Subcellular Biochemistry

Volume 13
Fluorescence Studies on Biological Membranes

A Continuation Order Plan is available for this series. A continuation order will bring delivery of each new volume immediately upon publication. Volumes are billed only upon actual shipment. For further information please contact the publisher.

Subcellular Biochemistry

Volume 13
Fluorescence Studies on Biological Membranes

Edited by

H. J. Hilderson
University of Antwerp
Antwerp, Belgium

Series Editor

J. R. Harris
North East Thames Regional Transfusion Centre
Brentwood, Essex, England

PLENUM PRESS • NEW YORK AND LONDON

The Library of Congress cataloged the first volume of this title as follows:

Sub-cellular biochemistry.

   London, New York, Plenum Press.
        v. illus. 23 cm. quarterly.
      Began with Sept. 1971 issue. Cf. New serial titles.
      1. Cytochemistry – Periodicals. 2. Cell organelles – Periodicals.
QH611.S84   574.8'76   73-643479

ISBN-13: 978-1-4613-9361-0   e-ISBN-13: 978-1-4613-9359-7
DOI: 10.1007/978-1-4613-9359-7

This series is a continuation of the journal *Sub-Cellular Biochemistry,*
Volumes 1 to 4 of which were published quarterly from 1972 to 1975

© 1988 Plenum Press, New York
A Division of Plenum Publishing Corporation
233 Spring Street, New York, N.Y. 10013
Softcover reprint of the hardcover 1st edition 1988

Contributors

Edward Blatt   Division of Applied Organic Chemistry, CSIRO, Melbourne, Victoria 3001, Australia

Robert Blumenthal   National Cancer Institute, National Institutes of Health, Bethesda, Maryland 20892

Nor Chejanovsky   Institute of Life Sciences, The Hebrew University of Jerusalem, Jerusalem 91904, Israel

Jeffrey Clarke   Division of Molecular Biology and Biochemistry, School of Basic Life Sciences, University of Missouri, Kansas City, Missouri 64110-2499

Hugo Depauw   RUCA–Laboratory for Human Biochemistry, University of Antwerp, B2020 Antwerp, Belgium

Marc De Wolf   RUCA–Laboratory for Human Biochemistry, University of Antwerp, B2020 Antwerp, Belgium

Wilfried Dierick   RUCA–Laboratory for Human Biochemistry, University of Antwerp, B2020 Antwerp, Belgium

Jacques Gallay   Laboratoire pour l'Utilisation du Rayonnement Electromagnétique (L.U.R.E.)–CNRS-CEA-MEN, Université Paris-Sud, 91405 Orsay-Cedex, France

Jose-Carlos Garcia-Borron   Division of Molecular Biology and Biochemistry, School of Basic Life Sciences, University of Missouri, Kansas City, Missouri 64110-2499

Jose-Manuel Gonzalez-Ros   Division of Molecular Biology and Biochemistry, School of Basic Life Sciences, University of Missouri, Kansas City, Missouri 64110-2499

Josef Gut   Stanford University, Stanford, California 94025

Herwig Hilderson   RUCA–Laboratory for Human Biochemistry, University of Antwerp, B2020 Antwerp, Belgium

Akira Ikegami   Institute of Physical and Chemical Research, Wako-shi, Saitama 351-01, Japan

Kazuhiko Kinosita, Jr.   Institute of Physical and Chemical Research, Wako-shi, Saitama 351-01, Japan

Tsutomu Kouyama   Institute of Physical and Chemical Research, Wako-shi, Saitama 351-01, Japan

Barbara C. Kunz   National Institutes of Health, Bethesda, Maryland 20205

Albert Lagrou   RUCA–Laboratory for Human Biochemistry, University of Antwerp, B2020 Antwerp, Belgium

Joseph R. Lakowicz   Department of Biological Chemistry, School of Medicine, University of Maryland, Baltimore, Maryland 21201

Abraham Loyter   Institute of Life Sciences, The Hebrew University of Jerusalem, Jerusalem 91904, Israel

Marino Martinez-Carrion   Division of Molecular Biology and Biochemistry, School of Basic Life Sciences, University of Missouri, Kansas City, Missouri 64110-2499

Ofer Nussbaum   Institute of Life Sciences, The Hebrew University of Jerusalem, Jerusalem 91904, Israel

P. Proulx   Department of Biochemistry, School of Medicine, Faculty of Health Sciences, University of Ottawa, Ottawa, Ontario K1H 8M5, Canada

Christoph Richter   Laboratory of Biochemistry, Swiss Federal Institute of Technology, Zürich CH-8092, Switzerland

William H. Sawyer   Russell Grimwade School of Biochemistry, University of Melbourne, Parkville, Victoria 3052, Australia

Elsie M. B. Sorensen   Department of Pharmacology and Toxicology, College of Pharmacy, University of Texas, Austin, Texas 78712

Hisako Urabe   Institute of Physical and Chemical Research, Wako-shi, Saitama 351-01, Japan

Wim J. van Blitterswijk   Division of Cellular Biochemistry, The Netherlands Cancer Institute, Antoni van Leeuwenhoek-Huis, 1066 CX Amsterdam, The Netherlands

B. Wieb Van der Meer   Department of Physics and Astronomy, Western Kentucky University, Bowling Green, Kentucky 42101

Guido Van Dessel   UIA–Laboratory for Pathological Biochemistry, University of Antwerp, B2020 Antwerp, Belgium

Michel Vincent Laboratoire pour l'Utilisation du Rayonnement Electromagnétique (L.U.R.E.)–CNRS-CEA-MEN, Université Paris-Sud, 91405
Orsay-Cedex, France

Preface

As stated by its first editor, Dr. D. B. Roodyn, the primary goal of the series *Subcellular Biochemistry* is to achieve an integrated view of the cell by bringing together results from a wide range of different techniques and disciplines.

This volume deals with the applications of fluorescence spectroscopy to membrane research. It seeks to present complementary biochemical and biophysical data on both the structure and the dynamics of biological membranes. Biophysics and biochemistry are improving more and more in their ability to study biomembranes, overlapping somewhat in this area and explaining the functioning of the whole cell in terms of the properties of its individual components. Therefore, we have brought together an international group of experts in order to report on and review advances in fluorescence studies on biological membranes, thereby highlighting subcellular aspects.

The first chapters present a critical evaluation of the current applications of dynamic and steady-state fluorescence techniques. Subsequent chapters discuss more specific applications in cells, biological membranes, and their constituents (lipids, proteins).

This volume opens with a chapter by B. Wieb Van der Meer addressing two questions: (1) What are the relevant parameters that describe membrane structure and dynamics? (2) What are the various fluorescence techniques for measuring these parameters? After updating our knowledge of the theoretical aspects and providing a critical evaluation of examples giving insight into the physical state, composition, and functions of membranes, the author concludes that fluorescence can often only demonstrate a correlation between a physical and a functional parameter. By showing the way of thinking that must be followed by everyone working in this area, this chapter becomes a must for those starting research in this field.

If, however, the reader is discouraged by a more physical treatment of the subject, he or she is referred to Chapter 2, by Kazuhiko Kinosita, Jr., and Akira Ikegami, dealing with the optical anisotropy decay method (in particular, fluorescence depolarization and absorption and phosphorescence anisotropy decay). As a less theoretical and more visual description of the interactions in biomembranes, DNA, and actin and myosin filaments, this chapter can serve as an introduction. The aim of the authors is not to be exhaustive but rather to explain what can be learned from optical anisotropy decay. According to them, it is a powerful means of studying the structure of a key molecule incorporated into a higher-order structure. They conclude that "molecular machines" working in living organisms appear to be highly flexible and that the whole machine, as well as its parts, undergoes continual fluctuations.

Time-resolved data are usually obtained by measurements of the time-dependent emission from samples excited with brief pulses of light. In Chapter 3, Joseph R. Lakowicz describes an alternative method in which the time-resolved emission is determined from the frequency response of the emission to intensity-modulated light. The examples include the resolution of closely spaced lifetimes, measurement of complex decays of fluorescence anisotropy, and calculation of time-resolved emission spectra from the wavelength-dependent frequency of the sample. A new application is described by which the distribution of distances between two sites on a flexible molecule can be recovered. According to the author, correlation times as short as 8 psec can be measured.

An excellent and profound review of the use of probes to study the effects of sterols and unsaturation on the lipid order parameter in biomembranes has been written by Michel Vincent and Jacques Gallay. In Chapter 4, the authors focus on the use of pulse fluorescence anisotropy techniques to investigate the effect of cholesterol and its derivatives on membrane lipid order and dynamics. Fluorescence lifetimes are extremely sensitive to the environment, but until recently they were very difficult to interpret with certainty. Very recent analysis methods (the maximum entropy method) are expected to allow a precise description of the multiexponential behavior of the total fluorescence decay in terms of a continuous distribution of excited-state lifetimes. The authors believe that the advent of such a powerful tool calls for a systematic reevaluation of probe behavior in membranes and other systems.

Fluorescence polarization results also provide valuable insights into ion–membrane interactions. Based on a recognition of the inherent limitations of fluorescence polarization procedures, Elsie M. B. Sorensen summarizes in Chapter 5 a portion of the scientific data base on ion–membrane interactions. The prevalence of reported studies on calcium-induced alterations of membrane fluidity results in our having more data for calcium than for other ions. From this study ratios of ions appear to be important, as does ionic charge, ionic radius, and binding affinity to the components of the plasma membrane. The author concludes that much research remains to be done to elucidate the interactions of a complex mixture of extracellular ions with plasma membranes. Moreover, ionic interactions with the internal surface of the phospholipid bilayer might prove to be as important as those with the external surface.

In Chapter 6, Hugo Depauw *et al.* apply fluorescence polarization procedures to the study of the fluidity of thyroid plasma membranes, bringing together biophysical methods and subcellular biochemistry. The isolation of enriched plasma membrane fractions and subfractions is described. From reconstitution experiments, it becomes clear that both lipids and proteins display a major effect on thyroid membrane fluidity, S_{DPH} being mainly affected by neutral lipids and D_{diff} being more sensitive to proteins. The authors demonstrate that adenylate cyclase activity can be modulated by manipulating the plasma membrane composition (incorporation of phospholipids, gangliosides, dolichol and derivatives, membrane-perturbing drugs). In this respect they claim that there is not always a parallel effect of drugs on bilayer fluidity and adenylate cyclase stimulation.

In Chapter 7, Akira Ikegami *et al.* describe a spectroscopic analysis of the structure of bacteriorhodopsin using the fluorescence energy transfer technique and polarized resonance Raman scattering. The application of x-ray crystallographic analysis to the study of membrane proteins is restricted, since the crystallization of the proteins is prevented by their hydrophobic nature. The authors thus turned to these methods in a new attempt to determine the three-dimensional location and orientation of the retinal chromophore in intact purple membrane. They also estimated the location of Lys-41 using fluorescence depolarization and predicted from free-energy calculations seven amino acid sequences corresponding to seven transmembrane α-helixes. Finally, they estimated possible folding of the polypeptide chain of bacteriorhodopsin in the purple membrane. As the visual pigment rhodopsin is considered a prototype of a G-protein-linked receptor (J. L. Marx, 1987, *Science* **238**:615–616), the elucidation of its structure and function in different organs and membranes is now of major importance.

Another group of proteins studied in native membranes is the liver microsomal monoxygenase system (Chapter 8). Christoph Richter *et al.* give an extensive review of the literature and their own work. After highlighting the biophysical consequences of lipid peroxidation and the mobility of membrane-bound cytochrome P-450 (as studied by delayed fluorescence polarization and fluorescence recovery after photobleaching using fluorescently labeled P-450), the rotational mobility and structure of the free and membrane-bound molecule are described. The authors also focus on the structure of NADPH–cytochrome P-450 reductase and the interactions of cytochrome P-450 reductase, as well as on lipid–protein interactions as studied by DPH fluorescence anisotropy.

Chapter 9 by P. Proulx is devoted to some of the numerous applications of fluorescence spectroscopy to the study of prokaryotic membranes. Using a large number of examples the author shows how this technique, applied in a coordinated manner with other physical techniques, provides useful information on the physical state of the membrane and helps define the molecular interactions that form the basis for the membrane structure. Particular attention is given to experiments dealing with uptake of exogenous lipids. Fluorescence techniques seem to be quite useful in defining the mechanism of this event as well as in revealing the resulting changes in membrane structure.

The study of cytoskeletal protein interactions by fluorescence probe techniques is described in Chapter 10 by Edward Blatt and William H. Sawyer. The authors concentrate on the polymerization properties of actin and its interactions with other cytoskeletal proteins. A summary of the dynamic interactions of cytoskeletal components is given. The specific experimental details of the fluorescence probe technique are examined as applied to the study of cytoskeletal interactions, employing the fluorescence characteristics of either intrinsic or labeled proteins. A considerable number of labels are described. Finally, experimental results of two related luminescence techniques are discussed, especially as they are used to monitor long rotational motions.

The scope of Chapter 11, by Marino Martinez-Carrion et al., is to review the several novel applications of fluorescent probes to the study of the relevant aspects of acetylcholine receptor structure and function. The authors pioneered the application of a variety of pyrene derivatives and discuss the information obtained with such probes. The applicability of the probes in monitoring the properties of the lipid bilayer, the protein–lipid interface, the exocytoplasmic surface, and the endocellular side is described. Yet the chapter is not restricted to pyrene derivatives; information obtained by means of other fluorescent probes is discussed as well.

Chapter 12, by Wim J. van Blitterswijk, is devoted to membrane fluidity in normal and tumor cells. The quantitative contribution of individual lipids in the membrane to DPH-fluorescence polarization is outlined. According to the author the apparent existence of preferential affinities among certain types of membrane lipids has some important cell biological consequences. After reviewing alterations in membrane fluidity in lymphoid tumor cells and their physiological control, the author describes how fatty acid profiles of the membrane phospholipids can, to a certain extent, be modified by dietary means but with limited overall effect on membrane fluidity.

Chapter 13, by Nor Chejanovsky et al., gives a description of the availability of active-fusogenic fluorescently labeled enveloped viruses and fluorescence dequenching methods that make it possible to study at a molecular level the fusion of intact virions or reconstituted envelopes with various preparations. Supported by the most recent literature and by their own research, the authors are able to demonstrate that fusion of enveloped virions with liposomes or biological membranes requires a high percentage of cholesterol, that fusion with negatively charged liposomes does not reflect the biological activity of the viral envelope, and that sialoglycolipids must be present in PC/chol liposomes to allow expression of the viral lytic activity. The authors conclude that the question of whether or not viral binding (spike) proteins and their membrane receptors play an active role in the process of virus–membrane fusion is still a matter of debate.

Let me close this preface with two citations from within the book:

> Since many proteins are present in the cell in small amounts, the biochemist has resorted to microanalytical methods to obtain this information. Recent developments in instrumentation and computer technologies have greatly aided this process. Fluorescence spectroscopy is an important tool in such endeavors [Edward Blatt and William H. Sawyer].

> Fluorescence spectroscopy has proved to be a powerful tool in the study of the structure and function of biological membranes and their model systems. Its advantages are its high sensitivity, the low degree of membrane perturbation, and the favorable time scale, which allows one to observe a wide range of interesting molecular processes [B. Wieb Van der Meer].

I want to express my thanks to Dr. Marc De Wolf and Dr. Hugo Depauw for their advice in preparing this book. I wish all of you much enthusiasm and pleasure in reading it.

Herwig Hilderson

Antwerp, Belgium

Contents

Chapter 1
Biomembrane Structure and Dynamics Viewed by Fluorescence

B. Wieb Van der Meer

1. Introduction to Fluorescence 1
2. Dynamics and Structure of Membranes 4
   2.1. Lateral and Rotational Diffusion 5
   2.2. Orientational Order and Packing 7
   2.3. Asymmetry 9
   2.4. Lipid Domains 10
3. Fluorescence Techniques and What They Make Visible 10
   3.1. Fluorescence Depolarization 11
   3.2. Quenching 26
   3.3. Fluorescence Energy Transfer 28
   3.4. Fluorescence Recovery after Photobleaching (FRAP) 35
   3.5. Excimer Fluorescence 38
4. Summary and Conclusions 41
5. References 42

Chapter 2
**Dynamic Structure of Membranes and Subcellular Components
Revealed by Optical Anisotropy Decay Methods**

Kazuhiko Kinosita, Jr., and Akira Ikegami

1. Introduction 55
2. Optical Anisotropy Decay 56
   2.1. Principle of Optical Anisotropy Decay Method 56
   2.2. Experimental Techniques 59
   2.3. Information Contained in an Anisotropy Decay 63

2.4. Optical Anisotropy Decay as a Tool in Bioscience 65
3. Examples of Application 66
    3.1. Dynamic Structure of Membranes Probed by
         Diphenylhexatriene 66
    3.2. Protein Rotations in Membrane and on Membrane Surface . 72
    3.3. Internal Motion of DNA 75
    3.4. Internal Motion of Actin Filament 80
    3.5. Dynamic Structure of Myosin Filament 83
4. Concluding Remarks 85
5. References 86

Chapter 3
**Principles of Frequency-Domain Fluorescence Spectroscopy and
Applications to Cell Membranes**
Joseph R. Lakowicz

1. Introduction 89
2. Comparison of Time- and Frequency-Domain Measurements 90
    2.1. A First-Generation Frequency-Domain Fluorometer 93
    2.2. Resolution of a Two-Component Mixture 94
3. Theory of Frequency-Domain Fluorometry 95
    3.1. Decays of Fluorescence Intensity 95
    3.2. Decays of Fluorescence Anisotropy 97
4. Intensity Decays of DPH-Labeled Membranes 98
5. Anisotropy Decays of Labeled Membranes 100
    5.1. Hindered Rotations of Diphenylhexatriene 100
    5.2. Anisotropic Rotations of Perylene 101
6. Time-Resolved Emission Spectra 103
    6.1. Calculation of Time-Resolved Emission Spectra 107
    6.2. Time-Resolved Emission Centers of Gravity and Spectral
         Half-Widths 108
    6.3. Time-Resolved Spectral Data for Patman-Labeled
         Membranes 109
7. Energy Transfer in Membranes 111
    7.1. Distribution of Distances in a Covalently Linked
         Donor–Acceptor Pair 113
8. A 2-GHz Frequency-Domain Fluorometer 116
    8.1. Picosecond Resolution of Tyrosine Intensity and Anisotropy
         Decays 118
    8.2. Measurement of a 8-psec Correlation Time 120
9. Future Developments 121

10. Summary 122
11. References 123

Chapter 4
Time-Resolved Fluorescence Depolarization Techniques in Model Membrane Systems: Effect of Sterols and Unsaturations
Michel Vincent and Jacques Gallay

1. Introduction 127
2. Intrinsic Motional Properties of Some Widely Used Fluorescent
   Probes 129
   2.1. Motional Characteristics 129
   2.2. Excited-State Characteristics 133
3. Sterol–Phospholipid Interactions in Model Membranes 134
   3.1. Cholesterol–Phospholipid Interactions: Lecithin as Bilayer
        Matrix 134
   3.2. Cholesterol–Phospholipid Interactions: Phospholipids Other
        Than Lecithin as Bilayer Matrix 142
   3.3. Cholesterol Chemical Modification: Effect on Phospholipid
        Fatty Acyl Chains Order and Dynamics 144
4. Concluding Remarks 149
5. References 151

Chapter 5
Fluorescence Polarization to Evaluate the Fluidity of Natural and Reconstituted Membranes
Elsie M. B. Sorensen

1. Introduction 159
   1.1. Aims and Scope of This Chapter 159
   1.2. Mechanism of Action and Biological Significance of
        Fluorescence Polarization Measurements of Membrane
        Fluidity 163
2. Methodology 172
   2.1. Theory of Fluorescence Polarization for Ion–Membrane
        Measurements 172
   2.2. Probe–Membrane Interactions 177
   2.3. Probe–Ion Interactions 180
3. Current Advancements in the Measurement of Ion–Membrane
   Interactions Using Fluorescence Polarization 180

3.1. Natural Membranes 180
3.2. Reconstituted Membranes 182
4. Critical Evaluation of the Significance of Ion–Membrane
    Measurements 183
    4.1. Advantages of Fluorescence Polarization for Evaluation of
         Ion–Membrane Interactions 183
    4.2. Limitations of Fluorescence Polarization for Measurement
         of Ion–Membrane Interactions 184
    4.3. Substantiation of the Fluorescence Polarization
         Measurements of Ion–Membrane Interactions 185
5. Concluding Remarks 185
6. References 186

Chapter 6
Fluidity of Thyroid Plasma Membranes

Hugo Depauw, Marc De Wolf, Guido Van Dessel, Herwig Hilderson, Albert Lagrou, and Wilfried Dierick

1. Introduction 193
2. Thyroid Plasma Membranes 194
    2.1. Enriched Plasma Membrane Fractions 195
    2.2. Chemical Characterization of Purified Plasma Membranes .. 197
    2.3. Enzymic Characterization of Purified Plasma Membranes .. 197
    2.4. Subfractionation of Thyroidal Plasma Membranes 198
    2.5. Characterization of Thyroid Plasma Membrane
         Subfractions 200
3. Fluidity of Thyroid Plasma Membranes 200
    3.1. Fluidity Measurements 201
    3.2. Fluidity of Thyroid Subcellular Fractions 201
    3.3. Fluidity of a P_2 Fraction in Reconstituted Thyroid
         Plasma Membranes 202
    3.4. Fluidity Characteristics of Plasma Membrane Subfractions . 209
4. Modulation of the Adenylate Cyclase Activity by Manipulating
    the Plasma Membrane Composition 209
    4.1. Incorporation of Phospholipids 210
    4.2. Incorporation of Gangliosides 214
    4.3. Incorporation of Dolichol and Dolichyl Derivatives 218
    4.4. Addition of Membrane-Perturbing Drugs 223
5. Involvement of Membrane Fluidity on Human Normal and
    Pathological Thyroid Glands 231

6. References 233

Chapter 7
Spectroscopic Analysis of the Structure of Bacteriorhodopsin
Akira Ikegami, Tsutomu Kouyama, Hisako Urabe, and
Kazuhiko Kinosita, Jr.

1. Introduction 241
2. Principle of the Fluorescence Energy Transfer Technique 242
3. Three-Dimensional Disposition of the Retinal Chromophore in
   the Purple Membrane 244
   3.1. In-Plane Location 244
   3.2. Transmembrane Location 246
   3.3. Orientation of the Molecular Plane 249
4. In-Plane Location of NBD (7-Chloro-4-Nitrobenzo-2-Oxa-1,3-
   Diazole) Bound to Lys-41 in the Purple Membrane 250
5. Conformational Prediction of Bacteriorhodopsin Molecule 254
6. References 255

Chapter 8
**Structure and Dynamics of the Liver Microsomal Monoxygenase
System**
Christoph Richter, Josef Gut, and Barbara C. Kunz

1. Introduction 259
   1.1. General Structure of Biological Membranes 259
   1.2. Peroxidation of Membrane Lipids 262
   1.3. Microsomal Monoxygenase 263
2. Membrane Dynamics and Order Studied by Fluorescence 265
   2.1. Biophysical Consequences of Lipid Peroxidation 265
   2.2. Mobility of Membrane-Bound Cytochrome P-450 267
   2.3. Rotational Mobility of Cytochrome P-450 in Peroxidized
        Rat Liver Microsomes 269
   2.4. Structure of Free and Membrane-Bound Cytochrome P-450 270
   2.5. Structure of NADPH–Cytochrome P-450 Reductase 272
   2.6. Interaction of Cytochrome P-450 and Its Reductase in
        Membranes 272
   2.7. Lipid–Protein Interactions Studied by DPH Fluorescence
        Anisotropy 273

3. References 274

Chapter 9
Fluorescence Studies on Prokaryotic Membranes
P. Proulx

1. Introduction 281
2. Fluorescent Probes 282
3. Structural Aspects of Bacterial Membranes 286
    3.1. Outer Membrane of Gram-Negative Bacteria 286
    3.2. Molecular Interactions 286
    3.3. Phase Transitions and Homeoviscous Adaptation 292
    3.4. Effects of Alcohols 293
    3.5. Permeability of the Outer Membrane to Hydrophobic
         Substances 295
    3.6. Membrane-Potential-Related Permeability Changes 297
    3.7. Factors Increasing Cell Resistance and Membrane Stability   299
4. Periplasm 301
5. Incorporation of Exogenous Lipids into Prokaryotic
    Membranes 301
    5.1. Gram-Negative Bacteria 301
    5.2. Other Bacteria 304
    5.3. Effect of Lipid Uptake on Membrane Function 307
    5.4. Interactions with Vehicle Liposomes 307
6. Concluding Remarks 308
7. References 308

Chapter 10
**The Study of Cytoskeletal Protein Interactions by Fluorescence Probe
Techniques**
Edward Blatt and William H. Sawyer

1. Introduction 323
2. The Cytoskeleton 324
    2.1. Organization of Cytoskeletal Proteins 324
    2.2. Assembly of Actin Filaments 326
3. Fluorescence Probe Techniques 327
    3.1. Introduction 327
    3.2. Energy Transfer 328
    3.3. Fluorescence Enhancement 335

3.4. Anisotropy 339
3.5. Fluorescence Photobleaching Recovery 342
3.6. Quenching 345
3.7. Pressure Relaxation 349
4. Alternative Luminescence Techniques 349
4.1. Introduction 349
4.2. Transient Absorption Anisotropy 351
4.3. Phosphorescence 352
5. Summary and Future Prospects 353
6. References 355

Chapter 11
Fluorescent Probes for the Acetylcholine Receptor Surface Environments

Marino Martinez-Carrion, Jeffrey Clarke, Jose-Manuel Gonzalez-Ros, and Jose-Carlos Garcia-Borron

1. Introduction 363
2. An Overview of AchR Properties 364
2.1. Structural Characteristics of *Torpedo californica* AchR 364
2.2. Ligand Binding and Pharmacological Properties 365
3. PTSA: A Probe for Measuring AchR-Mediated Ionic Fluxes in the Physiological Time Scale 367
3.1. Stopped-Flow Assays for AchR-Mediated Ionic Fluxes 367
3.2. Stopped-Flow Assay with PTSA and Thallous Ion 369
4. Pyrene-1-Sulfonyl Azide (PySA): A Probe for the Study of the AchR–Lipid Interface 370
4.1. Optical Properties of PySA and Its Photoproducts 370
4.2. Labeling of the AchR 371
4.3. Applications of PySA to the Study of AchR Structure and Function 376
5. Pyrene Maleimide (PM): The Labeling of a Functionally Relevant Sulfhydryl Group 380
5.1. Labeling of Solubilized Receptor 381
5.2. Labeling of Native Membranes 382
6. State and Organization of the Lipid Bilayer in AchR Membranes 384
6.1. Probing of AchR Membranes with Pyrene 384
6.2. Fluidity of the AchR Membranes as Probed by DPH and TMA-DPH 386
7. Summary 387

8. References 387

Chapter 12
**Structural Basis and Physiological Control of Membrane Fluidity in
Normal and Tumor Cells**
Wim J. van Blitterswijk

1. Introduction 393
2. Quantitative Contribution of Individual Types of Lipid to
    Membrane Fluidity 395
    2.1. Cholesterol, Sphingomyelin, and Fatty Acyl (Un)saturation 395
    2.2. Cell Biological Implications of Preferential Interactions
          between Individual Lipids 397
3. Alterations in Membrane Fluidity in Lymphoid Tumor Cells 399
    3.1. Tumor Cell Type and Location 399
    3.2. Plasma Lipoproteins and Cholesterol Biosynthesis 400
4. Effects of Dietary Lipids on Membrane Fluidity 403
5. References 410

Chapter 13
**Fusion of Enveloped Viruses with Biological Membranes:
Fluorescence Dequenching Studies**
Nor Chejanovsky, Ofer Nussbaum, Abraham Loyter, and
Robert Blumenthal

1. Introduction 415
2. Receptors for Enveloped Viruses 416
    2.1. Myxoviruses 417
    2.2. Paramyxoviruses 418
    2.3. Togaviruses 418
    2.4. Rhabdoviruses 419
    2.5. Retroviruses 419
    2.6. Herpesviruses 420
    2.7. Other Enveloped Viruses 420
3. Interaction of Enveloped Viruses with Receptor-Depleted
    Cells 420
    3.1. Use of Antimembrane Antibodies or Polypeptide Hormones
          to Mediate Virus Attachment 421
    3.2. Implantation of Receptors or Binding Proteins for
          Enveloped Virions into Recipient Cell Membranes 421

4. Theoretical Aspects of the Use of Fluorescence Dequenching to Measure Viral Fusion 422
5. Fusion of Enveloped Viruses with Animal Cells and Biological Membranes: Studies with Intact Virions 429
6. Use of Fluorescent Dequenching Methods to Study Fusion of Enveloped Viruses with Biological Membranes Lacking Virus Receptors 434
7. Role of Viral Glycoproteins in the Process of Virus Membrane Fusion: Studies with Reconstituted Viral Envelopes 435
8. Fusion of Enveloped Viruses with Negatively Charged and Neutral Liposomes 439
9. Role of Conformational Changes and Cooperativity of Viral Proteins in Mediating Membrane Fusion 444
10. Conclusions 449
11. References 450

Index 457

Chapter 1

Biomembrane Structure and Dynamics Viewed by Fluorescence

B. Wieb Van der Meer

1. INTRODUCTION TO FLUORESCENCE

Fluorescence is the phenomenon that certain molecules emit light with longer wavelength than the light with which they were illuminated. Such molecules are called fluorophores. Incident photons can excite a fluorophore from its electronic ground state, S_0, to a vibrational level of the electronic state S_1 or S_2. Vibrational energy is lost thermally after excitation within picoseconds, and the molecule drops to the ground vibrational state of the excited electronic state. The molecule then returns to one of the levels of S_0 after a short period of the order of nanoseconds, by emission of fluorescent light. This process

---

Abbreviations used in this chapter: DPH, 1,6-diphenyl-1,3,5-hexatriene; DMPC, dimyristoyl-phosphatidylcholine; DPPC, dipalmitoylphosphatidylcholine; ANS, 1-anilinonaphthalene-8-sulfonate; NBD, 7-nitro-2,1,3-benzoxadiazol-4-yl; SUV, small unilamellar vesicles; LUV, large unilamellar vesicles; MLV, multilamellar vesicles; BLM, black lipid membranes; FET, fluorescence energy transfer; *trans*-PnA, 9,11,13,15-all-*trans*-octadecatetraenoic acid; *cis*-PnA, 9,11,13,15-*cis, trans, trans, cis*-octadecatetraenoic acid; ROS, rod outer segment; TNP-PE, *N*-(trinitrophenyl)dimyristoylphosphatidylethanolamine; DDA, dansyldodecylamine; *n*-AO, *n*-(9-anthroyloxy) fatty acids; $Tb(DPA)_3^{3-}$, chelate of one terbium ion (Tb^{3+}) and three dipicolinate ions (DPA^{2-}); EPE, eosin phosphatidylethanolamine; FRAP, fluorescence recovery after photobleaching; FPR, fluorescence photobleaching recovery.

B. Wieb Van der Meer   Department of Physics and Astronomy, Western Kentucky University, Bowling Green, Kentucky 42101.

is depicted in Figure 1. The absorption and emission of light is also illustrated in the Jablonski diagram of Figure 2, where the energy levels are depicted as vertical lines with energy increasing to the right. [Solvent relaxation and intersystem crossing can also be included in such diagrams (Lakowicz, 1983) but are not shown here.] The positions of the S_1 and S_2 bands in the Jablonski diagram correspond to maxima in the absorption spectrum, which is a plot of the absorbance versus wave number, proportional to transition energy (see the upper left part of Figure 2) or wavelength (see the upper right-hand side of Figure 2). The corresponding fluorescence spectra, plots of fluorescence intensity versus wave number (lower left of Figure 2) or wavelength (lower right of Figure 2), are also shown.

Fluorescence can also be presented as a kinetic scheme (Eq. 1):

$$A + h\nu_E \longrightarrow A^* \begin{array}{c} \xrightarrow{k_f} A + h\nu_F \\ \xrightarrow{k_i} A \end{array} \tag{1}$$

where A represents a molecule in the ground state and A^* denotes an excited fluorophore; h is Planck's constant and $h\nu$ represents a quantum of radiation

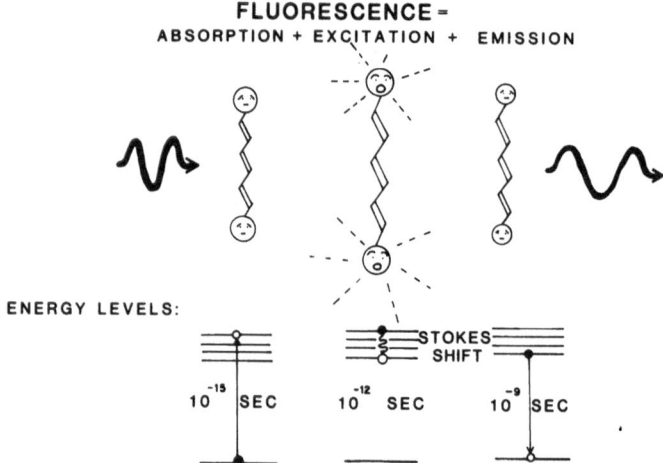

FIGURE 1. Schematic diagram of the fluorescence process, in which absorption and excitation take place, followed by emission. The photon that is absorbed has a shorter wavelength than the emitted photon. The molecule resembles 1,6-diphenyl-1,3,5-hexatriene. Possible configurational changes upon excitation are greatly exaggerated. The transitions from the ground state to higher energy levels and back to the ground state are also indicated, as well as the corresponding time scales. The Stokes shift is the difference in wave number between absorption and fluorescence.

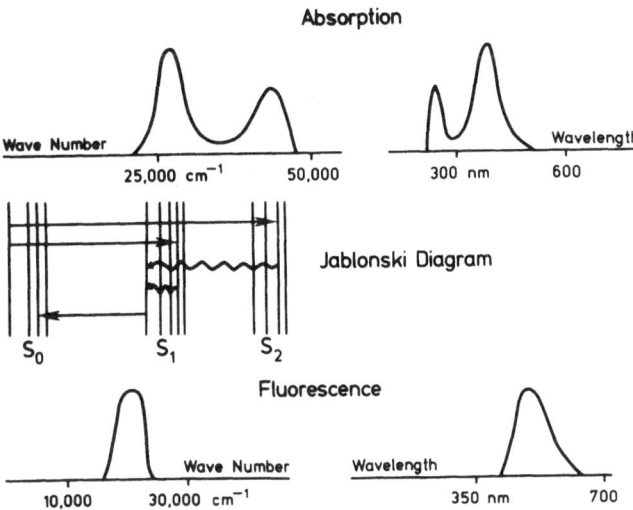

FIGURE 2. A Jablonski diagram is shown, together with corresponding absorption and fluorescence spectra. The vertical lines denote energy levels. The straight arrows represent transitions with radiation (absorption to the right, emission to the left) and the wavy arrows represent radiationless transitions. The vibrational levels of the ground state form the S_0 band. Similarly, the vibrational levels of the first and second electronically excited states form the bands S_1 and S_2, respectively. The locations of the S_1 and S_2 bands correspond to absorption peaks in the absorption spectrum, shown above the Jablonski diagram, where the absorbance is plotted versus wave number. The diagram can only give the locations of the peaks, not their relative heights. The same absorption spectrum plotted versus wavelength is also shown. The possible band of radiation transitions to S_0 corresponds with a peak in the fluorescence spectrum, which is a plot of fluorescence intensities versus wave number. This spectrum is shown under the Jablonski diagram. The same spectrum plotted versus wavelength is shown as well.

(photon); ν_E and ν_F are wave numbers of the excitation and the fluorescence, respectively; k_f and k_i are the radiative and nonradiative rate constants for depopulation of the excited state. The quantum yield, Q, is the ratio of the number of photons emitted to the number of photons absorbed and is given by

$$Q = \frac{k_f}{k_f + k_i} \tag{2}$$

The lifetime of the excited state or the fluorescence lifetime, which is an indication of the average time spent in the excited state before emitting a photon, equals

$$\tau = \frac{1}{k_f + k_i} \tag{3}$$

Only a few basic concepts of fluorescence spectroscopy have been discussed above. A number of review articles and books are available that treat the various aspects of fluorescence spectroscopy in more detail (Weber, 1969a; Udenfriend, 1969; Berlman, 1971; Pesce et al., 1971; Chen and Edelhoch, 1975; Weber, 1976; Lakowicz, 1983; Jameson, 1984; Haugland, 1985). Instrumentation is discussed by Lakowicz (1983).

Fluorescence spectroscopy has proved to be a powerful tool in the study of the structure and function of biological membranes and their model systems. Its advantages are its high sensitivity, the low degree of membrane perturbation, and the favorable time scale, which allows one to observe a wide range of interesting molecular processes.

In this chapter a number of structural and dynamical parameters of membranes will be discussed and the fluorescence techniques for measuring these parameters explained. A number of interesting examples of fluorescence studies will be given as well.

2. DYNAMICS AND STRUCTURE OF MEMBRANES

The generally accepted model of membrane structure is the *fluid mosaic model*, proposed by Singer and Nicholson (1972). In this model the membrane is pictured as a sea of lipids, in which proteins are floating like icebergs (Singer, 1975). The lipids form a bilayer. Nonbilayer lipid structures may transiently exist as well, however (De Gier et al., 1985). Proteins are either bound to the membrane surface (peripheral proteins) or bound to the hydrophobic core of the bilayer as well as to the polar surface (integral proteins). The model emphasizes the fluidity of the structure (Singer, 1975); the molecules have a great deal of freedom to move laterally and to rotate or wobble (Cherry, 1979). Beautiful cartoons are available illustrating the fluid mosaic structure (Singer, 1975). Proteins form the major component of all biomembranes, from 50 to 70% by weight, with the exception of myelin, in which the lipids account for nearly 90% of the weight (Singer, 1975). Proteins perform most of the biological functions of the membrane, like transport, enzymatic activity, and immunorecognition, while lipids are believed to be mainly responsible for its fluidity, although they have other functions as well (Pagano and Sleight, 1985).

Membrane fluidity is a widely used concept in membrane research (Shinitzky, 1984). However, the term lacks a precise definition. This ambiguity is due to the fact that the membrane differs from an ordinary isotropic fluid such as an oil, where the motional freedom of particles dispersed in the fluid can be described with one physical parameter, the fluidity (= 1/viscosity). A membrane, on the other hand, is an anisotropic, two-dimensional fluid and the lateral and rotational mobilities of proteins and lipids cannot be expressed solely in terms of one parameter. Rather, a set of physical parameters is needed to quantitatively describe such motions. These parameters are discussed below. Other factors that complicate a precise definition of the physical state of a membrane are related to the rich lipid polymorphism (Van der Meer, 1984) and the coupling between the plasma membrane and the glycocalix and the cytoskeleton (McCloskey and Poo, 1984; Van der Meer, 1984).

2.1. Lateral and Rotational Diffusion

Brownian motion is the random movement of a particle due to exchange of thermal energy with its environment. The mean square of the distance traveled in such a random motion is proportional to the time. The lateral diffusion coefficient (or constant), D_L, is the mean square displacement per second [except for a factor of 2 or 4, depending on the dimensionality of the system; Berg (1983)]. The commonly used unit is cm²/sec. Similarly, the mean square of an angular displacement is proportional with time and a rotational diffusion coefficient (constant) is the mean square angular variation per second [again, except for a factor of 2 or 4; Kinosita et $al.$ (1977)]. The unit is sec^{-1} (frequency). A particle with cylindrical symmetry has two rotational constants: D_R, for rotations about the symmetry axis, and D_W, for wobbling of the symmetry axis around a perpendicular axis. It is important to distinguish two cases:
1. Diffusion without wobbling ($D_W = 0$). This case applies to integral membrane proteins, which are known to have a fixed orientation (Cherry, 1979), and perhaps, as a first approximation, to rigid lipids like cholesterol or lipids with two chains like phosphatidylcholine.
2. Diffusion with wobbling ($D_W \neq 0$). This case applies to molecules that do not have a fixed orientation like the lipophilic probe diphenylhexatriene.

2.1.1. Diffusion without Wobbling

Saffman and Delbrück (1975) have introduced a hydrodynamic model for diffusion of a particle in a membrane. The particle is depicted as a cylinder with radius a and length h, which is assumed to be equal to the membrane thickness. The cylinder is restricted to move laterally in the plane of the membrane and to rotate around its axis only ($D_w = 0$). The viscosity for motions in the plane of the membrane is denoted by η. They find for D_R

$$D_R = \frac{kT}{4\pi a^2 h \eta} \tag{4}$$

where k is Boltzmann's constant and T is the absolute temperature. The lateral diffusion constant cannot be derived without referring to the viscosities w_1

and w_2 of the bathing fluids above and below the membrane. Saffman and Delbrück (1975) have derived the following expression for D_L:

$$D_L = \frac{kT}{4\pi\eta h} [\ln(2/\epsilon) - 0.5772] \tag{5}$$

where $\epsilon = (w_1 + w_2)a/\eta h$. An essential assumption in the theory of Saffman and Delbrück is that ϵ is much smaller than unity, that is, $w_1 + w_2$ is much smaller than η. Hughes *et al.* (1982) have solved this diffusion model for all values of ϵ. They showed that there is a unique relation between ϵ and the parameter p, defined as

$$p = \frac{D_L}{a^2 D_R} \tag{6}$$

Table I illustrates this relation between ϵ and p. Hughes *et al.* argue that the interaction between hydrophilic parts of the protein and the glycocalix could result in ϵ being of the order unity (Hughes *et al.*, 1982).

It follows from the Saffman–Delbrück equations that D_L is rather insensitive to the radius of the particle, while D_R depends strongly on the particle radius. Consequently, protein clustering (dimerization or oligomerization) could be detectable from rotational diffusion, but invisible in a measurement of lateral diffusion. Peters and Cherry (1982) have shown that knowledge of both the lateral and the rotational diffusion constants enables one to determine not only the membrane viscosity but also the radius of the particle. Biological implications of protein diffusion have been discussed by McCloskey and Poo (1984) and by Axelrod (1983).

2.1.2. Diffusion with Wobbling

Shinitzky and Yuli (1982) have discussed a hydrodynamic model for lipidlike probes in a membrane. They consider two components of the viscosity tensor, one for resistance to motion parallel to the membrane plane and another for viscous drag along the membrane normal. However, to the best of our knowledge, generalized Saffman–Delbrück equations for D_L, D_R,

Table I
Values for ϵ and p^a

ϵ	p
0.001	7.6
0.01	4.7
0.1	2.7
1	1.5
10	0.6

[a] From Hughes *et al.* (1982).

and D_W in terms of the viscosity components, ϵ, and the dimensions of the particle are not available at present. One could speculate that such equations should predict a relation between the orientation of the particle and the anisotropy of the viscosity tensor (Van der Meer, 1984).

For lateral diffusion of lipidlike molecules in a membrane, a number of lattice models are available. The diffusion limit model (Förster, 1969), the random walk model (Galla *et al.*, 1979; Galla and Hartmann, 1980), and the milling crowd model (Eisinger *et al.*, 1986) have been reviewed by Eisinger *et al.* (1986) and are discussed in Section 3.5. A modified percolation theory (Saxton, 1982) and effective medium theories (Saxton, 1982; O'Leary, 1987) are also available. Diffusion measured over distances large compared to a typical protein diameter (larger than 10 nm, say) can be significantly slower than diffusion over a few nanometers (Eisinger *et al.*, 1986; O'Leary, 1987). This difference is due to obstructed diffusion and is discussed in Section 3.4.

Models for rotational diffusion in membranes have been discussed by Kinosita *et al.* (1977, 1984), Lipari and Szabo (1980, 1981), Zannoni (1981), Zannoni *et al.* (1983), Van der Meer *et al.* (1984), and Szabo (1984). See also Section 3.1. Rotational diffusion, especially wobbling, is closely coupled to orientational order.

2.2. Orientational Order and Packing

The membrane is an ordered fluid, where orientation perpendicular to the membrane plane is more probable than in-plane. It can be assumed that the membrane normal is an axis of symmetry and that the membrane molecules can be considered essentially cylindrically symmetric. As a result, the orientational distribution function for a lipid or a lipidlike probe will depend only on the angle θ between the membrane normal and the molecular axis. This distribution function $f(\theta)$ can be uniquely determined, if all the order parameters $\langle P_{2L} \rangle$ of all ranks (all L) are known. These order parameters are defined as

$$\langle P_{2L} \rangle = \int_0^\pi P_{2L}(\cos \theta) f(\theta) \sin \theta \, d\theta \tag{7}$$

with

$$L = 1: \quad P_2(\cos \theta) = (3\cos^2\theta - 1)/2 \tag{8a}$$

$$L = 2: \quad P_4(\cos \theta) = (35\cos^4\theta - 30\cos^2\theta + 3)/8 \tag{8b}$$

where P_{2L} are the Legendre polynomials of rank $2L$ and $\langle P_2 \rangle$ and $\langle P_4 \rangle$ are the second and fourth rank orientational order parameters, respectively. Here it is assumed that the asymmetry (see Section 2.3) can be ignored; that is, it is assumed that all $\langle P_{2L+1} \rangle$ vanish, $L = 0,1,2, \ldots$ and that $\langle P_2 \rangle$ and $\langle P_4 \rangle$ can be measured (Ameloot *et al.*, 1984). Order parameters of higher rank have not been measured as yet. However, knowledge of $\langle P_2 \rangle$ and/or $\langle P_4 \rangle$ allows for estimating the distribution function. Having $\langle P_4 \rangle$ in addition to $\langle P_2 \rangle$ gives much more insight into the orientational distribution than having $\langle P_2 \rangle$ alone. To illustrate this point, consider Figure 3 and the following examples. Suppose that $\langle P_2 \rangle$ is positive. That would indicate that most of the molecules would have a θ between 0° and 55°. If, however, one would also know that $\langle P_4 \rangle$ is positive, then we can see from Figure 3 that most of the θ values would lie between 0° and 30°, while a negative $\langle P_4 \rangle$ would indicate in that case that the majority of θ values would fall in the range from 30° to 55°. As another example compare an isotropic distribution ($f = \frac{1}{2}$) with a distribution having a peak at 55°. Both models would yield $\langle P_2 \rangle = 0$, but the distributions could be readily distinguished, if $\langle P_4 \rangle$ would be known, because the former model would correspond with $\langle P_4 \rangle = 0$ and the latter with $\langle P_4 \rangle = -0.36$. The significance of the orientational order parameters $\langle P_2 \rangle$ and $\langle P_4 \rangle$ of fluorophores in membranes has been discussed by Pottel *et al.* (1986).

Packing is the number of molecules per unit area. If the orientational order is high, the lipid chains are extended and allow neighboring molecules to approach each other closely; on the other hand, if the order is low, the chains have irregular conformation and thus occupy a larger volume or area.

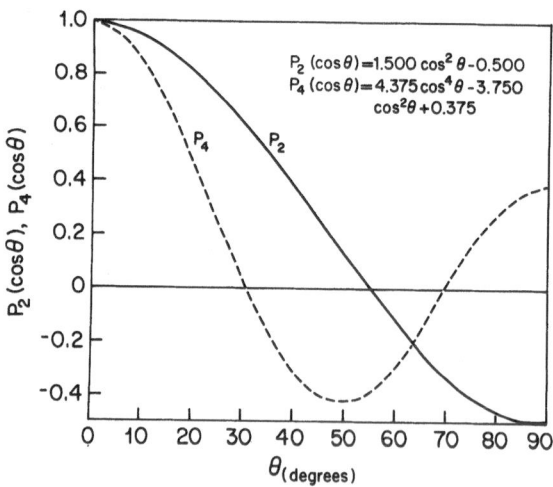

FIGURE 3. The second (P_2) and the fourth (P_4) Legendre polynomial of cos θ as a function of θ.

Therefore, packing must be related to orientational order (Van der Meer, 1984). This connection has been demonstrated by Mely *et al.* (1975), who have shown that the order parameter profiles for two lamellar systems with the same area per polar head, but with different hydration and temperature, are virtually identical. This relation between packing and order has also been emphasized by molecular-statistical theories of lipid chain ordering (Marčelja, 1974; Gruen, 1980). Lipid packing plays an important role in lipid protein interactions (Israelachvili *et al.*, 1980; Van der Meer, 1984). The less fluid (more dense) a bilayer the less easily a protein may be expected to pack into it. Thus, addition of cholesterol yields a tighter and more ordered membrane (Demel *et al.*, 1967; Kawato *et al.*, 1978) and this increase in packing can lead to an increased exposure of proteins (Borochov and Shinitzky, 1976). A similar increase in protein exposure has been observed upon lowering the temperature (Armond and Staehelin, 1979). Models for this vertical displacement of membrane proteins have been discussed by Van der Meer (1984). It is also noteworthy that almost all the proteins of the erythrocyte membrane are embedded in the inner monolayer, which is more fluid than the outer monolayer (Bretscher, 1973). Lipid–protein interactions have been reviewed by Israelachvili *et al.* (1980) and Devaux and Seignereut (1985).

2.3. Asymmetry

Biological membranes are asymmetric; the protein and lipid composition of the inner monolayer differs from that of the outer monolayer. Membrane asymmetry has been reviewed by Rothman and Lenard (1977), Op den Kamp (1979, 1981), and Storch and Kleinfeld (1985). The protein asymmetry is absolute; there is no transbilayer movement or "flip-flop" on proteins. Lipid flip-flop is slow; on the order of 8–16 hr at room temperature in erythrocytes and model membranes (Kornberg and McConnell, 1971; Seigneuret and Devaux, 1984). In erythrocytes, where the differences between inner and outer monolayers are known in detail, the aminophospholipids are located predominantly at the inside and phosphatidylcholine and sphingomyelin at the outside (Op den Kamp, 1979). Results of a number of groups suggest that this asymmetry is maintained by ATP-dependent transporter proteins (flipases), not only in erythrocytes (Seigneuret and Devaux, 1984; Daleke and Huestis, 1985) but also in rat liver microsomes (Backer and Dawidowicz, 1985; Bishop and Bell, 1985) and cultured fibroblasts (Martin and Pagano, 1987). Flipflop related shape changes have been observed in erythrocytes (Seigneuret and Devaux, 1984; Daleke and Huestis, 1985). Seigneuret and Devaux (1984) explained the shape changes they observed as follows. The outer to inner surface ratio is higher for the echinocyte than for the discocyte and the transport of aminophospholipids they observe could reduce this ratio. As a result, this transport could induce the transition from echinocyte to discocyte. This explanation is consistent with the bilayer coupling hypothesis of Sheetz and Singer (1974). However, the cytoskeleton could also play a role in determining the erythrocyte shape. Op den Kamp *et al.* (1985) argue that in membrane areas where the bilayer and cytoskeleton are uncoupled, the phospholipids may no longer experience the constraints of the cytoskeleton and could increase their rate of flip-flop.

2.4. Lipid Domains

A biological membrane contains many different species of lipids and proteins. The question of whether or not certain membrane components will aggregate is decided by a balance between energy and entropy (Israelachvili *et al.*, 1980). The entropy of mixing always favors randomization of the molecules; the energy of interactions between certain membrane components may favor aggregation. If interactions are not too selective, the entropy of mixing wins out and the membrane components remain randomly dispersed. If, on the other hand, a particularly strong interaction exists between certain types of molecules, the energy will dominate and phase separation into domains, for example, fluid and solid domains, will occur (Israelachvili *et al.*, 1980). Lipids in the fluid state generally mix homogeneously in the membrane (Shimshick and McConnell, 1973; Mabrey and Sturtevant, 1976). In a few cases, however, immiscibility between fluid-phase lipids has been observed (Wu and McConnell, 1975). Segregation between anionic and neutral lipids can be induced by calcium ion binding (Ohnishi and Ito, 1973; Van Dijck *et al.*, 1976) and polylysine (Hartmann *et al.*, 1977). Phase separation into fluid and solid lipid domains is a general feature of lipid mixtures (Shimshick and McConnell, 1973; Taylor *et al.*, 1973; Hui and Parsons, 1975; Mabrey and Sturtevant, 1976; Luna and McConnell, 1978). Phase diagrams of lipid mixtures have been discussed by Lee (1977). Quantitative descriptions of the lateral membrane organization have been introduced by Freire and Snyder (1980a, 1980b, 1982). The existence of domains in membranes is expected to have biological implications. Domains may play a role, for example, in surface receptor capping (Klausner *et al.*, 1980), immunorecognition (Ruysschaert *et al.*, 1977), diffusion-mediated trapping (McCloskey and Poo, 1984), and the regulation of transport to coated pits (Eisinger and Halperin, 1986).

3. FLUORESCENCE TECHNIQUES AND WHAT THEY MAKE VISIBLE

A selection of a few fluorescence techniques that are frequently used in membrane research is discussed here. Other important techniques are measurement of membrane potential using fluorescent probes (Waggoner, 1979, 1985, 1986; London *et al.*, 1986), visualization of calcium concentrations (Campbell, 1983; Tsien *et al.*, 1984), solvent relaxation studies (Stubbs *et al.*, 1985), imaging and quantitative fluorescence microscopy (Arndt-Jovin *et al.*, 1985; Bright and Taylor, 1986; Webb and Gross, 1986), and total internal reflection fluorimetry (Axelrod *et al.*, 1984) with interesting applications to studies of immunorecognition processes (Brian and McConnell, 1984; Watts *et al.*, 1984).

3.1. Fluorescence Depolarization

When fluorophores embedded in membranes are illuminated by polarized light, the fluorescence they emit will also be polarized. As will be explained below, fluorescence emitted by molecules that rotate very rapidly and without restriction will be completely depolarized, whereas fluorophores that are strongly hindered in their rotational motion will emit fluorescence with a marked polarization. The polarization reflects the degree of hindrance exerted by the neighboring molecules in the membrane and depends on fluidity parameters.

The experiment is schematically shown in Figure 4. A beam of light from a lamp or a laser is directed through a monochromator or filter to select a narrow wavelength band and through a polarizer, which transmits vertically polarized light. This light enters the sample containing the membranes labeled with a small amount of probe molecules (typically, the probe to lipid ratio is about 1 : 500). Fluorescence emitted in a direction perpendicular to the incoming beam is measured. One observes the vertical component, I_V, measured with a polarizer parallel to the polarization direction of the exciting beam, and the horizontal component, I_H, measured with a polarizer perpendicular to that direction. Monochromators or cutoff filters are used in the fluorescent beams to select the fluorescence wavelength range of interest and to block scattered light. The degree of polarization is commonly expressed as polarization, P, or as fluorescence anisotropy, r:

$$P = \frac{I_V - I_H}{I_V + I_H} \tag{9}$$

$$r = \frac{I_V - I_H}{I_V + 2I_H} \tag{10}$$

It follows that P and r are closely related:

$$P = \frac{3r}{2 + r} \tag{11}$$

$$r = \frac{2P}{3 - P} \tag{12}$$

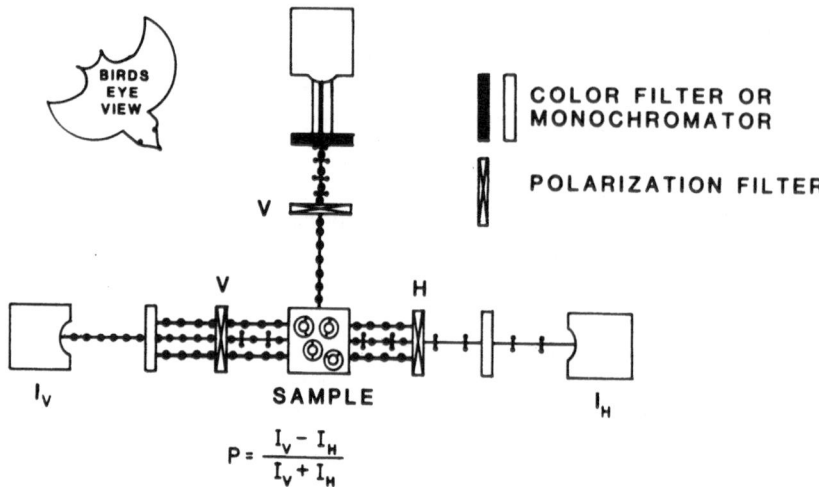

$$P = \frac{I_V - I_H}{I_V + I_H}$$

FIGURE 4. A schematic diagram of a fluorescence depolarization experiment. It is a bird's eye view on the instrumentation. The lamp (at the top of the diagram) emits a beam of radiation, a mixture of light of several colors. One color (wavelength band) is transmitted through a color filter or monochromator. This light, which may be considered as a mixture of vertically and horizontally polarized light (represented by a line with dots and bars), is headed toward a polarization filter, labeled V, which transmits only vertically polarized light (represented by a line with dots). This beam enters the sample cuvette containing a suspension of membranes, which have been labeled with a fluorophore. The membrane suspension will scatter the vertically polarized light, but this scattered light (lines with dots) will be blocked by the filters placed in the emission beams. The fluorescence emitted by the fluorophores can again be considered as a mixture of vertically and horizontally polarized light (lines with dots and bars). This radiation is detected by photomultiplier tubes at the end of the emission beams. The emission beam to the left of the sample contains a vertical polarization filter, which transmits vertically polarized light only (lines with dots). A cutoff filter (or monochromator) blocks the scattered light having the same color as the exciting light and allows only the fluorescence to enter the photomultiplier tube, which registers the intensity of the fluorescence with vertical polarization (I_V). The emission beam to the right of the sample is led through a horizontal polarization filter, which transmits horizontally polarized light only (lines with bars). A cutoff filter (or monochromator) blocks any possible scattered light and transmits the fluorescence into the photomultiplier tube, which measures the intensity of the fluorescence with horizontal polarization (I_H).

Here it is assumed that I_H is corrected for differences in detection for vertically and horizontally polarized light; that is (Chen and Bowman, 1965),

$$I_V = I_{VV} \tag{13}$$

$$I_H = \frac{I_{VH} I_{HV}}{I_{HH}} \tag{14}$$

where I_{VV}, I_{VH}, I_{HV}, and I_{HH} are intensities measured with the excitation and emission polarizers either vertical or horizontal as specified below:

	Excitation polarizer	
	Vertical	Horizontal
Emission polarizer vertical	I_{VV}	I_{HV}
Emission polarizer horizontal	I_{VH}	I_{HH}

The concept of photoselection, which is very helpful in understanding the principle of the technique, is depicted in Figure 5 for the rodlike probe DPH (1,6-diphenyl-1,3,5-hexatriene). This probe can absorb ultraviolet light (absorption peak near 360 nm), if the electric vector of the light is parallel to the long axis of the molecule. Probes oriented perpendicular to the polar-

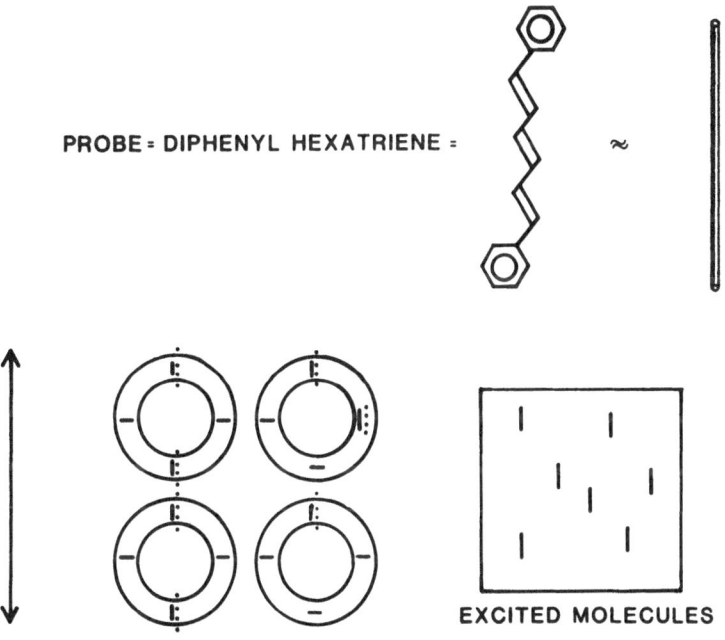

FIGURE 5. The concept of photoselection is illustrated. As an example, the probe 1,6-diphenyl-1,3,5-hexatriene has been chosen. This molecule can be considered as a rod, with absorption and emission moments aligned with its axis. When a suspension of membranes, labeled with these little rods, is illuminated with vertically polarized light, only those fluorophores that are oriented vertically (or at a small angle with the vertical) can be excited. The probes that have been photoselected in this way have been marked with dots and also grouped together in the lower right part of the figure. It should be noted that the alignment of this distribution of photoselected probes has been exaggerated somewhat; the probability of excitation is proportional to $\cos^2\alpha$, where α is the angle between the vertical and the axis of the probe.

ization direction of the incoming light cannot be excited. Consequently, a selection of excited molecules is created in the sample, oriented parallel to the polarization direction of the exciting light. The probability of absorbing a photon is proportional to $\cos^2\alpha$, where α is the angle between the absorption transition moment of the probe (in DPH along the long axis) and the electric vector of the exciting light. The alignment of the photoselected molecules is somewhat exaggerated in Figure 5. The light emitted by a fluorophore is completely polarized along the direction of its emission transition moment [along the long axis in the case of DPH; see Vos *et al.* (1983), however]. The intensity of the vertical component of the fluorescence emitted by such rodlike probes is proportional to the sum of the vertical projections squared and that of the horizontal component is proportional to the sum of the horizontal projections squared. The photoselected group of excited rodlike fluorophores, created at time zero by a flash of polarized light, will undergo changes due to rotational diffusion and also due to fluorescence, that is, to the decay of the excited state. These changes are depicted in Figure 6.

The effect of the rotational motion is that the initial alignment will decrease with time, so that, if this mechanism would act alone, I_V would decrease and I_H would increase. The effect of the decay of the excited state is that the group of excited molecules becomes smaller, so that both com-

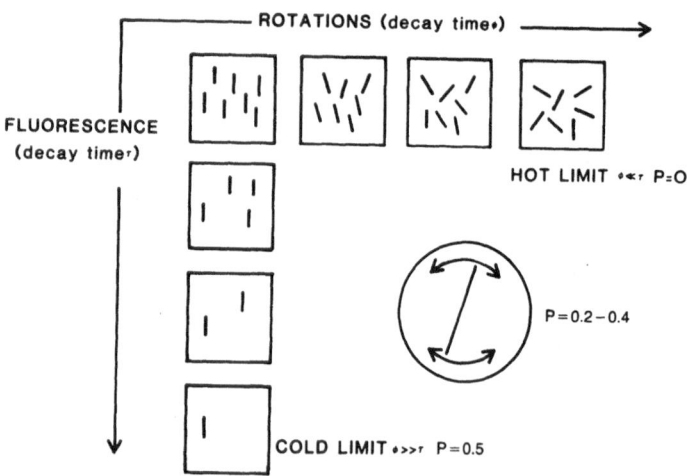

FIGURE 6. An illustration of the two main causes of change in the group of photoselected probes. In the hot limit, where the change due to reorientations dominates, the polarization is equal to 0. In the cold limit, on the other hand, the change due to fluorescence dominates and the polarization (in the case of rodlike probes) is 0.5. Rodlike probes embedded in membranes wobble around preferred orientations and in most cases the polarization is in the range 0.2–0.4.

ponents tend to fade away. Assuming that the rotational decay from the originally aligned distribution to the more random equilibrium distribution could be described by one decay time ϕ and that the decay of the excited state could be characterized by one decay time τ, we can distinguish two limiting cases:

1. The hot limit, $\phi << \tau$, where rotational reorientations are unrestricted and very fast compared to the fluorescence lifetime. In this limit the original alignment is completely lost before the fluorescence decay begins and, as a consequence, the polarization will be zero.
2. The cold limit, $\phi >> \tau$, where the rotational reorientations are very slow compared to the fluorescence lifetime, so that fluorescence decay will be completed before the diffusive reorientations begin. In this case the polarization turns out to be equal to 0.5 in the case of parallel emission and absorption transition moments.

As will be discussed below, rodlike probes like DPH reveal a decay behavior for fluorescence and rotations that is somewhere between those limiting cases. The rotational diffusion of such probes in biological membranes and most model membranes is restricted and fast, and the steady-state polarization values are generally in the range 0.2–0.4.

Several authors have discussed the theory of fluorescence depolarization in membrane suspensions (Kinosita *et al.*, 1977; Lipari and Szabo, 1980, 1981; Zannoni, 1981; Zannoni *et al.*, 1983; Szabo, 1984; Van der Meer *et al.*, 1984). This theory applies to a dilute suspension of labeled membranes obeying the following conditions:

1. Low turbidity, so that depolarization resulting from scattering of excited or emitted light is negligible. [Turbidity corrections have been discussed by Eisinger and Flores (1985) and Lentz *et al.* (1979).]
2. Ideal wavelength isolation, that is, directly scattered light or Raman scatter (Jameson, 1984) is blocked and its contributions to the intensity and to r are negligible. If the contribution of scattered light is not negligible, one can correct for it by considering the signals from the unlabeled sample. In that way one can control for possible autofluorescence as well.
3. Low fluorophore/lipid ratio so that depolarization due to self-quenching by fluorescence energy transfer can be ignored (Weber, 1969b).

Under these conditions fluorescence is a two-step process resulting from absorption of a photon followed by emission from the excited state (see Figure 1). If we assume these processes to be independent, we can write the contribution from one fluorophore excited at time 0 to the fluorescence emitted at time t as a product:

$$I(t) \sim P_a(0)F(t)P_e(t) \qquad (15)$$

where $P_a(0)$ = the probability that the molecule is excited at time 0
$\qquad\qquad$ = $\cos^2\alpha(0)$, where $\alpha(0)$ is the angle between the excitation po-
$\qquad\qquad$ larization and the absorption moment at time 0;
$\qquad F(t)$ = the probability that the molecule is still excited at time t
$\qquad\qquad$ = $(1/\tau)\exp(-t/\tau)$, if the decay is single exponential;
$\qquad P_e(t)$ = the probability that the photon emitted at time t will pass
$\qquad\qquad$ through the emission polarizer
$\qquad\qquad$ = $\cos^2\theta(t)$, where $\theta(t)$ is the angle between the emission polar-
$\qquad\qquad$ ization and the emission moment at time t.

It is convenient to describe first the case of fluorescence following excitation by a very short pulse at time zero, because other cases of interest, like continuous excitation (steady-state experiments) and sinusoidal excitation (phase and modulation measurements) can be treated as a deconvolution of this case. Consider a fluorescence depolarization experiment, where a very short pulse of plane polarized light with a polarization parallel to the z axis travels along the laboratory y axis and excites the sample at the origin. Fluorescence emitted in the direction of the x axis is observed through a polarizer either parallel to the z axis (I_V) or perpendicular to the z axis (I_H). The intensities I_V and I_H can then be expressed as

$$I_V = cQ\langle\cos^2\alpha(0)\cos^2\epsilon(t)\rangle F(t) \qquad (16)$$
$$I_H = cQ\langle\cos^2\alpha(0)\cos^2 n(t)\rangle F(t) \qquad (17)$$

where Q is the quantum yield, c is a geometrical factor, and the angles $\alpha(0)$, $\epsilon(t)$, and $n(t)$ are defined as follows:

$\qquad \alpha(0)$ = angle between the z axis and the absorption moment at time
$\qquad\qquad$ zero;
$\qquad \epsilon(t)$ = angle between the z axis and the emission moment at time t;
$\qquad n(t)$ = angle between the y axis and the emission moment at time t.

The brackets $\langle\ \ \rangle$ denote an ensemble average over all molecular orientations up to time t. The fluorescence anisotropy can be calculated from Eqs. (16), (17), and (10):

$$r(t) = T/N$$
$$T = \langle \cos^2\alpha(0)\{\cos_2\epsilon(t) - \cos^2 n(t)\}\rangle \qquad (18)$$
$$N = \langle \cos^2\alpha(0)\{\cos^2\epsilon(t) + 2\cos^2 n(t)\}\rangle$$

By using the fact that the membrane sample as a whole (not the membrane itself) is isotropic, it can be shown (Zannoni, 1981; Zannoni et al., 1984; Szabo, 1984) that $r(t)$ reduces to

$$r(t) = 0.4\langle P_2(\cos\theta_{0t})\rangle \qquad (19)$$

where P_2 is the second Legendre polynomial, $P_2(x) = (3x^2 - 1)/2$, and θ_{0t} is the angle between the absorption transition moment at time zero and the emission transition moment at time t.

The average of $P_2(\cos\theta_{0t})$, $\langle P_2(\cos\theta_{0t})\rangle$, depends on the reorientation dynamics of the probe in the membrane and is an intricate function of time, diffusion constants, and orientational order parameters (Zannoni, 1981; Szabo, 1984). However, a relatively simple expression can be derived, if one assumes that $r(t)$ can be approximated as an exponential plus a constant (Heyn, 1979; Jähnig, 1979; Van der Meer et al., 1984):

$$r(t) = r_\infty + (r_0 - r_\infty)\exp(-t/\phi) \qquad (20)$$

where ϕ is the rotational correlation time, r_0 is the fundamental anisotropy, and r_∞ is the limiting anisotropy. The fundamental anisotropy r_0 is the fluorescence anisotropy immediately after excitation with a very short flash (time width much smaller than ϕ and τ). Equation (20) yields a satisfactory description of the anisotropy decay for a number of probes in a wide variety of biological and model membranes (Kinosita et al., 1984). This decay behavior is illustrated in Figure 7.

From Eq. (19) it follows that

$$r_0 = 0.4P_2(\cos\theta_0) \qquad (21)$$

where θ_0 is the angle between absorption and emission moment in the fluorophore. The limiting anisotropy, r_∞, is the $r(t)$ for $t >> \phi$ and can be evaluated readily, if one assumes that the orientational distribution function of the probe is cylindrically symmetric with respect to the membrane normal

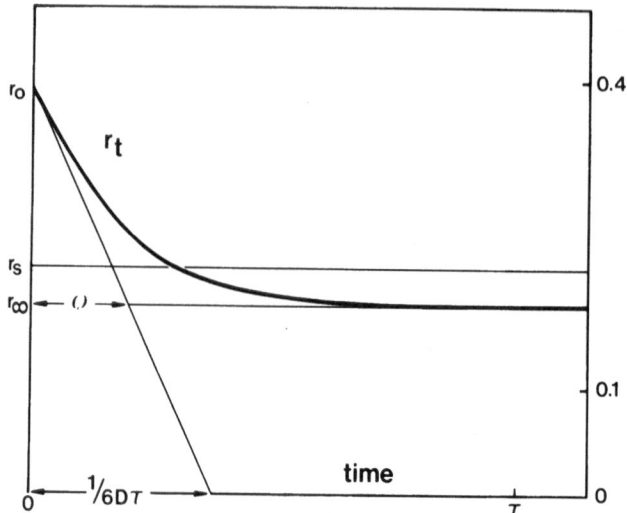

FIGURE 7. The fluorescence anisotropy decay as a function of time $[r_t = r(t)]$ following a very short flash illumination at time zero. In the particular example shown, the fundamental anisotropy (r_0) is 0.380, the steady-state fluorescence anisotropy (r_s) is 0.213, the limiting anisotropy (r_∞) equals 0.180, the fluorescence lifetime (τ) has been taken as the time unit, the rotational correlation time (ϕ) is 0.2, and the wobbling diffusion constant (D) equals $0.44/\tau$. From Van der Meer *et al.* (1986), with permission from the publisher and authors.

and that the probes themselves are effectively cylindrically symmetric as well (Zannoni, 1981; Szabo, 1984). In that case,

$$r_\infty = 0.4 P_2(\cos \theta_a) P_2(\cos \theta_e) \langle P_2 \rangle^2 \tag{22}$$

where θ_a and θ_e are the angles between the axis of the probe and the absorption moment and the emission moment, respectively, and $\langle P_2 \rangle$ is the second rank orientational order parameter of the probe (Van der Meer *et al.*, 1984). If it further can be assumed that the time dependence of $r(t)$ originates from a rotational diffusion process (which means that the probe molecules reorient through small angular steps and not by strong collisions with the surrounding lipids), then the time derivative of $r(t)$ immediately after flash excitation is proportional to a rotational diffusion constant. In the case of rodlike probes this derivative is proportional to D_W, the wobbling diffusion constant, which refers to reorientations of the long molecular axes and is in general different from D_R, the rotational diffusion coefficient for reorientations around the long axis. If the absorption and/or emission moment is aligned with the long molecular axis, ϕ will depend only on D_W. If both absorption and emission moments are at an angle with respect to the long molecular axis, $r(t)$ cannot be described by Eq. (20). However, if D_R and D_W do not differ too much, this equation could still be a reasonable approximation, and ϕ is related to the average rotational diffusion constant (Van der Meer et al., 1986):

$$\phi = \frac{r_0 - r_\infty}{6Dr_0} \tag{23}$$

where D is the wobbling diffusion constant in case of rodlike probes or else the average rotational diffusion coefficient.

The fluorescence anisotropy, $r(t)$, is thus, within the approximation of Eq. (20), determined by two fluidity-related parameters: the orientational order parameter $\langle P_2 \rangle$, which is inversely related to the amplitude of random librationals, and the rotational diffusion coefficient D, which reflects the frequency of these motions.

The steady-state fluorescence anisotropy r_s can be calculated from Eq. (20) as a time average (Heyn, 1979; Jähnig, 1979):

$$r_s = r_\infty + r_f \tag{24}$$

with

$$r_f = \frac{r_0 - r_\infty}{1 + \tau/\phi} \tag{25}$$

where τ is the fluorescence lifetime and it is assumed that the decay of the total fluorescence is single exponential. It follows that in the cold limit, $\tau/\phi \ll 1$, introduced above, $r_s = r_0$, with a maximal value of 0.4, corresponding to a maximal P value of 0.5, for $\theta_0 = 0°$. In the hot limit, $\tau/\phi \gg 1$, one would expect the motion to be unrestricted, so that $\langle P_2 \rangle = 0$ and therefore $r_s = r_\infty = 0 = P$ in that case.

It is clear that one single r_s measurement for a particular membrane cannot yield values for both parameters D and $\langle P_2 \rangle$ at the same time. However, it has been shown for many membranes and a number of fluorescent probes that a strong correlation exists between r_s and r_∞ (Van Blitterswijk et al., 1981; Pottel et al., 1983; Van der Meer et al., 1986). This correlation can be quantitatively formulated as follows. Compare two systems labeled with the same probe and having the same r_s value, of which one is a membrane suspension with restricted probe rotation, $r_s = r_f + r_\infty$, and the other is an isotropic reference oil with anisotropy r_s ($r_\infty = 0$). In the membrane, r_f is inversely related to the (average) rotational diffusion constant D and, similarly, in the isotropic oil, r_s is inversely related to D_i, the rotational diffusion constant in this system. Because r_f is smaller than r_s, D must be larger than D_i, whereas for D_i, Perrin's equation holds (Perrin, 1936; Shinitzky and Barenholz, 1978). Therefore, $6D\tau$ must be larger than $r_0/r_s - 1$; that is,

$$6D\tau = r_0/r_s - 1 + m \qquad (26)$$

where m has a positive value. The parameter m expresses the difference between the rotational diffusion of the probe in the membrane and that in the isotropic reference oil. Equation (26) is the extended Perrin equation, introduced by Van der Meer *et al.* (1986). Rearranging Eqs. (24), (25), and (26) gives the following $r_\infty - r_s$ relation (Van der Meer *et al.*, 1986):

$$r_\infty = \frac{r_0 r_s^2}{r_0 r_s + (r_0 - r_s)^2/m} \qquad (27)$$

This relation is not one unique relation but rather represents a set of $r_\infty(r_s)$ functions. A few examples are shown in Figure 8. A wide range of $r_\infty(r_s)$ data have been fitted to Eq. (27) (Van der Meer *et al.*, 1986). An example of such a fit for DPH data in liposomes of saturated lipids is shown in Figure 9.

It turns out that the m value depends on the type of membrane; for

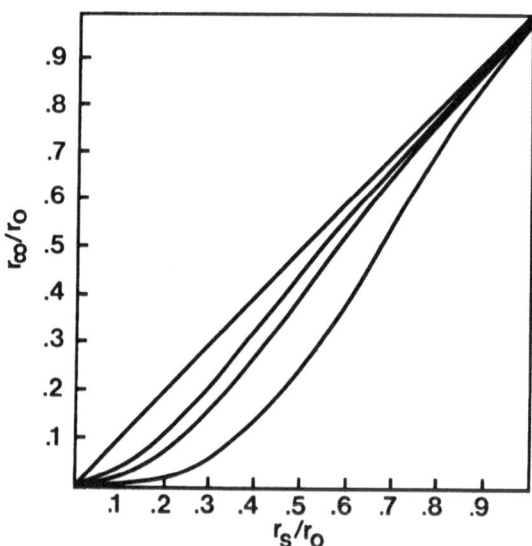

FIGURE 8. Four examples of plots of r_∞/r_0 versus r_s/r_0, derived from Eq. (27), for $m = \infty$, 4, 2, and 0.5, from the upper to the lowest curve, respectively. From Van der Meer *et al.* (1986), with permission from the publisher and authors.

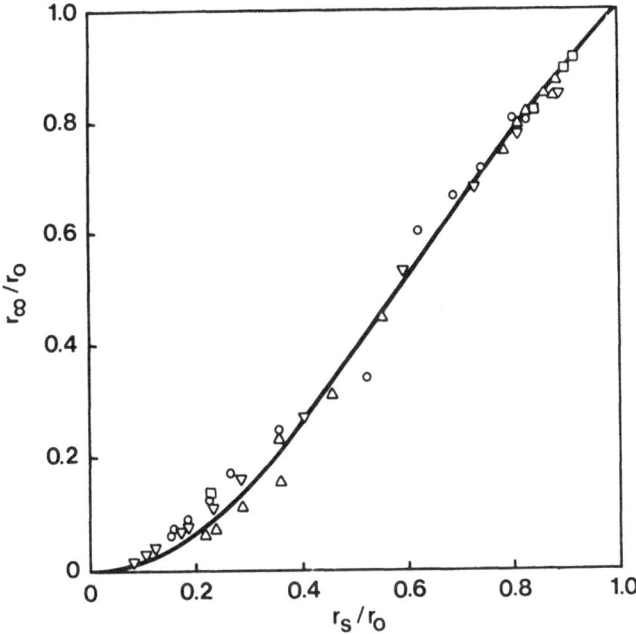

FIGURE 9. The best fit to Eq. (27) for literature data, yielding $m = 1.7$, of DPH in liposomes of dipalmitoylphosphatidylcholine [□, Stubbs *et al.* (1981); ○, Lakowicz *et al.* (1979); △, Kawato *et al.* (1977)] and of the same probe in liposomes of dimyristoylphosphatidylcholine [▽, Lakowicz and Prendergast (1978)]. The data correspond to a range of temperatures between 5 and 60°C. From Van der Meer *et al.* (1986), with permission from the publisher and authors.

unsaturated lipids $m(\text{DPH}) = 0.6$, for saturated lipids $m(\text{DPH}) = 1.7$, and for a wide range of biological membranes the m value equals 1.7 in the case of DPH (Van der Meer *et al.*, 1986). This variability leads to uncertainties in the estimations for D and $\langle P_2 \rangle$ from Eqs. (26) and (27), respectively. It has been shown that the estimation of $\langle P_2 \rangle$ for DPH is accurate for high r_s and the estimation of D is reliable only for low r_s values (Pottel *et al.*, 1983; Van der Meer *et al.*, 1986). The microviscosity interpretation of steady-state fluorescence depolarization data from DPH-like probes has been introduced by Shinitzky and co-workers (Shinitzky *et al.*, 1981; Shinitzky and Barenholz, 1974, 1978). At that time the fluorescence anisotropy in membranes was believed to behave as in an isotropic reference oil and the r_∞ contribution was assumed to be zero. Later, when the time dependence of the fluorescence anisotropy could actually be measured, it turned out that it contained a very noticeable r_∞ component (Chen *et al.*, 1977; Dale *et al.*, 1977; Kawato *et al.*, 1977). The r_∞ contribution to DPH data of r_s was shown to be dominant for biological membranes (Van Blitterswijk *et al.*, 1981). Shinitzky and Yuli introduced the concept of "true microviscosity," which is proportional to the rotational diffusion constant D defined above (Shinitzky and Yuli, 1982). This parameter was also obtained from steady-state r_s data but was corrected for the r_∞ contribution, using the apparent correlation between r_∞ and r_s (Van Blitterswijk *et al.*, 1981). It has been shown, however, that DPH r_s data for biological membranes fall in a region where the estimation of the structural order parameter is accurate, but that of the true microviscosity is not reliable (Pottel *et al.*, 1983; Van der Meer *et al.*, 1986). Therefore, the microviscosity interpretation of steady-state fluorescence anisotropy data using DPH and similar probes should be abandoned, because it only considers the rate of rotational reorientations and erroneously neglects the major contribution to the steady-state fluorescence anisotropy, that is, the limiting anisotropy r_∞, determined exclusively by membrane structural order.

Although Eq. (20) forms a good first approximation to the fluorescence anisotropy of rodlike probes in membranes, high-quality time-resolved data reveal some deviations from this time dependence (Ameloot *et al.*, 1984). These discrepancies may indicate that the approximation in deriving Eq. (20) from Eq. (19) has been too crude. An improved approximation has been proposed (Van der Meer *et al.*, 1984; Szabo, 1984) based on decomposing the correlation function $\langle P_2(\cos \theta_{0t}) \rangle$ into its components and interpolating for each component the short-time behavior with the long-time tail, again assuming that the reorientation process is diffusive. The result for a fluorophore with effective cylindrical symmetry that has the emission and/or the absorption moment aligned with its symmetry axis (like DPH) is

$$r(t) = r_0[\langle P_2 \rangle^2 + \beta_0 \exp(-\alpha_0 t) + 2\beta_1 \exp(-\alpha_1 t) + 2\beta_2 \exp(-\alpha_2 t)] \quad (28)$$

where α_m and β_m ($m = 1, 2, 3$) are given by

$$
\begin{align}
\beta_0 &= 0.2 + 2\langle P_2 \rangle/7 + 18\langle P_4 \rangle/35 - \langle P_2 \rangle^2 & (29a)\\
\alpha_0 &= 6D(0.2 + \langle P_2 \rangle/7 - 12\langle P_4 \rangle/35)/\beta_0 & (29b)\\
\beta_1 &= 0.2 + \langle P_2 \rangle/7 - 12\langle P_4 \rangle/35 & (29c)\\
\alpha_1 &= 6D(0.2 + \langle P_2 \rangle/14 + 8\langle P_4 \rangle/35)/\beta_1 & (29d)\\
\beta_2 &= 0.2 - 2\langle P_2 \rangle/7 + 3\langle P_4 \rangle/35 & (29e)\\
\alpha_2 &= 6D(0.2 - \langle P_2 \rangle/7 - 2\langle P_4 \rangle/35)/\beta_2 & (29f)
\end{align}
$$

where D is the wobbling diffusion constant, and $\langle P_2 \rangle$ and $\langle P_4 \rangle$ are the second rank and fourth rank orientational order parameters, respectively, defined above in Eq. (7). The original papers can be consulted for a treatment of a more general case and for details of the derivations (Van der Meer *et al.*,

1984; Szabo, 1984). Application of the $r(t)$ model of Eq. (28) to time-resolved data allows for measuring D, $\langle P_2 \rangle$, and $\langle P_4 \rangle$.

Ameloot *et al.* (1984) have applied the model in Eq. (28) to the analysis of time-resolved fluorescence anisotropy measurements for DPH embedded in small unilamellar vesicles of dimyristoylphosphatidylcholine (DMPC) and dipalmitoylphosphatidylcholine (DPPC). Their results are plotted in Figure 10. These results are not in agreement with the cone model, introduced by Kinosita *et al.* (1977) and discussed by Lipari and Szabo (1981). Especially above the chain-melting transition temperature T_m, a dramatic discrepancy between the data and the $\langle P_2 \rangle$ and $\langle P_4 \rangle$ values predicted by the cone model was observed. The results suggest that while below T_m most of the probes are wobbling around the membrane normal, above T_m about 50% of the probes are parallel to the plane of the membrane and about 50% are more or less perpendicular to the membrane (Ameloot *et al.*, 1984). This interpretation is depicted in Figure 11.

Wang *et al.* (1986) have studied essentially the same systems using the multifrequency phase and modulation technique [Ameloot *et al.* (1984) used the pulse technique] and analyzed the results with a different model. Their data were consistent with two components with different reorientation behavior. One component had a long lifetime (10 nsec below T_m and 7 nsec above T_m) and a restricted rotation that could be described by Eq. (20) (with r_∞ different from zero) and the other had a short lifetime (3 nsec) and an unrestricted rotation ($r_\infty = 0$) (Wang *et al.*, 1986). Although this interpretation differs from the one discussed above, the conclusion about the overall orientational distribution for the probes is essentially similar in the two studies; namely, that it is a bimodal distribution with two maxima, near 0° and 90°, respectively. Results from a recent phase and modulation fluorometry study are in agreement with the results and the interpretation of Ameloot *et al.* (1984) (Pottel *et al.*, 1987; Van der Meer *et al.*, 1987).

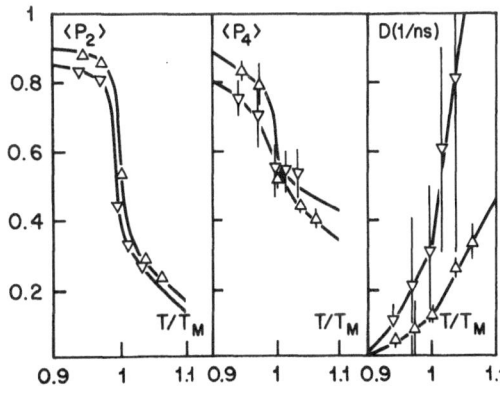

FIGURE 10. Data for $\langle P_2 \rangle$, $\langle P_4 \rangle$, and D for DPH in DMPC and DPPC as a function of temperature (Ameloot *et al.*, 1984). ($\langle P_2 \rangle$, 2nd rank orientational order parameter; $\langle P_4 \rangle$, 4th rank orientational order parameter; D, wobbling diffusion constant (1/nanoseconds); T, temperature; T_M, transition temperature; $\triangle$, DMPC = dimyristoylphosphatidylcholine; $\triangledown$, DPPC = dipalmitoylphosphatidylcholine.)

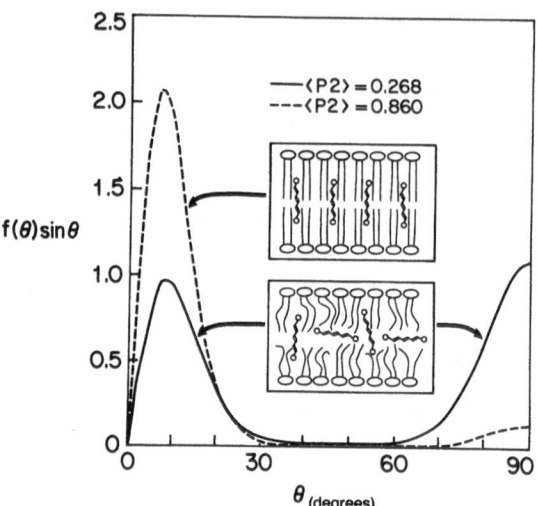

FIGURE 11. Distribution functions for DPH in a lipid vesicle below the chain-melting temperature (dashed line, corresponding to $\langle P_2 \rangle = 0.860$ and $\langle P_4 \rangle = 0.79$) and above the phase transition (continuous line, corresponding to $\langle P_2 \rangle = 0.268$ and $\langle P_4 \rangle = 0.52$). Cartoons of these distributions have also been indicated in the figure with exaggerated probe concentrations; the real lipid/probe ratio was about 500. Redrawn from Ameloot et al. (1984).

Fluorescence depolarization in membranes is an extremely broad area of research and so far we have shown very little of it. The remainder of this section is used to sketch a map of this area and to point out a few directions.

Fluorescence depolarization studies of membranes can be classified in a number of ways, according to (1) technique (time-resolved or steady-state), (2) sample (membrane suspensions or oriented membranes), (3) membrane type (biological or model membranes, intact cells), (4) probe (e.g., DPH, perylene, parinaric acid), or (5) type of research (from physical structure of liposomes to medical applications).

(1) **Technique.** Steady-state fluorescence depolarization was developed by Perrin (1936) and introduced to biochemistry by Weber (1952) and to membrane research by Shinitzky et al. (1971). In this technique the illumination and, as a consequence, the fluorescence are constant in time. Initially, the data were analyzed in terms of microviscosity. This analysis was based on the comparison of the depolarization of a probe in a membrane with that in a reference oil (Shinitzky et al., 1971; Shinitzky and Barenholz, 1974, 1978). However, experiments of Hare and Lussan with a number of different reference oils revealed some ambiguity of the method (Hare and Lussan, 1977). Later, when fast time-resolved techniques became available, it became clear that the Perrin equation did not hold for fluorescence depolarization of probes in membranes (Chen et al., 1977; Dale et al., 1977; Kawato et al., 1977). The assumption that this

equation was applicable to membranes was fundamental in the microviscosity interpretation. It therefore became clear that the microviscosity concept should be abandoned, at least in the interpretation of fluorescence depolarization data of DPH and similar rodlike probes in membranes [see discussion above and, e.g., the following references: Kinosita et al. (1977), Heyn (1979), Jähnig (1979), Dale (1983), Hare (1983), and Sklar (1984)].

Time-resolved fluorescence anisotropy has been widely applied to membrane probe studies [see discussion above; for a review see, e.g., Dale (1983)]. There are two types of time-resolved technique: the pulse method and the harmonic response method. In the former the sample is illuminated by a series of short pulses (pulse width of the order of a nanosecond, repetition rate about 0.01–1 MHz) and the fluorescence response is measured using a single photon-counting apparatus and a multichannel analyzer (Cundall and Dale, 1983). In the harmonic response method or phase and modulation fluorometry, the sample is illuminated by light that is sinusoidally modulated in intensity at high frequencies, typically in the megahertz range. The fluorescence will then also be modulated at the same frequency but will be shifted in phase and demodulated with respect to the exciting light. The phase shift and demodulation are measured, preferably at more than a few frequencies [for reviews see Gratton et al. (1984) and Jameson et al. (1984)]. Time-resolved fluorescence studies give not only information on the time dependence of the fluorescence anisotropy but also on fluorescence lifetimes (Wolber and Hudson, 1981; Dale, 1983; Kinosita et al., 1984; Parasassi et al., 1984b). Measurement of fluorescence lifetimes has important biological applications, for example, in the detection of lipid domains (Klausner et al., 1980; Parasassi et al., 1984a; Barrow and Lentz, 1985).

(2) Sample. In the majority of studies the sample is an aqueous suspension of membranes, but in a number of papers oriented membranes have been used (Badley et al., 1973; Kooyman et al., 1981; Van der Meer et al., 1982; Vos et al., 1983; Van de Ven and Levine, 1984; Vogel and Jähnig, 1985). These membrane systems consist of a stack of hydrated lipid multilayers sandwiched between two cover slides. Interesting results can be obtained when the angles between the membrane normal and the excitation and/or the emission polarization are varied. Such measurements may be called angle-resolved fluorescence depolarization experiments (Vos et al., 1983). At least three different geometries have been used (Kooyman et al., 1981; Vos et al., 1983; Vogel and Jähnig, 1985). Care must be taken to avoid artifacts and it is not clear if in all studies the optical corrections have been applied (Van der Meer et al., 1982) and defects have been avoided (Asher and Pershan, 1979; Vogel and Jähnig, 1985). As an alternative to angle-resolved studies, one can measure the depolarization at different regions of a curved membrane having different membrane orientation with respect to the optical axis of the system (Yguerabide and Stryer, 1971; Andrich and Vanderkooi, 1976; Axelrod, 1979).

(3) **Membrane Type.** Membrane suspensions can be either artificial (model) membranes [small or large unilamellar lipid vesicles (SUV or LUV), multilamellar vesicles (MLV), or black lipid membranes (BLM)] or biological membranes (e.g., erythrocyte ghosts or plasma membranes isolated from other tissue) or intact cells (e.g., see Shinitzky and Barenholz, 1978; Dale, 1983; Shinitzky, 1984). Suspensions of MLV, biomembranes, and intact cells scatter light. This scattering gives rise to artifacts, which can be corrected, however (Lentz et al., 1979; Eisinger and Flores, 1985). Fluorescence depolarization from lipophilic probes incorporated in whole cells represents a measure of the average fluidity and order of all lipids present in a cell and thus does not exclusively monitor the cell surface membrane (Van Hoeven et al., 1979; Bouchy et al., 1981). For amphiphilic probes, the situation is different (Kleinfeld, 1985; Martin and Pagano, 1987). An interesting quenching method to resolve the fluorescence anisotropy contribution from the plasma membrane and endomembranes has been proposed by Grunberger et al. (1982). It may not be entirely clear why this method has not been applied since.

(4) **Probe.** The most frequently used fluorescence depolarization probe is DPH. Many other probes are available (Shinitzky and Barenholz, 1978; Haugland, 1985). Examples are ANS [1-anilinonaphthalene-8-sulfonate (Slavik, 1982)], perylene (Zannoni et al., 1983; Chong et al., 1985a), anthroyloxy fatty acids (Thulborn and Sawyer, 1978; Vincent et al., 1982), and parinaric acid (Sklar et al., 1979), which can be used to study domains in membranes (see Section 3.3.1).

(5) **Type of Research.** Fluorescence depolarization has been used to study the physical state of membranes (e.g., see Dale, 1983; Zannoni et al., 1983), the chemical factors determining polarization (e.g., see Van Blitterswijk et al., 1987; Van Blitterswijk, Chapter 12 in this volume), biological functions of membranes (e.g., see Nathan et al., 1979; De Laat and Van der Saag, 1982; Chong et al., 1985b), and medical implications of membrane fluidity (e.g., see Shinitzky, 1984).

3.2. Quenching

When certain substances are added to a fluorescent sample, the fluorescence intensity decreases in a concentration-dependent way. This process is called quenching and substances that can induce this decrease are called quenchers. A wide variety of quenchers are available (Lakowicz, 1983). Examples are oxygen, iodine, and acrylamide. Fluorescence quenching data are commonly presented in quenching plots as shown in Figure 12. Quenching can result from collisional encounters between fluorophore and quencher. This process is called collisional or dynamic quenching. Another process is static quenching, which is due to complex formation. It is important to distinguish between the two types of quenching.

In dynamic quenching the fluorescence intensity is decreased, because fluo-

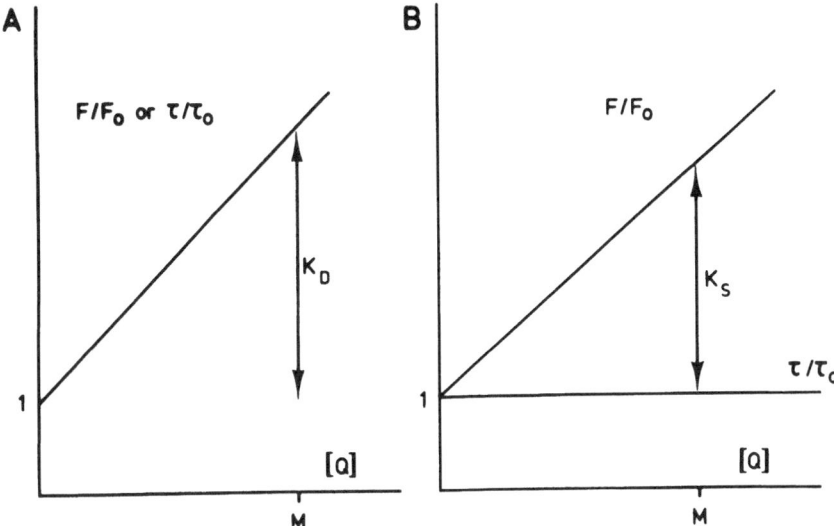

FIGURE 12. Illustrative quenching plots in the case of dynamic (A) and static (B) quenching. The unit of K_D and K_S is 1/M (M = Molar).

rophores in the excited state collide with quenchers and lose their excitation energy in the collision without emission of light. The quencher must diffuse to the fluorophore during the lifetime of the excited state. Therefore, the probability for quenching (and thus the constant K_D, see below) is proportional to the lifetime of the excited state and to the sum of the diffusion coefficients of fluorophore and quencher. Dynamic quenching is described by the Stern–Volmer equation:

$$F_0/F = \tau_0/\tau = 1 + K_D[Q] \tag{30}$$

In this equation F_0 and F are the fluorescence intensities in the absence and presence of quencher, respectively, τ_0 and τ are the lifetimes in the absence and presence of quencher, respectively, $[Q]$ is the concentration of quencher, and K_D is the dynamic or Stern–Volmer quenching constant.

In static quenching, a quencher–fluorophore complex is formed, which is nonfluorescent. The quencher acts to reduce the level of free fluorophore that can be excited, and hence the fluorescence intensity decreases. Any remaining fluorophore that is excited, however, will give the unquenched lifetime (see Figure 12B). Static quenching is described by

$$F_0/F = 1 + K_S[Q] \tag{31}$$

where K_S is the static quenching constant, which is equal to the equilibrium constant for the fluorophore–quencher complex (Lakowicz, 1983).

If both static and dynamic processes occur at the same time, the Stern–Volmer equation may be modified to

$$F_0/F = (1 + K_D[Q])(1 + K_S[Q]) \qquad (32)$$

This equation predicts an upward curvature of F_0/F in a plot versus quencher concentration.

There are a number of applications to membrane research. Fluorescence quenching by spin probes and various ions has been used to determine the transverse location of fluorescent probes, such as the n-(9-anthroyloxy) fatty acids and pyrene derivatives (Thulborn and Sawyer, 1978; Haigh et al., 1979; Luisetti et al., 1979; London and Feigenson, 1981; Chalpin and Kleinfeld, 1983; Blatt and Sawyer, 1985; Chattopadhyay and London, 1987). Leto et al. (1980) have studied the mechanism of cytochrome b5 intervesicle exchange by measuring quenching of its tryptophan fluorescence by brominated lipids. Also using tryptophan quenching by brominated lipids, Everett et al. (1986) have shown that the tryptophan in the membrane-binding domain of cytochrome b5 is located 0.7 nm below the surface of the membrane vesicles. In this study, asymmetric vesicles were used where the inner and outer monolayers contained different amounts of brominated lipids (Everett et al., 1986). Determination of the degree of exposure of membrane proteins to the aqueous phase is another possible application (Shinitzky and Rivnay, 1977). The binding of fatty acids to lipid bilayers has been evaluated by measuring fluorescence quenching by fatty acids and their brominated analogues (Froud et al., 1986). The quenching of fluorescence of n-(9-anthroyloxy) fatty acids and other probes by different ubiquinone homologues and analogues has been exploited to assess the localization and lateral mobility of the quinones in lipid bilayers of model and mitochondrial membranes (Fato et al., 1986). Deleers et al. (1986) have used quenching of NBD-labeled lipids in vesicles to monitor the lipid phase separation induced by zinc and calcium ions. And a method for calculating membrane permeability and surface potential from ionic quenching data has also been presented (Morris et al., 1985).

3.3. Fluorescence Energy Transfer

Fluorescence energy transfer (FET) is the transfer of excited-state energy from a donor (D) to an acceptor (A) whose absorption spectrum overlaps with the fluorescence spectrum of the donor. This transfer does not occur by the transmission of a photon but is the result of resonance dipole–dipole interactions between the donor and the acceptor (Förster, 1948, 1949, 1959; Stryer, 1978; Lakowicz, 1983; Blumberg, 1985). Fluorescence energy transfer (FET) is observed as a reduction of donor fluorescence in the presence of acceptor, or as the increased emission of the acceptor when excited at the absorbance of the donor (see Figure 13). When a cuvette containing donor and acceptor is illuminated by light at a wavelength where the donor absorbs and donor molecules are excited, a competition process begins between energy decaying as donor fluorescence and energy transferring to the acceptor. This dual decay can be described by the following reaction scheme:

$$
\begin{array}{ccc}
D + A + h\nu_E \longrightarrow & D^* + A \xrightarrow{\ k_T\ } & D + A^* \\
& \downarrow k_D & \downarrow \\
& A + D + h\nu_{FD} & D + A + h\nu_{FA} \quad (33)
\end{array}
$$

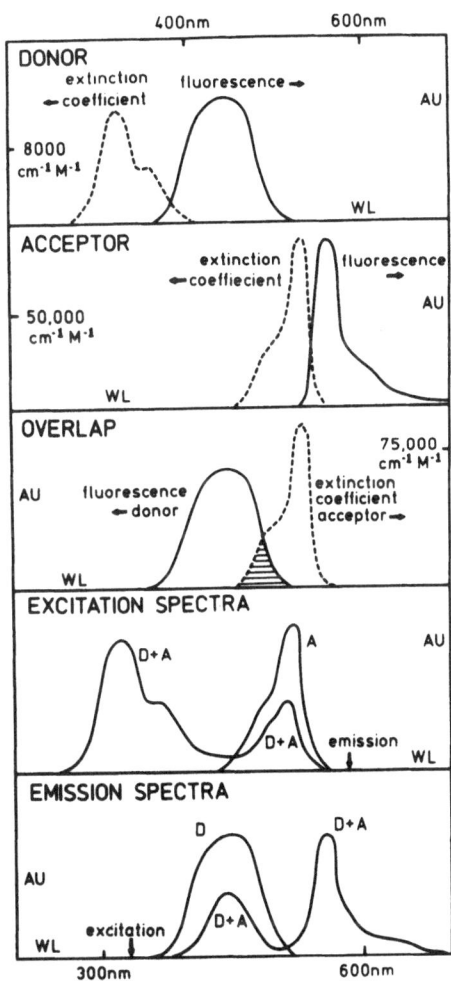

FIGURE 13. Donor and acceptor spectra illustrating fluorescence energy transfer. AU stands for arbitrary units and WL for wavelength. D + A in the plots of excitation spectra refers to the case of excess donor and A means without donor. D + A in the plots of emission spectra refers to the case of excess acceptor and D means without acceptor.

where k_D is the radiational decay rate of the donor in the absence of acceptor and k_T is the rate of energy transfer; ν_E, ν_{FD}, and ν_{FA} are wave numbers, ν_E of the donor excitation, ν_{FD} of the donor fluorescence, and ν_{FA} of the acceptor fluorescence (sensitized emission). It is assumed here that radiationless decay can be ignored. These decay modes for A* and D* could also be taken into account in Eq. (33), similarly as indicated in Eq. (1). The fraction of absorbed photons that are transferred to the acceptor is called the efficiency of transfer, E:

$$E = k_T/(k_T + k_D) \tag{34}$$

This parameter is the most readily accessible measure of transfer. It can be obtained from the fluorescence quantum yield in the presence (Q_{DA}) and in the absence of acceptor (Q_D):

$$E = 1 - Q_{DA}/Q_D \tag{35}$$

The transfer efficiency is very sensitive to the distance R between donor and acceptor. According to theory (Förster, 1948) and experiment (Stryer, 1978) this distance dependence is given by

$$E = R_0^6/(R_0^6 + R^6) \tag{36}$$

which follows from

$$k_T = (R_0/R)^6 \tag{37}$$

where R_0, the distance at which the transfer efficiency is 50%, is given by

$$R_0 = 979(JK^2Q_Dn^{-4})^{1/6} \text{ nm} \tag{38}$$

In this expression J denotes the spectral overlap integral between acceptor absorbance and donor emission (Förster, 1948; Stryer, 1978), K^2 is the orientation factor (Dale et al., 1979a; Eisinger et al., 1981; Blumberg, 1985), and n is the refractive index of the medium.

Equations (34), (36), and (37) apply only to the case of donor–acceptor pairs that are at a fixed distance. Examples are proteins or peptides labeled with two different chromophores (Stryer, 1978). A fixed distance results in a single transfer rate, and the decay of the emission will therefore be a single exponential. In FET experiments in membranes the donor–acceptor distance is not fixed and the fluorescence decay deviates from monoexponential decay. In such measurements there is a distribution of donor–acceptor pairs. In membranes, the dependence of the transfer efficiency on distance parameters is more intricate than the relation of Eq. (36) and has been studied by several groups (Tweet et al., 1964; Shaklai et al., 1977; Fung and Stryer, 1978;

Estep and Thompson, 1979; Koppel *et al.*, 1979; Wolber and Hudson, 1979; Dewey and Hammes, 1980: Snyder and Freire, 1982). It can be derived in the following steps:

1. Write down the expression for the fluorescence decay for a particular donor–acceptor configuration by summing over the transfer rates for transfer from one donor to surrounding acceptors.
2. Calculate the average decay by averaging over all possible configurations.
3. Obtain the steady-state fluorescence (Q_{DA}, Q_D) by integrating over time.
4. Calculate the transfer efficiency from Q_{DA} and Q_D using Eq. (35).

The orientation factor K^2 complicates the determination of distance parameters. In the dynamic regime, where reorientational relaxation is fast compared to the fluorescence and transfer rates, bounds for the average value of K^2 can be obtained from polarization measurements (Dale *et al.*, 1979a,b; Eisinger *et al.*, 1981; Blumberg, 1985).

A few illustrative examples of FET in membranes are discussed in the following sections.

3.3.1. Detection of Domains in Membranes

Sklar *et al.* (1979) have studied the lipid organization in the retinal rod outer segment (ROS) plasma membrane using the fluorescence properties of *trans*-parinaric acid and *cis*-parinaric acid (*trans*-parinaric acid = *trans*-PnA = 9,11,13,15-all-*trans*-octadecatetraenoic acid; *cis*-parinaric acid = *cis*-PnA = 9,11,13,15-*cis,trans,trans,cis*-octadecatetraenoic acid). ROS membranes are rich in the photopigment rhodopsin. In rats whose photoreceptor phospholipid was modified by dietary manipulation, it was found that the electroretinogram (a measure of photoreceptor function) depends on the position and total number of double bonds in the dietary supplements (Wheeler *et al.*, 1975). These studies suggest that membrane lipids play a role in photoreceptor function.

The results of Sklar *et al.* (1979) form an important contribution to the understanding of the lipid organization in these membranes. They have combined the information from FET, polarization, and quantum yield measurements. The polarization of the parinaric acid probes is very sensitive to domain formation. *Trans*-PnA partitions preferentially into lipid solid phase where it exhibits a strongly enhanced quantum yield; the polarization of this probe is responsive to as little as 1% solid phase. *Cis*-PnA distributes more equally between coexisting solid and fluid lipid phase (Sklar *et al.*, 1979). The polarization data suggest that solid and fluid lipid phases coexist at low temperatures (around 5°C), while the membrane is homogeneously fluid at high temperatures (around 35°C). FET can provide some insight as to how large the rigid patches are and if rhodopsin prefers the fluid phase or not.

The emission spectra of *cis*-PnA and *trans*-PnA overlap the absorption spectra of retinal (free retinal or rhodopsin-bound retinal). FET can therefore occur from PnA (donor) to retinal (acceptor). In ROS phospholipids with added retinal, the energy transfer efficiency from *trans*-PnA to retinal decreases strongly over the temperature associated with the phospholipid lateral phase separation. This result can be explained by the partition of *trans*-PnA into rigid lipid patches as they form and the exclusion of retinal from these patches. Preliminary calculations suggest that solid phase patches must be at least twice the R_0 distance for this system or about 7 nm, and probably much larger, in order to account for the reduced energy transfer at low temperature.

Tryptophan is a fluorescence donor to either *cis*-PnA or *trans*-PnA (R_0 equals 2.5–3.0 nm). In ROS membranes, the energy transfer efficiency between rhodopsin tryptophan and *trans*-PnA is reduced at temperatures where rigid domains are present in the membrane; the energy transfer to *cis*-PnA is virtually temperature independent.

Sklar *et al.* study not only the ROS membranes themselves but also lipid mixtures that mimic the composition of the outer and inner monolayers. The picture that emerges from their results and the literature data they quote is the following:

At 5°C rigid patches exist in the ROS membrane of about 7 nm in diameter. *Trans*-PnA is included and rhodopsin is excluded and *cis*-PnA is virtually unaffected by these rigid domains. About 15% of the lipids contribute to the domains; in the inner monolayer these lipids are mainly disaturated phosphatidylcholines plus cholesterol and in the outer monolayer primarily monosaturated phophatidylethanolamines and phosphatidylserines. Clusters of phospholipids may exist at 15°C but are small compared to the patches at 5°C. Around 35°C the membrane is fluid and homogeneous.

3.3.2. Tryptophan Imaging of Membrane Proteins

Determination of the molecular structure of membrane proteins is an important problem in molecular biology. Tertiary structure information originates most often from diffraction studies of crystalline proteins. Mapping of tryptophan residues can be achieved by FET between tryptophans and membrane probes (Kleinfeld, 1985). This approach could be essential for determining the tertiary structure in the case that crystallization would not be feasible for the membrane protein studied.

Two groups have performed FET measurements for determining the location of the fluorescent tryptophan (Trp-109) of cytochrome b5 reconstituted in lipid vesicles (Fleming *et al.*, 1979; Kleinfeld and Lukacovic, 1985).

Cytochrome b5 is a membrane protein that plays an important role in the electron transfer processes on the surface of the endoplasmic reticulum. Structural data indicate that the protein is composed of a heme containing a polar segment located above the plane of the membrane and a nonpolar segment buried halfway into the membrane. This nonpolar peptide (NPP) can be isolated by trypsin digestion (Fleming *et al.*, 1979).

Fleming *et al.* have studied FET from the donor Trp-109 in cytochrome b5 or in its nonpolar peptide, reconstituted into lipid vesicles, to the acceptors [(N-(trinitrophenyl)dimyristoylethanolamine (TNP-PE, R_0 = 3.2 nm) or dansyldodecylamine (DDA, R_0 = 2.6 nm)], embedded in the lipid bilayer (Fleming *et al.*, 1979). The matrix lipid in the vesicles was formed by dimyristoylphosphatidylcholine (95%) and dimyristoylphosphatidylethanolamine (5%) and the lipid/protein ratio was 200–600. The acceptor/lipid ratio was between 0.2 and 2%. The acceptor chromophores were shown to be placed at the head group region of the outer monolayer (Fleming *et al.*, 1979). Energy transfer efficiencies were obtained from the FET-induced decrease of the tryptophan fluorescence. The results were analyzed in terms of the theory of Koppel and co-workers (Koppel *et al.*, 1979). Orientational factors were taken into account by measuring axial depolarization factors. However, the orientational factor for TNP-PE could not be obtained in this way because this chromophore is nonfluorescent and an assumption was made for its orientation factor. This assumption has been criticized (Blumberg, 1985). The conclusion of their study is that the tryptophan residue (Trp-109) is located 2–3 nm below the outer membrane surface, so that it sits essentially in the middle of the bilayer.

Kleinfeld and Lukacovic have studied FET from the same donor (Trp-109 in cytochrome b5 or in its nonpolar peptide), reconstituted into sonicated vesicles of dimyristoylcholine, to the series of acceptors n-(9-anthroyloxy) fatty acids (n-AO), with n = 2, 3, 6, 7, 9, 10, 12, or 16 (R_0 = 2.4 nm) (Kleinfeld and Lukacovic, 1985). The n-AO were embedded in the outer monolayer of the membrane. The anthroyloxy fluorophore is located 1.9 nm above the membrane midplane for n = 2 and 0.3 nm above this plane for n = 16. These authors measured also the FET from the tryptophan donor to the heme moiety of cytochrome b5 (R_0 = 2.9 nm) and from the AO donors to the heme acceptor (R_0 = 3.6–4.0 nm). Transfer efficiencies were obtained from the decrease in the tryptophan fluorescence in the case of FET from tryptophan to n-AO and from tryptophan to heme and from the decrease in AO fluorescence for FET from AO to heme. In the case of FET from tryptophan to AO, the energy transfer efficiencies can be obtained either from the tryptophan quenching or from the FET-sensitized AO emission. These two efficiencies were reported to be in excellent agreement.

The conclusion from this study is that the tryptophan residue is located at 2 nm from the membrane surface, in agreement with the results of Fleming and co-workers (Fleming et al., 1979; Kleinfeld and Lukacovic, 1985). In addition, Kleinfeld and Lukacovic find that the effective protein radius of the hydrophilic segment is 2.2 nm, the radius of the hydrophobic part is 1.4 nm, and the heme is located 1.5 nm above the membrane surface.

3.3.3. Diffusion-Enhanced FET

If the donor-acceptor distance can change during the excited-state life-time of the donor by diffusion, FET can be enhanced (Haas et al., 1978; Thomas et al., 1978; Katchalski-Katzir and Steinberg, 1981). The parameter determining the degree of this enhancement is

$$Z = D\tau_0/s^2 \tag{39}$$

where D is the sum of the diffusion coefficients of the donor and acceptor, τ_0 is the donor lifetime in the absence of transfer, and s is the mean donor–acceptor distance.

Three diffusion regimes can be distinguished (Thomas et al., 1978):

1. $Z < < 1$, the static limit, where the transfer is low and constant (i.e., there is essentially no variation with diffusion).
2. $Z \approx 1$, the intermediate range, where the efficiency is sensitive to diffusion.
3. $Z > > 1$, the rapid diffusion limit, where the efficiency approaches a maximal value and becomes again independent of diffusion.

Since distances of interest are in the range of 1–10 nm and the diffusion coefficient of a typical fluorophore in aqueous solvents is of the order of 10^{-6} cm^2/sec, the donor lifetime in the case of maximally enhanced FET ($Z >> 1$) should be several orders of magnitude above the conventional nano-second time range. Thomas et al. (1978) used the fluorophore Tb^{3+} ion chelated to dipicolinate [Tb(DPA)$_3{}^{3-}$], which has a lifetime close to 2 msec in aqueous solutions. They measured the FET from the donor [Tb(DPA)$_3{}^{3-}$] trapped within vesicles to eosin phosphatidylethanolamine (EPE) embedded in the vesicle membrane. The vesicles were small liposomes of egg phos-phatidylcholine containing between 0.06 and 0.3% EPE. The transfer effi-ciency was shown to be very sensitive to both the acceptor/phospholipid ratio and the distance of closest approach between donor and acceptor, in excellent agreement with theory (Thomas et al., 1978). The distance of closest ap-proach between the inner vesicle surface and the EPE chromophore, measured with this technique, is 1 nm (Thomas et al., 1978).

Thomas and Stryer (1982) have applied diffusion-enhanced FET to elucidate the position of the retinal chromophore of rhodopsin in retinal disc membranes. They find that the distance of closest approach between terbium dipicolinate and retinal is 2.2 nm from the intradisc surface and 2.8 nm from the external surface of the disc membrane. They also estimate the precision of the method to be of the order of 0.1–0.2 nm. However, factors such as irregularities of the membrane surfaces and electrostatic interactions can worsen the precision somewhat (Thomas and Stryer, 1982). Applications of diffusion-enhanced FET are also discussed by Yeh and Meares (1980), Meares and Rice (1981), and Isaacs *et al.* (1986).

FET has also been used as a means of measuring membrane fusion (Keller *et al.*, 1977; Gibson and Loew, 1979; Struck *et al.*, 1981), lipid–protein interactions (Kimelman *et al.*, 1979), surface density (Fung and Stryer, 1978), and protein–protein interactions (Fernandez and Berlin, 1976).

3.4. Fluorescence Recovery after Photobleaching (FRAP)

The principle of a FRAP experiment is as follows. The membrane sample is labeled with fluorescent molecules and is located in the sample holder of a fluorescence microscope. A small spot (about 1–10 μm in diameter) is illuminated by a beam of light of low intensity, exciting the fluorophores in that spot. The intensity of the emitted fluorescence is a measure for the number of fluorophores present in the spot. By exposing this spot suddenly to a very brief pulse of high-intensity light, a portion of the fluorophores in that region will be bleached irreversibly. Subsequently, one monitors the same spot with the low-intensity light. A gradual increase of the fluorescence is now observed as bleached and unbleached fluorophores randomize their positions by diffusion. This FRAP process is illustrated in Figure 14.

As indicated in Figure 14, two parameters can be derived from the fluorescence recovery curves—D_L, the lateral diffusion coefficient of the mobile fluorophores, and R, the fraction of fluorophores that are mobile on the experimental time scale (Axelrod *et al.*, 1976). D_L reflects a dynamic property of the mobile molecules, while R relates to the organization of the marker in the membrane (Yechiel *et al.*, 1985).

The bleached region is not necessarily a spot. In fact, often other bleaching patterns are used, each topology focusing on one particular aspect of the structure or dynamics of the membrane system studied. Periodic pattern bleaching (Smith and McConnell, 1978; Koppel and Sheetz, 1983) measures dynamics averaged over large areas of one or more regions. Localized line or spot bleaching (Axelrod *et al.*, 1976; Koppel, 1979) yields spatially resolved information, allowing a comparison between the properties of different

B. Wieb Van der Meer

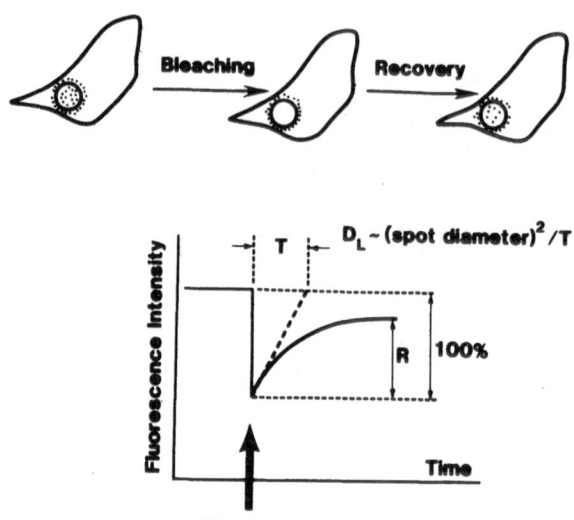

FIGURE 14. Illustration of the technique, which is commonly referred to by the abbreviation FRAP—fluorescence recovery (or redistribution) after photobleaching—used here, or FPR—fluorescence photobleaching recovery. Fluorophores in a membrane (e.g., the plasma membrane of a cell) are illuminated with a weak-intensity beam in a spot. Next, these fluorophores are bleached by a short high-intensity pulse. This spot is again illuminated by the low-intensity beam, immediately after the flash, and the recovery of fluorescence intensity as a result of exchanging places of bleached and unbleached fluorophores by lateral diffusion is measured. The fluorophores can be labeled proteins or lipids. Two parameters can be obtained from the recovery curves: the lateral diffusion constant D_L and the recovery R.

regions and different locations within the same region. Bleaching whole sections of the surface with a scanning slit (Koppel *et al.*, 1986) yields information on interregional transport.

FRAP has made an enormous contribution to our present knowledge about the mobility of proteins and lipids in membranes and has provided insight into the biological functions of this mobility. This progress has been clearly described in various excellent reviews and research papers; for example, see the papers by Peters *et al.* (1974), Poo and Cone (1974), Koppel et al. (1976), McCloskey and Poo (1984), Jacobson *et al.* (1976), Cherry (1979), Axelrod (1983), Devaux and Seineuret (1985), Edidin (1981), Edidin *et al.* (1976), Webb *et al.* (1982), and the literature cited therein. Here it will suffice to quote some general results and mention a few illustrative examples.

Typical D_L values for proteins in biological membranes fall in the range of 10^{-10}–10^{-12} cm^2/sec, about 1000–100,000 times slower than an average soluble protein in aqueous solution. On the other hand, isolated integral proteins diffuse much faster, 10^{-8} cm^2/sec, and with nearly complete recovery when incorporated into liposomes composed of fluid phase phospholipids. The extent of recovery, R, for proteins in biological membranes is in general less than 100%. Fluorescent lipid analogues are generally rapidly mobile (D_L of the order of 10^{-8} cm^2/sec) in both model and biological membranes. Recovery is usually high for lipid diffusion in both systems.

The wealth of lipid and protein diffusion data available now enables us to perform a critical evaluation of the validity of the diffusion theories that have been developed for membranes so far. As discussed in Section 2.1, we have the hydrodynamic model introduced by Saffman and Delbrück (1975) and extended by Hughes et al. (1982), and on the other hand we have a random walk model (Galla et al., 1979), a milling crowd model (Eisinger et al., 1986), a modified percolation theory (Saxton, 1982), and effective medium theories (Saxton, 1982; O'Leary, 1987). Convincing evidence is available that for large proteins embedded in lipid model membranes the diffusion behavior is in agreement with the hydrodynamic model (Peters and Cherry, 1982; Vaz et al., 1984). To check if for such systems the assumption holds that the viscosity of the aqueous phase is much smaller than that of the membrane (Saffman and Delbrück, 1975) or not (Hughes et al., 1982), it suffices to evaluate the parameter P, defined above in Eq. (6). Experimental data for D_R and D_L of bacteriorhodopsin reconstituted into dimyristoylphosphatidylcholine at a low protein/lipid ratio of 0.5% (Peters and Cherry, 1982) yield:

$$D_L = 1.4 \ \mu m^2/sec \ (24.5°C) \text{ and } 2.4 \ \mu m^2/sec \ (28.5°C)$$

$$D_R = 64,000/sec^{-1} \ (24.5°C) \text{ and } 79,000 \ sec^{-1} \ (28.5°C) \text{ and } a = 2 \ nm$$

These data give $p = 5.5$ at 24.5°C and $p = 7.6$ at 28.5°C, consistent with the assumption made by Saffman and Delbrück (1975) that the membrane viscosity is much higher than the bathing viscosities ($\epsilon << 1$, see Table I). Moreover, the actual D_L values for various large membrane proteins (with a radius larger than that of the lipids) and the weak dependence on radius (Vaz et al., 1984) as well as the dependence on aqueous phase viscosity (Peters and Cherry, 1982) are in excellent agreement with the equations of Saffman and Delbrück (1975). The viscosity of a fluid bilayer is of the order of 0.6–1.5 poise (Peters and Cherry, 1982; Vaz et al., 1984). When the proteins are present in higher concentrations and the fractional membrane area occupied by proteins approaches 0.5 (which it is in erythrocytes), there is a drastic reduction in both the protein and lipid lateral diffusion (Peters and Cherry, 1982; Vaz et al., 1984; Eisinger et al., 1986; O'Leary, 1987). This reduction is directly dependent on the amount of protein in the membrane and can be explained as obstructed diffusion (Eisinger et al. 1986; O'Leary, 1987).

For estimating the parameter p for lipidlike probes, we can use $D_L <$
0.3×10^{-8} cm²/sec, obtained for dioctadecylindotricarbocyanine in lipid
bilayers (Fahey and Webb, 1978), $a = 0.7$ nm, and $D_R = 0.03$–0.3 nsec^{-1}
[(Ameloot et al., 1984); these values were actually for diphenylhexatriene
in lipid bilayers, however, this range is in agreement with $\tau D_R = 0.27$
(Axelrod, 1979) and τ being very short, 1–5 nsec (Haugland, 1985)]. These
values yield $p = 0.2$ to 0.02, clearly inconsistent with the theory of Saffman
and Delbrück, as discussed above. Moreover, the strong dependence of D_L
on the free area of lipidlike probes is not in agreement with the Saffman–
Delbrück equations (Vaz et al., 1984). Rather, the linear relation between ln
(D_L) and the free area of the probe fits the free volume model for diffusion
(Cohen and Thurnbull, 1959; Galla et al., 1979; Vaz et al., 1984).

3.5. Excimer Fluorescence

Pyrene and its derivatives have an interesting spectral property: the emission spectrum depends dramatically on the concentration of the fluorophore.
At low concentration one observes two peaks at 378 and 400 nm, but at
higher fluorophore concentration an additional broad peak appears around
470 nm (see Figure 15). Förster and his collaborators showed that the broad
470-nm peak is due to the formation of excited-state dimers (excimers =
excited-state dimers) and that the excimer production rate is diffusion limited
(Förster, 1969; and references therein). The kinetic processes involved can
be summarized in the following scheme:

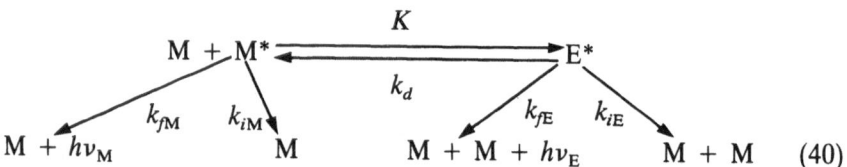

$$\tag{40}$$

where M denotes a monomer and M* an excited-state monomer, E* is an
excited state dimer (excimer); ν_M and ν_E are wave numbers for the monomer
and excimer emission, respectively; the decay constants k_{fE} and k_{fM} are the
excimer and monomer fluorescence decay parameters, respectively, and k_{iM}
and k_{fE} are the radiationless decay parameters for monomer and excimer,
respectively. The constant K is the rate constant for excimer formation and
k_d denotes the rate of regeneration of the excited monomer from the dissociation of excimer. K depends on the probe fraction x [= probe/(probe +
lipid)] and at low probe fractions K is proportional to x (Galla et al., 1979;
Galla and Hartmann, 1980; Eisinger et al., 1986). This decay scheme is more
complex than the ordinary scheme presented in the introduction. When an

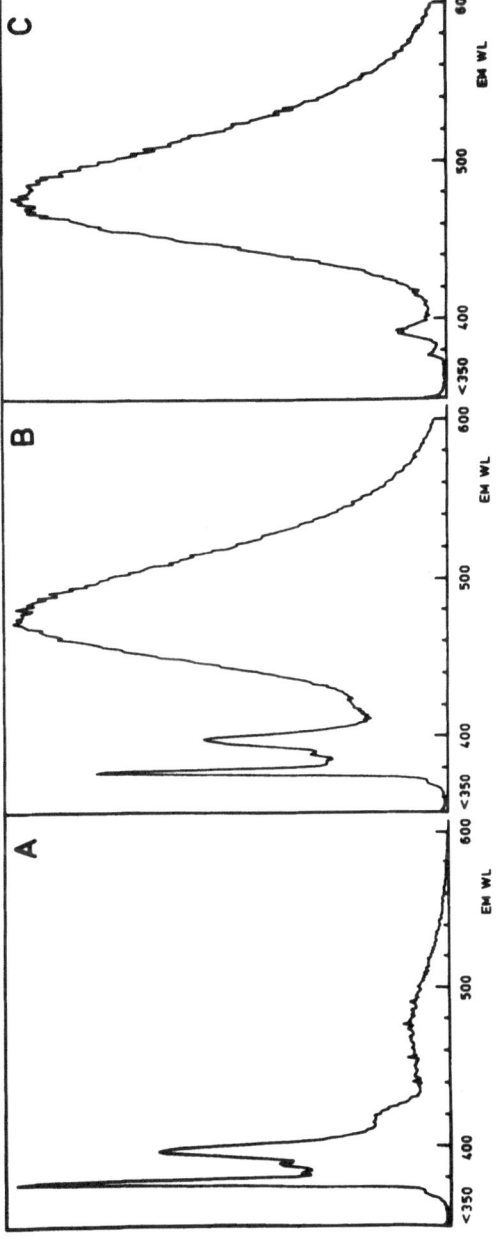

FIGURE 15. Emission spectra of a pyrene-labeled phosphatidylcholine [PyrPC = 3-palmitoyl-2-(1-pyrenedecanoyl)-1-α-phosphatidylcholine from Molecular Probes, OR] in lipid vesicles of PL [PL = PC/PS(95/5), PC = L-α-lecithin (egg), PS = phosphatidylserine (bovine brain), from Avanti Polar Lipids, AL], for (A) 2% PyrPC [PyrPC/PL(2/98)], (B) 16% PyrPC [PyrPC/PL(16/84)], and (C) 100% PyrPC, respectively.

ordinary fluorophore is excited, only one thing can happen; sooner or later it will lose its excitation energy (with or without emitting radiation). When a pyrene derivative is excited, however, one of two things can occur; either it meets another pyrene fluorophore in its ground state and interacts with it forming an excimer and subsequently decays from this state, or it does not interact and decays as a monomer. The monomeric and excimeric quantum yields, ϕ_M and ϕ_E, evaluated from this scheme, read (Eisinger et al., 1986):

$$\phi_M = \frac{\phi_M{}^{max}}{1 + D^*(x)} \tag{41}$$

$$\phi_E = \frac{\phi_E{}^{max} D^*(x)}{1 + D^*(x)} \tag{42}$$

with

$$\phi_M{}^{max} = k_{fM}\tau_M \tag{43}$$

$$\tau_M = \frac{1}{k_{fM} + k_{iM}} \tag{44}$$

$$\phi_E{}^{max} = \frac{k_{fE}\tau_E}{1 - \tau_E k_d} \tag{45}$$

$$\tau_E = \frac{1}{k_{fE} + k_{iE} + k_d} \tag{46}$$

$$D^*(x) = (1 - \tau_E k_d)\tau_M K \tag{47}$$

K is proportional to the jump frequency (Galla et al., 1979) or the rate of neighbor–neighbor exchanges (Eisinger et al., 1986) and is therefore proportional to the lateral diffusion coefficient. However, it also depends on the probe fraction x. This x dependence can be evaluated from various models, reviewed by Eisinger et al. (1986):

1. In the *diffusion limit model*, K is linearly proportional to x. This is analogous to Förster's three-dimensional diffusion model (Förster, 1969) and is expected to hold only for very low probe fractions.
2. In the *random walk model* (Galla et al., 1979; Galla and Hartmann, 1980), K is proportional to $(1/x)\ln(2/x)$. This model is based on the random walk of a probe on points of a lattice with a particular jump frequency.
3. In the *milling crowd model* (Eisinger et al., 1986), K is also a monotonically increasing function of x, which can be evaluated by a Monte Carlo simulation. In this model probes migrate by changing places with neighboring

membrane lipids at a particular frequency. Here K depends on an additional parameter as well, the excimerization probability between neighboring probes.

Excimer fluorescence has been used to study lateral diffusion (Galla and Sackmann, 1974; Vanderkooi and Callis, 1974; Galla *et al.*, 1979; Galla and Hartmann, 1980; Galla and Luisetti, 1980; Eisinger *et al.*, 1986; Jones and Lentz, 1986). The lateral diffusion coefficient D_L for lipidlike pyrene derivatives measured with the excimer technique is of the order of 10^{-8} to 10^{-7} cm²/sec in fluid bilayers (Galla *et al.*, 1979; Galla and Hartmann, 1980), in agreement with FRAP lateral diffusion data for similarly sized probes in fluid bilayers (Vaz *et al.*, 1984). However, excimer D_L values for lipid probes in erythrocyte membranes [10^{-8} to 10^{-7} cm²/sec (Galla and Luisetti, 1980; Eisinger *et al.*, 1986)] are considerably faster than FRAP D_L values for lipid probes ($2-8 \times 10^{-9}$ cm²/sec) in intact red cell membranes (Bloom and Webb, 1983) or their ghosts (Kapitza and Sackmann, 1980; Thompson and Axelrod, 1980). It should be noted that the excimer technique measures short-range diffusion over nanometers, while FRAP measures long-range diffusion over micrometers. Peters and Cherry (1982) have shown that the long-range diffusion coefficient of a lipid probe in a lipid bilayer decreases by an order of magnitude when the bacteriorhodopsin/lipid ratio increases from 0 to 33%. This decrease and the discrepancy between long- and short-range D_L values in erythrocyte membranes can be explained by stating that FRAP yields protein-obstructed diffusion coefficients, while the excimer technique measures local lateral diffusion, which is similar to that of pure lipid bilayers (Eisinger *et al.*, 1986; O'Leary, 1987).

Excimer fluorescence can also be used to study interbilayer and transbilayer lipid exchange (Galla and Hartmann, 1980; Galla and Luisetti, 1980; Roseman and Thompson, 1980; Pownall *et al.*, 1982; Frank *et al.*, 1983; Massey *et al.*, 1984; Wong *et al.*, 1984; Homan and Pownall, 1986) and the lateral organization of phospholipids (Galla and Sackmann, 1975; Haberer *et al.*, 1982; Chong and Thompson, 1985; Somerharju *et al.*, 1985; Wiener *et al.*, 1985; Hresko *et al.*, 1986, 1987; Jones and Lentz, 1986).

4. SUMMARY AND CONCLUSIONS

In this chapter we have discussed the application of fluorescence spectroscopy to membrane research. Fluorescence depolarization has been given special attention because of personal bias, not because it is the most important technique. Basic concepts in fluorescence have been explained very briefly in the Introduction. A number of books and review articles on fluorescence are available and have been cited here.

Which aspects of structure and dynamics can be made visible by fluorescence? That is the question we have addressed in this chapter. It seemed necessary to cut this question into two parts: What are the relevant parameters that describe the membrane structure and dynamics and what are the various fluorescence techniques for measuring those parameters? These questions are dealt with in Sections 2 and 3 of this chapter, respectively.

Fluorescence has contributed to the understanding of biomembranes. Examples have been given in this chapter of studies that have given insight into the physical state, the composition, and functions of membranes. Fluorescence can often only demonstrate a correlation between a physical and a functional parameter. The technique becomes more powerful, however, when applied in combination with relevant biological methods.

ACKNOWLEDGMENTS. Thanks are due to Dr. Dave Jameson for stimulating discussions. The expert assistance of Dick Irish with the preparation of a number of figures is gratefully acknowledged.

5. REFERENCES

Ameloot, M., Hendricks, H., Herreman, W., Pottel, H., Van Cauwelaert, F., and Van der Meer, B. W., 1984, Effect of orientational order on the decay of the fluorescence anisotropy in membrane suspensions. Experimental verification on unilamellar vesicles and lipid/α-lactalbumin complexes, *Biophys. J.* **46** : 247–256.

Andrich, M. P., and Vanderkooi, J. M., 1976, Temperature dependence of 1, 6-diphenyl-1, 3, 5-hexatriene fluorescence in phospholipid artificial membranes, *Biochemistry* **15** : 1257–1261.

Armond, P. A., and Staehelin, L. A., 1979, Lateral and vertical displacement of integral membrane proteins during lipid phase transition in *Anacystis nidulans, Proc. Natl. Acad. Sci. U.S.A.* **76**: 1901–1905.

Arndt-Jovin, D. J., Robert-Nicaud, M., Kaufman, S. J., and Jovin, T. M., 1985, Fluorescence digital imaging microscopy (DIM) in cell biology, *Science* **230** : 247–256.

Asher, S. A., and Pershan, P. S., 1979, Alignment and defect structures in oriented phosphatidylcholine multilayers, *Biophys. J.* **27** : 393–422.

Axelrod, D., 1979, Carbocyanine dye orientation in red cell membrane studied by microscopic fluorescence polarization, *Biophys. J.* **26** : 667–674.

Axelrod, D., 1983, Lateral motion of membrane proteins and biological function, *J. Membr. Biol.* **75** : 1–10.

Axelrod, D., Koppel, D. E., Schlessinger, J., Elson, E., and Webb, W. W., 1976, Mobility measurement by analysis of fluorescence photobleaching recovery kinetics, *Biophys. J.* **16** : 1055–1069.

Axelrod, D., Burghardt, T. P., and Thompson, N. L., 1984, Total internal reflection fluorescence, *Annu. Rev. Biophys. Bioeng.* **13** : 247–268.

Backer, J. M., and Dawidowicz, E. A., 1985, Membrane assembly: Evidence for a phospholipid "Flipase" in rat liver microsomes, *J. Cell Biol.* **101** : 261a.

Badley, R. A., Martin, W. G., and Schneider, H., 1973, Dynamic behavior of fluorescent probes in lipid bilayer model membranes, *Biochemistry* **12** : 268–275.

Barrow, D. A., and Lentz, B. R., 1985, Membrane structural domains. Resolution limits using diphenylhexatriene fluorescence decay, *Biophys. J.* **48** : 221–234.

Berg, H. C., 1983, *Random Walks in Biology*, Princeton University Press, Princeton, NJ.

Berlman, I. B., 1971, *Handbook of Fluorescence Spectra of Aromatic Molecules*, Academic Press, New York.

Bishop, W. R., and Bell, R. M., 1985, Assembly of the endoplasmatic reticulum phospholipid bilayer: The phosphatidylcholine transporter, *Cell* **42** : 51–60.

Blatt, E., and Sawyer, W. H., 1985, Depth-dependent fluorescent quenching in micelles and membranes, *Biochim. Biophys. Acta* **822** : 43–62.

Bloom, J. A., and Webb, W. W., 1983, Lipid diffusibility in the intact erythrocyte membrane, *Biophys. J.* **42** : 295–305.

Blumberg, W. E., 1985, Fluorescence energy transfer as a structural probe in membranes and membrane-bound proteins, in *Physical Methods on Biological Membranes and Their Model Systems* (F. Conti, W. E. Blumberg, J. de Gier, and F. Pocchiari, eds.), pp. 95–122, Plenum Press, New York.

Borochov, H., and Shinitzky, M., 1976, Vertical displacement of membrane proteins mediated by changes in microviscosity, *Proc. Natl. Acad. Sci. U.S.A.* **73** : 4526–4530.

Bouchy, M., Donner, M., and André, J. C., 1981, Evolution of fluorescence polarization of diphenylhexatriene during the labelling of living cells, *Exp. Cell Res.* **133** : 39–46.

Bretscher, M. S., 1973, Membrane structure: Some general principles, *Science* **181** : 622–629.

Brian, A. A., and McConnell, H. M., 1984, Allogenic stimulation of cytotoxic T-cells by supported planar membranes, *Proc. Natl. Acad. Sci. U.S.A.* **81** : 6159–6163.

Bright, G. R., and Taylor, D. L., 1986, Imaging at low light level in fluorescence microscopy, in *Applications of Fluorescence in the Biomedical Sciences* (D. L. Taylor, A. S. Waggoner, R. F. Murphy, F. Lanni, and R. R. Birge, eds.), pp. 257–288, Alan R. Liss, New York.

Campbell, A. K., 1983, *Intracellular Calcium*, Wiley, New York.

Chaplin, D. B., and Kleinfeld, A. M., 1983, Interaction of quenchers with the *n*-(9-anthroyloxy) fatty acid membrane probes, *Biochim. Biophys. Acta* **731** : 465–474.

Chattopadhyay, A., and London, E., 1987, Parallax method for direct measurement of membrane penetration depth utilizing fluorescence quenching by spin-labeled phospholipids, *Biochemistry* **26** : 39–45.

Chen, R. F., and Bowman, R. L., 1965, Fluorescence polarization: Measurement with ultraviolet-polarizing filters in a spectrophoto fluorometer, *Science* **147** : 729–732.

Chen, L. A., Dale, R. E., Roth, S., and Brand, L., 1977, Nanosecond time-dependent fluorescence depolarization of diphenylhexatriene in dymiristoyl-lecithin vesicles and the determination of "microviscosity," *J. Biol. Chem.* **252** : 2163–2169.

Chen, R. F., and Edelhoch, H., 1975, *Biochemical Fluorescence*, Vols. 1 and 2, Marcel Dekker, New York.

Cherry, R. J., 1979, Rotational and lateral diffusion of membrane proteins, *Biochim. Biophys. Acta* **559** : 289–327.

Chong, P. L.-G., and Thompson, T. E., 1985, Oxygen quenching of pyrene-lipid fluorescence in phospatidylcholine vesicles, *Biophys. J.* **47** : 613–621.

Chong, P. L.-G., Van der Meer, B. W., and Thompson, T. E., 1985a, The effects of pressure and cholesterol on rotational motions of perylene in lipid bilayers, *Biochem. Biophys. Acta* **813**: 253–265.

Chong, P. L.-G., Fortes, P. A. G., and Jameson, D. M., 1985b, Mechanisms of inhibition of (Na, K)-ATPase by hydrostatic pressure studied with fluorescent probes, *J. Biol. Chem.* **260**: 14484–14490.

Cohen, M. H., and Thurnbull, D., 1959, Molecular transport in liquids and glasses, *J. Chem. Phys.* **31** : 1164–1169.

Cundall, R. B., and Dale, R. E., 1983, *Time-Resolved Fluorescence Spectroscopy in Biochemistry and Biology*, Plenum Press, New York.

Dale, R. E., 1983, Membrane structure and dynamics by fluorescence probe depolarization kinetics, in *Time-Resolved Fluorescence Spectroscopy in Biochemistry and Biology* (R. B. Cundall and R. E. Dale, eds.), pp. 555–612, Plenum Press, New York.

Dale, R. E., Chen, L. A., and Brand, L., 1977, Rotational relaxation of the microviscosity probe diphenylhexatriene in paraffin oil and egg lecithin vesicles, *J. Biol. Chem.* 252 : 7500–7510.

Dale, R. E., Eisinger, J., and Blumberg, W. E., 1979a, Orientational freedom of molecular probes—the orientation factor in intramolecular energy transfer, *Biophys. J.* 26 : 161–194.

Dale, R. E., Eisinger, J., and Blumberg, W. E., 1979b, Correction: Orientational freedom of molecular probes—the orientation factor in intramolecular energy transfer, *Biophys. J.* 30 : 365.

Daleke, D. L., and Huestis, W. H., 1985, Incorporation and translocation of aminophospholipids in human erythrocytes, *Biochemistry* 24 : 5406–5416.

De Gier, J., Van Echteld, C. J. A., Killian, J. A., De Kruijff, B., Mandersloot, J. G., Noordam, P. C., Van der Steen, A. T. M., and Verkleij, A. J., 1985, Lateral and polymorphic phase transitions in relation to the barrier function of a lipid membrane, in *Physical Methods on Biological Membranes and Their Model Systems* (F. Conti, W. E. Blumberg, J. De Gier, and F. Pocchiari, eds.), pp. 81–93, Plenum Press, New York.

De Laat, S. W., and Van der Saag, P. T., 1982, The plasma membrane as a regulatory site in growth and differentiation of neuroblastoma cells, *Int. Rev. Cytol.* 74 : 1–75.

Deleers, M., Servais, J. P., and Wulfert, E., 1986, Micromolar concentrations of Zn^{2+} potentiates Ca^{2+}-induced phase separation of phosphatidyl serine containing liposomes, *Biochim. Biophys. Res. Commun.* 136 : 476–481.

Demel, R. A., Van Deenen, L. L. M., and Pethica, B. A., 1967, Monolayer interactions of phospholipids and cholesterol, *Biochim. Biophys. Acta* 135 : 11–19.

Devaux, P. F., and Seigneuret, M., 1985, Specificity of lipid–protein interactions as determined by spectroscopic techniques, *Biochim. Biophys. Acta* 822 : 63–125.

Dewey, T. G., and Hammes, G. G., 1980, Calculation of fluorescence resonance energy transfer on surfaces, *Biophys. J.* 32 : 1023–1036.

Edidin, M., 1981, Molecular motions and membrane organization and function, in *Membrane Structure* (J. B. Finean and R. H. Michell, eds.), pp. 37–82, Elsevier/North-Holland Biomedical Press, Amsterdam.

Edidin, M., Zagayanski, Y., and Lardner, T. J., 1976, Measurement of membrane protein lateral diffusion in single cells, *Science* 191 : 466–468.

Eisinger, J., and Flores, J., 1985, Fluorometry of turbid and absorbant samples and the membrane fluidity of intact erythrocytes, *Biophys. J.* 48 : 77–84.

Eisinger, J., and Halperin, B. I., 1986, Effects of lateral variation in membrane diffusibility and solubility on the transport of membrane components, *Biophys. J.* 50 : 513–521.

Eisinger, J., Blumberg, W. E., and Dale, R. E., 1981, Orientational effects in intra- and intermolecular long range excitation energy transfer, *Ann. N.Y. Acad. Sci.* 366 : 155–175.

Eisinger, J., Flores, J., and Petersen, W. P., 1986, A milling crowd model for local and long-range obstructed lateral diffusion. Mobility of excimeric probes in the membrane of intact erythrocytes, *Biophys. J.* 49 : 987–1001.

Estep, T. N., and Thompson, T. E., 1979, Energy transfer in lipid bilayers, *Biophys. J.* 26 : 195–208.

Everett, J., Zlotnick, A., Tennyson, J., and Holloway, P. W., 1986, Fluoresence quenching of cytochrome b5 in vesicles with an asymmetric transbilayer distribution of brominated phosphatidylcholine, *J. Biol. Chem.* 261 : 6725–6729.

Fahey, P. F., and Webb, W. W., 1978, Lateral diffusion in phospholipid bilayer membranes and multilamellar liquid crystals, *Biochemistry* 17 : 3046–3053.

Fato, R., Battino, M., Degli-Esposti, M., Parenti-Castelli, G., and Lenaz, G., 1986, Determination of partition and lateral diffusion coefficients of ubiquinones by fluorescence quenching of *n*-(9-anthroyloxy) stearic acids in phospholipid vesicles and mitochondrial membranes, *Biochemistry* 25 : 3378–3390.

Fernandex, S. M., and Berlin, R. D., 1976, Cell surface distribution of lectin receptors determined by resonance energy transfer, *Nature* 264 : 411–415.

Fleming, P. J., Koppel, D. E., Lau, A. L. Y., and Strittmatter, P., 1979, Intramembrane position of the fluorescent tryptophanyl residue in membrane-bound cytochrome b5, *Biochemistry* 18 : 5458–5464.

Förster, T., 1948, Zwischenmoleculare energiewanderung und fluoreszenz, *Ann. Phys.* 2 : 55–75.

Förster, T., 1949, Experimentelle und theoretische untersuchung des zwischenmolekularen übergangs vom electronenanregungsenergie, *Z. Naturforsch.* 4a : 322–327.

Förster, T., 1959, Transfer mechanisms of electronic excitation, *Discuss. Faraday Soc.* 27 : 7–17.

Förster, T., 1969, Excimers, *Angew. Chem. Int. Ed.* 8 : 333–343.

Frank, A., Barenholz, Y., Lichtenberg, D., and Thompson, T. E., 1983, Spontaneous transfer of sphingomyelin between phospholipid bilayers, *Biochemistry* 22 : 5647–5651.

Freire, E., and Snyder, B., 1980a, Estimation of the lateral distribution of molecules in two-component lipid bilayers, *Biochemistry* 19 : 88–94.

Freire, E., and Snyder, B., 1980b, Monte Carlo studies of the lateral organization of molecules in two-component lipid bilayers, *Biochim. Biophys. Acta* 600 : 643–654.

Freire, E., and Snyder, B., 1982, Quantitative characterization of the lateral distribution of membrane proteins within the lipid bilayer, *Biophys. J.* 37 : 617–624.

Froud, R. J., East, J. M., Rooney, E. K., Lee, A. G., 1986, Binding of long-chain alkyl derivatives to lipid bilayers and to $(Ca^{2+}\text{-}Mg^{2+})$-ATPase, *Biochemistry* 25 : 7535–7544.

Fung, B. K. K., and Stryer, L., 1978, Surface density determination in membranes by fluorescence energy transfer, *Biochemistry* 17 : 5241–5248.

Galla, H.-J., and Sackmann, E., 1974, Lateral diffusion in the hydrophobic region of membranes: Use of pyrene excimers as optical probes, *Biochim. Biophys. Acta* 339 : 103–115.

Galla, H.-J., and Sackmann, E., 1975, Chemically induced phase separation in mixed vesicles containing phosphatidic acid. An optical study, *J. Am. Chem. Soc.* 97 : 4114–4120.

Galla, H.-J., and Hartmann, W., 1980, Excimer-forming lipids in membrane research, *Chem. Phys. Lipids* 27 : 199–219.

Galla, H.-J., and Luisetti, J., 1980, Lateral and transversal diffusion and phase transitions in erythrocyte membranes. An excimer fluorescence study, *Biochim. Biophys. Acta* 596 : 108–117.

Galla, H.-J., Hartman, W., Theilen, U., and Sackmann, E., 1979, On two-dimensional passive random walk in lipid bilayers and fluid pathways in biomembranes, *J. Mol. Biol.* 48 : 215–236.

Gibson, G. A., and Loew, L., 1979, Phospholipid vesicle fusion monitored by fluorescence energy transfer, *Biochem. Res. Commun.* 88 : 135–140.

Gratton, E., Jameson, D. M., and Hall, R. D., 1984, Multifrequency phase and modulation fluorometry, *Annu. Rev. Biophys. Bioeng.* 13 : 105–124.

Gruen, D., 1980, A statistical mechanical model of the lipid bilayer above its phase transition, *Biochim. Biophys. Acta* 595 : 161–183.

Grunberger, D., Haimowitz, R., and Shinitzky, M., 1982, Resolution of plasma membrane lipid fluidity in intact cells labelled with diphenylhexatriene, *Biochim. Biophys. Acta* 688 : 764–774.

Haas, E., Katchalski-Katzir, E., and Steinberg, I. Z., 1978, Brownian motion of the ends of oligopeptide chains in solution as estimated by energy transfer between chain ends, *Biopolymers* 17 : 11–31.

Haberer, K., Pfistere, M., and Galla, H.-J., 1982, Virus capping on mycoplasma cells and its effect on membrane structure, *Biochim. Biophys. Acta* **688** : 720–726.

Haigh, E. A., Thulborn, K. R., and Sawyer, W. H., 1979, Comparison of fluorescence energy transfer and quenching methods to establish the position and orientation of components within the transverse plane of the lipid bilayer: Application to the gramicidin A–bilayer interaction, *Biochemistry* **18** : 3525–3532.

Hare, F., 1983, Simplified derivation of angular order and dynamics of rodlike fluorophores in models and membranes. Simultaneous estimation of the order and fluidity parameters for diphenylhexatriene by only coupling steady-state illumination polarization and lifetime of fluorescence, *Biophys. J.* **42** : 205–218.

Hare, F., and Lussan, G., 1977, Variations in microviscosity values induced by different rotational behavior of fluorescent probes in some aliphatic solvents, *Biochim. Biophys. Acta* **467** : 262–272.

Hartmann, W., Galla, H.-J., and Sackmann, E., 1977, Direct evidence of charge-induced lipid domain structure in model membranes, *FEBS Lett.* **78** : 169–172.

Haugland, R. P., 1985, *Handbook of Fluorescent Probes and Research Chemicals*, Molecular Probes, Junction City, OR.

Heyn, M. P., 1979, Determination of lipid order parameters and rotational correlation times from fluorescence depolarization experiments, *FEBS Lett.* **108** : 359–364.

Homan, R., and Pownall, H. J., 1986, Dependence of transbilayer diffusion of pyrene labeled phospholipids on head group composition, *Biophys. J.* **49** : 517a.

Hresko, R. C., Sugar, I. P., Barenholz, Y., and Thompson, T. E., 1986, Lateral distribution of a pyrene-labeled phosphatidylcholine in phosphatidylcholine bilayers: Fluorescence phase and modulation study, *Biochemistry* **25** : 3813–3823.

Hresko, R. C., Sugar, I. P., Barenholz, Y., and Thompson, T. E., 1987, The lateral distribution of pyrene-labeled sphingomyelin and glucosylceramide in phosphatidylcholine bilayers, *Biophys. J.* **51** : 725–734.

Hughes, B. D., Pailthorne, B. A., White, L. R., and Sawyer, W. H., 1982, Extraction of membrane microviscosity from translational and rotational diffusion coefficients, *Biophys. J.* **37** : 673–676.

Hui, S. W., and Parsons, D. F., 1975, Direct observation of domains in wet lipid bilayers, *Science* **190** : 383–384.

Isaacs, B. S., Husten, E. J., Esmon, C. T., and Johnson, A. E., 1986, A domain of membrane-bound blood coagulation factor Va is located far from the phospholipid surface. A fluorescence energy transfer measurement, *Biochemistry* **25** : 4958–4969.

Israelachvili, J. N., Marčelja, S., and Horn, R. G., 1980, Physical principles of membrane organization, *Q. Rev. Biophys.* **13** : 121–200.

Jacobson, K., Wu, E., and Poste, G., 1976, Measurement of the translational motion of concanavalin A in glycerol–saline solutions and on the cell surface, *Biochim. Biophys. Acta* **433** : 215–222.

Jähnig, F., 1979, Structural order of lipids and proteins in membranes: Evaluation of fluorescence anisotropy data, *Proc. Natl. Acad. Sci. U.S.A.* **76** : 6361–6365.

Jameson, D. M., 1984. Fluorescence: Principles, methodologies, and applications, in *Fluorescein Hapten: An Immunological Probe* (E. W. Voss, Jr., ed.), pp. 23–48, CRC Press, Boca Raton, FL.

Jameson, D. M., Gratton, E., and Hall, R. D., 1984, The measurement and analysis of heterogeneous emissions by multifrequency phases and modulation fluorometry, *Appl. Spectrosc. Rev.* **20** : 55–106.

Jones, M. E., and Lentz, B. R., 1986, Phospholipid lateral organization in synthetic membranes as monitored by pyrene-labeled phospholipids: Effects of temperature and prothrombin fragment 1 binding, *Biochemistry* **25** : 567–574.

Kapitza, H. G., and Sackman, E., 1980, Local measurement of lateral motion in erythrocyte membranes by photobleaching technique, *Biochim. Biophys. Acta* **595** : 56–64.

Katchalski-Katzir, E., and Steinberg, I. Z., 1981, Study of conformation and intramolecular motility of polypeptides in solution by a novel fluorescence method, *Ann. N.Y. Acad. Sci.* **366** : 44–61.

Kawato, S., Kinosita, K., Jr., and Ikegami, A., 1977, Dynamic structure of lipid bilayers studied by nanosecond fluorescence techniques, *Biochemistry* **16** : 2319–2324.

Kawato, S., Kinosita, K., Jr., and Ikegami, A., 1978, Effect of cholesterol on the molecular motion in the hydrocarbon region of lecithin bilayers studied by nanosecond fluorescence techniques, *Biochemistry* **17** : 5026–5033.

Keller, P. M., Person, S., and Snipes, W., 1977, A fluorescence enhancement assay of cell fusion, *J. Cell Sci.* **28** : 167–173.

Kimelman, D., Tecoma, E. S., Wolber, P. K., Hudson, B. S., Wickner, W. T., and Simoni, R. D., 1979, Protein–lipid interactions: Studies of the M13 coat protein in dimyristoylphosphatidylcholine vesicles using parinaric acid, *Biochemistry* **18** : 5874–5880.

Kinosita, K., Jr., Kawato, S., and Ikegami, A., 1977, A theory of fluorescence polarization decay in membranes, *Biophys. J.* **20** : 289–305.

Kinosita, K., Jr., Kawato, S., and Ikegami, A., 1984, Dynamic structure of biological and model membranes: Analysis by optical anisotropy decay measurement, *Adv. Biophys.* **17** : 147–203.

Klausner, R. D., Kleinfeld, A. M., Hoover, R. L., and Karnovsky, M. J., 1980, Lipid domains in membranes, *J. Biol. Chem.* **255** : 1286–1295.

Kleinfeld, A. M., 1985, Tryptophan imaging of membrane proteins, *Biochemistry* **24** : 1874–1882.

Kleinfeld, A. M., and Lukacovic, M. F., 1985, Energy-transfer study of cytochrome b5 using the anthroyloxy fatty acid membrane probes, *Biochemistry* **24** : 1883–1890.

Kooyman, R. P. H., Levine, Y. K., and Van der Meer, B. W., 1981, Measurement of second and fourth rank order parameters by fluorescence polarization experiments in a lipid membrane system, *Chem. Phys.* **60** : 317–326.

Koppel, D. E., 1979, Fluorescence redistribution after photobleaching: A new multi-point analysis of membrane translational dynamics, *Biophys. J.* **28** : 281–291.

Koppel, D. E., and Sheetz, M. P., 1983, A localized pattern photobleaching method for the concurrent analysis of rapid and slow diffusion processes, *Biophys. J.* **43** : 175–181.

Koppel, D. E., Axelrod, D., Schlessinger, J., Elson, E. L., and Webb, W. W., 1976, Dynamics of fluorescence marker concentration as a probe of mobility, *Biophys. J.* **16** : 1315–1327.

Koppel, D. E., Fleming, P. J., and Strittmatter, P., 1979, Intramembrane positions of membrane-bound chromophores determined by excitation energy transfer, *Biochemistry* **18** : 5450–5457.

Koppel, D. E., Primakoff, P., and Myles, D. G., 1986, Fluorescence photobleaching analysis of cell surface regionalization, in *Applications of Fluorescence in the Biomedical Sciences* (D. L. Taylor, A. S. Waggoner, R. F. Murphy, F. Lanni, and R. R. Birge, eds.), pp. 477–497, Alan R. Liss, New York.

Kornberg, R. D., and McConnell, H. M., 1971, Inside–outside transitions of phospholipids in vesicle membranes, *Biochemistry* **10** : 1111–1120.

Lakowicz, J. R., 1983, *Principles of Fluorescence Spectroscopy*, Plenum Press, New York.

Lakowicz, J. R., and Prendergast, F. G., 1978, Quantitation of hindered rotations of diphenylhexatriene in lipid bilayers by differential polarized phase fluorometry, *Science* **200** : 1399–1401.

Lakowicz, J. R., Prendergast, F. G., and Hogen, D., 1979, Differential polarized phase fluorometric investigations of diphenylhexatriene in lipid bilayers. Quantitation of hindered depolarizing rotations, *Biochemistry* **18** : 508–519.

Lee, A. G., 1977, Lipid phase transitions and phase diagrams. II. Mixtures involving lipids, *Biochim. Biophys. Acta* **472** : 285–344.

Lentz, B. R., Moore, B. M., and Barrow, D. A., 1979, Light-scattering effects in the measurement of membrane microviscosity with diphenylhexatriene, *Biophys. J.* **25** : 489–494.

Leto, T. L., Roseman, M. A., and Holloway, P. W., 1980, Mechanism of exchange of cytochrome b5 between phosphatidylcholine vesicles, *Biochemistry* **19** : 1911–1916.

Lipari, G., and Szabo, A., 1980, Effects of librational motion on fluorescence depolarization and nuclear magnetic resonance relaxation in macromolecules and membranes, *Biophys. J.* **30** : 489–506.

Lipari, G., and Szabo, A., 1981, Pade approximants to correlation functions for restricted rotational diffusion, *J. Chem. Phys.* **75** : 2971–2976.

London, E., and Feigenson, G. W., 1981, Fluorescence quenching in model membranes: An analysis of the local phospholipid environments of diphenylhexatriene and gramicidin A', *Biochim. Biophys. Acta* **649** : 89–97.

London, J. A., Zecevic, D., Loew, L. M., Orbach, H. S., and Cohen, L. B., 1986, Optical measurement of membrane potential in simple and complex nervous systems, in *Applications of Fluorescence in the Biomedical Sciences* (D. L. Taylor, A. S. Waggoner, R. F. Murphy, F. Lanni, and R. R. Birge, eds.), pp. 423–448, Alan R. Liss, New York.

Luisetti, J., Höhwald, H., and Galla, H. J., 1979, Monitoring the location profile of fluorophores in phosphatidyl bilayers by the use of paramagnetic quenching, *Biochim. Biophys. Acta* **552** : 519–530.

Luna, E. J., and McConnell, H. M., 1978, Multiple phase equilibria in binary mixtures of phospholipids, *Biochim. Biophys. Acta* **509** : 462–473.

Mabrey, S., and Sturtevant, J. M., 1976, Investigation of phase transitions of lipids and lipid mixtures by high sensitivity differential scanning calorimetry, *Proc. Natl. Acad. Sci. U.S.A.* **73** : 3862–3866.

Marčelja, S., 1974, Chain ordering in liquid crystals. II. Structure of bilayer membranes, *Biochim. Biophys. Acta* **367** : 165–176.

Martin, O. A., and Pagano, R. E., 1987, Transbilayer movement of fluorescent analogs of phosphatidylserine and phosphatidylethanolamine at the plasma membrane of cultured cells, *J. Biol. Chem.* **262** : 5890–5898.

Massey, J. B., Hickson, D., She, H. S., Sparrow, J. T., Via, D. P., Gotto, A. M., and Pownall, H. J., 1984, Measurement and prediction of the rates of spontaneous transfer of phospholipids between plasma lipoproteins, *Biochim. Biophys. Acta* **794** : 274–280.

McCloskey, M., and Poo, M.-M., 1984, Protein diffusion in cell membranes: Some biological implications, *Int. Rev. Cytol.* **87** : 19–81.

Meares, S. M., and Rice, L. S., 1981, Diffusion-enhanced transfer shows accessibility of ribonucleic acid polymerase inhibitor binding sites, *Biochemistry* **20** : 610–617.

Mely, B., Charvolin, J., and Keller, P., 1975, Disorder of lipid chains as a function of their lateral packing in lyotropic liquid crystals, *Chem. Phys. Lipids* **15** : 161–169.

Morris, S. J., Bradley, D., and Blumenthal, R., 1985, The use of cobalt ions as a collisional quencher to probe surface charge and stability of fluorescently labeled bilayer vesicles, *Biochim. Biophys. Acta* **818** : 365–372.

Nathan, I., Fleischer, G., Livne, A., Dvilansky, A., and Parola, A. H., 1979, Membrane microenvironmental changes during activation of human blood platelets by thrombin. *J. Biol. Chem.* **254** : 9822–9828.

Ohnishi, S., and Ito, T., 1973, Clustering of lecithin molecules in phosphatidylserine membranes induced by calcium ion binding to phosphatidylserine, *Biochem. Biophys. Res. Commun.* **51** : 132–138.

O'Leary, T. J., 1987, Lateral diffusion of lipids in complex biological membranes, *Proc. Natl. Acad. Sci. U.S.A.* **84** : 429–433.

Op den Kamp, J. A. F., 1979, Lipid asymmetry in membranes, *Annu. Rev. Biochem.* **48** : 47–71.

Op den Kamp, J. A. F., 1981, The asymmetric architecture of membranes, in *Membrane Structure* (J. B. Finean and R. H. Michell, eds.), pp. 83–126, Elsevier/North-Holland Biomedical Press, Amsterdam.

Op den Kamp, J. A. F., Roelofsen, B., and Van Deenen, L. L. M., 1985, Structural and dynamic aspects of phosphatidylcholine in the human erythrocyte membrane, *Trends Biochem. Sci.* **10** : 320–323.

Pagano, R. E., and Sleight, R. G., 1985, Emerging problems in the cell biology of lipids, *Trends Biochem. Sci.* **10** : 421–425.

Parasassi, T., Conti, F., Glaser, M., and Gratton, E., 1984a, Detection of phospholipid phase separation. *J. Biol. Chem.* **259** : 14011–14017.

Parasassi, T., Conti, F., and Gratton, E., 1984b, Study of heterogeneous emission of parinaric acid isomers using multifrequency phase fluorometry, *Biochemistry* **23** : 5660–5664.

Perrin, F., 1936, Mouvement brownian d'un ellipsoide (II). Rotation libre et dépolarisation des fluorescences. Translation et diffusion de molécules ellipsoides, *J. Phys.* **7** : 1–11.

Pesce, A. J., Rosen, C. G., and Pasby, T. L., 1971, *Fluorescence Spectroscopy: An Introduction for Biology and Medicine*, Marcel Dekker, New York.

Peters, R., and Cherry, R. J., 1982, Lateral and rotational diffusion of bacteriorhodopsin in lipid bilayers: Experimental test of the Saffman–Delbrueck equations, *Proc. Natl. Acad. Sci. U.S.A.* **79** : 4317–4321.

Peters, R., Peters, J., Tews, K. H., and Bahr, W., 1974, A microfluorimetric study of translational diffusion in erythrocyte membranes, *Biochim. Biophys. Acta* **367** : 282–295.

Poo, M.-M., and Cone, R. A., 1974, Lateral diffusion of rhodopsin in the photoreceptor membrane, *Nature* **247** : 438–441.

Pottel, H., Van der Meer, B. W., and Herreman, W., 1983, Correlation between the order parameter and the steady-state fluorescence anisotropy of 1,6-diphenyl-1,3,5-hexatriene and an evaluation of membrane fluidity, *Biochim. Biophys. Acta* **730** : 181–186.

Pottel, H., Herreman, W., Van der Meer, B. W., and Ameloot, M., 1986, On the significance of the fourth-rank orientational order parameter of fluorophores in membranes, *Chem. Phys.* **102** : 37–44.

Pottel, H., Van der Meer, B. W., Herreman, W., and Depauw, H., 1987, A new approach to polarization studies using phase and modulation fluorometry. II. Experiments with 1,6-diphenyl-1,3,5-hexatriene in lipid bilayers, *Eur. Biophys. J.* **15** : 47–58.

Pownall, H. J., Hickson, D., Gott, A. M., and Massey, J. B., 1982, Kinetics of spontaneous and plasma-stimulated sphingomyelin transfer, *Biochim. Biophys. Acta* **712** : 169–176.

Roseman, M. A., and Thompson, T. E., 1980, Mechanism of the spontaneous transfer of phospholipids between bilayers, *Biochemistry* **19** : 439–444.

Rothman, J. E., and Lenard, J., 1977 Membrane asymmetry, *Science* **195** : 743–753.

Ruysschaert, J. M., Tenenbaum, A., Berliner, C., and Delmelle, M., 1977, Correlation between lateral lipid phase separation and immunological recognition in sensitized liposomes, *FEBS Lett.* **81** : 406–410.

Saffman, P. G., and Delbrück, M., 1975, Brownian motion in biological membranes, *Proc. Natl. Acad. Sci. U.S.A.* **72** : 3111–3113.

Saxton, M. J., 1982, Lateral diffusion in an archipelago. Effects of impermeable patches on diffusion in a cell membrane, *Biophys. J.* **39** : 165–173.

Seigneuret, M., and Devaux, P. F., 1984, ATP-dependent asymmetric distribution of spin-labeled phospholipids in the erythrocyte membrane: Relation to shape changes, *Proc. Natl. Acad. Sci. U.S.A.* **81** : 3751–3755.

Shaklai, N., Yguerabide, J., and Ranney, H. M., 1977, Interaction of hemoglobin with red blood cell membranes as shown by a fluorescent chromophore, *Biochemistry* **16** : 5585–5592.

Sheetz, M. P., and Singer, S. J., 1974, Biological membranes as bilayer couples. A molecular mechanism of drug–erythrocyte interactions, *Proc. Natl. Acad. Sci. U.S.A.* **71** : 4457–4461.

Shimshick, E. J., and McConnell, H. M., 1973, Lateral phase separation in phospholipid membranes, *Biochemistry* **12** : 2351–2360.

Shinitzky, M., 1984, *Physiology of Membrane Fluidity*, Vol. 1 and 2, CRC Press, Boca Raton, FL.

Shinitzky, M., and Barenholz, Y., 1974, Dynamics of the hydrocarbon layer in liposomes of lecithin and sphingomyelin containing dicetylphosphate, *J. Biol. Chem.* **249** : 2652–2657.

Shinitzky, M., and Barenholz, Y., 1978, Fluidity parameters of lipid regions determined by fluorescence polarization, *Biochim. Biophys. Acta* **515** : 367–394.

Shinitzky, M., and Rivnay, B., 1977, Degree of exposure of membrane proteins determined by fluorescence quenching, *Biochemistry* **16** : 982–986.

Shinitzky, M., and Yuli, I., 1982, Lipid fluidity at the submacroscopic level; determination by fluorescence polarization, *Chem. Phys. Lipids* **30** : 261–282.

Shinitzky, M., Dianoux, A. C., Gitler, C., and Weber, G., 1971, Microviscosity and order in the hydrocarbon region of micelles and membranes determined with fluorescent probes. I. Synthetic micelles, *Biochemistry* **10** : 2106–2113.

Singer, S. J., 1975, Architecture and topography of biologic membranes, in *Cell Membranes. Biochemistry, Cell Biology & Pathology* (G. Weissman and R. Claiborne, eds.), pp. 35–44, HP Publishing, New York.

Singer, S. J., and Nicholson, G. L., 1972, The fluid mosaic model of the structure of cell membranes, *Science* **175** : 720–731.

Sklar, L. A., 1984, Fluorescence polarization studies of membrane fluidity: Where do we go from here?, in *Biomembranes*, Vol. 12 (M. Kates and L. A. Manson, eds.), pp. 99–131, Plenum Press, New York.

Sklar, L. A., Miljanich, G. P., Bursten, L. S., and Dratz, E. A., 1979, Thermal lateral phase separations in bovine retinal rod outer segment membranes and phospholipids as evidenced by parinaric acid fluorescence polarization and energy transfer, *J. Biol. Chem.* **254** : 9583–9597.

Slavik, J., 1982, Anilinonaphthalene sulfonate as a probe of membrane composition and function, *Biochim. Biophys. Acta* **694** : 1–25.

Smith, B. A., and McConnell, H. M., 1978, Determination of molecular motion in membranes using periodic pattern photobleaching, *Proc. Natl. Acad. Sci. U.S.A.* **75** : 2759–2763.

Snyder, B., and Freire, E., 1982, Fluorescence energy transfer in two dimensions: Numeric solutions for random and nonrandom distributions, *Biophys. J.* **40** : 137–148.

Somerharju, P. J., Virtanen, J. A., Eklund, K. K., Vainio, P., and Kinnunen, P. K. J., 1985, 1-Palmitoyl-2-pyrenedecanoyl glycerophospholipids as membrane probes: Evidence for regular distribution in liquid–crystalline phosphatidylcholine vesicles, *Biochemistry* **24** : 2773–2781.

Storch, J., and Kleinfeld, A. M., 1985, The lipid structure of biological membranes, *Trends Biochem. Sci* **10** : 418–421.

Struck, D. K., Hoekstra, D., and Pagano, R. E., 1981, Use of resonance energy transfer to monitor membrane fusion, *Biochemistry* **20** : 4093–4099.

Stryer, L., 1978, Fluorescence energy transfer as a spectroscopic ruler, *Annu. Rev. Biochem.* **47** : 819–846.

Stubbs, C. D., Kouyama, T., Kinosita, K., Jr., and Ikegami, A., 1981, Effect of double bonds on the dynamic properties of hydrocarbon region of lecithin bilayers, *Biochemistry* **20** : 4257–4262.

Stubbs, C. D., Meech, S. R., Lee, A. G., and Phillips, D., 1985, Solvent relaxation in lipid bilayers with dansyl probes, *Biochim. Biophys. Acta* **815** : 351–360.

Szabo, A., 1984, Theory of fluorescence depolarization in macromolecules and membranes, *J. Chem. Phys.* **81** : 150–167.

Taylor, J. A. G., Mingins, J., Pethica, B. A., Tan, B. Y. J., and Jackson, C. M., 1973, Phase changes and mosaic formation in single and mixid phospholipid monolayers at the oil–water interface, *Biochim. Biophys. Acta* **323** : 175–180.

Thomas, D. D., and Stryer, L., 1982, Transverse location of the retinal chromophore of rhodopsin in rod outer segment disc membranes, *J. Mol. Biol.* **154** : 145–157.

Thomas, D. D., Carlsen, W. F., and Stryer, L., 1978, Fluorescence energy transfer in the rapid diffusion limit, *Proc. Natl. Acad. Sci. U.S.A.* **75** : 5746–5750.

Thompson, N. L., and Axelrod, D., 1980, Reduced lateral mobility of fluorescent lipid probes in cholesterol-depleted erythrocyte membranes, *Biochim. Biophys. Acta* **597** : 155–165.

Thulborn, K. R., and Sawyer, W. H., 1978, Properties and the location of a set of fluorescent probes sensitive to the fluidity gradient of the lipid bilayer, *Biochim. Biophys. Acta* **511** : 125–140.

Tsien, R. Y., Pozzan, T., and Rink, T. J., 1984, Measuring and manipulating cytosolic Ca^{2+} with trapped indicators, *Trends Biochem. Sci.* **9** : 263–266.

Tweet, A. G., Bellamy, W. D., and Gaines, G. L., 1964, Fluorescence quenching and energy transfer in monomolecular films containing chlorophyll, *J. Chem. Phys.* **41** : 2068–2077.

Udenfriend, S., 1969, *Fluorescence Assay in Biology and Medicine*, Vols. 1 and 2, Academic Press, New York.

Van Blitterswijk, W. J., Van Hoeven, R. P., and Van der Meer, B. W., 1981, Lipid structural order parameters (reciprocal of fluidity) in biomembranes derived from steady-state fluorescence polarization measurements, *Biochim. Biophys. Acta* **644** : 323–332.

Van Blitterswijk, W. J., Van der Meer, B. W., and Hilkmann, H., 1987, Quantitative contributions of cholesterol, the individual classes of phospholipids and their degree of fatty acyl (un)saturation to membrane fluidity measured by fluorescence polarization, *Biochemistry* **26** : 1746–1756.

Vanderkooi, J. M., and Callis, J. B., 1974, Pyrene. A probe of lateral diffusion in the hydrophobic regions of membranes, *Biochemistry* **73** : 4000–4006.

Van der Meer, B. W., 1984, Physical aspects of membrane fluidity, in *Physiology of Membrane Fluidity* (M. Shinitzky, ed.), pp. 53–71, CRC Press, Boca Raton, FL.

Van der Meer, B. W., Kooyman, R. P. H., and Levine, Y. K., 1982, A theory of fluorescence depolarization in macroscopically ordered membrane systems, *Chem. Phys.* **66** : 39–50.

Van der Meer, B. W., Pottel, H., Herreman, W., Ameloot, M., Hendrickx, H., and Schröder, H., 1984, Effect of orientational order on the decay of the fluorescence anisotropy in membrane suspensions. A new approximate solution of the rotational diffusion equation, *Biophys. J.* **46** : 515–523.

Van der Meer, B. W., Van Hoeven, R. P., and Van Blitterswijk, W. J., 1986, Steady-state fluorescence polarization data in membranes. Resolution into physical parameters by an extended Perrin equation for restricted rotation of fluorophores, *Biochim. Biophys. Acta* **854** : 38–44.

Van der Meer, B. W., Pottel, H., and Herreman, W., 1987, A new approach to polarized fluorescence using phase and modulation fluorometry. I. Theory with reference to hindered and anisotropic rotations, *Eur. Biophys. J.* **15** : 35–45.

Van de Ven, M. J. M., and Levine, Y. K., 1984, Angle resolved fluorescence depolarization of macroscopically ordered bilayers of unsaturated lipids, *Biochim. Biophys. Acta* **777** : 283–296.

Van Dijck, P. W. M., De Kruijff, B., Verkleij, A. J., Van Deenen, L. L. M., De Gier, J., and Demel, R. A., 1976, The preference of cholesterol for phosphatidylcholine in mixed phosphatidylcholine–phosphatidylethanolamine bilayers, *Biochim. Biophys. Acta* **455** : 576–587.

Van Hoeven, R. P., Van Blitterswijk, W. J., and Emmelot, P., 1979, Fluorescence polarization measurements on normal and tumour cells and their corresponding plasma membranes, *Biochim. Biophys. Acta* **551** : 44–54.

Vaz, W. L. C., Goodsaid-Zalduondo, F., and Jacobson, K., 1984, Lateral diffusion of lipids and proteins in bilayer membranes, *FEBS Lett.* **174** : 199–208.

Vincent, M., De Forestra, B., and Alfsen, A., 1982, Nanosecond fluorescence anisotropy decays of *n*-(9-anthroyloxy) fatty acids in dipalmitoylphosphatidylcholine vesicles with regard to isotropic solvents, *Biochemistry* **21** : 708–718.

Vogel, H., and Jähnig, F., 1985, Fast and slow orientational fluctuations in membranes, *Proc. Natl. Acad. Sci. U.S.A.* **82** : 2029–2033.

Vos, M. H., Kooyman, R. P. H., and Levine, Y. K., 1983, Angle resolved fluorescence depolarization experiments on oriented lipid membrane systems, *Biochem. Biophys. Res. Commun.* **116** : 462–468.

Waggoner, A. S., 1979, Dye indicators of membrane potential, *Annu. Rev. Biophys. Bioeng.* **8** : 47–68.

Waggoner, A. S., 1985, Dye probes of cell organelle, and vesicle membrane potentials, in *The Enzyme of Biological Membranes* (A. Martonesi, ed.), pp. 313–331, Plenum Press, New York.

Waggoner, A. S., 1986, Fluorescent probes for analysis of cell structure function, and health by flow and imaging cytometry, in *Applications of Fluorescence in the Biomedical Sciences* (D. L. Taylor, A. S. Waggoner, R. F. Murphy, F. Lanni, and R. R. Birge, eds.), pp. 3–28, Alan R. Liss, New York.

Wang, S., Glaser, M., and Gratton, E., 1986, Frequency domain fluorescence studies of rotational motions of DPH in multilamellar vesicles, *Biophys. J.* **49** : 307a.

Watts, T. H., Brian, A. A., Kappler, J. W., Marrach, P., and McConnell, H. M., 1984, Antigen presentation by supported planar membranes containing affinity-purified I-A, *Proc. Natl. Acad. Sci. U.S.A.* **81** : 7564–7568.

Webb, W. W., and Gross, D., 1986, Patterns of individual molecular motions deduced from fluorescent image analysis, in *Applications of Fluorescence in the Biomedical Sciences* (D. L. Taylor, A. S. Waggoner, R. F. Murphy, F. Lanni, and R. R. Birge, eds.), pp. 405–422, Alan R. Liss, New York.

Webb, W. W., Barak, L. S., Tank, D. W., and Wu, E.-S., 1982, Molecular mobility on the cell surface, *Biochem. Soc. Symp.* **46** : 191–205.

Weber, G., 1952, Polarization of the fluorescence of macromolecules. I. Theory and experimental method, *Biochem. J.* **51** : 145–155.

Weber, G., 1969a, Fluorescence methods in kinetic studies, *Methods Enzymol.* **16** : 380–394.

Weber, G., 1969b, Polarization of the fluorescence of solutions, in *Fluorescence and Phosphorescence Analysis* (E. Hercules, ed.), pp. 217–240, Interscience, New York.

Weber, G., 1976, Practical applications and philosophy of optical spectroscopic probes, *Horizons Biochem. Biophys.* **2** : 163–198.

Wheeler, T. G., Benolken, R. M., and Anderson, R. E., 1975, Visual membranes: Specificity of fatty acid precursors for the electric response to illumination, *Science* **188** : 1312–1314.

Wiener, J. R., Pal, R., Barenholz, Y., and Wagner, R. R., 1985, Effect of the vesicular stomatitus virus matrix protein on the lateral organization of lipid bilayers containing phosphatidyl glycerol: Use of fluorescent phospholipid analogues, *Biochemistry* **24** : 7651–7658.

Wolber, P. K., and Hudson, B. C., 1979, An analytical solution to the Förster energy transfer problem in two dimensions, *Biophys. J.* **28** : 197–210.

Wolber, P. K., and Hudson, B. C., 1981, Fluorescence lifetime and time-resolved polarization anisotropy studies of acyl chain order and dynamics in lipid bilayers, *Biochemistry* **20** : 2800–2810.

Wong, M., Brown, R. E., Barenholz, Y., and Thompson, T. E., 1984, Glycolipid transfer protein from bovine brain, *Biochemistry* **23** : 6498–6505.

Wu, S. H.-W., and McConnell, H. M., 1975, Phase separations in phospholipid membranes, *Biochemistry* **14** : 847–854.

Yechiel, E., Barenholz, Y., and Henis, Y. I., 1985, Lateral mobility and organization of phospholipids and proteins in rat myocyte membranes. Effects of aging and manipulation of lipid composition, *J. Biol. Chem.* **260** : 9132–9136.

Yeh, S. M., and Meares, C. F., 1980, Characterization of transferrin metal-binding sites by diffusion-enhanced energy transfer, *Biochemistry* **19** : 5057–5062.

Yguerabide, J., and Stryer, L., 1971, Fluorescence spectroscopy of an oriented model membrane, *Proc. Natl. Acad. Sci. U.S.A.* **68** : 1217–1221.

Zannoni, C., 1981, A theory of fluorescence depolarization in membranes, *Mol. Phys.* **42** : 1303–1320.

Zannoni, C., Arcioni, A., and Cavatorta, P., 1983, Fluorescence depolarization in liquid crystals and membrane bilayers, *Chem. Phys. Lipids* **32** : 179–250.

Chapter 2

Dynamic Structure of Membranes and Subcellular Components Revealed by Optical Anisotropy Decay Methods

Kazuhiko Kinosita, Jr., and Akira Ikegami

1. INTRODUCTION

Molecules in living organisms, as well as those in dead matter, undergo continual, irregular motions. It is thermal fluctuations, or the Brownian motions.

The optical anisotropy decay method is a means of visualizing the Brownian motion of molecules. The method, applied to biology, reveals how the molecules composing living things move about in complex organized structures. Looking at the molecular motion is not simply a biophysicist's fun. The thermal motions provide many important clues toward the understanding of life at the molecular level.

Through the molecular motion one sees, first of all, structure: the structure of the molecule itself and/or of the surroundings. Currently, the major tools in structural biology at the molecular level are x-ray crystallography and electron microscopy, which are unsurpassed in the information content. Analysis of thermal motion, as explained below, complements these tech-

Kazuhiko Kinosita, Jr., and Akira Ikegami   Institute of Physical and Chemical Research, Wako-shi, Saitama 351-01, Japan.

niques in that it provides information about the structure in an environment where the biological molecule works. The optical anisotropy decay method, in particular, has a high sensitivity and focuses on a selected molecular species in a complex biological system. The method is therefore a powerful means of studying the structure of a key molecule incorporated in a higher-order structure.

Second, the structural information reported by the thermal motion is necessarily a dynamic one: flexibility or mobility is directly expressed in the thermal motion. Changes in conformation can also be followed by time-resolved measurements.

Third, it should not be overlooked that biological reactions rely heavily on thermal motions. Encounter between an enzyme and substrate is an obvious example, which may appear trivial. In membranes or cells, however, the motions are often hindered and impeded, possibly under biological controls. Catalysis is by itself a thermal reaction. Even in an energy-consuming reaction, most steps are thermally activated ones. In the photochemical reaction of rhodopsin and bacteriorhodopsin, for example, only the first step is activated by light; subsequent steps do not proceed at low temperature.

In what follows, we first discuss briefly the principle of the optical anisotropy decay method and then describe several applications chosen from our recent work. Our aim is to explain, in simple words, what can be learned from an optical anisotropy decay. We avoid the discussion of technical details and problems, for which the readers are referred to the references cited and those therein. This chapter is not meant to be a review. Although each topic contains a few references as an access to a fuller explanation, those by no means constitute a complete set. Many important contributions are left out.

2. OPTICAL ANISOTROPY DECAY

2.1. Principle of Optical Anisotropy Decay Method

The optical anisotropy decay method is based on the work of Perrin (1936) and Weber (1953), who showed that optical anisotropy such as the polarization of fluorescence reflects the rotational mobility of the dye molecules. Later, introduction of the decay method has enabled time-resolved measurement of the rotational motion (e.g., see Yguerabide, 1972; Rigler and Ehrenberg, 1973; Wahl, 1975; Kinosita et al., 1984b).

First, we take the example of fluorescence depolarization, through which one sees the rotational motion of fluorescent molecules. Imagine that the circles in Figure 1 represent fluorescent molecules. Each molecule has a unique axis fixed to it, represented by the bar; light polarized parallel to the axis is absorbed by the molecule while perpendicularly polarized light does not interact with the molecule. In an ordinary sample contained in a cuvette, molecules adopt random orientations by virtue of the thermal motion (Figure 1A). Now illuminate the sample, at time 0, with pulsed light that is vertically polarized. Then, only those molecules that happen to be oriented more or less vertically at this instant can absorb the light and get excited (thick molecules in Figure 1B). The excited molecule will sooner or later emit a photon as fluorescence and become deexcited. The light emitted by each molecule is completely polarized, again along the unique axis (the bar in the figure). At time 0, therefore, the fluorescence from the whole sample is strongly polarized along the vertical direction: the intensity of fluorescence observed through a vertically oriented polarizer is much higher than the intensity through a horizontal polarizer.

As time passes the orientations of the excited molecules become random due to the thermal motion (heavy circles in Figure 1C). Fluorescence at this stage is depolarized. The degree of polarization is expressed in terms of a quantity called anisotropy, which is a function of the ratio between the vertically and horizontally polarized intensity components. The anisotropy is maximal at the instant of the pulsed illumination and becomes zero for completely random orientations. Thus, the rotational Brownian motion is visualized as the decay of anisotropy of the optical signal (fluorescence). A fast decay reflects a fast motion, and a slow decay a slow motion. If the anisotropy does not decrease below a certain value, it implies that the rotational motion is restricted to a certain angular range.

In Figure 1, from B to C, the number of excited molecules decreases. When an excited molecule emits a photon, the molecule is deexcited; it never emits a second photon unless reexcited by absorbing light. The average lifetime of the excited state, or the fluorescence lifetime since this is the time during which one can observe fluorescence, is at most 100 nsec for ordinary

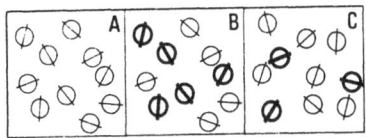

FIGURE 1. Principle of the optical anisotropy decay method. Molecules shown immediately before (A), immediately after (B), and some time after (C) illumination with a vertically polarized light pulse. Heavy circles represent molecules in the excited state.

fluorescent molecules (1 nsec $= 10^{-9}$ sec). Therefore, the motion that can be studied by the fluorescence anisotropy decay is a fast motion occurring in the nanosecond range. Fortunately, the thermal rotation of proteins and nucleic acids in aqueous environment occurs in this time range. The motion of these biological macromolecules is visualized by fixing a suitable fluorescent probe (dye) to the target molecule.

Slower motions, such as the rotation of proteins in a membrane, which takes place in the microsecond range, can be visualized with a dye that emits phosphorescence or delayed fluorescence (Jovin *et al.*, 1981). The lifetime of these emissions reaches milliseconds at room temperature in the absence of oxygen. Anisotropy of the emission reflects the rotational Brownian motion of the dye (hence of the host molecule that rigidly holds the dye molecule) as in the fluorescence anisotropy decay. Such dyes are called triplet probes (since it is the excited triplet state that is given the long lifetime).

Another method is the absorption anisotropy decay. A dye molecule that has been excited by absorbing light does not absorb light (of the same wavelength as the first one) again until it returns to the ground state. Thus, if one illuminates the sample in Figure 1A with a polarized light pulse of high intensity, the molecules drawn heavy in Figure 1B and 1C become "transparent," resulting in an absorbance loss. The change in the absorbance of the sample is detected by passing a weak light beam through the sample and monitoring the transmitted intensity. The absorbance change is anisotropic: the transmittance is higher for a vertically polarized beam than for a horizontally polarized beam. This absorption anisotropy also decays with time (from Figure 1B to 1C) as the dye molecules rotate. The triplet probe above was first introduced by Cherry and co-workers as the absorption anisotropy probe (Cherry, 1979). The long lifetime of the excited (triplet) state of the probe has enabled extensive studies of slow rotational motion, mainly of membrane proteins.

Of the variety of methods, the fluorescence anisotropy decay is suited for measuring rotations in the nanosecond or faster time range. The absorption method, though less sensitive, covers a wide time range from nanosecond to second. High sensitivity in the micro- to millisecond range is obtained by the phosphorescence or delayed fluorescence methods, or by a fluorescence depletion method which is a high-sensitivity version of the absorption method (Johnson and Garland, 1981). The absorption method in particular requires only an absorbance change upon light absorption; light emission from the excited state is not necessary. Thus, any sample that undergoes a photochemical reaction can be studied by the absorption anisotropy decay method. For example, rhodopsin in retina is bleached by light. Using this reaction, Cone (1972) detected the rotation of this protein in the disc membrane, the first successful observation of the rotation of a membrane protein.

2.2. Experimental Techniques

Here we illustrate only the basic principle. For details, see, for example, the following: for fluorescence, Yguerabide (1972), Wahl (1975), Badea and Brand (1979), Kinosita (1983), Lakowicz (1983), and Meech *et al.* (1984); for triplet signals, Cherry (1979), Jovin *et al.* (1981), Chan and Austin (1984), and Thomas (1986).

The optical diagram of our apparatus for phosphorescence and absorption anisotropy decay measurements is shown in Figure 2A. In the case of phosphorescence, we illuminate the sample with a pulsed light from a dye laser through a vertically oriented polarizer. The phosphorescence emission at right angles to the excitation beam is passed through the polarizers A_1 and A_2, deflected by the mirrors M_1 and M_2, and detected by the photomultiplier tubes PM_1 and PM_2. By setting A_1 vertical and A_2 horizontal the two polarized components of the phosphorescence can be measured simultaneously. The time courses of the two signals are recorded on a signal averager. The quality of the signals is improved by averaging over many excitation pulses. Finally, the phosphorescence anisotropy decay is calculated from the two signals.

In absorption measurements the polarizers A_1 and A_2 and the mirrors M_1 and M_2 are removed. The sample is exposed to a weak monitor beam of constant intensity from the halogen lamp. A desired wavelength component of the monitor beam is selected by the monochromator. The vertically and horizontally polarized components of the beam are separated by the beam-splitting polarizer and detected by the photomultiplier tubes PM_3 and PM_4. Changes in the beam intensity accompanying the laser illumination are amplified and fed to the signal averager.

Time resolution of our apparatus, determined by the electronics shown in Figure 2B, is 1 μsec. With a triplet probe such as eosin, which has a lifetime of a few milliseconds, rotational motions in the entire microsecond range can be examined. Techniques for achieving a higher resolution are discussed by Chan and Austin (1984). Measurements under a microscope have also been reported (Johnson and Garland, 1981).

Optical setup for fluorescence measurement is essentially the same as that for phosphorescence measurement. Time resolution of less than 1 nsec is obtained by detecting individual photons of fluorescence: the time lapse between the pulsed excitation and the arrival of a single photon at the photomultiplier tube is measured precisely by a special circuit. As a reference the apparatus that we have been using is shown in Figure 3.

To observe the rotational motion of a biological macromolecule, one needs a probe (dye) on the molecule. Ideally, the probe should be attached rigidly and specifically, via a chemical bond and/or adsorption, to a single

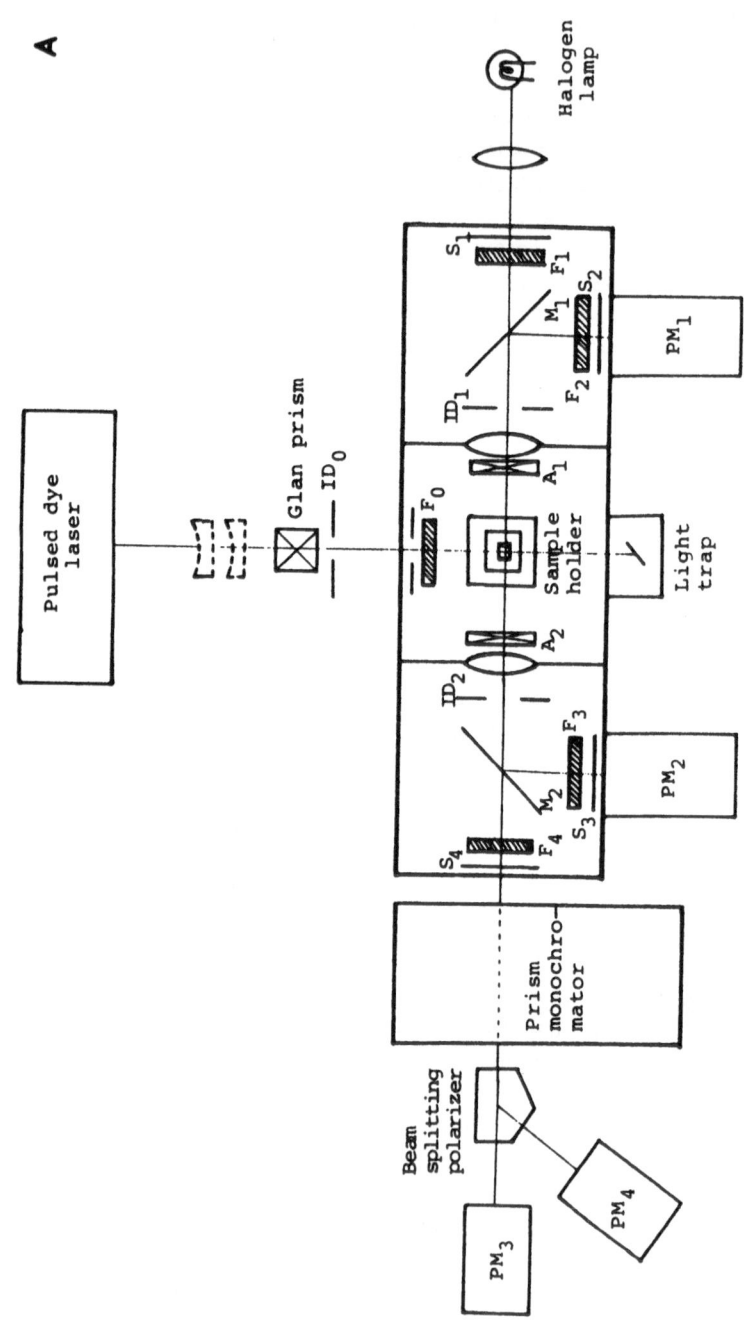

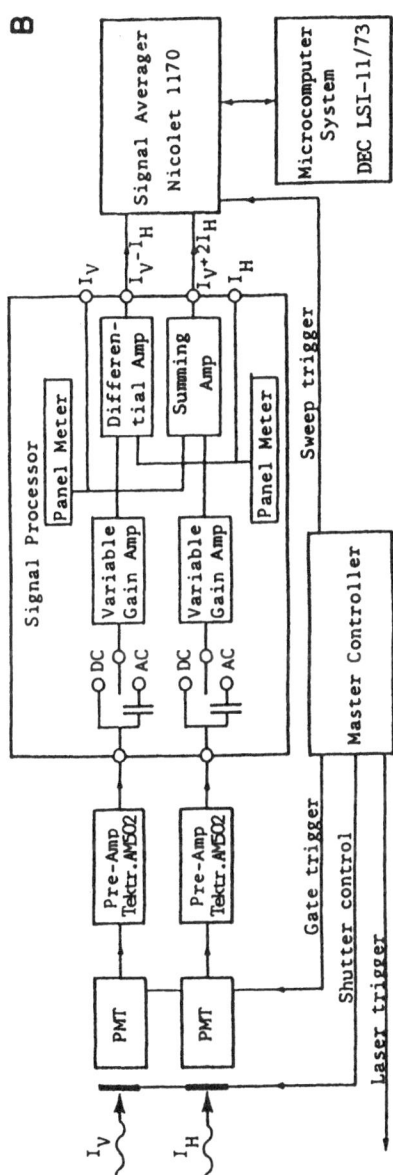

FIGURE 2. Optical (A) and electrical (B) diagrams of the apparatus for phosphorescence and absorption anisotropy decay measurement. In (A), ID, iris diaphragm; S, shutter; F, optical filter; M, mirror; A, polarization analyzer; PM, photomultiplier tube. In (B), I_V and I_H, vertically and horizontally polarized components of emitted (phosphorescence) or transmitted (absorption) light; PMTs, a pair of photomultiplier tubes corresponding to PM_1 and PM_2 (phosphorescence) or PM_3 and PM_4 (absorption) in (A). In phosphorescence measurement, the photomultiplier tubes set at a high sensitivity are gated off during the laser illumination. The phosphorescence anisotropy (as well as fluorescence anisotropy) is defined as $(I_V - I_H)/(I_V + 2I_H)$. The absorption anisotropy is defined similarly except that the intensity I is replaced by the absorbance change ΔA. Diagram (A) is from Kinosita et al. (1984a).

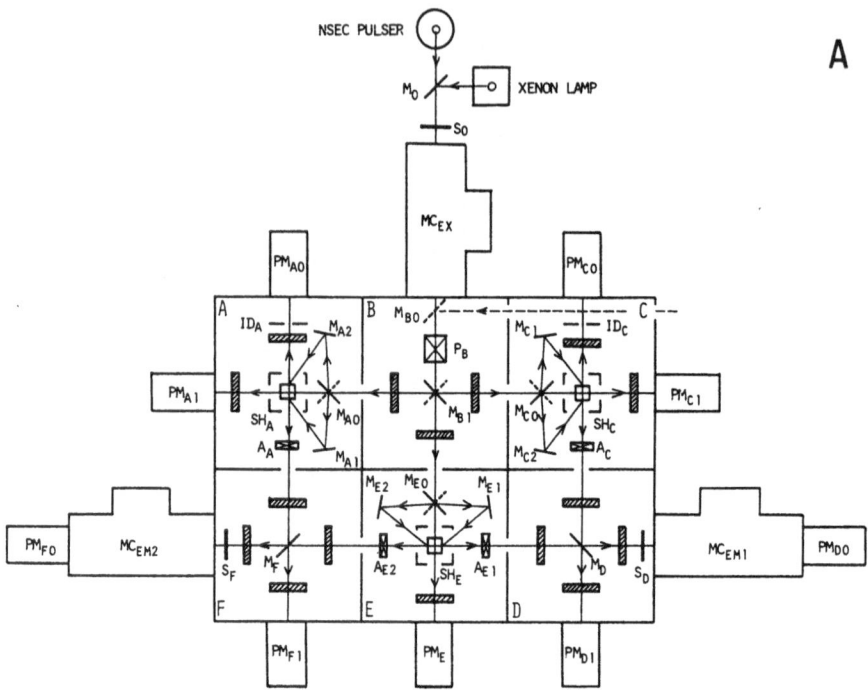

FIGURE 3. Optical (A) and electrical (B) diagrams of the apparatus for fluorescence anisotropy decay measurement. In (A), M, mirror; S, shutter; MC, monochromator; P, polarizer; A, polarization analyzer; SH, sample holder; PM, photomultiplier tube; ID, iris diaphragm. The nanosecond light pulser at the top provides a light pulse of duration 0.8 nsec. The light is monochromatized by MC_{EX}, polarized by P_B, and focused on the sample in chamber E. The temperature-controlled sample holder can accommodate sample cuvettes of various sizes, including dewars, and is equipped with a magnetic stirrer. Emitted fluorescence is decomposed into the polarized components by A_{E1} and A_{E2} and detected by PM_{D1} and PM_{F1} or optionally by PM_{D0} and PM_{F0} through MC_{EM1} and MC_{EM2}. Additional mirrors in chamber E allow front surface illumination. By placing a light-scattering solution in chambers A and C, the time profile of the excitation light pulse can be monitored through the path B→A→F or B→C→D. The xenon lamp at the top is for steady-state measurements. In this case the mirror M_{B1} deflects part of the excitation light to a reference sample (quantum counter) in chamber A. Fluorescence from the reference is monitored by PM_{A0} and is used for the correction for the variation of excitation light intensity. Most of the optical components are driven electrically under computer control. In (B), PMR and PML, either of the pair PM_{F1} and PM_{D1} or PM_{F0} and PM_{D0} in (A); TAC, time-to-amplitude converter; PHA, pulse height analyzer; numbers with an asterisk indicate modules made by Ortec. Fluorescence from the sample is detected as single photons by PMR and PML. The electrical signal corresponding to the detection of a photon is fed to the stop input of TAC. The start input synchronous with the excitation light pulse is generated in the nanosecond light pulser. The TAC converts the time lapse between the start and stop signals into

B

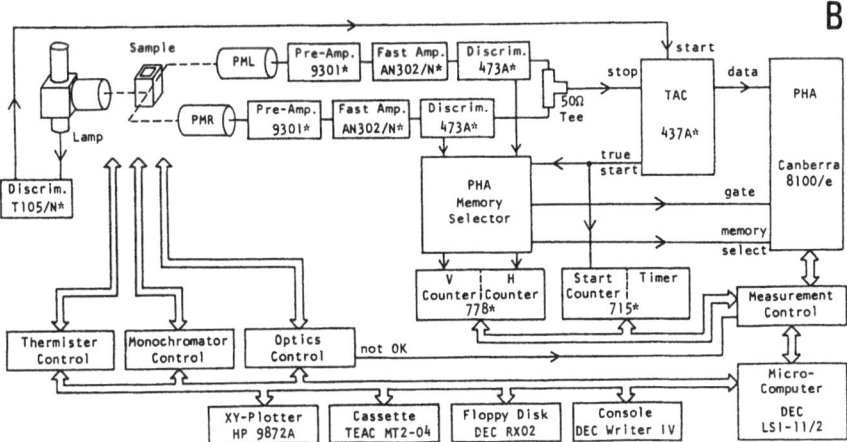

FIGURE 3. a voltage signal, which in turn is digitized and accumulated in the PHA. After many excitations, the data in PHA are the number of occurrences versus the time lapse, that is, the fluorescence intensity versus the time after excitation. The dual electrical paths to the stop input allow the simultaneous measurement of the vertical and horizontal polarization components. The PHA memory selector judges which of the two PMs has detected the photon and issues a proper instruction to PHA. When signals from the two PMs overlap, the event is discarded. At intervals the roles of the two paths, vertical and horizontal, are interchanged by rotating the analyzers A_{E1} and A_{E2} so as to cancel the slight mismatch between the two. The optics control watches all the optical components. If any of them deviates from the assigned position (not-OK signal), the measurement is immediately blocked and a mending cycle is initiated by the computer.

site in the target molecule. Finding an appropriate probe and the method of specific attachment is the key to successful measurement of anisotropy decay. When available, an intrinsic probe, such as the retinal chromophore in rhodopsin, often proves to be satisfactory.

2.3. Information Contained in an Anisotropy Decay

The motion that one sees via the anisotropy decay is the thermal motion of molecules. In principle, "biological" motions such as the conformational change of a protein in a biological reaction should also be visualized. In fact, however, these motions are usually buried under the continual thermal motion. How then does the anisotropy decay method contribute to biological science?

The data in the anisotropy measurement are given in terms of anisotropy versus time (Figure 4). Usually, the ordinate is in the log scale. The decrease of the anisotropy with time implies the change from the state B to C in Figure 1.

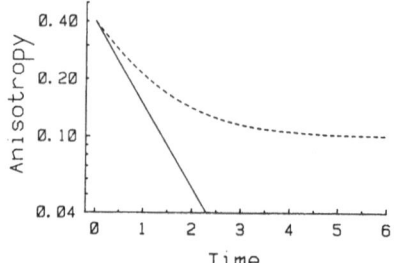

FIGURE 4. Typical anisotropy decay curves. Solid curve, free rotation of globular particles; dashed curve, angle-restricted rotation in or on a structure.

The solid line in Figure 4 is the curve expected for a globular protein undergoing thermal rotation in water. The (log of) anisotropy falls linearly with time toward zero, that is, toward the completely random orientations. The slope is proportional to the rate of rotational motion, as is evident from Figure 1. The rate, or the rotational diffusion coefficient D, is inversely proportional to the cube of the diameter L of the rotating molecule:

$$D = kT/\pi\eta L^3 \tag{1}$$

where k is the Boltzmann constant, T the absolute temperature, and η the viscosity of the surrounding medium (in this case water). Since T and η are experimental quantities, one can estimate the size of the molecule L from the rate of anisotropy decay. Conversely, by putting a probe molecule of known size in cell membrane, for example, the viscosity of the membrane is estimated. Thus, the optical anisotropy decay reveals, first of all, the size of the rotating object or the viscosity of the medium.

The important point is the cubic dependence of the rotational rate on the size. Complex formation between two protein molecules, for example, will be detected as a marked slowing down of the rotational motion. Conformational change of a molecule will also be detected if it is accompanied by a change in the diameter L. [When the shape of the molecule is not spherical, L in Eq. (1) stands, very roughly, for the largest diameter.]

A curve like the dashed one in Figure 4 is often observed for a probe dye in a supramolecular structure such as a membrane or muscle. The anisotropy levels off at a finite value instead of falling toward zero. The data imply that the orientations of the molecules do not become random: the angular range of the rotational motion is restricted. Such is expected when the molecule carrying the probe dye is connected with a "spring" to a larger structure, or when the molecule is trapped in a gap between "walls." The harder the spring or the narrower the gap, the higher the final plateau level of the anisotropy. Thus, from the final anisotropy value one sees how the molecule is incorporated in the supramolecular structure. This is the second role of the anisotropy decay method.

Even when the angular range of rotational motion is restricted, the initial slope of the anisotropy decay is proportional to the rotational rate D in Eq. (1). The initial slope is determined solely by the size of the rotating molecule and the viscosity of the surrounding medium; the spring constant or the gap width does not affect the initial slope.

In summary, the initial slope of the anisotropy decay reflects the rate of rotational motion and the final plateau level the angular range. The rate tells one the size of the molecule (or the medium viscosity) and the range the organization of the supramolecular architecture. The thermal motion elucidates structure. [For more rigorous discussion, see, e.g., Kinosita *et al.* (1984b) and references therein.]

2.4. Optical Anisotropy Decay as a Tool in Bioscience

Below we list the major characteristics of this tool.

1. Visualization of thermal motion. The information the thermal motion conveys has already been discussed above. Examples will be given in the next section.

2. Visualization of rotational motion. The two types of thermal motion, translational (fluctuation in position) and rotational (fluctuation in orientation), can be studied by a variety of physical methods. Some sense both types of motion. The anisotropy decay reports only on rotation and hence is free from separation problems. The rate of rotation is by far more sensitive to the size of the molecule than the rate of translational motion: the latter is inversely proportional to the first power of the diameter whereas the rotational rate is inversely proportional to the third power.

3. Visualization only of the target molecules. Of the variety of molecules in a sample, only those with a probe dye give rise to the anisotropy decay. This tool is therefore particularly suited for the analysis of complex supramolecular architectures. Different parts of the structure may be stained with different probes, each giving specific information.

4. High sensitivity. For the measurement of emission anisotropy decay, 1 ml of a sample with a dye concentration of the order of 10–100 nM is sufficient. Single cell measurement under a microscope is also possible. Absorption anisotropy requires a dye concentration about 100 times as high. Yet the sensitivity is higher than many other methods.

5. Broad time range. Calculation based on Eq. (1) shows that the time by which a protein molecule of diameter 6 nm changes its orientation by 45° in water is, on the average, 25 nsec. For a dye molecule of diameter 1 nm

this number is 0.1 nsec. In cell membranes for which the effective viscosity is about 100 times as high as that of water (see below), the rotational motions are 100 times as slow. All these motions are within the range of the anisotropy decay method.

    6. Time-resolved measurement. The time course of the rotational process is followed by the anisotropy decay method. Thus, distinction between the rate and range is straightforward. This is not the case with time-averaged techniques. For example, with the magnetic resonance spectroscopy one may see that the rotational "mobility" of a molecule is low. However, whether the low mobility implies slow motion, restriction in the angular range, or both is not immediately obvious. In contrast, coexistence of mobile and immobile components is easily detected with magnetic resonance, whereas this case will give rise to an anisotropy decay similar to the dashed curve in Figure 4. The distinction between the two-component case and the restricted rotation is not straightforward in the anisotropy decay. The two methods complement each other, optical on the time axis and magnetic on the component axis.

3. EXAMPLES OF APPLICATION

3.1. Dynamic Structure of Membranes Probed by Diphenylhexatriene

3.1.1. Diphenylhexatriene as a Probe of Lipid Dynamics

    The fluorescent dye 1,6-diphenyl-1,3,5-hexatriene, introduced by Shinitzky and Barenholz (1974), has been widely used as a probe of lipid dynamics in biological and model membranes. As is seen in Figure 5, it is an approximately rod-shaped molecule with a thickness similar to that of a fatty acyl chain of a lipid molecule. The dye taken up by a membrane presumably enters in-between lipid molecules and replaces an acyl chain. The motion of the dye molecule will reflect the lipid chain dynamics.

    Fluorescence anisotropy decay of diphenylhexatriene in various membranes was always biphasic as the dashed curve in Figure 4. The rod-shaped molecule "wobbled" in a restricted angular range rather than undergoing free rotation. This is expected in the lipid bilayer, since the lipid chains have a tendency to lie parallel to each other. Whether the tendency is strong or weak is judged from the final plateau level of the anisotropy decay. A high level points to a high degree of orientational order. From the initial part of the anisotropy decay, in contrast, the rate of wobbling is estimated. The higher the rate, the lower the effective viscosity of the membrane interior.

    The topic of lipid dynamics, particularly that probed by diphenylhexa-

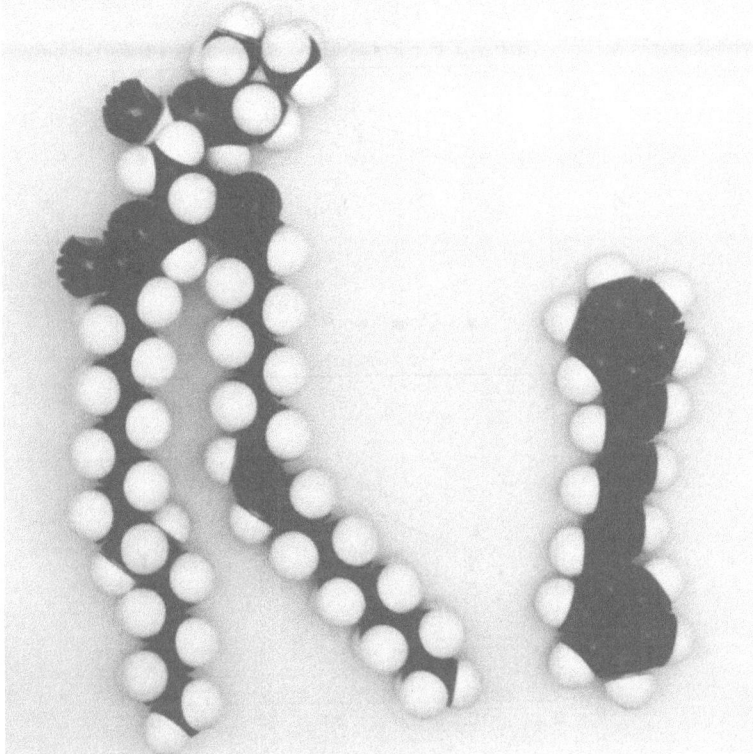

FIGURE 5. Space-filling models of a phospholipid (1-palmitoyl-2-oleoylphosphatidylcholine) and 1,6-diphenyl-1,3,5-hexatriene. In the lipid model, chain 2 (at right) is bent at the position of the double bond. For chain 1 the straight configuration is most stable. Here a kink is introduced in chain 1 to illustrate the chain flexibility.

triene, is also dealt with by other authors of this volume. The following are excerpts from the studies made in our laboratory (Ikegami *et al.*, 1982; Kinosita *et al.*, 1984b).

3.1.2. Lipid Bilayers

Liposomal membranes made of pure phospholipid undergo an order to disorder (crystalline to liquid crystalline) phase transition at a temperature characteristic of the lipid species. The changes in the lipid chain dynamics accompanying the transition, as probed by diphenylhexatriene, are shown in Figure 6 for a saturated phospholipid with a transition temperature of 41°C (solid circles). At low temperatures the wobbling motion of the probe molecule was slow and restricted to a narrow angular range. Above the transition

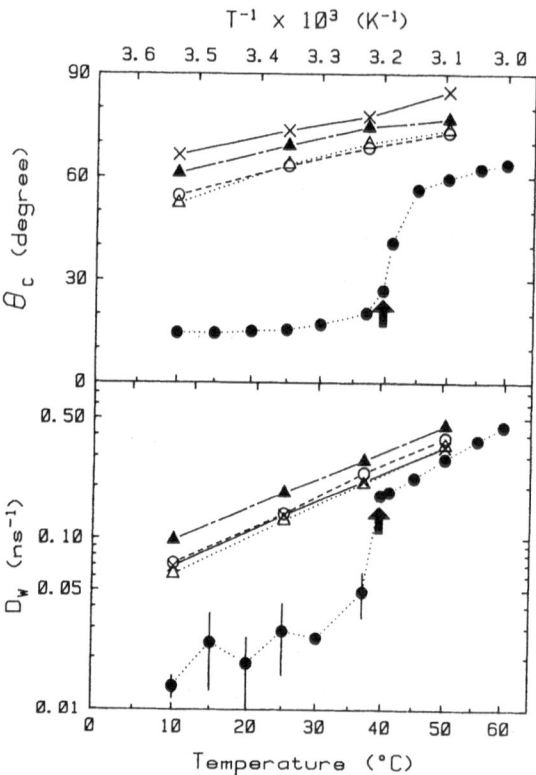

FIGURE 6. Temperature dependence of the rate and range of wobbling motion of diphenyl-hexatriene in pure phospholipid membranes. D_W, the rate of wobbling (wobbling diffusion coefficient); θ_C, the range of wobbling (half-angle of the cone); ●, dipalmitoylphosphatidylcholine; ○, 1-palmitoyl-2-linoleoylphosphatidylcholine; △, 1-palmitoyl-2-oleoylphosphatidylcholine; ▲, 1-palmitoyl-2-arachidonoylphosphatidylcholine; ×, dioleoylphosphatidylcholine. Adapted from Stubbs *et al.* (1981).

temperature, in contrast, the wobbling rate was an order of magnitude higher. The effective viscosity of the membrane interior was 0.15 poise at 50°C (only 30 times the viscosity of water at this temperature), compared to 2.7 poise at 10°C (200 times the viscosity of water). The wobbling range at high temperatures was quite wide, which is consistent with the laterally expanded state of the bilayer above the phase transition (Trauble and Haynes, 1971). An interesting finding was that the increase in the wobbling rate at the transition preceded that in the wobbling range (arrows in Figure 6). Fluctuation in the ordered array of lipid chains at low temperature is activated to a high level before the array melts into the disordered liquid-crystalline state.

Symbols other than the solid circles in Figure 6 are for unsaturated phospholipids, for which the transition temperatures are all below 0°C. The degree of unsaturation was quite different among the four lipids: one double bond in chain 2 ($\triangle$), two double bonds in chain 2 ($\bigcirc$), four double bonds in chain 2 ($\blacktriangle$), and one double bond each in chains 1 and 2 ($\times$). Yet the wobbling parameters were rather similar to each other. The first double bond introduced in a phospholipid molecule renders the bilayer in the disordered state at physiological temperatures. The second and subsequent double bonds make only a minor contribution to the dynamic structure of the membrane. The importance of the first double bond in a molecule was also confirmed in mixtures of unsaturated and saturated phospholipids. The primary factor that determines the lipid chain dynamics is the percentage of unsaturated phospholipid molecules in the membrane, not the percentage of unsaturated fatty acyl chains or the average number of double bonds per molecule.

Cholesterol, one of the major components of biological membranes, exerted pronounced effects on lipid chain dynamics. In the disordered state above the phase transition, addition of cholesterol greatly reduced the wobbling range. The range approached the value in the ordered state as the cholesterol content was increased from 0 to 40 mol %. The wobbling rate, in contrast, remained close to that of the disordered state, although a slight decrease was observed. An opposite effect was found in the ordered state. There the already narrow wobbling rate did not respond to the addition of cholesterol. Instead, the wobbling rate increased and approached the value in the disordered state. Thus, cholesterol, a molecule with a rigid steroid nucleus, tends to realize a state with a narrow wobbling range (characteristic of the ordered state) and a high wobbling rate (disordered state), irrespective of the starting state. The state at high cholesterol content is reminiscent of the state of pure phospholipid at the onset of phase transition.

3.1.3. Lipid–Protein Interaction

The effects of membrane proteins on lipid chain dynamics appear to be qualitatively similar to the effects of cholesterol. In the presence of cytochrome oxidase, a typical membrane protein, the wobbling range of diphenylhexatriene was narrow and the wobbling rate was high, whether the membrane was above or below the transition temperature. Both proteins and cholesterol may be regarded as rigid particles. Insertion of the rigid particles in the fluid bilayer above the transition will naturally reduce the wobbling range. In the bilayer in the ordered state, the lipid chains are closely packed, allowing only cooperative motions. At the interface between lipid chains and the particle surface, the cooperativity may well be disrupted, leading to a high-frequency (but narrow) wobbling.

At the time we were studying the lipid–protein interaction, the dynamics of the boundary lipid, those lipid molecules that are in contact with membrane proteins, was still controversial. Earlier ESR studies had suggested that the boundary lipid was immobile, distinct from the highly mobile bilayer lipid away from protein molecules. NMR studies, in contrast, did not detect any immobile component. NMR further indicated that the average angular range of the chain motion slightly increased upon insertion of membrane proteins into lipid bilayers.

Our fluorescence anisotropy results are shown in Figure 7B. The wobbling motion of diphenylhexatriene was severely restricted in angular range in the boundary region (curve b), compared to the motion in the bilayer region (curve a). The restriction was consistent with the ESR result; above. The anisotropy decay, however, also revealed a small-amplitude wobbling with a rate of the same order of magnitude as that in the bilayer region (curve b).

Lipid chains are constrained on the irregular protein surface by the pressure from neighboring chains and/or by possible interactions between the lipid head group and the membrane protein. The chains pressed against the irregular surface do not have a large freedom of reorientation. Yet they undergo a rapid (nanosecond) thermal wobbling. Occasionally, a chain will acquire a sufficient thermal energy to overcome the constraint and depart from the protein surface. If the chains go into and out of the boundary region once in microseconds, the boundary and bilayer chains are not resolved by NMR, which averages out all motions faster than a millisecond. For NMR all chains appear to be highly mobile. On the nanosecond time scale for which ESR

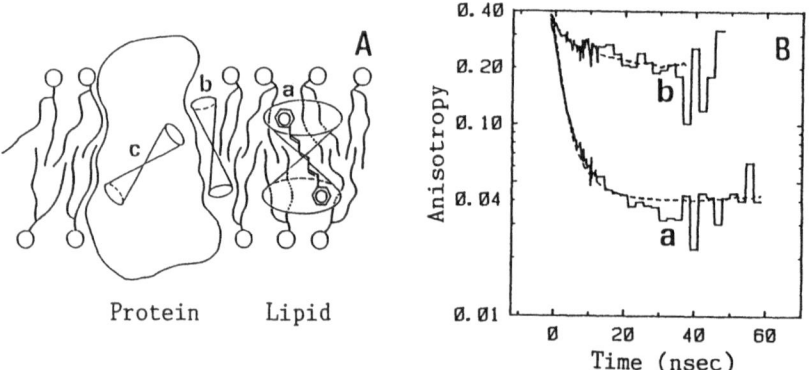

FIGURE 7. The effect of membrane protein on the lipid chain dynamics. (A) Schematic diagram with cones representing the angular range of diphenylhexatriene wobbling. (B) Fluorescence anisotropy decays of diphenylhexatriene in pure phospholipid (dimyristoylphosphatidylcholine) membrane (a), and in the same phospholipid membrane containing cytochrome oxidase at such a concentration that most lipid chains are in contact with the protein surface (b). Adapted from Kinosita *et al.* (1981).

and fluorescence are sensitive, in contrast, the motion in the bilayer part is relatively free, whereas the motion of the boundary chains is severely restricted. On the irregular protein surface, the axis of the restricted wobbling may well differ from site to site, reflecting the surface structure. This situation is illustrated in Figure 7A where the axis and angular range of wobbling are represented by a cone. The bilayer cone a and the boundary cones b and c are for nanosecond motions. For NMR, there is only one wide vertical cone, which is a superposition of all nanosecond cones. Thus, the overall angular range on the NMR time scale is greater in the presence of the membrane protein than in its absence. The controversy is settled.

3.1.4. Biological Membranes

Investigation of biological membranes from several sources has suggested that the lipid chain dynamics is explained, basically, by a simple formula: a biological membrane ~ unsaturated phospholipid + protein + cholesterol. Most phospholipid molecules in biological membranes contain at least one double bond. The chain dynamics is therefore expected to be of the type characteristic of unsaturated phospholipids. In fact, the temperature dependencies of both the rate and range of the wobbling motion of diphenylhexatriene in biological membranes were quite similar to those for unsaturated lipids in Figure 6. The major difference between the pure unsaturated phospholipids and biological membranes was the smaller angular range in the latter. The difference is accounted for by the effect of protein and cholesterol. As discussed above, these molecules act as range reducers in phospholipids above the phase transition. The effect appeared to be additive.

Of course, the above formula is only a first-order approximation referring to average motional properties in a membrane. Microheterogeneity may well exist both in the plane of the membrane and in the vertical direction, as suggested, for example, by the distinction between the boundary and bilayer lipids. Various membrane processes will respond to either the average or local molecular dynamics depending on the spatial and temporal scales of the reaction. The reader is referred to other chapters in this volume.

3.1.5. Steady-State Measurement

Direct measurement of an optical anisotropy decay requires a pulsed light source and a fast detection system. Indirect information, however, is obtained by the steady-state measurement of the anisotropy of the optical signal under constant illumination. Steady-state measurement of fluorescence anisotropy is quite popular, since it is easy to perform and since the high precision in the steady-state measurement allows the detection of small changes in the motional properties.

The steady-state fluorescence anisotropy is a weighted average of the time-dependent anisotropy decay that would be obtained after a pulsed excitation. Essentially, the decay over the time period equal to the fluorescence lifetime of the probe is averaged. If the decay to the final plateau level is much faster than the lifetime, then the steady-state anisotropy reflects the plateau level or the angular range of the rotational motion. If, on the other hand, the decay is slow compared to the lifetime, then the steady-state value reflects the initial slope of the decay or the rate of motion.

For the diphenylhexatriene in lipid environment, the anisotropy decay in most cases is substantially faster than the fluorescence lifetime. The steady-state anisotropy of diphenylhexatriene fluorescence therefore mainly reflects the wobbling range of the probe in the membrane. A high anisotropy points to a narrow range and a low anisotropy a wide range. Such interpretation, however, is not always warranted. Direct confirmation by time-resolved measurement is desirable. For more quantitative discussion, see Kinosita *et al.* (1984b).

3.2. Protein Rotations in Membrane and on Membrane Surface

3.2.1. In-Plane Rotation of Membrane Proteins

A transmembrane protein, like cytochrome oxidase, that vertically spans the bilayer membrane rotates mainly in the plane of membrane. Flip-flop across the bilayer hardly takes place. Since those portions of the protein molecule that protrude into the aqueous phase are hydrophilic, pulling them into the hydrophobic core of the membrane requires a lot of energy. In-plane rotation, in contrast, requires little energy since it preserves the hydrophobic contact between the lipid chains and the body of the protein.

Since the size of a protein molecule is more than an order of magnitude larger than the size of diphenylhexatriene, protein rotation in membrane occurs in the microsecond time range. Hence, its detection requires an optical probe with a long lifetime. The triplet probes with a chemical handle for conjugation have been introduced for this purpose. In addition, many membrane proteins contain a chromophore that serves as an intrinsic probe.

Figure 8 shows an example in which the rotational motion of a protein bacteriorhodopsin was examined. This protein, a typical transmembrane protein of molecular weight about 26,000, has retinal as a rigidly bound internal chromophore. Light absorption by the chromophore initiates a cycle of reaction, during which the purple color of the chromoprotein disappears. The large absorbance change is an excellent signal for the anisotropy decay measurement.

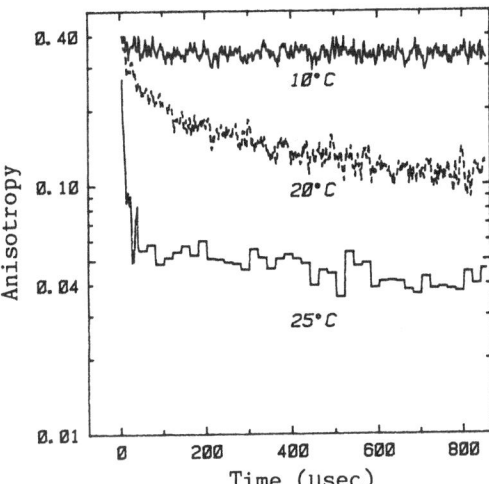

FIGURE 8. Absorption anisotropy decay arising from the in-plane rotation of bacteriorhodopsin in a phospholipid (dimyristoylphosphatidylcholine) membrane (K. Kinosita, Jr., and A. Ikegami, unpublished data). From the final plateau level of the anisotropy decay at the high temperature where the protein rotates in the monomeric form, the angle between the membrane normal and the axis of the chromophore, retinal, can be estimated (Heyn *et al.*, 1977).

In the experiment in Figure 8, bacteriorhodopsin was put in a phospholipid membrane. At 25°C, where the lipid was in the disordered state, the anisotropy decayed to a low constant value in several microseconds. The in-plane rotation of monomeric bacteriorhodopsin in the bilayer lipid, in which the effective viscosity is of the order of 1 poise, is expected to occur in this time range. The finite plateau level is explained by the restriction in the rotational mode, that is, by the suppression of the flip-flop rotation. As the temperature was lowered beyond the transition temperature of this lipid of 23°C, the protein motion greatly slowed down. At 10°C the protein was almost immobile. There is evidence that bacteriorhodopsin tends to aggregate and form patches in lipid membranes below the phase transition. The extreme slowing down of the rotational motion must therefore in large part be due to the increase in the size of the rotating unit.

That the anisotropy decay is sensitive to the association state of membrane proteins was put forward by Cherry and co-workers and has been widely evidenced. In biological membranes the final level of anisotropy decay is often high, which is taken to mean the presence of an immobile population. One reason is that the membrane is crowded with many proteins to the extent that nonspecific protein microaggregates form. In fact a high final level, for the anisotropy of cytochrome oxidase, in mitochondrial membrane was shown to decrease upon dilution of the membrane with exogenous lipid (Kawato *et al.*, 1982). In plasma membranes the immobilization also results from the interaction with cytoskeleton (Nigg and Cherry, 1980). Specific interaction between membrane proteins, the formation of one-to-one complex, has also

74   Kazuhiko Kinosita, Jr., and Akira Ikegami

been revealed (Gut *et al.*, 1983). The anisotropy decay method has proved to be an excellent tool for studies of protein–protein interactions in biological membranes.

3.2.2. Wobbling Motion of Antibody on the Membrane Surface

Complex formation on the membrane surface, between a membrane component (e.g., a receptor or transporter protein) and an external ligand or substrate, is the initial event in many biological processes. Elucidation of the dynamic structure of the complex is a key to the understanding of the molecular process. Individual components can be studied by the anisotropy decay method. In addition, the method may reveal the presence, degree, or site of flexibility in the complex. Whether the complex (or part of it) is buried in the membrane or is held in the aqueous phase can also be inferred from the rotational rate.

As an example we studied the interaction between an antibody and hapten on the membrane surface. To observe the rotational motion, an antibody against a fluorescent hapten was prepared. The small hapten molecule (▲ in Figure 9A) was bound rigidly by the Y-shaped antibody (Figure 9A-a) as well as by its isolated arm portion (F_{ab} fragment, Figure 9A-b). Fluorescence anisotropy decays for these complexes in solution are shown in curves a and b in Figure 9B. The anisotropy fell toward zero, indicating free rotation. The rotational motion of the whole antibody (curve a) was slower than that of the

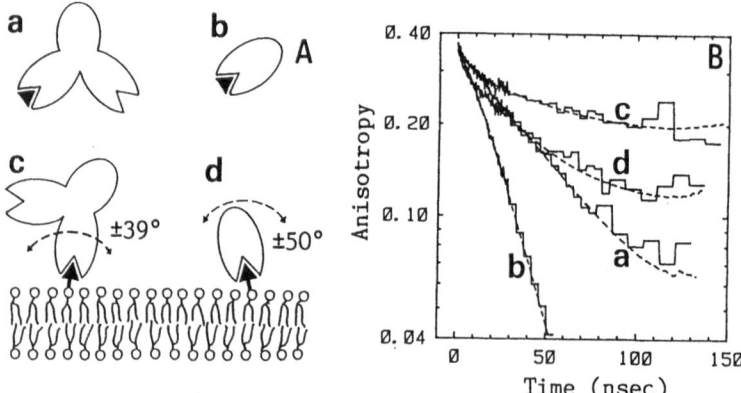

FIGURE 9. Rotational motion of antibody in solution and on membrane surface. (A) Schematic diagram. ▲, *N*-iodoacetyl-*N'*-(5-sulfo-1-naphthyl)ethylenediamine (1,5-IAEDANS) as a fluorescent hapten; Y-shaped body in a, antibody (immunoglobulin G) raised against the hapten; ellipsoid in b, F_{ab} fragment of the antibody; in c and d the hapten was conjugated to phosphatidylethanolamine and mixed with dipalmitoyl- or dimyristoylphosphatidylcholine to form bilayer membranes. (B) Fluorescence anisotropy decays of the hapten in the configurations shown in (A). Adapted from Osada *et al.* (1984).

arm portion (curve b), reflecting the difference in the molecular size. The size difference alone, however, would predict a much slower decay for the whole antibody. The relatively fast decay as observed indicates, as is known, that the antibody molecule is flexible: reorientation of the fluorescent dye (hapten) was accelerated by internal rotational motions (e.g., the arm against the rest).

To model a cell surface antigen, the fluorescent hapten was planted on the surface of a lipid bilayer membrane (Figure 9A-c,d). Both the antibody and isolated arm rigidly held the hapten, which was covalently linked to the head group of a lipid molecule. The linkage produced a large effect on the late portion of the anisotropy decays (Figure 9B-c,d): the anisotropy remained at a high level, indicating a restriction of the rotation angle. The initial decay rate, however, was not different from the rate of free rotation in solution. The friction against the rotational motion [the viscosity η in Eq. (1)] still comes from water. The picture that has emerged is shown in Figure 9A-c,d. The antibody does not penetrate into the membrane interior. It moves about in the aqueous phase and is "reflected" by the membrane surface. As expected from the size difference, the motion of the whole antibody is more restricted in angular range than the motion of the isolated arm. The initial decay rates in Figure 9B also suggest that the arm (or the whole antibody) on the membrane surface still rotated around its own center. If the hapten had been firmly attached on the membrane surface and served as a pivot, the decay would have been much slower (due to the larger friction for the rotation around the pivot than for the rotation around the molecular center). The hapten on the membrane must have an appreciable freedom of translational motion, which presumably comes from the flexible linkage between the hapten dye and the lipid head.

In the above experiment the antibody was in large excess of the hapten. The dynamics thus refers to the antibody with one arm binding the hapten and the other arm free. When hapten on the membrane is in excess, the nanosecond rotational motion of the antibody in the one-to-one complex will help the free arm bind a second hapten. The dynamics of the two-to-one complex on the membrane is yet to be analyzed.

3.3. Internal Motion of DNA

3.3.1. Long DNA in Solution

Double-stranded DNA in solution normally takes the form of right-handed double helix (B-form DNA). A fluorescent dye ethidium, with a planar structure, fits into the narrow spacing between neighboring nucleotide base pairs, which form the ladderlike backbone of the helix. Fluorescence anisotropy of the intercalated ethidium reports the thermal motion of the DNA helix.

The anisotropy decay for a filamentous structure, like the DNA helix, is best understood by referring to the model in Figure 10: small cylinders connected by springs. Suppose that the reporter dye is fixed on one of the cylinders. Initially, the anisotropy decays due to the rotation of this cylinder. The rotation is rapid since the cylinder is small, but the motion is soon arrested by the springs holding the cylinder. The anisotropy decreases only slightly. Then comes the concerted and therefore much slower motion of three cylinders, the dye-carrying one together with the neighboring two, leading to a further decrease of anisotropy. This is followed by a concerted motion of five, then seven, and so on, cylinders. In this way, the rate of anisotropy decay decreases rapidly with time. The expected decay is in the form shown in Figure 11 (dashed line a). [For a rigorous theory, see Allison and Schurr (1979).] The noisy solid line is experimental, obtained for B-form DNA in solution. Of the two kinds of motion, bending (reorientation of the filament axis) and twisting (rotation around the filament axis), the bending motion is much slower since friction from water is larger. What one sees in the anisotropy decay is therefore mainly the twisting motion. Curve a in Figure 11 represents the twisting motion of DNA.

FIGURE 10. A model of an elastic filament.

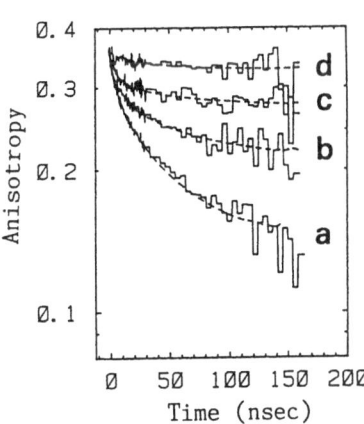

FIGURE 11. Fluorescence anisotropy decays of ethidium intercalated in DNA. a, Calf thymus DNA in solution (Ashikawa *et al.*, 1983); b, DNA in the head of immature sperm (late spermatid) from boar (Ashikawa *et al.*, 1987); c, DNA in chicken erythrocyte nucleus (Ashikawa *et al.*, 1985); d, DNA in the head of mature sperm (cauda spermatozoon) (Ashikawa *et al.*, 1987). Zigzag solid lines are experimental. Smooth dashed lines are theoretical curves for free twisting (a) or for twisting of a short stretch of filament with both ends fixed (b–d).

The overall decay rate is a function of the spring constant. The decay is slower for stronger springs, since the decrease of anisotropy in each step above is then smaller. From the anisotropy decay, one can therefore estimate the rigidity of the filamentous structure. The torsional (twisting) rigidity of B-form DNA, estimated from the data, is about one-thousandth the rigidity of an iron filament of the same thickness. Expressed differently, the neighboring base pairs in the DNA helix thermally rotate over each other by as much as $\pm 5°$ on the average. The DNA helix is flexible.

Yet another way of expressing the rigidity is in terms of the persistence length. It is defined as such that the mutual direction, either bending-wise or twisting-wise, between two short segments in the filament becomes random when the two are separated by more than the persistence length along the filament. This length for twisting is about 50 nm for B-form DNA. The value agrees with the persistence length for bending estimated by other means. The DNA helix happens to be equally flexible bending-wise and twisting-wise.

A particular sequence of DNA forms a left-handed helix (Z-form DNA) in high salt solutions. The anisotropy for Z-form DNA decayed much faster than that for B-form DNA. The estimated torsional rigidity was several times smaller than that for B-form DNA (Ashikawa *et al.*, 1984a).

3.3.2. Chromatin

In the nucleus of eukaryotic cells DNA exists as chromatin, a "beads-on-a-string" structure. The bead, nucleosome bead, is a complex of histone proteins around which the DNA is wound. The DNA extends to the next bead with a short connecting stretch called linker DNA. The entire string is packed in a helical fashion, forming a "solenoid."

The dynamics of DNA on the surface of the histone complex was investigated by intercalating the dye ethidium in DNA in isolated nucleosome beads. Analysis of the fluorescence anisotropy decay indicated that the rigidity of the wound DNA was not very different from that of free DNA and that the DNA had a considerable motional freedom in the nucleosome bead. Since the histone core and the wound DNA appear to be in close contact only on a few points (Kornberg and Klug, 1981), angle-restricted twisting of intermediary stretches is not surprising. The motional freedom indicated in the fluorescence, however, was more than that: some flexibility even in the contact regions was suggested.

When ethidium was added to the whole chromatin, the dye was intercalated primarily in the linker portion. The early portion of the anisotropy decay was rather similar to that for free DNA, indicating a similar rigidity, as may be expected. The late portion, in contrast, was quite high, reflecting the constraint of motion imposed by the beads on both ends of the linker DNA. A very slow decay was seen in the late portion, presumably arising from a concerted motion of the linker and neighboring beads. The slow decay was sensitive to the higher-order structure of the chromatin: the decay became faster in a low-salt solution where the solenoidal chromatin melted into an extended beads-on-a-string form. Measurement on intact nuclei produced a decay curve (Figure 11, curve c) indistinguishable from that for the isolated chromatin in the solenoidal form.

The higher-order structure of chicken erythrocyte chromatin was more stable than that of calf thymus chromatin. The anisotropy decay in the late portion was slower in chicken erythrocyte chromatin than in calf thymus chromatin, both for the solenoidal and extended forms. Furthermore, solenoidal form of chicken chromatin was more resistant to low-salt conditions. In chicken erythrocyte the DNA is dormant: it is no longer duplicated or transcribed. The immobilization, or stabilization, of the higher-order structure in chicken chromatin may be related to the loss of activity.

A class of histone, histone H1, is important in maintaining the higher-order structure. In chicken erythrocyte, histone H5 plays the role. When chicken erythrocyte chromatin was depleted of the protein, the anisotropy decay in the late portion became notably faster.

3.3.3. DNA in Phage Head

Double-stranded DNA in bacteriophages such as λ or T4 is closely packed in the phage head. Structural studies have indicated that the DNA helices are aligned parallel with each other with an interhelix distance almost equal to the diameter of a hydrated DNA helix.

The consequence of the packaging on the torsional dynamics was investigated on suspensions of phages stained with ethidium (Ashikawa *et al.*, 1984b). In the bacteriophage λ, an anisotropy decay close to curve c in Figure 11 was observed, indicating a restriction in the twist angle. Presumably, motion of the DNA strand was hindered at many points by neighboring strands. Stretches in between were relatively free, although the ends were fixed, giving rise to the observed small-amplitude decay of anisotropy. The motion of the stretches, however, was not as fast as would be expected in water: the initial portion of the anisotropy decay in the phage head was somewhat slower than that in the erythrocyte nucleus in curve c. Friction by neighboring strands may account for the slower motion.

If the suppression of DNA motion really resulted from the dense packing, it would be relieved by decreasing the density. This was in fact the case: in a λ mutant whose DNA content was deficient by 18%, the anisotropy decay was significantly enhanced particularly in the late portion. Also, when soluble DNA was collapsed artificially with ethanol to almost the same density as in λ wild, an anisotropy decay very similar to that in the phage was observed.

In the bacteriophage T4, the ethidium anisotropy decay was much more extensive than in λ wild, although the DNA in T4 was also fully packed. The reason turned out to be the difference in the chemical structure of DNA: in T4 DNA, cytosine residues are glucosylated. A T4 mutant having normal cytosine residues gave an anisotropy decay similar to that in λ wild. When isolated DNA was examined in solution, the rigidity of the glucosylated DNA from T4 wild was similar to or even slightly higher than that of the normal DNA from λ or the T4 mutant. Thus, the enhanced mobility of the DNA in T4 wild phage should be ascribed to an alteration in the DNA–DNA interaction, or the packaging mode, induced by the glucosylation.

3.3.4. DNA in Sperm Head

Dramatic changes in chromatin structure take place in the differentiation stage of sperm production. A diffuse chromatin characteristic of a genetically active cell is transformed into a highly condensed inactive state. In this course the histone proteins interacting with DNA are replaced with protamine.

The change in DNA mobility is also dramatic. In the nuclei of immature sperm, the ethidium anisotropy was seen to decay extensively as shown in curve b in Figure 11. The sperm were at a late stage of differentiation and their nucleus had already acquired a nonround shape characteristic of the mature sperm. In the nuclei, however, the exchange of histone with protamine was still in progress. The DNA mobility in these differentiating sperm was much higher than that in the chicken erythrocyte (curve c). In the mature sperm, in contrast, the decay was almost absent (curve d). Presumably, all stretches of DNA were complexed with protamine, forming a tightly condensed chromatin.

Under a microscope the nuclei of the immature and mature sperm were indistinguishable. The size, shape, and ethidium stain pattern were exactly alike. The difference in the mobility cannot be attributed to different packing densities.

From the comparison of various chromatins, including the case of sperm above, we have an impression that functionally dormant DNA tends to be thermally motionless; or, functional activity requires thermal mobility. Whether this is a general rule or not remains to be seen.

3.4. Internal Motion of Actin Filament

3.4.1. Naked F-Actin

Actin is a globular protein with a molecular weight of about 42,000. At physiological ionic strength it polymerizes into a double-helical filament called F-actin. Actin filaments, together with myosin filaments, form the contractile apparatus of muscle. These proteins are also implicated in various forms of motility in nonmuscle cells.

The twisting motion of DNA takes place in the nanosecond time range (Figure 11). The motion of F-actin is expected in the microsecond range, since the radius of F-actin, 5 nm, is about four times as large as that of DNA. (The filament rigidity, as well as the friction from water, is an increasing function of the radius.) To study the filament dynamics, therefore, F-actin was labeled with the triplet dye eosin, which gave an excited-state lifetime of a few milliseconds at room temperature.

The labeled F-actin in solution gave an anisotropy decay in microseconds, which was similar to curve a in Figure 11 in that the decay rate decreased continuously with time. Analysis revealed that F-actin is torsionally very flexible. The rigidity of F-actin against twist was found to be as much as an order of magnitude smaller than that against bend estimated by other means. (The torsional persistence length was about 500 nm whereas the bending persistence length is several micrometers.) The anisotropic rigidity of F-actin contrasts with the case of DNA, for which twisting and bending rigidities are similar to each other. Electron microscopy has also indicated a high torsional flexibility of F-actin (Egelman *et al.*, 1982).

In the double-helical structure of F-actin, two neighboring actin monomers on one strand are separated by 5.5 nm along the filament axis. During thermal motion these monomers are twisted over each other by $\pm 6°$ on the average. At a radius of 3 nm from the filament axis, the twist induces a displacement of 0.3 nm. If actin monomers in the filament are rigid, bonds between neighboring monomers will be broken. (Note that instantaneous displacement in the thermal motion will occasionally exceed the average value above.) Either the actin monomers are by themselves very flexible, and/or the interaction between neighboring monomers allows many twisted arrangements separated by low-energy barriers. The fluctuation angle between neighboring base pairs of DNA is also as large as $\pm 5°$. In terms of displacement, however, the fluctuation in DNA is modest.

3.4.2. Regulated F-Actin

In skeletal muscle, regulatory proteins tropomyosin and troponin decorate F-actin, forming the *thin filament*. These proteins modify the myosin–actin interaction in response to calcium ions.

The anisotropy decay for the regulated F-actin, or the thin filament, was several-fold slower than the decay for naked F-actin. Since the regulatory proteins do not increase the filament radius greatly, the regulated F-actin must be torsionally stiffer than naked F-actin. The increase in rigidity, however, is rather small. Regulated F-actin is also more flexible against twist than against bend. The form of anisotropy decay also suggested that the length of torsional unit, the cylinder in Figure 10, is larger in the regulated F-actin. In naked F-actin, the unit is small and probably corresponds to an actin monomer. In regulated F-actin, in contrast, a long stretch containing many monomers rotates as a unit. The long tropomyosin molecules, which lie parallel to the filament, may be responsible for this effect. The results are shown diagrammatically in Figure 12. Calcium ions, to which the thin filament responds, did not affect the anisotropy decay in both the presence and absence of the regulatory proteins.

3.4.3. Interaction with Myosin

Myosin is a Y-shaped molecule with two globular heads and a long rod portion. In muscle, the rods are bundled parallel to each other to form myosin filament or *thick filament*. The head portions protrude from the filament backbone into water and interact with the thin filament. The head alone, cut out by an enzyme, can interact with actin.

Myosin–actin interaction was studied with the eosin-labeled F-actin and the isolated myosin heads. When the heads were added to regulated F-actin in the absence of ATP (the condition that favors the formation of actin–myosin complex), the anisotropy slowed down in proportion to the amount

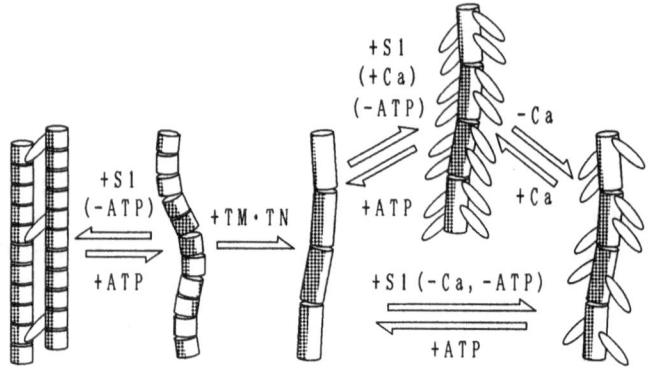

FIGURE 12. A highly schematic diagram representing the torsional dynamics of actin filament. The filament made of short cylinders represents naked F-actin, while that of long cylinders the regulated F-actin, a complex of F-actin with the regulatory proteins tropomyosin (TM) and troponin (TN). The ellipsoids represent the head of myosin (S1, or myosin subfragment 1). Based on Yoshimura *et al.* (1984) and Mihashi *et al.* (1988) [a preliminary report appears in Kinosita *et al.* (1985)].

added. Calculation showed that the slowdown is explained by the increase in the effective radius of the filament due to the bound heads. Regulated F-actin decorated with myosin heads is still torsionally flexible.

The effect of myosin heads was modulated by calcium. Removing calcium ions from the solution, either before or after the addition of the heads, partially reversed the reduction in the decay rate. Apparently, the affinity of regulated actin for myosin is reduced in the absence of free calcium ions (Figure 12).

When ATP was added to the mixture of regulated F-actin and myosin heads, the original anisotropy decay for regulated F-actin alone was restored whether calcium was present or not. ATP is a dissociating agent for actin–myosin complex. In the absence of calcium ions muscle relaxes (myosin heads are detached from the thin filament). The restoration of the original anisotropy decay above is consistent with the relaxation. When both ATP and calcium ions are present, muscle contracts. The anisotropy decays under relaxing and contracting conditions were indistinguishable, suggesting apparently the same torsional rigidity. Whether the thin filament in contracting muscle really remains as flexible as that in relaxed muscle, however, has to be answered by the measurement on the whole muscle, since the protein concentration in the measurement *in vitro* was too low to simulate the situation in muscle.

A quite different result was obtained when myosin heads were added to naked F-actin. The anisotropy decay became flat (no decay) at a head/actin molar ratio as small as 0.05. The result may be explained by the formation of actin bundles in which the myosin head cross-links actin filaments (Ando and Scales, 1985). In regulated F-actin, the regulatory proteins presumably blocked the head binding site(s) other than the genuine one; cross-linking was thus inhibited.

3.5. Dynamic Structure of Myosin Filament

3.5.1. Myosin Filament in Solution

Contraction of muscle results from the sliding of actin and myosin filaments over each other. The molecular basis of this sliding motion is not fully understood yet. The prevailing model has been one in which the myosin head protruding from the myosin filament attaches to the actin filament and rotates in such a way as to pull the two filaments against each other. Recent experiments, for example, one by Yanagida *et al.* (1985), cast a serious doubt on the rotating–head model. Resolution calls for more experimental facts.

Myosin head, on which actin-binding and ATP-hydrolyzing sites reside, is undoubtedly a key part of the contractile apparatus. The thermal motion of the head was revealed in the anisotropy decay of the triplet probe eosin tightly conjugated to the head (Figure 13).

In Figure 13A is shown the anisotropy decay for a solution of individual myosin heads obtained by cutting away the rod portion enzymatically. The heads rotated freely. The decay curve indicated that the head is rigid and is approximated by a prolate ellipsoid of revolution with a size between 16 nm × 4.7 nm and 17 nm × 4.5 nm.

When myosin filaments in solution were examined, the curve shown in Figure 13B was obtained. Since the myosin filament is thicker than the actin filament, the microsecond decay in the figure is too fast to be accounted for by the motion of the filament backbone. The decay must have arisen from

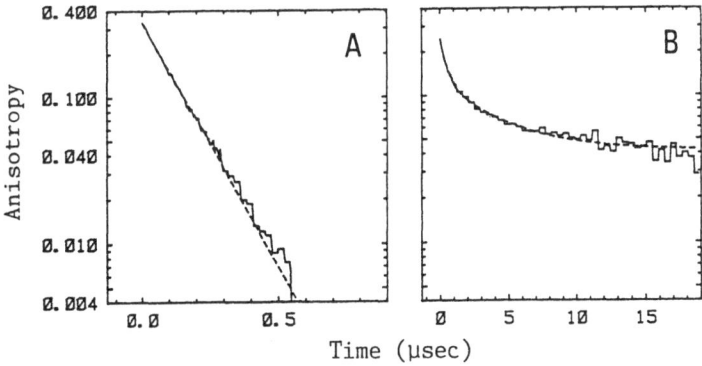

FIGURE 13. The absorption anisotropy decays of eosin bound to the head of myosin. (A) Solution of isolated heads (myosin subfragment 1). (B) Suspension of myosin filaments made *in vitro* of purified myosin. Adapted from Kinosita *et al.* (1984a).

the motion of the protruding heads relative to the filament backbone. The motion of the heads connected to the backbone is naturally restricted in angular range, as is suggested by the decay curve. The initial decay rate in Figure 13B is lower than the rate in Figure 13A by a factor of about 3.5. (Note the difference in time scales.) The reduction in the rate is accounted for by postulating a swivel at the head–rod junction. The rotation about the swivel is slower than the rotation about the center of the head just by the observed factor above. This one-swivel model, however, predicts an aniso-tropy curve consisting of a submicrosecond decay followed immediately by a constant phase. The observed decay in Figure 13B exhibits an additional component with a decay time of a few microseconds before the final plateau level is reached. The slow component must reflect the rotational motion of an object larger than the head, that is, a combined rotation of the head and something attached to it. A natural candidate is a part of the rod portion next to the head–rod junction. In this way a model depicted in Figure 14 has been reached.

The model places a second swivel joint in the rod portion at 14 nm from the head–rod junction. The experiment does not exclude an alternative pos-sibility that a portion of the rod is bent uniformly rather than at the unique site. Then the estimated length of 14 nm should be understood as the effective length of the distributed flexibility. In any case the bend of the rod near the head–rod junction is a new finding. Previous studies have shown the presence of possibly flexible sites that are highly susceptible to enzymatic attacks. These sites, however, are remote from the head–rod junction and are distinct from the flexibility site inferred from the anisotropy decay.

3.5.2. Myofibril

A muscle fiber is a bundle of myofibrils. The myofibril is a thin thread that contains the contractile apparatus. Whether the above model obtained in solution is applicable to myosin in the contractile apparatus was investigated

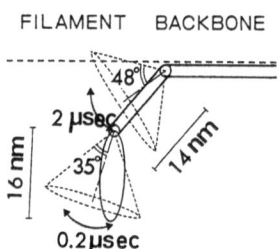

FILAMENT   BACKBONE

FIGURE 14. A model of head dynamics on a myosin filament. Only one of the two heads of myosin is shown. Adapted from Kinosita *et al.* (1984a).

in a suspension of myofibrils in which myosin heads were selectively labeled with the eosin probe (Ishiwata *et al.*, 1987).

The anisotropy decay in relaxed myofibrils, in the presence of ATP and in the absence of calcium ions, was quite similar to the one in Figure 13B. The scheme in Figure 14 applies, except for slight differences in the parameter values, when myosin is detached from actin. The major consequence of this finding is that the myosin heads detached from actin fluctuate translationally, by virtue of the two-swivel structure, over a distance of a few tens of nanometers in a matter of a few microseconds. The gap between myosin and actin filaments is about 20 nm. Thus, the heads make continual searches for a binding site(s) on the actin filament from various directions and with different orientations. The search frequency is enormous. A myosin head interacting with actin hydrolyzes ATP only 10 times a second.

In the absence of ATP where the heads were expected to attach to the actin filament, the anisotropy did not decay in the time range of 1 msec. Attached heads were immobile. Apparently, the result is not consistent with the finding that the regulated actin filament remains torsionally flexible in the presence of attached myosin heads. Actin was mobile while the attached myosin was not. The discrepancy is explained by the fact that an actin filament in muscle is surrounded by several myosin filaments from which many heads protrude. When all heads attach, the actin filament cannot be twisted even though it is flexible as the solution study has indicated.

The head dynamics during contraction has to be investigated on a muscle fiber with fixed ends, since myofibrils contract quickly and irreversibly when ATP is added in the presence of calcium ions. We have not completed the analysis. Preliminary data suggest, however, that the behavior of heads when detached from actin is qualitatively similar to that in the relaxed myofibrils above. How an attached head pulls, or is pulled by, the actin filament is not known yet.

4. CONCLUDING REMARKS

Dynamic structures of various biological molecules and supramolecular systems have been elucidated by the optical anisotropy decay method. The "molecular machines" working in living organisms appear highly flexible. The whole machine as well as its parts undergoes continual fluctuations. Without the fluctuation, the biomolecular machine probably will not work.

The fluctuation is important in two ways: one is to drive the molecular machines properly, while the other is to reveal the architecture of the machine via, for example, the anisotropy decay.

The first applications of the time-resolved optical anisotropy decay method to biological systems appeared in the 1960s. With the development of techniques both experimental and theoretical, the area of applications has expanded from simple determinations of rotational rates to analyses of complex supramolecular structures. Currently, efforts are being made toward the application of the method to oriented samples, as is also discussed in several other chapters in this volume.

Measurement on an oriented sample adds a new class of information: the direction of rotation with respect to the structure. In Figure 14, for example, two cones are depicted that represent the rotation ranges determined from the anisotropy decay. The directions of the cone axes, however, are drawn arbitrarily since the measurement on suspension did not give the information. Measurements on an oriented muscle fiber, in which the myosin filaments are aligned parallel, will yield the required information.

The added information implies an increased number of parameters. Experimental determination as well as theoretical analysis is naturally more complicated. We believe that the development of simplifying procedures, appropriate to each system studied, is the key to successful analyses. Progress is being made in many laboratories including our own.

ACKNOWLEDGMENTS. The optical system in Figure 3A was built by Mr. Y. Sawada, Mr. T. Kishimoto, and Mr. T. Nagamura, who were then at Union Giken Co., Ltd. The system in Figure 2A was made by Sigma Koki Co., Ltd. Many of the parts were also made at the Technology Division of the Institute of Physical and Chemical Research. Mr. T. Hakamada and his colleagues at Hamamatsu Photonics Co., Ltd. gave us valuable suggestions on the use of photomultiplier tubes. The work described here would have been impossible without these supports.

The work is a product of fruitful collaborations with many colleagues. In particular, we thank Dr. I. Ashikawa, Dr. S. Ishiwata, Dr. S. Kawato, Dr. T. Kouyama, Dr. K. Mihashi, and Dr. M. Nakanishi for critically reading the manuscript.

This work was supported by Special Coordination Funds for Promoting Science and Technology and a research grant for Solar Energy-Photosynthesis given by the Agency of Science and Technology of Japan, and by Grants-in-Aid from the Ministry of Education, Science and Culture of Japan.

5. REFERENCES

Allison, S. A., and Schurr, J. M., 1979, Torsion dynamics and depolarization of fluorescence of linear macromolecules. I. Theory and application to DNA, *Chem. Phys.* **41** : 35–59.

Ando, T., and Scales, D., 1985, Skeletal muscle myosin subfragment-1 induces bundle formation by actin filaments, *J. Biol. Chem.* **260** : 2321–2327.

Ashikawa, I., Kinosita, K., Jr., Ikegami, A., Nishimura, Y., Tsuboi, M., Watanabe, K., Iso, K., and Nakano, T., 1983, Internal motion of deoxyribonucleic acid in chromatin. Nanosecond fluorescence studies of intercalated ethidium, *Biochemistry* **22** : 6018–6026.

Ashikawa, I., Kinosita, K., Jr., and Ikegami, A., 1984a, Dynamics of Z-form DNA, *Biochim. Biophys. Acta* **782** : 87–93.

Ashikawa, I., Furuno, T., Kinosita, K., Jr., Ikegami, A., Takahashi, H., and Akutsu, H., 1984b, Internal motion of DNA in bacteriophages, *J. Biol. Chem.* **259** : 8338–8344.

Ashikawa, I., Kinosita, K., Jr., Ikegami, A., Nishimura, Y., and Tsuboi, M., 1985, Increased stability of the higher order structure of chicken erythrocyte chromatin: Nanosecond anisotropy studies of intercalated ethidium, *Biochemistry* **24** : 1291–1297.

Ashikawa, I., Kinosita, K., Jr., Ikegami, A., and Tobita, T., 1987, Changes of the DNA packaging mode during boar sperm maturation, *Biochim. Biophys. Acta* **908** : 263–267.

Badea, M. G., and Brand, L., 1979, Time-resolved fluorescence measurements, *Methods Enzymol.* **61** : 378–425.

Chan, S. S., and Austin, R. H., 1984, Laser photolysis in biochemistry, *Methods Biochem. Anal.* **30** : 105–139.

Cherry, R. J., 1979, Rotational and lateral diffusion of membrane proteins, *Biochim. Biophys. Acta* **559** : 289–327.

Cone, R. A., 1972, Rotational diffusion of rhodopsin in the visual receptor membrane, *Nature New Biol.* **236** : 39–43.

Egelman, E. H., Francis, N., and DeRosier, D. J., 1982, F-actin is a helix with a random variable twist, *Nature* **298** : 131–135.

Gut, J., Richter, C., Cherry, R. J., Winterhalter, K. H., and Kawato, S., 1983, Rotation of cytochrome P-450. Complex formation of cytochrome P-450 with NADPH-cytochrome P-450 reductase in liposomes demonstrated by combining protein rotation with antibody-induced cross-linking, *J. Biol. Chem.* **258** : 8588–8594.

Heyn, M. P., Cherry, R. J., and Müller, U., 1977, Transient and linear dichroism studies on bacteriorhodopsin: Determination of the orientation of the 568 nm all-*trans* retinal chromophore, *J. Mol. Biol.* **117** : 607–620.

Ikegami, A., Kinosita, K., Jr., Kouyama, T., and Kawato, S., 1982, Structure and dynamics of biological membranes studied by nanosecond fluorescence spectroscopy, in *Structure, Dynamics, and Biogenesis of Biomembranes* (R. Sato and S. Ohnishi, eds.), pp. 1–32, Japan Scientific Societies Press, Tokyo.

Ishiwata, S., Kinosita, K., Jr., Yoshimura, H., and Ikegami, A., 1987, Rotational motions of myosin heads in myofibril studied by phosphorescence anisotropy decay measurements, *J. Biol. Chem.* **262** : 8314–8317.

Johnson, P., and Garland, P. B., 1981, Depolarization of fluorescence depletion. A microscopic method for measuring rotational diffusion of membrane proteins on the surface of a single cell, *FEBS Lett.* **132** : 252–256.

Jovin, T. M., Bartholdi, M., Vaz, W. L. C., and Austin, R. H., 1981, Rotational diffusion of biological macromolecules by time-resolved delayed luminescence (phosphorescence, fluorescence) anisotropy, *Ann. N.Y. Acad. Sci.* **366** : 176–196.

Kawato, S., Lehner, C., Müller, M., and Cherry, R. J., 1982, Protein–protein interactions of cytochrome oxidase in inner mitochondrial membranes. The effect of liposome fusion on protein rotational mobility, *J. Biol. Chem.* **257** : 6470–6476.

Kinosita, K., Jr., Kawato, S., Ikegami, A., Yoshida, S., and Orii, Y., 1981, The effect of cytochrome oxidase on lipid chain dynamics. A nanosecond fluorescence depolarization study, *Biochim. Biophys. Acta* **647** : 7–17.

Kinosita, K., Jr., 1983, Nanosecond fluorometry, in *Fluorometry. Applications to Biological Science* (K. Kinosita, Jr. and K. Mihashi, eds.), pp. 99–159, Japan Scientific Societies Press, Tokyo (in Japanese).

Kinosita, K., Jr., Ishiwata, S., Yoshimura, H., Asai, H., and Ikegami, A., 1984a, Submicrosecond and microsecond rotational motions of myosin head in solution and in myosin synthetic filaments as revealed by time-resolved optical anisotropy decay measurements, *Biochemistry* **23** : 5963–5975.

Kinosita, K., Jr., Kawato, S., and Ikegami, A., 1984b, Dynamic structure of biological and model membranes: Analysis by optical anisotropy decay measurement, *Adv. Biophys.* **17** : 147–203.

Kinosita, K., Jr., Mihashi, K., Ishiwata, S., Yoshimura, H., Nishio, T., Asai, H., and Ikegami, A., 1985, Optical anisotropy decay studies of the dynamics of muscle proteins, in *Actin: Structure and Functions (Proceeding of the 11th Taniguchi International Symposium, Division of Biophysics)* (T. Yanagida, ed.), pp. 32–45, The Taniguchi Foundation, Kyoto.

Kornberg, R. D., and Klug, A., 1981, The nucleosome, *Sci. Am.* **244** : 52–64.

Lakowicz, J. R., 1983, *Principle of Fluorescence Spectroscopy*, Plenum Press, New York.

Meech, S. R., Stubbs, C. D., and Phillips, D., 1984, The application of fluorescence decay measurements in studies of biological systems, *IEEE J. Quantum Electr.* **QE-20** : 1343–1352.

Mihashi, K., Ooi, A., Suzuki, N., Yoshimura, H., and Kinosita, K., Jr., 1988, Dynamic polarity of F-Actin and its Ca-dependent regulation under the influence of tropomyosin-troponin system, *Life Sci. Adv.* **7** (in press).

Nigg, E. A., and Cherry, R. J., 1980, Anchorage of a band 3 population at the erythrocyte cytoplasmic membrane surface: Protein rotational diffusion measurements, *Proc. Natl. Acad. Sci. U.S.A.* **77** : 4702–4706.

Osada, H., Nakanishi, M., Tsuboi, M., Kinosita, K., Jr., and Ikegami, A., 1984, Rotational dynamics of immunoglobulins with fluorescent haptens on a membrane surface, *Biochim. Biophys. Acta* **773** : 321–324.

Perrin, F., 1936, Mouvement Brownien d'un ellipsoide (II). Rotation libre et dépolarisation des fluorescenses. Translation et diffusion de molécules ellipsoidales, *J. Phys. Radium* **7** : 1–11.

Rigler, R., and Ehrenberg, M., 1973, Molecular interactions and structure as analyzed by fluorescence relaxation spectroscopy, *Q. Rev. Biophys.* **6** : 139–199.

Shinitzky, M., and Barenholz, Y., 1974, Dynamics of the hydrocarbon layer in liposomes of lecithin and sphingomyelin containing dicetylphosphate, *J. Biol. Chem.* **249** : 2652–2657.

Stubbs, C. D., Kouyama, T., Kinosita, K., Jr., and Ikegami, A., 1981, Effect of double bonds on the dynamic properties of the hydrocarbon region of lecithin bilayers, *Biochemistry* **20** : 4257–4262.

Thomas, D. D., 1986, Rotational diffusion of membrane proteins, in *Techniques for the Analysis of Membrane Proteins* (C. I. Ragan and R. J. Cherry, eds.), pp. 377–431, Chapman and Hall, London.

Trauble, H., and Haynes, D. H., 1971, The volume change in lipid bilayer lamellae at the crystalline–liquid crystalline phase transition, *Chem. Phys. Lipids* **7** : 324–335.

Wahl, P., 1975, Nanosecond pulse fluorimetry, *New Tech. Biophys. Cell Biol.* **2** : 233–285.

Weber, G., 1953, Rotational Brownian motion and polarization of the fluorescence of solutions, *Adv. Protein Chem.* **8** : 415–459.

Yanagida, T., Arata, T., and Oosawa, F., 1985, Sliding distance of actin filament induced by a myosin crossbridge during one ATP hydrolysis cycle, *Nature* **316** : 366–369.

Yguerabide, J., 1972, Nanosecond fluorescence spectroscopy of macromolecules, *Methods Enzymol.* **26** : 498–578.

Yoshimura, H., Nishio, T., Mihashi, K., Kinosita, K., Jr., and Ikegami, A., 1984, Torsional motion of eosin-labeled F-actin as detected in the time-resolved anisotropy decay of the probe in the sub-millisecond time range, *J. Mol. Biol.* **179** : 453–467.

Chapter 3

Principles of Frequency-Domain Fluorescence Spectroscopy and Applications to Cell Membranes

Joseph R. Lakowicz

1. INTRODUCTION

Fluorescence spectroscopy is often used to study the dynamic and hydrody-namic properties of proteins, membranes, and nucleic acids (Cundall and Dale, 1980; Lakowicz, 1983, 1986; Visser, 1985; Demchenko, 1986). More recently, the sensitivity of fluorescence detection, and advances in two-di-mensional detectors have resulted in increased emphasis on the use of fluo-rescence microscopy to obtain a more detailed understanding of cellular phenomena (Taylor *et al.*, 1986). An unfavorable characteristic of fluores-cence is the relatively low degree of specificity. The emission spectra of fluorophores often overlap on the wavelength scale, and the emission spectra of different fluorophores are often similar in shape. A further complication is that the emission may be complex due to the presence of multiple envi-ronments in a membrane, several fluorophores in a macromolecule, or the intrinsically complex emission of macromolecules or even simple molecules like tryptophan.

Joseph R. Lakowicz   Department of Biological Chemistry, School of Medicine, University of Maryland, Baltimore, Maryland 21201.

The complex emission from biological samples can sometimes be resolved into the underlying components by measurement of time-resolved data. This is because each fluorescent substance displays one or more characteristic decay times, and the time-resolved emission is usually the sum of the emission from the individual components. If the time-resolved emission can be recovered with adequate precision, then it is possible to use the data to determine the amounts of each substance in the sample. For instance, suppose the sample contains two fluorescent substances with decay times τ_1 and τ_2. Then, following a δ-function pulse excitation, the intensity decays as

$$I(t) = \sum_{i=0}^{2} \alpha_i \, e^{-t/\tau_i} \tag{1}$$

The fractional intensity of each component is

$$f_i = \frac{\alpha_i \, \tau_i}{\sum_i \alpha_i \, \tau_i} \tag{2}$$

It is generally difficult to obtain data to resolve multiexponential decays and hence mixtures of fluorophores. This is because of the time scale of fluorescence. Most decay times are in the range from 1 to 100 nsec, with the average being near 10 nsec. Resolution of complex or multiexponential decays requires data with a high signal-to-noise ratio, which is difficult to obtain on this time scale. Additionally, resolution of a sum of exponentials is difficult, even with a good signal-to-noise ratio, because the parameters (α_i and τ_i) are highly correlated.

In this chapter we describe a new method for resolving the emission from complex samples, called frequency-domain fluorometry. At present, this new instrumentation exists in only a few laboratories. Nonetheless, the technique is technologically simple, and instrumentation and software are now commercially available. The method appears to be reliable, to provide good resolution of rapid and/or complex decays, and to be mostly free of systematic errors.

2. COMPARISON OF TIME- AND FREQUENCY-DOMAIN MEASUREMENTS

The objective of both time- and frequency-domain fluorometry is to recover the parameters describing the time-resolved decay. Suppose the sample is excited with a pulse of light. If the emission decays with a single decay time (Figure 1, solid line), it is easy to measure the decay time with good accuracy. The more difficult task is recovery of multiple decay times, which is illustrated for two widely spaced decay times in Figure 1 (2 and 20 nsec, dashed line). As the two decay times become more closely spaced, it becomes increasingly difficult to distinguish a double exponential decay law from a single exponential with an intermediate decay time. Even if the decay can be recognized as a double exponential, the values of closely spaced decay times are often rather uncertain. One requires a high signal-to-noise ratio, or equivalently a large number of photons, to recover the decay times with reasonable confidence.

At present, the most widely used method for recovering complex decays is time-correlated single photon counting (O'Connor and Phillips, 1984). In this technique the sample is excited with a brief flash of light, followed by measurement of the arrival time of the first emitted photon following the excitation pulse. The time distribution of arrival times represents the time-resolved emission. To obtain an accurate representation of the emission it is necessary to collect less than 1 photon per 20 pulses. The commonly used flash lamp sources have repetition rates near 20 kHz with pulse widths near 2 nsec. Hence, photon acquisition rates are usually near 1 kHz and long data acquisition times are needed to build up statistical accuracy in the data. Additionally, the relatively wide pulses make it difficult to observe processes that occur in 1 nsec or less, since these events are mostly completed during the excitation. Both problems are now being remedied by the use of high repetition rate lasers, which provide pulse rates near 1 MHz and pulse widths near 10 psec (Visser, 1985). With these laser sources recovery of complex decays can be obtained with just minutes of data acquisition.

The light source and measurements are different for frequency-domain

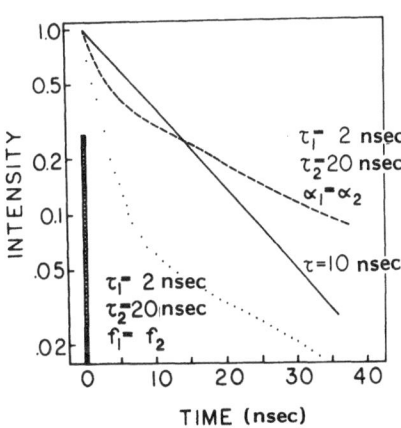

FIGURE 1. Time-domain measurement of fluorescence decays. The sample is excited with a brief pulse of light. The solid line shows the intensity decay for a single decay time of 10 nsec. The dashed line shows the decay for $\tau_1 = 2$ nsec, $\tau_2 = 20$ nsec, $\alpha_1 = \alpha_2 = 0.5$ [Eq. (1)].

fluorometry. The pulsed source is replaced with an intensity-modulated light source. The modulation frequency ($F = \omega/2\pi$) is varied over the widest possible range, with the frequencies being comparable to the decay rates (1/τ) of the emission. Because of the time lag between absorption and emission, the emission is delayed in time relative to the modulated excitation (Figure 2). The solid line in Figure 2 represents the data expected for a single exponential decay of 10 nsec. The delay is described as the phase shift (ϕ_ω), which increases from 0° to 90° with increasing frequency. The finite time response of the sample also results in demodulation of the emission by a factor (m_ω), which decreases from 1.0 to 0.0 with increasing frequency. The phase angle and modulation, measured over a wide range of frequencies, constitute the frequency response of the emission.

The frequency response is determined by the number of decay times displayed by the sample. If the decay is a single exponential (Figure 2, solid line), the frequency response is simple. If the sample displays two decay times (2 and 20 nsec, Figure 2, dashed lines), a more complex response is observed. The frequency-dependent phase and modulation data are used to determine the lifetimes (τ_i) and amplitudes (α_i) that characterize the fluorescence decay.

Surprisingly, phase shift measurements were the first type of data used to determine fluorescence lifetimes (Gaviola, 1926). However, this method did not gain popularity because it was not possible to measure the entire frequency response of the sample. Instead, the commercial phase fluorometers, which have been available since 1975, only allowed phase and modulation measurements at one to three frequencies, spaced from 6 to 30 MHz. In spite of efforts in several laboratories (Jameson and Weber, 1981; Lak-

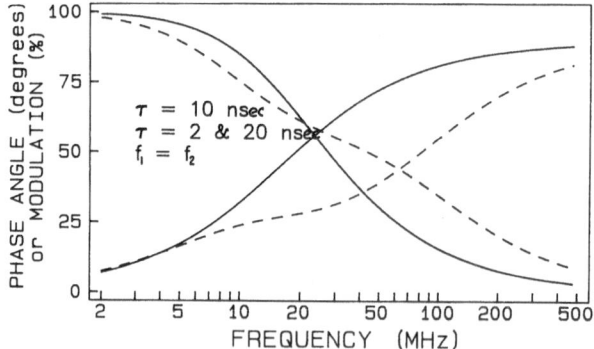

FIGURE 2. Frequency-domain measurements of fluorescence decays. The solid line is for a single decay time of 10 nsec, and the dashed line is for a double-exponential decay, $\tau_1 = 2$ nsec, $\tau_2 = 20$ nsec, $f_1 = f_2 = 0.5$ [Eq. (2)]. The phase angles increase and the modulation decreases with increasing modulation frequency.

owicz and Cherek, 1981; Weber, 1981), the information content of the limited data could not generally support the multicomponent analyses (Jameson and Gratton, 1983). This situation has now changed, and a wide and continuous range of modulation frequencies is now available.

2.1. A First-Generation Frequency-Domain Fluorometer

A schematic of a frequency-domain fluorometer is shown in Figure 3 (Gratton and Limkemann, 1983; Lakowicz and Maliwal, 1985). This design is similar to that of the commercially available instruments and is rather similar to a standard fluorometer. The main differences are the laser light source, the light modulator, and the associated radio-frequency electronics. Until recently, it was thought to be difficult to obtain wide-band light modulation unless the light was highly collimated. Continuously variable frequency modulation is possible with electro-optic modulators, whereas the acousto-optic modulators usually operate at discrete resonances over a limited range of frequencies. Additionally, many electro-optic modulators have long narrow apertures. Hence, only laser sources seemed practical for use with electro-optic modulators. It is now known that laser sources can easily be modulated to 200 MHz using one of several electro-optic modulators. Surprisingly, it is now possible to modulate arc lamps to 200 MHz (Gratton, 1986).

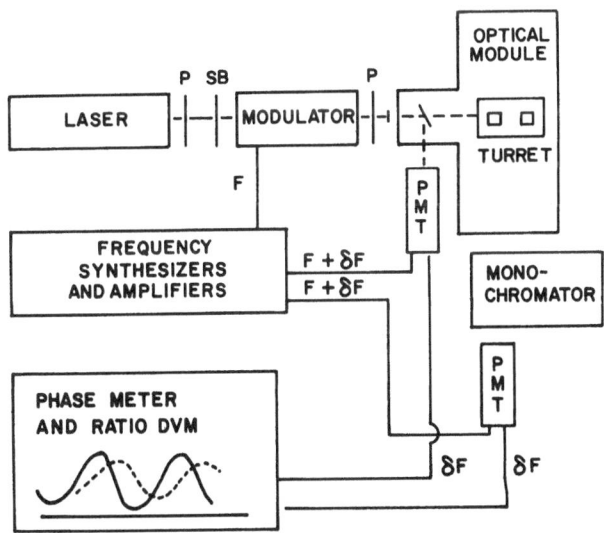

FIGURE 3. Schematic of the variable-frequency phase-modulation fluorometer. P, polarizer; SB, Soleil–Babinet compensator; F, frequency; δF, cross-correlation frequency; PMT, photomultiplier tube; DVM, digital voltmeter.

The next apparent difficulty is measurement of the phase angle and modulation at high frequencies. The measurements appear more difficult when one realizes that resolution of multiexponential decays requires accuracy greater than 0.5° in phase and 0.5% in modulation. In fact, the measurements are surprisingly easy and free of interference because of cross-correlation detection. The gain of the detector is modulated at a frequency offset $(F + \delta F)$ from that of the modulated excitation (F). The difference in frequency $(\delta F = 25 \text{ Hz})$ contains the phase and modulation information. Hence, at all modulation frequencies the phase and modulation can be measured at the low cross-correlation frequency (δF) with a zero-crossing detector and a ratio digital voltmeter. The use of cross-correlation detection results in the rejection of harmonics and other sources of noise. The cross-correlation method is surprisingly robust. The harmonic content (frequency components) of almost any excitation profile can be used if it contains frequency components that are synchronized with the detector. Both pulsed lasers (Lakowicz *et al.*, 1986b) and synchrotron radiation (Gratton and Lopez-Delgato, 1980) have been used as modulated light sources. Since pulse lasers provide harmonic content to many gigahertz, the bandwidth of the frequency-domain instruments is now limited by the detector and not the modulator. By use of a fast microchannel plate (MCP) photomultiplier tube (PMT), the frequency range has now been extended to 2 GHz (Lakowicz *et al.*, 1986b) as will be described later.

2.2. Resolution of a Two-Component Mixture

It is instructive to examine data for a mixture of fluorophores. An example is shown in Figure 4, which shows the frequency response of a mixture of POPOP (1.37 nsec) and 9-CA (9-cyanoanthracene, 12.1 nsec). The fre-

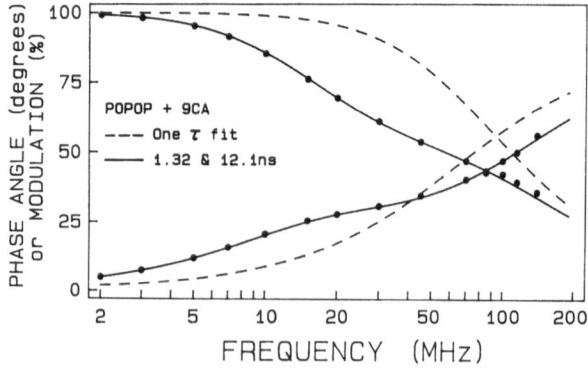

FIGURE 4. Frequency response for a two-component mixture of POPOP and 9-cyanoanthracene. The dashed line shows the best single-component fit and the solid line the best two-decay-time fit.

quency response is complex, reflecting the multiple decay times. Shoulders are seen for both the phase and modulation profiles near 30 MHz. At this frequency the contribution of the 9-CA to the data is diminishing because its modulated amplitude is decreasing. The presence of multiple decay times is immediately evident from an attempt to fit the data to a single decay time. The best fit, shown as the dashed line, is obviously inadequate. In contrast, the data can be explained by a curve with decay times of 1.3 and 12.1 nsec (solid line). We note that the resolution shown in Figure 4 is a simple one because the decay times are different by nearly 10-fold. In other studies we demonstrated that more closely spaced lifetimes can be determined, as can the decay times of three-component mixtures (Gratton *et al.*, 1984b; Lakowicz and Maliwal, 1985).

3. THEORY OF FREQUENCY-DOMAIN FLUOROMETRY

3.1. Decays of Fluorescence Intensity

At present, the frequency-domain data are mostly analyzed by the method of nonlinear least squares (Bevington, 1969; Gratton *et al.*, 1984a; Lakowicz *et al.*, 1984c). The measured data are compared with values predicted from a model, and the parameters of the model are varied to yield the minimum deviations from the data. It is possible to predict the phase and modulation values for any assumed decay law. Suppose the time-resolved decay can be described by a sum of exponentials

$$I(t) = \sum_i \alpha_i \, e^{-t/\tau_i} \tag{3}$$

The frequency-domain data can be calculated from the sine and cosine transforms of $I(t)$:

$$N_\omega = \frac{\int_0^\infty I(t) \sin \omega t \, dt}{\int_0^\infty I(t) dt} \tag{4}$$

$$D_\omega = \frac{\int_0^\infty I(t) \cos \omega t \, dt}{\int_0^\infty I(t) dt} \tag{5}$$

where ω is the circular modulation frequency ($2\pi \times$ frequency in Hz). Even if the decay is more complex than a sum of exponentials, it is generally adequate to approximate the decay by such a sum. If needed, nonexponential decay laws can be transformed numerically. For a sum of exponentials the transforms are

$$N_\omega = \left(\sum_i \frac{\alpha_i \omega \tau_i^2}{1 + \omega^2 \tau_i^2} \right) / \left(\sum_i \alpha_i \tau_i \right) \tag{6}$$

$$D_\omega = \left(\sum_i \frac{\alpha_i \tau_i}{1 + \omega^2 \tau_i^2} \right) / \left(\sum_i \alpha_i \tau_i \right) \tag{7}$$

The frequency-dependent values of the phase angle ($\phi_{c\omega}$) and the demodulation ($m_{c\omega}$) are

$$\tan \phi_{c\omega} = N_\omega / D_\omega \tag{8}$$

$$m_{c\omega} = \left[N_\omega^2 + D_\omega^2 \right]^{1/2} \tag{9}$$

The parameters (α_i and τ_i) are varied to yield the best fit between the data and the calculated values, as indicated by a minimum value for the goodness-of-fit parameter χ_R^2,

$$\chi_R^2 = \frac{1}{\nu} \sum_\omega \left(\frac{\phi_\omega - \phi_{\omega c}}{\delta \phi} \right)^2 + \frac{1}{\nu} \sum_\omega \left(\frac{m_\omega - m_{\omega c}}{\delta m} \right)^2 \tag{10}$$

where ν is the number of degrees of freedom, and $\delta \phi$ and δm are the uncertainties in the phase and modulation values, respectively. The subscript c is used to indicate calculated values based on the assumed parameters of the decay (α_i and τ_i). The correctness of a model is judged by the values of χ_R^2. For an appropriate model and random noise, χ_R^2 is expected to be near unity. If χ_R^2 is greater than unity, one should consider whether the χ_R^2 value is adequate to reject the model. Rejection is judged from the probability that random noise could be the origin of the value of χ_R^2 (Bevington, 1969; Taylor, 1982). For instance, a typical frequency-domain measurement in this laboratory contains phase and modulation data at 25 frequencies. A double-exponential model contains three floating parameters (two τ_i and α_i), resulting in 47 degrees of freedom. A value of χ_R^2 equal to 2 is adequate to reject the model with a certainty of 99.9% or higher (Taylor, 1982).

In practice, the values of χ^2 change depending on the values of $\delta \phi$ and δm used in its calculation. For consistency we chose to use constant and frequency-dependent values of $\delta\phi = 0.2°$ and $\delta m = 0.005$. While the precise values may vary between experiments, the χ_R^2 values calculated in this way indicate to us the degree of error in a particular data set. The use of fixed values of $\delta\phi$ and δm does not introduce any ambiguity in the analysis, because it is the relative values of χ_R^2 that should be used in accepting or rejecting a model. Hence, we typically compare χ_R^2 for the one-, two-, and three-exponential fits. If χ_R^2 decreases two-fold or more as the model is incremented, then the data most probably justify inclusion of the additional decay time.

3.2. Decays of Fluorescence Anisotropy

It is also of interest to recover the fluorescence anisotropy of probes in membranes and for other biological macromolecules. These decays can potentially reveal the size, shape, and segmental mobility of the molecule under investigation (Belford *et al.*, 1972; Chuang and Eisenthal, 1972; Munro *et al.*, 1979; Lakowicz *et al.*, 1986a,c). In the case of membranes, the anisotropy decay can reveal the rate of probe rotation and the order parameters of the membranes (Kinosita *et al.*, 1977; Heyn, 1979). Suppose the sample is excited with a δ-function pulse of vertically polarized light. The decays of the parallel and perpendicular components of the emission are given by

$$I_\parallel(t) = \frac{1}{3} I(t) [1 + 2r(t)] \tag{11}$$

$$I_\perp(t) = \frac{1}{3} I(t) [1 - r(t)] \tag{12}$$

where $r(t)$ is the time-resolved anisotropy. Generally, $r(t)$ can be described as a multiexponential decay

$$r(t) = r_0 \sum_i g_i e^{-t/\theta_i} \tag{13}$$

where r_0 is the limiting anisotropy in the absence of rotational diffusion, θ_i the individual correlation times, and g_i the associated amplitudes.

In the time domain one measures the time-resolved decays of the polarized components of the emission [Eqs. (11) and (12)]. These decays are used to determine which anisotropy decay law is most consistent with the data. The frequency-domain measurements are different. The measured quantities are the phase angle difference (Δ_ω) between the perpendicular ($\phi\perp$) and parallel ($\phi_\parallel$) components of the modulated emission ($\Delta_\omega = \phi\perp - \phi_\parallel$) and the amplitude ratio (Λ_ω) of the parallel ($m_\parallel$) and perpendicular ($m\perp$) components of the modulated emission ($\Lambda = m_\parallel/m\perp$). These values are compared with those expected for an assumed anisotropy decay law. At present, we believe that the direct measurement of the difference signal in the frequency domain is an advantage over the use of a calculated difference for the time-domain measurements.

The expected values of Δ_ω ($\Delta_{c\omega}$) and Λ_ω ($\Lambda_{c\omega}$) can be calculated from the sine and cosine transforms of the individual polarized decays [Eqs. (11) and (12)]. The frequency-dependent values of Δ_ω and Λ_ω are given by

$$\Delta_{c\omega} = \arctan\left(\frac{D_\parallel N_\perp - N_\parallel D_\perp}{N_\parallel N_\perp + D_\parallel D_\perp}\right) \tag{14}$$

$$\Lambda_{c\omega} = \left(\frac{N_\parallel^2 + D_\parallel^2}{N_\perp^2 + D_\perp^2}\right)^{1/2} \tag{15}$$

where the N_i and D_i are calculated at each frequency. The parameters describing the anisotropy decay are obtained by minimizing the squared deviations between measured and calculated values, using an expression analogous to Eq. (10).

The meaning of the modulation ratio Λ_ω is more apparent when presented as a frequency-dependent anisotropy (Maliwal and Lakowicz, 1986; Maliwal et al., 1986) using

$$r_\omega = \frac{\Lambda_\omega - 1}{\Lambda_\omega + 2} \tag{16}$$

At low frequency the value of r_ω is equal to the steady-state anisotropy, and at the high frequency limit $r_\omega = r_0$.

4. INTENSITY DECAYS OF DPH-LABELED MEMBRANES

We now describe several applications of the frequency-domain method to model membranes. One of the most common applications is resolution of multiexponential decays. This application is illustrated in Figure 5 for diphenylhexatriene (DPH) in DPPC–melittin complexes (Faucon and Lakowicz, 1987). Melittin is an amphipathic peptide that interacts with and disrupts cell membranes. Upon binding to DPPC bilayers, the intensity decay of DPH becomes increasingly multiexponential. This is seen (Table I) by the increasing values of χ_R^2 for the single-exponential fits. As the amount of melittin is increased, χ_R^2 increases from 19 to 183. The value of $\chi_R^2 = 19$ found in the absence of protein indicates that the DPH intensity decay is more complex than a single exponential even in the absence of melittin. This short com-

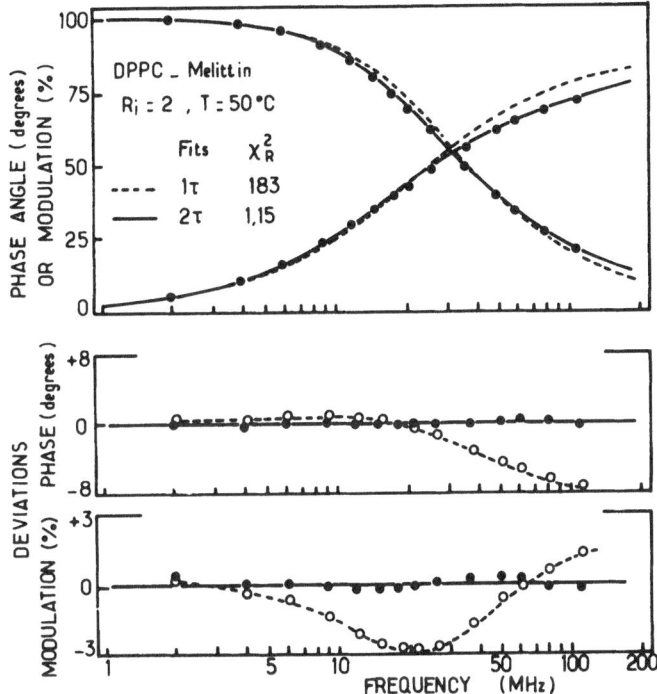

FIGURE 5. Frequency-domain intensity decays of DPH in DPPC–melittin complexes. The dashed line shows the best single-component fit and the solid line the best two-component fit. The lower panels show the deviation from the one- (○) and two-component (●) fits. From Faucon and Lakowicz (1987).

Table I
Intensity Decays of DPH in DPPC–Melittin
Complexes at 50°C

[Lipid]/[protein]	τ_i (nsec)	f_i	χ_R^2
∞	7.78	1.0	19
5	7.34	1.0	83
2	7.24	1.0	183
∞	0.29	0.01	
	7.91	0.99	2.0
5	1.34	0.04	
	7.79	0.96	1.9
2	1.53	0.08	
	8.07	0.92	1.2

ponent in pure DPPC bilayers has been observed previously (Parasassi *et al.*, 1984), but its origin is unclear. In each case the data can be adequately fit using the double-exponential model. As the melittin concentration is increased, the long decay time remains constant, but the amplitude of the short component progressively increases (Table I). It seems probable that the short component is due to DPH molecules, which are close to the melittin molecules or in regions of the bilayer that are perturbed by melittin.

It is instructive to examine the actual frequency-response data for DPH, which are shown in Figure 5. The data (solid dots) cannot be fit using a single-exponential model (dashed line). This is seen from the mismatch between the data and the calculated curve, and from the large and systematic deviations (lower panels). In contrast, the double-exponential model (solid line) provides a good fit to the data, and the deviations become small and randomly distributed. From these and other studies we know that the frequency-domain method provides good resolution of multiexponential decays. Hence, this method should be valuable in more detailed studies of the effects of membrane composition and temperature on the phase behavior of model membranes.

5. ANISOTROPY DECAYS OF LABELED MEMBRANES

5.1. Hindered Rotations of Diphenylhexatriene

It is well known that DPH does not rotate freely in model membranes (Dale *et al.*, 1977; Kawato *et al.*, 1977; Veatch and Stryer, 1977). Instead, DPH behaves predominantly as a hindered rotator for which the anisotropy decays to some limiting value r_∞:

$$r(t) = (r_0 - r_\infty) e^{-t/\theta} + r_\infty \tag{17}$$

It is probable that the actual DPH anisotropy decays are more complex than this, as has been suggested by theory (Kinosita *et al.*, 1977) and experimentation (Faucon and Lakowicz, 1987). Nonetheless, this simple model illustrates the use of frequency-domain measurements to recover a complex anisotropy decay.

In the frequency domain the measured quantities for an anisotropy decay are the differential polarized phase angle (Δ_ω) and the modulation ratio (Λ_ω) (Lakowicz and Prendergast, 1978). The differential phase angles are shown in Figure 6 for DPH in DPPC below (7°C) and above (39°C) its phase transition temperature. Below the transition temperature the phase angles are small. These data are for sonicated vesicles, and still smaller phase angles were found for unsonicated liposomes (Faucon and Lakowicz, 1987). The small phase angles are the result of the large value of r_∞, which uniformly suppresses the values of Δ_ω. The differential phase angles increase substantially at higher temperature, due to the larger amplitude of the DPH motions (smaller r_∞). The parameters

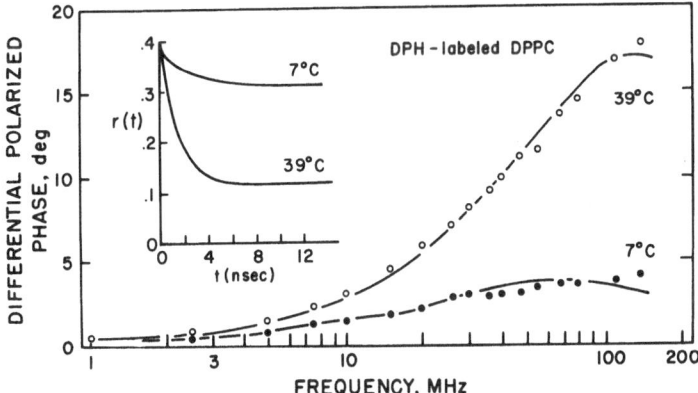

FIGURE 6. Frequency-dependent differential polarized phase angles of DPH-labeled DPPC vesicles. The insert shows the time-resolved anisotropies, which were derived from the frequency-domain data. From Lakowicz (1986).

(θ and r_∞) recovered from the analysis were used to reconstruct the time-resolved anisotropy decays, which are shown as an insert in Figure 6. Such anisotropy decays are easily recovered from the frequency-domain data, and the results are in good agreement with those from other laboratories (Dale *et al.*, 1977; Kawato *et al.*, 1977).

5.2. Anisotropic Rotations of Perylene

Our experience with the frequency-domain measurements suggested that it should be possible to recover more complex anisotropy decays. A multiexponential anisotropy decay is expected for perylene because its planar structure results in different rotational rates for the in-plane and out-of-plane motions (Mantulin and Weber, 1977; Barkley *et al.*, 1978). We tested the resolution of the method by examining perylene in propylene glycol at 28°C (Figure 7). At this temperature the correlation times were expected to be near 1. It was not possible to account for the data using a single correlation time, as is easily seen from the mismatch between the one correlation time fit (dashed line) and the data (solid dots). The data were adequately fit using two correlation times of 0.19 and 1.33 nsec, resulting in a 340-fold decrease in χ_R^2. This result indicates that we could resolve still more complex anisotropy decays, or a two-component decay for which the correlation times are shorter than 0.2 and 1.3 nsec. This capability is likely to be valuable for interpreting the details of anisotropy decay from more complex systems, such as the more complex decays predicted for hindered rotators in membranes (Kinosita *et al.*, 1977).

A final example of a complex anisotropy decay is shown in Figure 8 for the anisotropic molecule perylene in DMPC bilayers. Previous studies indicated

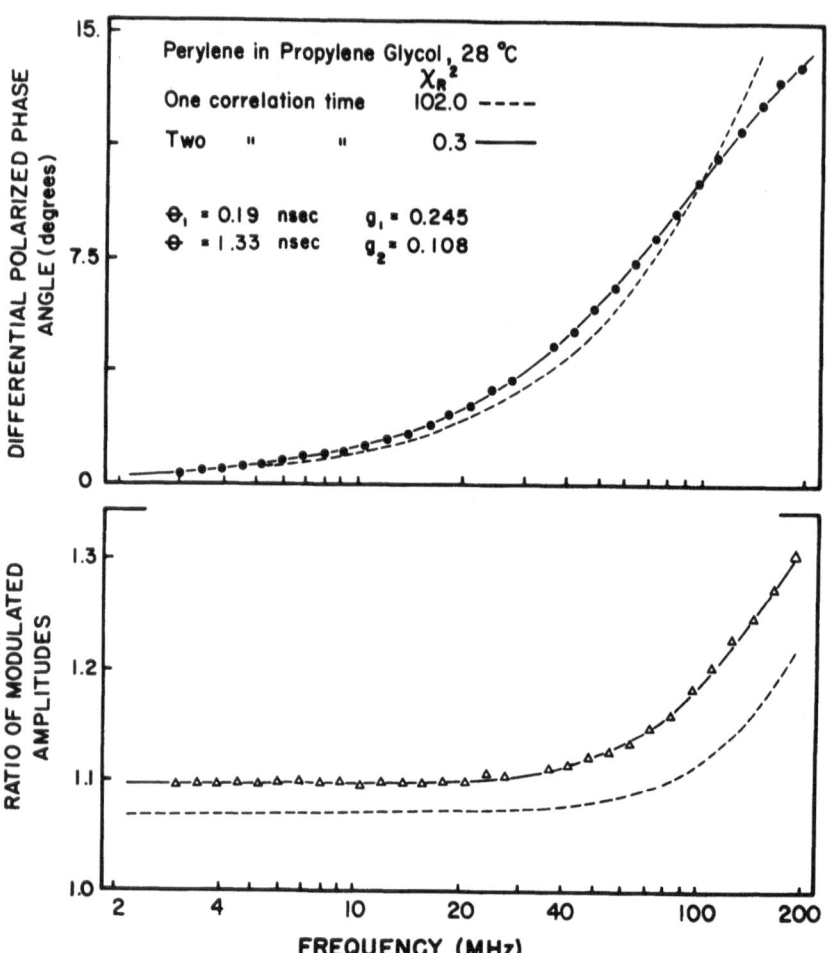

FIGURE 7. Resolution of a two-component anisotropy decay on the subnanosecond time scale (J. R. Lakowicz and B. Maliwal, unpublished data). The data are for perylene in propylene glycol at 28°C.

that below the transition temperature (<25°C) the rotational motions of perylene were hindered by the bilayer (Lakowicz and Knutson, 1980). Since perylene is itself asymmetric and displays two correlation times in isotropic solvents, we expected that three correlation times could be present in a membrane. The hindered model [Eq. (17)] was completely inadequate to account for the data ($\chi_R^2 = 106$). Even the fit with two correlation times is easily judged to be inadequate ($\chi_R^2 = 14.4$). The three-correlation-time fit provides a reasonable value of $\chi_R^2 = 0.5$, and reasonable correlation times (0.2, 1.4, and 47 nsec). We attribute the two shorter correlation times to rapid motions of perylene within

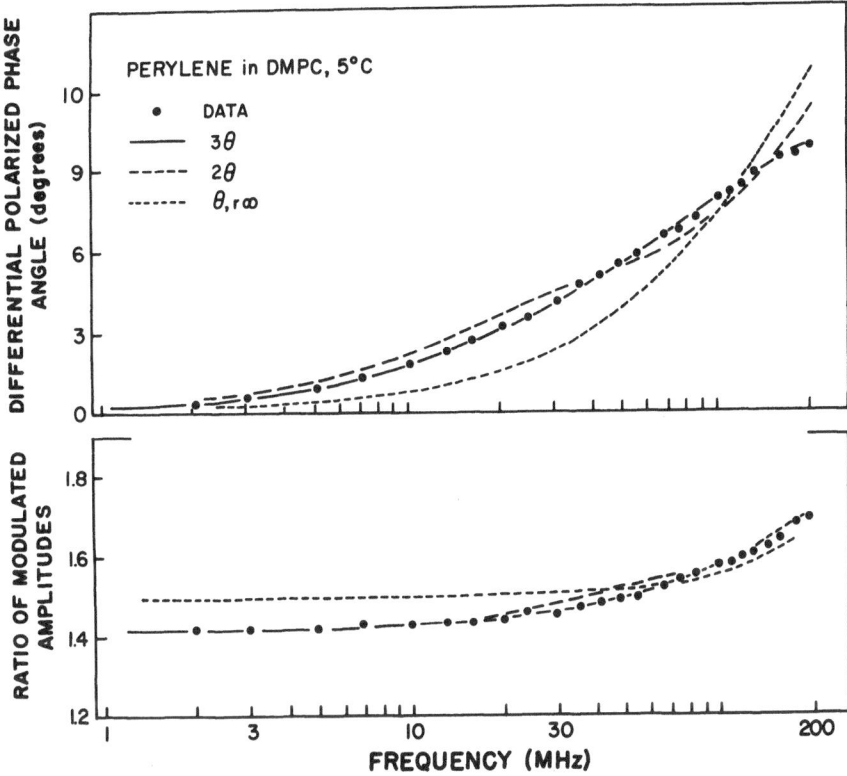

FIGURE 8. Three-component anisotropy decay of perylene in DMPC bilayers. The values of χ_R^2 for the three-correlation-time, two-correlation-time and hindered model fits are 0.5, 14.4, and 106, respectively. From Lakowicz *et al.* (1985).

the nonhindered environment and the 47-nsec correlation time to the hindered motion. It is remarkable that one could obtain a 28-fold decrease in χ_R^2 for distinguishing two versus three rotational correlation times. The results presented in Figures 7 and 8 demonstrate that the frequency-domain method provides remarkable resolution of complex decays of fluorescence anisotropy.

6. TIME-RESOLVED EMISSION SPECTRA

We now consider a different time-dependent process—time-dependent spectra shifts of fluorophores that are sensitive to the surrounding environment. A schematic representation of solvent relaxation is shown in Figure 9. Generally, the dipole moment of a fluorophore increases upon excitation. If the solvent is fluid, the rate of solvent relaxation is much more rapid than emission, and a red-shifted emission (R) is observed. If the solvent is cold and/or viscous, the rate

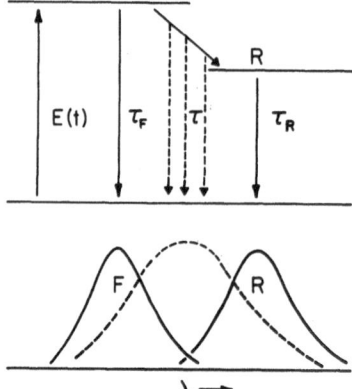

FIGURE 9. Schematic representation of solvent relaxation. The spectra for F and R represent the hypothetical spectra for the initially excited and relaxed states, respectively. The steady-state spectrum expected when $\tau = \tau_s$ is also shown (---). From Lakowicz (1983).

of solvent reorientation around the fluorophore is slow, and the emission spectrum shifts to shorter wavelengths (F). At any given wavelength the decay of fluorescence intensity is complex because the emission results from at least two states, and more probably a larger number of intermediate states. The decay may be estimated by a sum of exponential decays, but the actual decay is probably nonexponential. In spite of this complexity, it is possible to analyze the time-dependent spectral shifts using data observed in the frequency domain.

Time-resolved emission spectra are the spectra that would be observed at defined times following δ-pulse excitation. The determination of such detailed spectral information is generally regarded as a unique capability of the time-resolved method. If such spectra could be obtained from frequency-domain data, this would demonstrate the essential equivalence of the methods. This has been accomplished for labeled proteins and membranes (Lakowicz *et al.*, 1984a, 1984b). As an example we describe data for Patman, a lipidlike probe whose emission spectra are highly sensitive to the phase state of the bilayers (Lakowicz *et al.*, 1983).

The steady-state emission spectra of Patman-labeled vesicles are shown in Figure 10. Above the transition temperature of the DPPC vesicles, the emission spectrum of Patman shifts dramatically to longer wavelengths. For the DOPC vesicles, which are above their transition temperature even at 6°C, the emission spectrum is broad even at low temperature. At higher temperature, the emission spectrum becomes narrower and shifts to longer wavelengths. From earlier measurements, we know that these spectral shifts are due to time-dependent processes on the nanosecond time scale.

The time-dependent shifts can be quantified by measurements at various wavelengths across the emission spectrum. Wavelength-dependent phase and modulation data for Patman-labeled DPPC vesicles are shown in Figure 11. It

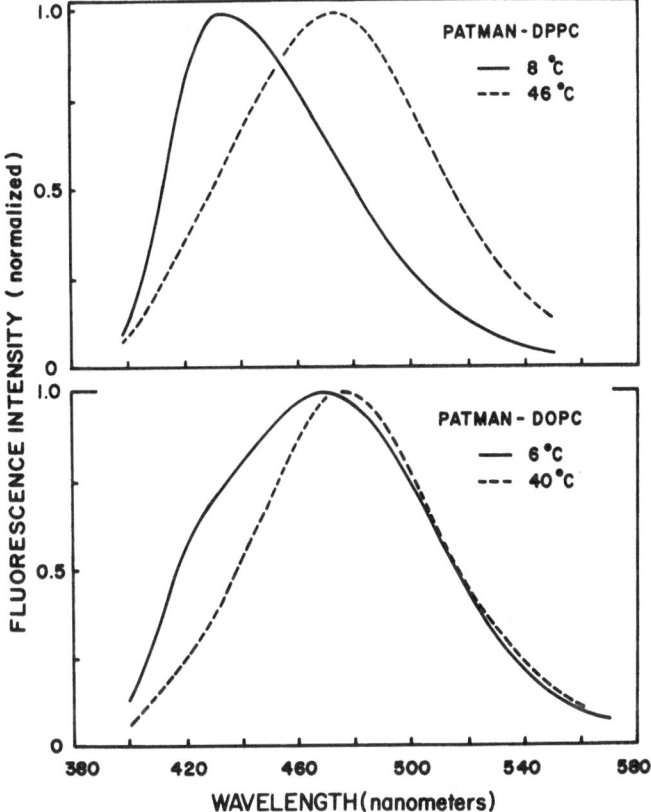

FIGURE 10. Steady-state emission spectra of Patman-labeled phospholipid vesicles. (*Top*) DPP vesicles at 8°C (—) and 46°C (---). (*Bottom*) DOPC vesicles at 6°C (—) and 40°C (---). From Lakowicz *et al*. (1984a).

is not practical to display the data at all emission wavelengths. Instead, we chose to show the data obtained at three emission wavelengths, located on the blue, center, and red sides of the emission. These data are characteristic of an excited-state reaction. That is, the phase angles increase and the modulation decreases with increasing emission wavelength. Furthermore, the data are unambiguous in demonstrating the occurrence of a time-dependent process that occurs subsequent to excitation. The apparent phase and modulation lifetimes at each frequency increase with the modulation frequency (data not shown), and the measured phase angles exceed 90°. This is not possible for a mixture of directly excited fluorophores (Lakowicz and Balter, 1982a). It is interesting to compare the results for Patman–DPPC at low and high temperatures (Figure 11). At high temperature and on the blue side of the emission, the phase angles are shorter,

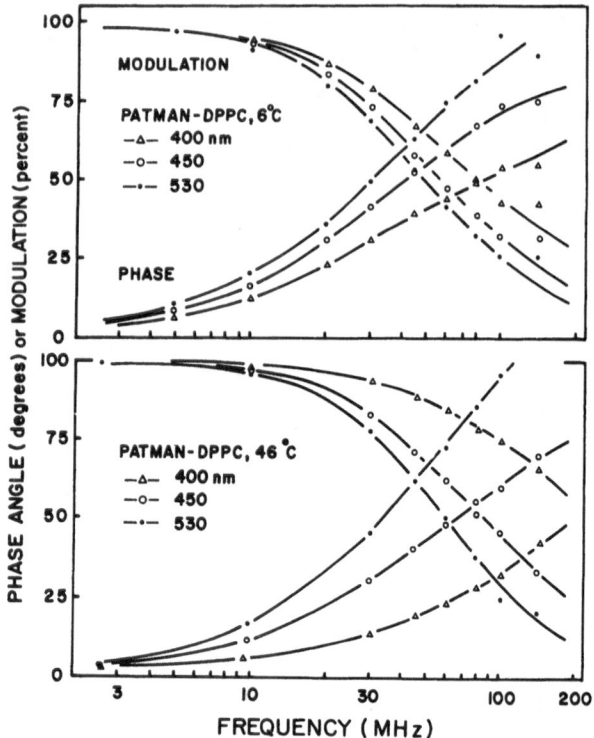

FIGURE 11. Frequency-dependent phase and modulation values for Patman-labeled DPPC vesicles at (*top*) 6°C and (*bottom*) 46°C. For both temperatures, data are shown for 400 (△), 450 (○), and 530 (●) nm. From Lakowicz *et al.* (1984a).

and the modulation values higher, than observed at lower temperature. This distinction reflects the shortened decay time on the blue side (400 nm) of the spectrum due to faster relaxation at the higher temperature. At longer emission wavelengths (530 nm), the data are more characteristic of an excited-state process, as is evident from the phase angles that exceed 90°.

The phase and modulation data can be used to derive the wavelength-dependent impulse response functions. These are the multiexponential intensity decays that would be observed for δ-function excitation and an infinitely fast detector. Typical impulse response functions for the Patman-labeled DPPC vesicles are summarized in Table II. The three decay times are similar at both 8°C and 46°C. The main distinction between these results is the larger contribution of terms with negative preexponential factors at the higher temperature. Such terms are characteristic of emission, which is formed by an excited-state process. At the longest emission wavelengths, these positive and negative terms are

Table II
Impulse Response Function for Patman-Labeled DPPC Vesicles[a]

Wavelength at 8°C (τ, nsec)	$\alpha_1(\lambda)$[b] (0.98 nsec)	$\alpha_2(\lambda)$ (3.74 nsec)	$\alpha_3(\lambda)$ (5.52 nsec)
400	1.15	0.60	0.25
410	0.87	0.74	0.39
430	0.21	1.02	0.77
450	−0.13	0.95	0.92
470	−0.40	0.73	0.87
490	−0.50	0.57	0.93
510	−0.61	0.33	1.06
530	−0.74	0.18	1.08

Wavelength at 46°C (τ, nsec)	$\alpha_1(\lambda)$ (0.95 nsec)	$\alpha_2(\lambda)$ (1.48 nsec)	$\alpha_3(\lambda)$ (3.94 nsec)
400	1.84	—	0.10
410	1.51	0.39	0.10
430	0.52	1.16	0.32
450	−0.56	0.91	0.53
470	−0.77	0.45	0.78
490	−0.85	0.14	1.01
510	−0.71	−0.20	1.09
530	−0.50	−0.42	1.07
550	−0.04	−0.90	1.06

[a]From Lakowicz *et al.* (1984a).
[b]The sum of the absolute values of $\alpha_i(\lambda)$ is set to 2.0 at each wavelength.

nearly equal in magnitude and opposite in sign. This is a characteristic of emission from the product of an excited-state process without substantial contribution from the emission due to the initially excited state (Laws and Brand, 1979).

6.1. Calculation of Time-Resolved Emission Spectra

Time-resolved emission spectra can be calculated from the properly normalized impulse response functions (Easter *et al.*, 1976). Specifically, the total intensity at each wavelength must be normalized to the steady-state intensity at this same wavelength [$F(\lambda)$]. Let

$$H(\lambda) = \frac{F(\lambda)}{\int_0^\infty I(\lambda,t)dt} \tag{18}$$

$$= \frac{F(\lambda)}{\alpha_1(\lambda)\,\tau_1 + \alpha_2(\lambda)\,\tau_2 + \alpha_3(\lambda)\,\tau_3} \tag{19}$$

Then the appropriately normalized functions are given by

$$I'(\lambda,t) = H(\lambda) \, I(\lambda,t) \tag{20}$$
$$= \sum_i \alpha_i' (\lambda)^{-t/\tau_i} \tag{21}$$

where $\alpha_i' (\lambda) = H(\lambda)\alpha_i(\lambda)$. By this procedure one forces the time-integrated intensity $(\Sigma_i \alpha_i \tau_i)$ to be equal to the steady-state intensity $[F(\lambda)]$. For any desired time (t_i), the time-resolved emission spectrum is obtained by plotting the values $I'(\lambda,t)$.

6.2. Time-Resolved Emission Centers of Gravity and Spectral Half-Widths

The time-dependent spectral shifts can be characterized by the time-dependent center of gravity, in wave numbers (cm^{-1} or kK), which is proportional to the average energy of the emission. The center of gravity of the emission is defined by

$$\bar{v}_{cg}(t) = \frac{\int_0^\infty I'(\bar{v},t) \, \bar{v} \, d\bar{v}}{\int_0^\infty I'(\bar{v},t)d\bar{v}} \tag{22}$$

where $I'(\bar{v},t)$ represents the number of photons per wave number interval. For equally spaced wavelength intervals, and using an emission monochromator with a constant bandpass in nanometers, the center of gravity in kK ($= 10^3 \, cm^{-1}$) is given by

$$\bar{v}(t) = 10,000 \, \frac{\sum_\lambda I'(\lambda,t)\lambda^{-1}}{\sum_\lambda I'(\lambda,t)} \tag{23}$$

The time-dependent spectral width $\Delta\bar{v}(t) \, cm^{-1}$ can reveal whether the spectral relaxation is best described by a continuous or stepwise process. We chose to characterize this parameter using

$$\Delta\bar{v}(t) = \frac{\int_0^\infty [\bar{v} - \bar{v}_{cg}(t)]^2 \, I'(\bar{v},t)d\bar{v}}{\int_0^\infty I'(\bar{v},d)d\bar{v}} \tag{24}$$

6.3.   Time-Resolved Spectral Data for Patman-Labeled Membranes

Time-resolved emission spectra of the Patman-labeled vesicles are shown in Figures 12 and 13. To facilitate the comparison of these spectra we chose to display, in each case, time-resolved spectra at 0.2, 2, and 20 nsec. For DPPC, below its transition temperature, there is only a modest red shift at 8°C, about 15 nm. At 46°C, a dramatic spectral shift is seen to occur between 0.2 and 20 nsec, about 50 nm.

Time-resolved spectra for DOPC vesicles are shown in Figure 13. In this case, a dramatically broadened spectrum is seen at 2 nsec 6°C. The spectral widths are obviously more narrow at 0.2 and 20 nsec, indicating that spectral relaxation has not proceeded substantially at 0.2 nsec and that it is mostly complete at 20 nsec. The wider spectrum seen at 2 nsec is probably the result of emission from two distinct states and is thus evidence for a two-state relax-

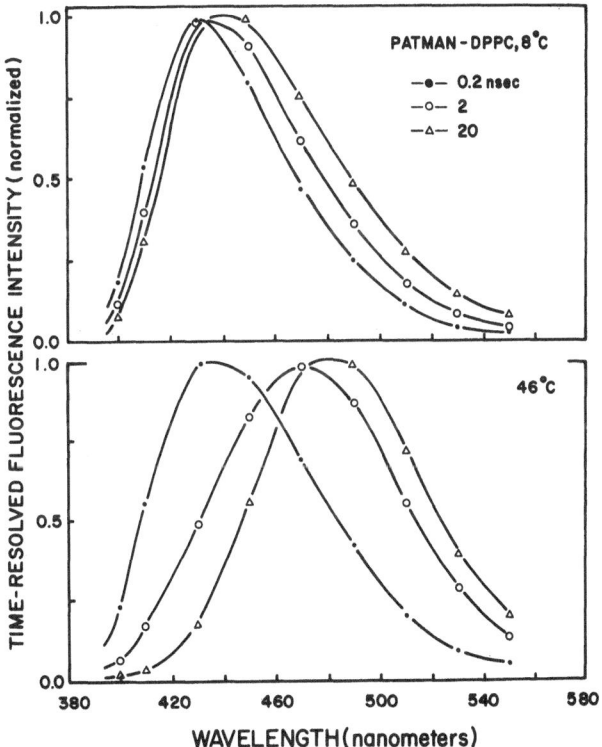

FIGURE 12. Time-resolved emission spectra of Patman-labeled DPPC vesicles at (*top*) 8°C and (*bottom*) 46°C. Time-resolved spectra are shown at 0.2 (●), 2 (○), and 20 (△) nsec. From Lakowicz *et al.* (1984a).

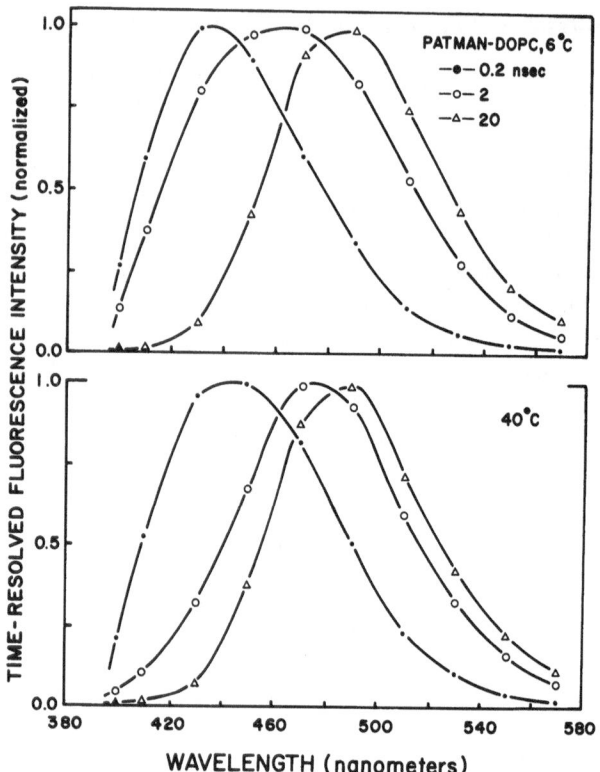

FIGURE 13. Time-resolved emission spectra of Patman-labeled DOPC vesicles at (*top*) 6°C and (*bottom*) 40°C. Time-resolved emission spectra are shown at 0.2 (●), 2 (○), and 20 nsec (△). From Lakowicz *et al.* (1984a).

ation process. At higher temperature, the spectral relaxation appears to be more rapid. This is evident from the nearly relaxed spectrum seen at 2 nsec at 40°C.

A more quantitative measure of the rates of spectral relaxation can be obtained from the time-dependent decay of the center of gravity $[\bar{\nu}_{cg}(t)]$, summarized in Figure 14. For DPPC, both the extent and rate of spectral relaxation are increased above the transition temperature. For the DOPC vesicles, a similar amount of energy (2.5 kK) is lost at low and high temperatures, and the rate of decay is increased at the higher temperature. It appears that spectral relaxation is more rapid in DOPC than in DPPC vesicles. The relaxation times appear to be near 2 nsec. If desirable, these time-dependent values of $\bar{\nu}_{cg}(t)$ could be used to calculate the decay law for the center of gravity. A preliminary examination of these data revealed that the decay is more complex than a single exponential.

We also examined the time-dependent changes in the spectral widths $[\Delta\bar{\nu}_{cg}(t)]$

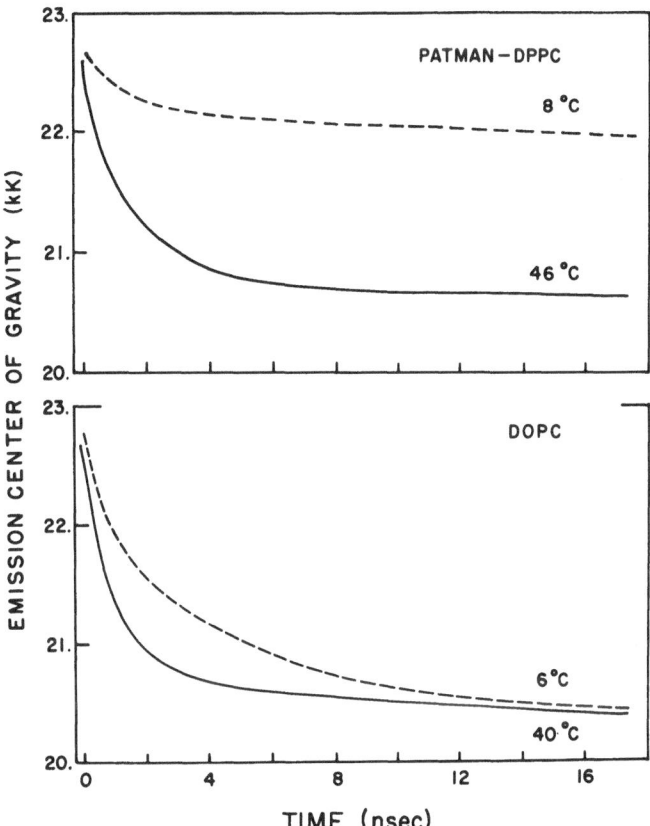

FIGURE 14. Time-resolved emission center of gravity of Patman-labeled vesicles. (*Top*) DPPC at 8°C (---) and 46°C (—). (*Bottom*) DOPC at 6°C (---) and 40°C (—). From Lakowicz *et al.* (1984a).

(Figure 15). In all cases, except for DPPC at 8°C, there is a substantial increase in spectral width at intermediate times. The increased width is the result of contributions from the initially excited and the relaxed states to the emission spectrum. At higher temperature, the increase in $\Delta\bar{\nu}(t)$ is of shorter duration, reflecting a more rapid decay to the equilibrium excited state. This transient increase in $\Delta\bar{\nu}(t)$ suggests that the spectral relaxation of Patman is best described by the two-state model.

7. ENERGY TRANSFER IN MEMBRANES

Fluorescence energy transfer is widely used to estimate the distance between sites on macromolecules (Steinberg, 1971; Stryer, 1978). Such measure-

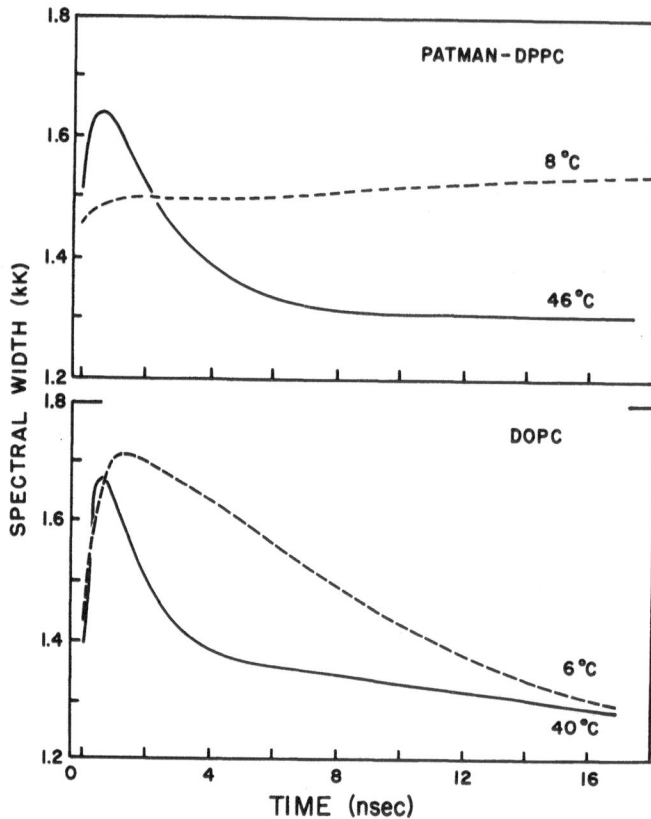

FIGURE 15. Time-resolved spectral width of Patman-labeled vesicles. (*Top*) DPPC at 8°C (---) and 46°C (—). (*Bottom*) DOPC at 6°C (---) and 40°C (—). From Lakowicz *et al.* (1984a).

ments have been widely applied to proteins and nucleic acids, which have a discrete structure and discrete distances. Energy transfer measurements have been used less frequently in membranes because single distances between sites may not exist. We now show how energy transfer can be used in the more difficult case where one wishes to recover a distribution of distances. This possibility is illustrated in Figure 16, which shows the intensity decay data of dansyl-PE, in the presence and absence of a membrane-bound acceptor (trinitrophenyl-PE, TNP-PE). In the absence of acceptor (Δ) the intensity decay is predominantly a single exponential, $\chi_R^2 = 1.2$. In the presence of acceptor the intensity decay becomes highly heterogeneous, as is seen from $\chi_R^2 = 220$, and the mismatch between the data (solid dots) and the best single-exponential fit. The increased heterogeneity is the result of a range of decay times. This dispersion of decay times is the result of a variety of donor–acceptor distances.

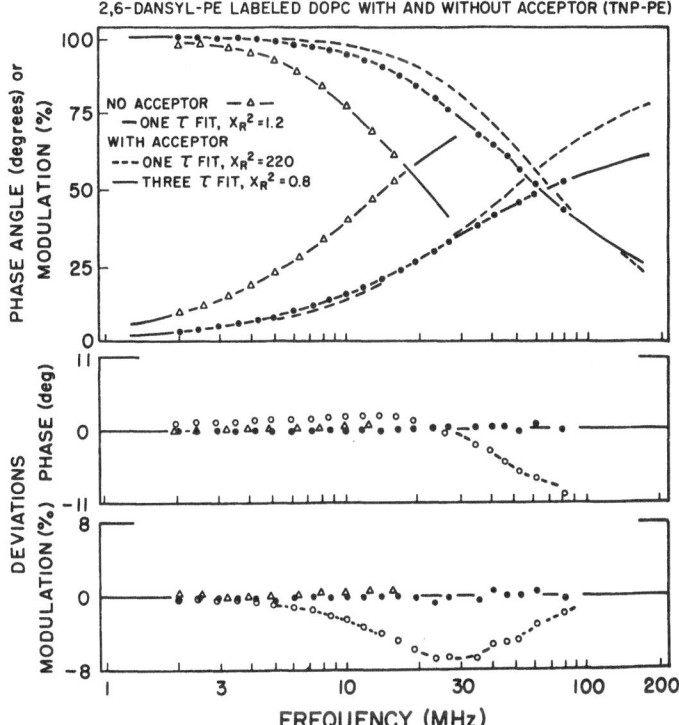

FIGURE 16. Intensity decay of dansyl-PE-labeled DOPC vesicles, with and without an acceptor (TNP-PE).

We are currently developing the software to analyze the increased dispersion in the frequency response in terms of the spatial distribution of acceptors around the donor. Once these techniques are developed it should be possible to use the frequency response of membrane-bound donors to detect processes that cause nonrandom acceptor distributions, such as phase separations or boundary lipid around proteins.

7.1. Distribution of Distances in a Covalently Linked Donor–Acceptor Pair

At present, we can analyze the data from a donor–acceptor pair, if only one acceptor is present per donor (Lakowicz et al., 1987e). One such pair is illustrated in Figure 17. The donor is the indole ring and the acceptor is the dansyl moiety. The donor alone is provided by tryptamine myristic acid (TMA). To determine the distribution of distances we use the donor decay, which is

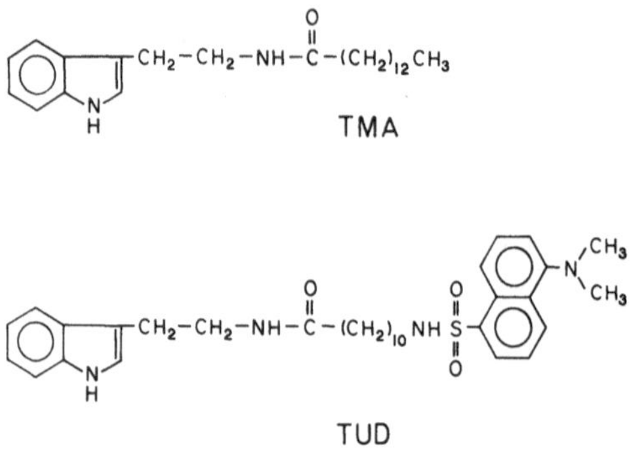

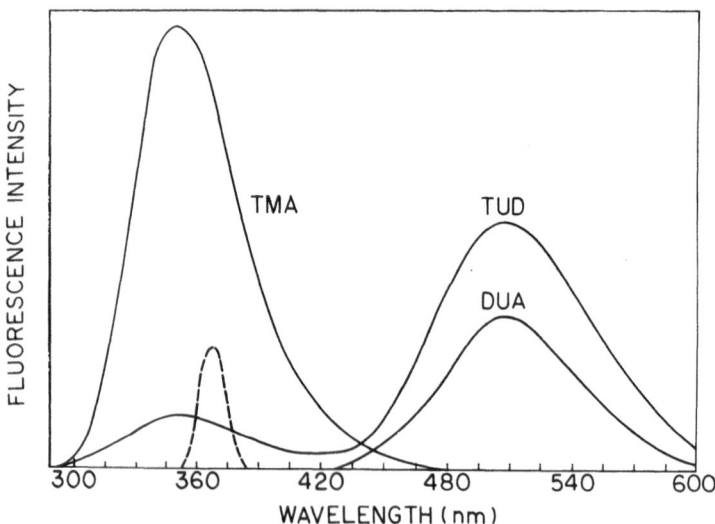

FIGURE 17. Donor–acceptor pair for distance distribution analysis.

isolated using an interference filter at 360 nm (Figure 18). These decays are measured in the presence and absence of energy transfer.

The frequency response of the donor emission is shown in Figure 19. In the absence of energy transfer (TMA) the emission is a single exponential, as is seen by the good match between the data (solid dots, left) and the best single-exponential fit. The intensity decay of the donor–acceptor pair is considerably more complex, as is seen from our inability to fit the data (open circles) to a single-exponential decay and by the elevated value of $\chi_R^2 = 1494$.

The data were then analyzed using algorithms that allow for a Gaussian

FIGURE 18. Emission spectra of TMA (donor), TUD (donor–acceptor), and DUA (acceptor) in propylene glycol at 20°C. The dashed line near 360 nm is the donor emission as observed through a 360-nm interference filter.

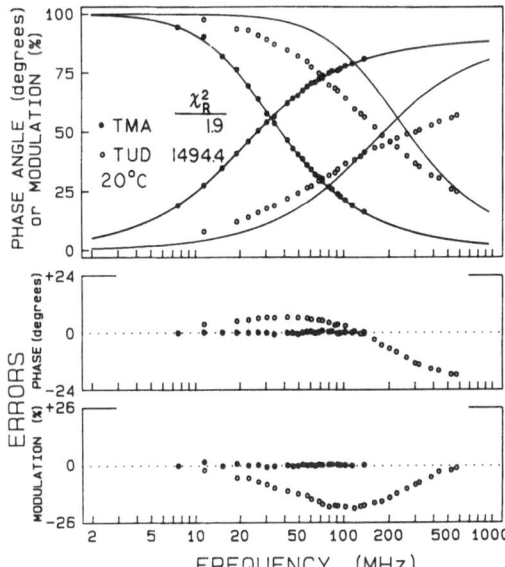

FIGURE 19. Frequency response of the donor (360 nm) emission without (●) and with (○) the covalently attached acceptor at 20°C. The solid lines show the best single-exponential fits to the data.

distribution of donor–acceptor distances (Figure 20). This model results in a good fit to the data, with χ_R^2 values near 5. While these χ_R^2 values are not as low as for a triple-exponential fit, this model has only two floating parameters (the average distance R and the half-width of the distribution HW), whereas the triple-exponential model contains five adjustable parameters (three τ_i and two α_i). Also, it is unlikely that the donor–acceptor distribution is described precisely by any of these models because the linker is not infinitely long (Flory, 1969). It is not possible to fit the data to a narrow range of distances. For instance, if the half-width is held fixed at 2 Å, the data cannot be fit, resulting in $\chi_R^2 = 1110$ (dashed line, Figure 20).

The distance distributions were recovered at several temperatures and by using different analytical forms of the distribution (indicated by r^0, r^1, and r^2). The range of distance distributions recovered with these various models, for the three temperatures, is shown in Figure 21. The recovered donor–acceptor distributions were similar at each temperature and for each model (r^0, r^1, and r^2). These results indicate that the donor–acceptor distance distribution is broad, and only limited variations in the distribution are consistent with the data.

The ability to recover a distribution of distances has numerous potential applications in biochemical research in addition to its use in studies of lipid distributions in membranes (Figure 16). For example, small peptides and proteins can exist in several conformations, whose relative populations are determined by the peptide backbone, structures of the side chains, and the rotational potential function. Determination of the end-to-end distributions for such pep-

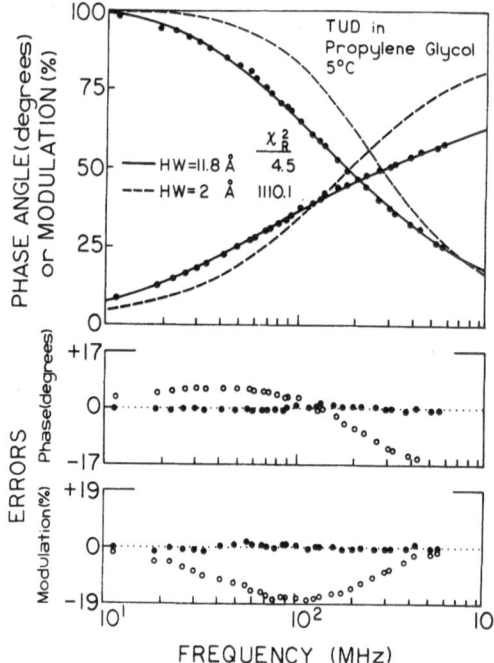

FIGURE 20. Distance distribution analysis of 360-nm emission from TUD at 5°C. The solid line shows the best fit to the data using the decay law of the TMA and a Gaussian distance distribution with $R = 11.8$ Å and HW = 14.8 Å. Also shown is the best fit when the half-width is fixed at 2 Å (---).

tides can be used for comparison of the experimental distributions with calculations based on theoretical models for the conformational potential functions. This technique can also be used to determine whether two sites on a protein are indeed at a single distance, or whether there exists a range of distances due to either multiple conformations or structural fluctuations around a mean distance.

8. A 2-GHz FREQUENCY-DOMAIN FLUOROMETER

In the frequency domain it is desirable to measure data over the widest possible range of frequencies. There were several limitations to the measurement at higher frequencies. The upper frequency limit of most photomultiplier tubes is near 200 MHz, and it is difficult to obtain useful intensity modulation above 200 MHz. Traveling wave modulators—with UV transmission, low half-wave voltages, and that work to several gigahertz—have been described (Peters, 1965; White and Chin, 1972; Kaminov, 1974). However, we were unable to use a commercially available version of these modulators, apparently because of their sensitivity to small temperature gradients, which cause drifts in transmission through the crossed polarizers.

The need for a light modulator was eliminated by using the intrinsic harmonic content of a laser pulse train (Figure 22). The light source consists of a

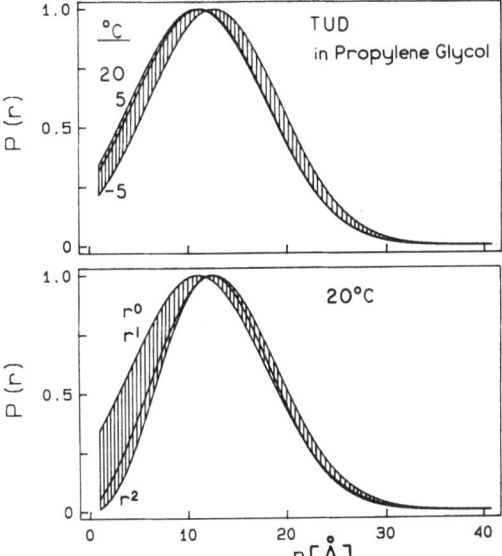

FIGURE 21. (*Top*) Distance distributions recovered at -5, 5, and 20°C for TUD in propylene glycol. (*Bottom*) Distance distributions found using Gaussian distribution multiplied by r^0, r^1, and r^2. The shaded region indicates the maximum range of $P(r)$ values found at 20°C using the different models.

mode-locked argon ion laser and a synchronously pumped and cavity-dumped dye laser. This source provides 5-psec pulses at a repetition rate of 4 MHz. In the frequency domain this source is intrinsically modulated to many gigahertz, as shown by the schematic Fourier transform (Figure 22). The idea of using the harmonic content of a pulse train was originally proposed by Merkelo and colleagues (1969) for a pulsed laser source and later by Gratton and co-workers for synchrotron radiation (1980, 1984). This source provides light that is intrinsically modulated at each integer multiple of the repetition rate, to about 50 GHz.

A second and significant advantage of this source is that it is rather easy to frequency-double the pulse train because of the high peak power. Additionally, the doubled output is used directly to excite the sample. It is no longer necessary to use an electro-optic modulator and nearly crossed polarizers, which result in a significant attenuation of the incident light (Lakowicz and Maliwal, 1985). During the past year we performed numerous measurements with the harmonics extending to 2 GHz. There is no detectable increase in noise at the highest frequencies, suggesting that there is no multiplication of phase noise at the higher harmonics. We are pleased with this approach and at present do not feel a need to identify other light modulators.

The second obstacle to higher-frequency measurements was the detector. The PMT in our original instrument (Figure 3) was replaced with a microchannel plate (MCP) PMT. These devices are 10- to 20-fold faster than a standard PMT (Kinosita and Kushida, 1985; Yamazaki *et al.*, 1985) and hence a bandwidth of 2 GHz was expected. As presently designed, the MCP PMTs do not allow

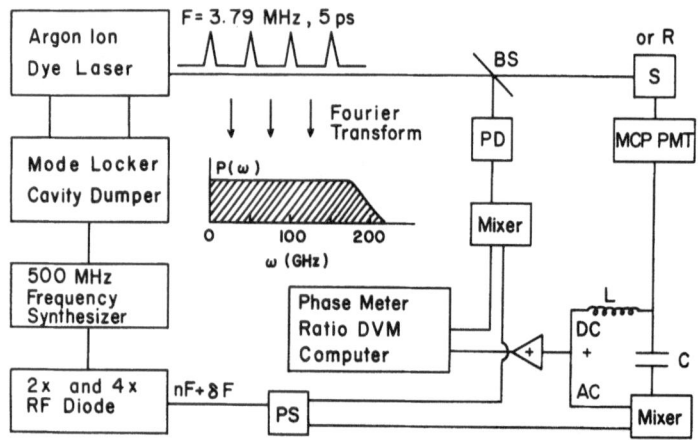

FIGURE 22. The 2-GHz frequency-domain fluorometer. PD, photodiode; PS, power splitter; MCP PMT, microchannel plate photomultiplier tube; BS, beam splitter; F, fundamental frequency of cavity-dumped dye laser output; δF, cross-correlation frequency of 25 Hz; n, number of the harmonic; S, sample; R, reference or scatter. From Lakowicz *et al.* (1986b).

internal cross-correlation, which is essential for an adequate signal-to-noise ratio. This problem was circumvented by designing an external mixing circuit, which preserved both the phase and the modulation data. With these modifications, data can now be obtained to 2 GHz (Lakowicz *et al.*, 1986b).

The light source is presently a mode-locked argon ion laser. The ion laser pumps a dual jet dye laser, whose output is cavity dumped at 3.79 MHz. The dye laser with R6G provides excitation wavelengths from 570 to 600 nm. For excitation with UV its output is frequency-doubled to 285–300 nm. The average UV power is near 0.5 mW, which is attenuated 50– to 100–fold prior to the sample. Alternatively, the beam diameter is expanded to about 5 mm using a negative lens. For our application the pulse width and shape are not critical, but the pulse width of the visible output of the dye laser was 5 psec or less.

8.1. Picosecond Resolution of Tyrosine Intensity and Anisotropy Decays

Relatively few time-resolved measurements have been performed on tyrosine-containing proteins. This is because most proteins also contain tryptophan, whose emission overwhelms that of the tyrosine residues and quenches their emission by energy transfer. Tyrosine fluorescence is difficult to measure because of its short lifetime, short absorption and emission wavelengths, and small Stokes shift between absorption and emission. Recently, the time-resolved emission from a variety of tyrosine-like compounds have been studied with synchrotron radiation (Laws *et al.*, 1986) and the time-resolved emission of histone H1 was examined with a pulsed laser source by Libertini and Small (1985).

We examined the intensity and anisotropy decays of several tyrosine compounds and peptides (Lakowicz *et al.*, 1986c, 1987a). One example is shown in Figure 23, which shows the frequency response of the tyrosine emission from oxytocin. This substance is a cyclic nonapeptide that contains a single tyrosine residue. The mean decay time for oxytocin is near 0.7 nsec. Even with this short decay time, the entire frequency response was measured, as seen by phase angles that extend to 70° and modulations to 20%. These data are adequate to support a three-exponential analysis. The apparent decay times are 80, 359, and 927 psec. To the best of our knowledge, this is the first resolution of three decay times that are all below 1 nsec.

The 2-GHz data also provide resolution of complex anisotropy decays on the picosecond time scale. Data for the anisotropy decay of oxytocin are shown in Figure 24. It is not possible to fit the data using a single correlation time (dashed lines, $\chi_R^2 = 292$). In contrast, a two-correlation-time fit provides a good fit, which is not improved by the use of a third correlation time. We believe the correlation times of 29 and 454 psec reflect local tyrosyl motions and overall rotational diffusion, respectively. It is important to note that the measurements to 2 GHz provide considerable information content beyond the previous 200-MHz limit. Data to 200 MHz would not display the shoulder seen at 600 MHz, which represents the transition from rotational diffusion to segmental motions.

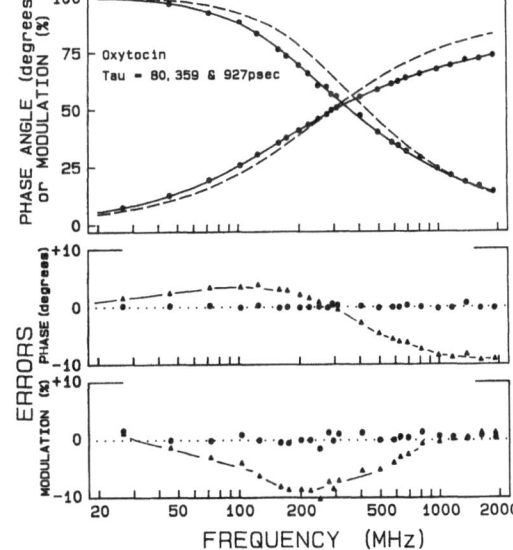

FIGURE 23. Frequency-domain data for the tyrosine intensity decay of oxytocin. The upper panel shows the data (●). The solid line is the best three-exponential fit and the dashed line the best one-exponential fit. The lower panels show the deviations between the data and the calculated values for the one (△) and three (●) decay time models. From Lakowicz *et al.* (1986c).

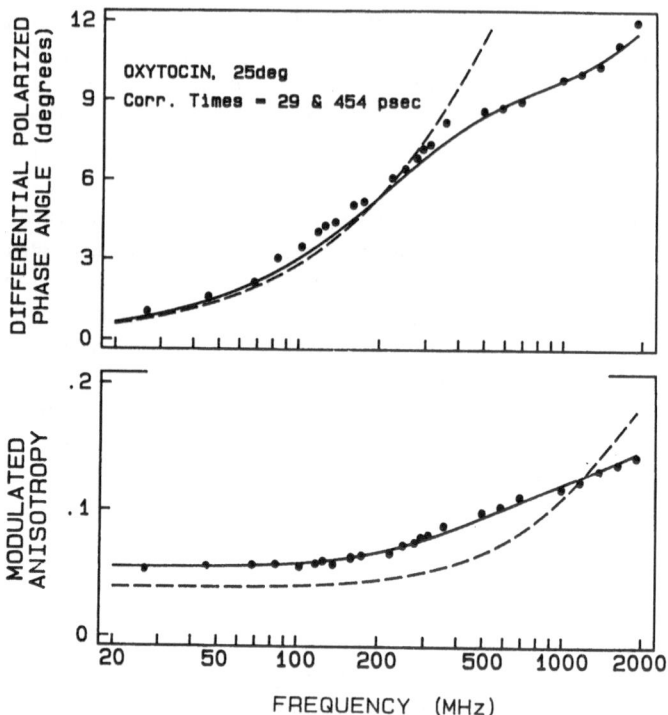

FIGURE 24. Frequency-domain data for the tyrosine anisotropy decay of oxytocin. The data (●) could not be fit using a single correlation time (---) but were adequately fit using two correlation times (—). From Lakowicz *et al.* (1986c).

8.2. Measurement of an 8-psec Correlation Time

As the result of the successful resolution of the oxytocin anisotropy decay, we question whether still shorter correlation times could be measured. This test was performed using indole as the fluorophore, to model the emission from proteins. The chosen solvent was methanol because of its low viscosity. To allow measurement to the instrumental limit of 2 GHz, we used acrylamide to decrease the mean decay times to near 0.18 nsec (Lakowicz *et al.*, 1987b). The temperature-dependent anisotropy data are shown in Figure 25. At −55°C the 145-psec rotational correlation time of indole is easily measured. Even for this short correlation time we were able to measure beyond the peak of the Lorentzian. As the temperature is increased the phase angles shift to higher frequencies. At 5°C the 28-psec correlation time yields a 9° phase angle shift, which is easily measured. To obtain a still faster rate of rotation, the sample was heated to 80°C in a pressurized vessel. At this temperature the apparent correlation time is 8 psec. Even for this short correlation time the uncertainty is about 1 psec. This ability to measure picosecond correlation times should result in a

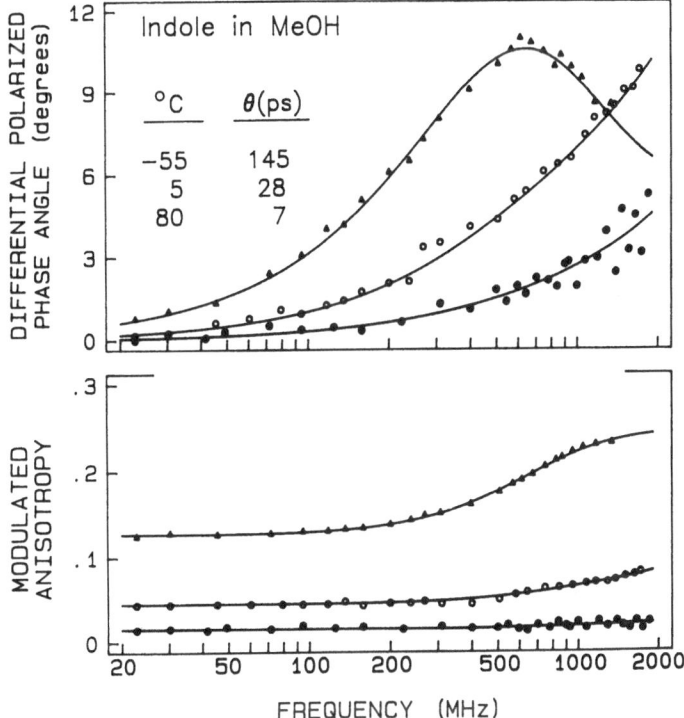

FIGURE 25. Anisotropy decays of indole in methanol at -55 ($\triangle$), 5 ($\circ$), and 80°C ($\bullet$). Acrylamide was used to decrease the decay times to about 0.18 nsec. From Lakowicz *et al.* (1987b).

comparison of experimental measures of protein dynamics with molecular dynamics simulations (Karplus, 1986).

9. FUTURE DEVELOPMENTS

Since the instrumentation for frequency-domain fluorometry has only appeared in the past four years, there has been little opportunity to determine the limits of resolution. To date, only a few instrumental configurations have been tested, and all data analysis has been done by the method of nonlinear least squares. It is probable that there will be rapid advances in both instrumentation and analysis.

Regarding instrumentation, several developments seem inevitable. At present, phase and modulation values are measured at one frequency at a time. However, light sources may be modulated simultaneously at many frequencies (Kaminov, 1974) or intrinsically modulated at many frequencies like a pulse laser. Simultaneous measurement at many frequencies will decrease the time required for data acquisition and possibly increase the accuracy. The sources for modulated excitation are likely to become less expensive. Arc lamps can now be used with electro-optic modulators (Gratton, 1986) and simple acousto-optic modulators can provide simultaneous modulation at several frequencies. Additionally, laser diodes may serve as an inexpensive source for the modulation excitation. Laser diodes can be modulated to about 10 GHz. Unfortunately, the emission of these sources is near 830 nm, but reports of frequency doubling their output are starting to appear. Hence, the instrumentation should become faster, simpler, and less expensive.

One can also expect improved resolution as the result of alternative means of data analysis. In this laboratory we are emphasizing multiple measurements, followed by simultaneous analysis of the multiple data sets. By variation of the excitation, emission wavelengths, or quencher concentration, it is often possible to vary the weight of various components in the data, in a manner that allows the decay times or correlation times to be linked across all the data (Gryczynski et al., 1987). In addition to global analysis of data obtained at several excitation wavelengths, we are using collisional quenching to decrease the decay times (Lakowicz et al., 1987d). The anisotropy measurements under these conditions are thus obtained with various mean decay times. Global analysis of these multiple data sets provides enhanced resolution of rapid and complex anisotropy decays (Lakowicz et al., 1987d). These studies resulted in the observation of transient effects in quenching, which are now known to be easily detected using the frequency-domain method (Joshi et al., 1987; Lakowicz et al., 1987). Importantly, these effects allow estimation of the diffusion coefficients of quenchers in proteins (Lakowicz et al., 1987c). Finally, we have not yet tested alternative analysis methods, such as factor analysis (Malinowski and Howery, 1980; Marchiarullo and Ross, 1985). Such methods may become desirable as the instrumentation is automated to provide rapid measurement of the wavelength-dependent frequency responses.

10. SUMMARY

Time-resolved fluorescence spectroscopy is often used for studies of biological macromolecules. Time-resolved data are usually obtained by measurements of the time-dependent emission from samples excited with brief pulses of light. In this chapter we describe an alternative method, in which the time-resolved emission is determined from the frequency response of the emission to intensity-modulated light. We describe the instrumentation for these measurements and the methods used to analyze the data. The examples include the resolution of closely spaced lifetimes, measurement of complex decays of fluorescence anisotropy, and calculation of time-resolved emission spectra from the wavelength-dependent frequency response of the sample. We also describe a new application, by which one can recover the distribution of distances between two sites on a flexible molecule. Finally, we describe our recent extension of the measurements to 2 GHz, which has allowed measurement of correlation times as short as 8 psec.

ACKNOWLEDGMENTS. This work was supported by grants DMB-8511065 and DMB-08502835 from the National Science Foundation and grants GM-29318 and GM-35154 from the National Institutes of Health. The author offers his special thanks to the National Science Foundation for supporting development of the frequency-domain method.

11. REFERENCES

Belford, G. G., Belford, R. L., and Weber, G., 1972, Dynamics of fluorescence polarization in macromolecules, *Proc. Natl. Acad. Sci. U.S.A.* **69** : 1392–1393.

Barkley, M. D., Kowalczyk, A. A., and Brand, L., 1978, Fluorescence decay studies of anisotropic rotations of small molecules, *J. Chem. Phys.* **75** : 3581–3593.

Bevington, P. R., 1969, *Data Reduction and Error Analysis for the Physical Sciences*, McGraw-Hill, New York.

Chuang, T. J., and Eisenthal, K. B., 1972, Theory of fluorescence depolarization by anisotropic rotational diffusion, *J. Chem. Phys.* **57** : 5094–5097.

Cundall, R. B., and Dale, R. E. (eds.), 1980, *Time-resolved fluorescence spectroscopy in biochemistry and biology*, Plenum Press, New York.

Dale, R. E., Chen, L. A., and Brand, L., 1977, Rotational relaxation of the "microviscosity" probe diphenylhexatriene in paraffin oil and egg lecithin vesicles, *J. Biol. Chem.* **252** : 7500–7510.

Demchenko, A. P., 1986, *Ultraviolet Spectroscopy of Proteins*, Springer-Verlag, Berlin.

Easter, J. H., De Toma, R. P., and Brand, L., 1976, Nanosecond time-resolved emission spectroscopy of a fluorescence probe adsorbed to L-α-egg lecithin vesicles, *Biophys. J.* **16**: 571–583.

Faucon, J. F., and Lakowicz, J. R., 1987, Anisotropy decay of diphenylhexatriene in melittin-phospholipid complexes by multi-frequency phase-modulation fluorometry, *Arch. Biochem. Biophys.* **252** : 245–258.

Flory, P., 1969, *Statistical Mechanics of Chain Molecules*, Interscience, New York.

Gaviola, Z., 1926, Ein Fluorometer. Apparat zur Messung von Fluoreszenzabklingungszeiten, *Z. Phys.* **42** : 853–861.

Gratton, E., 1986, ISS, Inc., Urbana, Illinois.

Gratton, E., and Limkemann, M., 1983, Picosecond fluorescence spectroscopy by time-correlated single-photon counting, *Biophys. J.* **44** : 315–324.

Gratton, E., and Lopez-Delgado, R., 1980, Measuring fluorescence decay times by phase-shift and modulation techniques using the high harmonic content of pulsed light sources, *Nuovo Cimento* **B56** : 110–124.

Gratton, E., James, D. M., Rosato, N., and Weber, G., 1984a, Multifrequency cross-correlation phase fluorometer using synchrotron radiation, *Rev. Sci. Instrum.* **55** : 486–494.

Gratton, E., Limkemann, M., Lakowicz, J. R., Maliwal, B. P., Cherek, H., and Laczko, G.,

1984b, Resolution of mixtures of fluorophores using variable-frequency phase and modulation data, *Biophys. J.* **46** : 479–486.

Gryczynski, I., Cherek, H., Laczko, G., and Lakowicz, J. R., 1987, Enhanced resolution of anisotropic rotational diffusion by multi-wavelength frequency-domain fluorometry and global analysis, *Chem. Phys. Lett.* **135** : 193–199.

Heyn, M. P., 1979, Determination of lipid order parameters and rotational correlation times from fluorescence depolarization experiments, *FEBS Lett.* **108** : 359–364.

Jameson, D. M., and Gratton, E., 1983, Analysis of heterogeneous fluorescence by multifrequency phase and modulation fluorometry, in *New Directions in Molecular Luminescence* (D. Eastwood, ed.), pp. 67–81, American Society for Testing and Materials, Philadelphia.

Jameson, D. M., and Weber, G., 1981, Resolution of the pH-dependent heterogeneous fluorescence decay of tryptophan by phase and modulation measurements, *J. Phys. Chem.* **85**: 953–958.

Joshi, N., Johnson, M. L., Gryczynski, I., and Lakowicz, J. R., 1987, Radiation boundary conditions in collisional quenching of fluorescence: Resolution by frequency-domain fluorometry, *Chem. Phys. Lett.* **135** : 200–207.

Karplus, M., 1986, Internal dynamics of proteins, in *Methods in Enzymology* (C. H. W. Hirs and S. N. Timasheff, eds.), Vol. 131, pp. 283–307, Academic Press, New York.

Kaminov, I. P., 1974, *An Introduction to Electro-optic Devices*, Academic Press, New York.

Kawato, S., Kinosita, K., and Ikegami, A., 1977, Dynamic structure of lipid bilayers studied by nanosecond fluorescence techniques, *Biochemistry* **16** : 2319–2324.

Kinosita, K., Kawato, S., and Ikegami, A., 1977, Construction of a nanosecond fluorometric system for applications to biological samples at cell or tissue levels, *Biophys. J.* **20** : 289–305.

Kinosita, S., and Kushida, T., 1985, Picosecond fluorescence spectroscopy by time-correlated single-photon counting, *Anal. Instrum.* **14** : 503–524.

Lakowicz, J. R., 1983, *Principles of Fluorescence Spectroscopy*, Plenum Press, New York.

Lakowicz, J. R., 1986, Fluorescence studies of the dynamics of proteins and membranes, in *Methods in Enzymology* (C. H. W. Hirs and S. N. Timasheff, eds.), Vol. 131, pp. 518–567, Academic Press, New York.

Lakowicz, J. R., and Balter, A., 1982, Theory of phase-modulation fluorescence spectroscopy for excited state processes, *Biophys. Chem.* **16** : 99–115.

Lakowicz, J. R., and Cherek, H., 1981, Fluorescence spectroscopic investigations of the dynamic properties of proteins, membranes and nucleic acids, *J. Biochem. Biophys. Methods* **5** : 19–35.

Lakowicz, J. R., and Knutson, J. R., 1980, Hindered depolarizing rotations of perylene in lipid bilayers; detection by lifetime-resolved anisotropy measurements, *Biochemistry* **19** : 905–911.

Lakowicz, J. R., and Maliwal, B. P., 1985, Construction and performance of a variable-frequency phase-modulation fluorometer, *Biophys. Chem.* **21**: 61–78.

Lakowicz, J. R., and Prendergast, F. G., 1978, Detection of hindered rotation of 1,6-diphenyl-1,3,5-hexatriene in lipid bilayers by differential polarized phase fluorometry, *Biophys. J.* **23** : 213–231.

Lakowicz, J. R., Bevan, D. R., Cherek, H., Balter, A., and Maliwal, B. P., 1983, Synthesis and characterization of a fluorescence probe of the phase transition and dynamic properties of membranes, *Biochemistry* **22** : 5714–5722.

Lakowicz, J. R., Cherek, H., Laczko, G., and Gratton, E., 1984a, Time-resolved fluorescence emission spectra of labeled phospholipid vesicles, as observed using frequency-domain fluorometry, *Biochim. Biophys. Acta* **777** : 183–193.

Lakowicz, J. R., Cherek, H., Maliwal, B., Laczko, G., and Gratton, E., 1984b, Determination of time-resolved fluorescence emission spectra and anisotropies of a fluorophore–protein complex using frequency-domain phase-modulation fluorometry, *J. Biol. Chem.* **259** : 10967–10972.

Lakowicz, J. R., Laczko, G., Cherek, H., Gratton, E., and Limkemann, M., 1984c, Analysis of fluorescence decay kinetics from variable-frequency phase shift and modulation data, *Biophys. J.* **46** : 463–467.

Lakowicz, J. R., Cherek, H., Maliwal, B. P., and Gratton, E., 1985, Time-resolved fluorescence anisotropies of diphenylhexatriene and perylene in solvents and lipid bilayers obtained from multifrequency phase and modulation fluorometry, *Biochemistry* **24** : 376–383.

Lakowicz, J. R., Laczko, G., Gryczynski, I., and Cherek, H., 1986a, Measurement of sub-nanosecond anisotropy decays of protein fluorescence using frequency-domain fluorometry, *J. Biol. Chem.* **261** : 2240–2245.

Lakowicz, J. R., Laczko, G., and Gryczynski, I., 1986b, A 2 GHz frequency-domain fluorometer, *Rev. Sci. Instrum.* **57** : 2499–2506.

Lakowicz, J. R., Laczko, G., and Gryczynski, I., 1986c, Picosecond resolution of oxytocin tyrosyl fluorescence by 2 GHz frequency-domain fluorometry, *Biophys. Chem.* **24** : 97–100.

Lakowicz, J. R., Laczko, G., and Gryczynski, I., 1987a, Picosecond resolution of tyrosine fluorescence and anisotropy decays by 2 GHz frequency-domain fluorometry, *Biochemistry* **26** : 82–90.

Lakowicz, J. R., Szmacinski, H., and Gryczynski, I., 1987b, Picosecond resolution of indole anisotropy decays and spectral relaxation by 2 GHz frequency-domain fluorometry, *Photochem. Photobiol.* **47** : 31–41.

Lakowicz, J. R., Joshi, N. B., Johnson, M. L., Szmacinski, H., and Gryczynski, I., 1987c, Diffusion coefficients of quenchers in proteins from transient effects in the intensity decays, *J. Biol. Chem.* **262** : 10907–10910.

Lakowicz, J. R., Cherek, H., Gryczynski, I., Joshi, N., and Johnson, M. L., 1987d, Enhanced resolution of fluorescence anisotropy decays by simultaneous analysis of progressively quenched samples; applications to anisotropic rotations and protein dynamics, *Biophys. J.* **51** : 755–768.

Lakowicz, J. R., Johnson, M. L., Wiczk, W., Bhat, A., and Steiner, R. F., 1987e, Resolution of a distribution of distances by fluorescence energy transfer and frequency-domain fluorometry, *Chem. Phys. Lett.* **138** : 587–593.

Lakowicz, J. R., Johnson, M. L., Gryczynski, I., Joshi, N., and Laczko, G., 1987f, Transient effects in fluorescence quenching measured by 2 GHz frequency-domain fluorometry, *J. Phys. Chem.* **91** : 3277–3285.

Laws, J. R., and Brand, L., 1979, Analysis of two-state excited state reactions. The fluorescence decay of 2-naphthol, *J. Phys. Chem.* **83** : 795–802.

Laws, W. R., Ross, J. B. A., Wyssbrod, H. R., Beechem, J. M., Brand, L., and Sutherland, J. C., 1986, Time-resolved fluorescence and ¹H NMR studies of tyrosine and tyrosine analogues: Correlation of NMR-determined rotamer populations and fluorescence kinetics, *Biochemistry* **25** : 599–607.

Libertini, L. J., and Small, E. W., 1985, The intrinsic tyrosine fluorescence of histone H1—steady-state and fluorescence decay studies reveal heterogeneous emission, *Biophys. J.* **47**: 765–772.

Malinowski, E. R., and Howery, D. G., 1980, *Factor Analysis in Chemistry*, Wiley-Interscience, New York.

Maliwal, B. P., and Lakowicz, J. R., 1986, Resolution of complex anisotropy decays by variable frequency phase-modulation fluorometry: A simulation study, *Biochim. Biophys. Acta* **873**: 161–172.

Maliwal, B. P., Hermetter, A., and Lakowicz, J. R., 1986, A study of protein dynamics from anisotropy decays obtained by variable frequency phase-modulation fluorometry: Internal motions of *N*-methylanthraniloyl melittin, *Biochim. Biophys. Acta* **873** : 173–181.

Mantulin, W. W., and Weber, G., 1977, Rotational anisotropy and solvent–fluorophore bonds; an investigation by differential polarized phase fluorometry, *J. Chem. Phys.* **66** : 4092–4099.

Marchiarullo, M. A., and Ross, R. T., 1985, Resolution of component spectra for spinach chloroplasts and green algae by means of factor analysis, *Biochim. Biophys. Acta* **807** : 52–63.

Merkelo, H. S., Hartman, S. R., Mar, T., Singhal, G. S., and Govindjee, 1969, Mode-locked lasers: Measurements of very fast radiative decay in fluorescent systems, *Science* **164** : 301–303.

Munro, I., Pecht, I., and Stryer, L., 1979, Subnanosecond motions of tryptophan residues in proteins, *Proc. Natl. Acad. Sci. U.S.A.* **761** : 156–60.

O'Connor, D. V., and Phillips, D., 1984, *Time Correlated Single Photon Counting*, Academic Press, New York.

Parasassi, T., Conti, F., Glaser, M., and Gratton, E., 1984, Detection of phospholipid phase separation, *J. Biol. Chem.* **259** : 14011–14017.

Peters, C. J., 1965, Gigacycle-bandwidth coherent-light traveling wave amplitude modulator, *Proc. IEEE* **53** : 455–460.

Steinberg, I. Z., 1971, Long-range nonradiative transfer of electronic excitation energy in proteins and polypeptides, *Annu. Rev. Biochem.* **40** : 83–114.

Stryer, L., 1978, Fluorescence energy transfer as a spectroscopic ruler, *Annu. Rev. Biochem.* **47** : 819–916.

Taylor, D. L., Waggoner, A. S., Murphy, R. F., Lanni, F., and Birge, R. R. (eds.), 1986, *Applications of Fluorescence in the Biomedical Sciences*, Alan R. Liss, New York.

Taylor, J. R., 1982, *An Introduction to Error Analysis, the Study of Uncertainties in Physical Measurements*, University Science Books, Mill Valley, CA.

Veatch, W. R., and Stryer, L., 1977, The dimeric nature of the Gramicidin A transmembrane channel: Conductance and fluorescence energy transfer studies of hybrid channels, *J. Mol. Biol.* **177** : 1109–1113.

Visser, A. J. W. G. (ed.), 1985, Special symposium on time-resolved fluorescence spectroscopy, in *Analytical Instrumentation*, Vol. 14, pp. 193–566, Marcel Dekker, New York.

White, G., and Chin, G. M., 1972, Traveling wave electro-optic modulators, *Opt. Commun.* **5**: 374–379.

Weber, G., 1981, Resolution of the fluorescence lifetimes in a heterogeneous system by phase and modulation measurements, *J. Phys. Chem.* **85** : 949–953.

Yamazaki, I., Tamai, N., Kume, H., Tsuchiya, H., and Oba, K., 1985, Microchannel-plate photomultiplier applicability to the time-correlated photon-counting method, *Rev. Sci. Instrum.* **56** : 1187–1194.

Chapter 4

Time-Resolved Fluorescence Depolarization Techniques in Model Membrane Systems

Effect of Sterols and Unsaturations

Michel Vincent and Jacques Gallay

1. INTRODUCTION

Application of time-resolved fluorescence depolarization techniques to the study of biological membrane structure and dynamics requires selective signal monitoring of specifically located intrinsic or extrinsic fluorophores. In such systems, at least to date, the fluorescence characteristics of the naturally occurring chromophores (e.g., aromatic amino acid residues, enzymatic reactional cofactors, polyenic fatty acids and sterols, chlorophyll, and retinal) are of limited use because of low and/or reabsorbed emitted intensities, low signal-to-noise ratio (e.g., light scattering), and so on.

To overcome this drawback, the probing of membrane systems by using extrinsic fluorescent molecules of high quantum yield and extinction coefficient has been widely developed since the pioneering work, in the time-resolved mode, of Wahl *et al.* (1971), but mostly at the expense of a precise

Michel Vincent and Jacques Gallay   Laboratoire pour l'Utilisation du Rayonnement Electromagnétique (L.U.R.E.) – CNRS-CEA-MEN, Université Paris–Sud, 91405 Orsay-Cedex, France.

knowledge of the location of the guest dye. This is especially true for the noncovalently bound fluorophores. A way to eliminate this problem is found in the study of well-defined artificial membrane systems containing a limited number of constituents but for which fluorescence signals are now relevant to some of the above-mentioned biological molecules.

No fundamental reason can be invoked to incorporate extrinsic fluorescent probes into membrane model systems. In fact, studies in well-defined artificial systems are expected to be of great help in the accurate interpretation of the probe fluorescent signals when incorporated into complex biological membranes. This will allow scientists to take advantage of the high sensitivity of this technique when applied to biological systems. The artificial systems (Bangham, 1981; Hope et al., 1986) encompass uni- or multilamellar vesicles as well as oriented mono- (Teissié, 1987) or multilayer (Mulders et al., 1986) or reverse micelles (Keh and Valeur, 1981; Visser et al., 1984; Nicot et al., 1985; Gallay et al., 1987; Vos et al., 1987). In fact, it must be emphasized that although also true for many other spectroscopic techniques used in biology, the signal intensity is mostly correlated in an opposite way with the amount of information expected from a given experiment. In each case, the most realistic optimization of these "conflicting" parameters has to be established.

As now demonstrated, biomembranes are characterized by the existence of a phospholipid bilayer matrix, in which proteins are immersed, can laterally diffuse, and/or rotate [for a recent review see Devaux and Seigneuret (1985)]. Some general trends are now well recognized: both lateral and transverse phospholipid distributions may not be random; gel state phospholipids segregate from liquid-crystalline ones (Shimshick and McConnell, 1973; Barenholz et al., 1976; Sklar et al., 1979); and formation of clusters may be induced by interaction with constituents such as proteins, cholesterol or derivatives, ligand receptors, drugs, or related compounds and ions. Lipid asymmetry has been emphasized (Op den Kamp, 1979) and has recently been shown to be ATP dependent in erythrocytes and platelets (Seigneuret and Devaux, 1984; Zachowski et al., 1986; Sune et al., 1987). The lipid bilayer, and especially the hydrophobic core of the acyl chains, may be physically characterized by what the physicochemist has called the *membrane fluidity* (Shinitzky et al., 1971). This term not only reflects the reciprocal of the viscosity (as in isotropic solvents) but covers many physical dynamic parameters: (1) rotational diffusion rate, (2) average orientation of the acyl chains or of their segments, and (3) lateral diffusion.

From a practical point of view, the lateral diffusion process occurs on a time scale that can differ by three to five orders of magnitude as compared to that of the rotational diffusion, typically in the nanosecond range. Moreover, there is no correlation between translational and rotational motions in biological membrane systems (Kleinfeld *et al.*, 1981). Consequently, only rotational diffusion rates and orientational distributions are accessible parameters for probes fluorescing with typical excited-state lifetimes ranging from 0.1 to 15 nsec. Whenever the probe's lateral diffusion cannot be monitored by collection of both vertically and horizontally polarized fluorescence decays of probes excited with vertically polarized light, phenomenologically physically related techniques such as fluorescence recovery after photobleaching (FRAP) (Owicki and McConnell, 1980) or kinetics of pyrene excimer formation remain available (Liu *et al.*, 1980).

In this chapter we focus on the use of pulse fluorescence anisotropy techniques on the effect of cholesterol—one of the two main effectors of the membrane structure with integral proteins—and its derivatives on membrane lipid order and dynamics. Recently, a theoretical model was proposed for the effects of impurities on biomembranes (O'Leary, 1983). O'Leary assumes a spatial perturbation of the membrane system by the external molecule leading to an increased fatty acyl chain disorder depending on both the size and shape of the impurity molecule. With regard to the chemical heterogeneity of the membrane matrix (e.g., fatty acyl chains composition, polar head groups, proteins) the universality of cholesterol as a regulator of the membrane "fluidity" has been stressed in eukaryotic systems (Bloch, 1983). Such a molecule has optimized structural properties for the efficiency of the interaction with the membrane components; any chemical modification results in a lowering and even a vanishing of the mother molecule efficiency.

2. INTRINSIC MOTIONAL PROPERTIES OF SOME WIDELY USED FLUORESCENT PROBES

2.1. Motional Characteristics

Besides its ability, in terms of sensitivity, to report the properties of its microenvironment, a fluorescent probe ideally must be well-localized and nonperturbing to its surrounding medium. Conditions of a minimal disturbance of the membrane dynamic properties by the probe, as well as a more precise knowledge of its location, have to be evaluated.

An example of improving the use of fluorescent probes is provided by the case of the most popular one: 1,6-diphenyl-1-3-5-hexatriene (DPH), initially introduced in the membrane research field by Shinitzky and Barenholz (1974, 1978) as a means of "microviscosity" measurement (Shinitzky and Yuli, 1982). The validity of this parameter has been widely discussed (Hare *et al.*, 1979; Van Blitterswijk *et al.*, 1981; Hare, 1983) and still remains a much debated question (see Van der Meer, Chapter 1 in this volume). We do not deal with this question. In early studies, DPH was used in membrane systems just as it is, that is, with a wide uncertainty in its transverse and lateral localization in the bilayer leaflet (Lentz *et al.*, 1976). A first step of improvement was to substitute a positively charged polar moiety on one of the phenyl rings (Prendergast *et al.*, 1981); this allowed a better alignment of the polyene moiety with the phospholipid fatty acyl chains. More recently, a substitution by a carboxyl group on the phenyl ring was performed. Depending on the nature of the surrounding phospholipid polar head groups, a differential anchoring of TMA-DPH or Prop-DPH can be expected. A subsequent acylation of the Prop-DPH on a lysophospholipid (Cranney *et al.*, 1983) gave rise to a probe expected to mimic closely a biological lecithin inside the membrane. The fluorescence signal can then be related to a specific entity, depending on the chemical structure of its polar head group. As a result of these chemical modifications leading ultimately to a biologically designed probe, the reported membrane lipid dynamics and ordering are strongly modified. Essentially, the order parameter reported by TMA-DPH is much higher than the one reported by DPH in the liquid-crystalline phase. This is observed either in models or in natural membranes and in their lipid extracts (Prendergast *et al.*, 1981; Stubbs *et al.*, 1984; De Foresta *et al.*, 1986; Illsley *et al.*, 1987). This is also the case for Prop-DPH in ghost erythrocyte membranes (F. Beaugé and J. Gallay, unpublished data). This can be accounted for by the existence of the high amphiphilic character of these two probes whatever the nature of the electrostatic charge. Both probes are likely to be held nearer than DPH to the membrane–water interface and report a higher lipid packing in this region. The effect is probably amplified since a fraction of DPH molecules is likely to be randomly distributed in the hydrocarbon chains as indicated by angle-resolved fluorescence depolarization (Vos *et al.*, 1983). The PC-DPH probe reports intermediate values of the order parameter (Stubbs *et al.*, 1984) that are consistent with the location of the fluorophore deeper in the bilayer hydrophobic core than Prop-DPH or TMA-DPH. ^{2}H-NMR has also reported a weaker lipid ordering in this membrane region (Seelig and Seelig, 1980). Ironically, and despite its intrinsic limitations, the usefulness of DPH has recently been rediscovered by ^{2}H-NMR (Kintanar *et al.*, 1986). Using deuterated DPH, these authors have found a good correlation between their data and those obtained by fluorescence measurements.

However, it must be noted that such an agreement is not always fulfilled between probe techniques and "nonperturbing" techniques. For instance, the protein effects on lipid ordering appear to be quite different when monitored by ESR or fluorescence and by ^{2}H-NMR or Raman spectroscopy (Seelig and Seelig, 1980; Jaehnig *et al.*, 1982). A rationale for an understanding of these discrepancies can be found in Jaehnig's paper. It has been proposed that Raman and ^{2}H-NMR spectroscopy report on the "conformational order" of the methylene segments of the acyl chain whereas DPH or rodlike probes such as parinaric acids report on the "rigid-body orientational order," which is a resultant of the orientation of the methylene segment along the stiff probe. According to this idea, the α-helical parts of the membrane protein would restrict the rigid-body orientational order of the acyl chains. On the other hand, the irregular surfaces of these α-helixes, at the scale of a methylene segment, may result in tilts along the acyl chains in proximity and would decrease the conformational order to which ^{2}H-NMR and Raman spectroscopies are sensitive.

Still in the series of the popular rodlike probes, *cis*- and *trans*-parinaric acids are conjugated polyenic, 18-carbon fatty acids. *Cis*-parinaric acid does not display a marked preferential partition into fluid or ordered lipid domains, whereas *trans*-parinaric acid preferentially partitions into gellike domains (Sklar *et al.*, 1977, 1979). In a similar way to DPH derivative probes, the parinaric acids can be bound covalently to a given phospholipid (Wolber and Hudson, 1981; Cranney *et al.*, 1983; Berkhout *et al.*, 1984; Stubbs *et al.*, 1984; Parente and Lentz, 1985).

The second major class of fluorescent probes is represented by the disk-shaped ones. Let us first emphasize that the major criticism directed against these probes—their large size—is counterbalanced by the intrinsic wealth of their depolarizing motions. Indeed, as pointed out by Weber (1971), three modes of rotation can be detected, depending on excitation wavelength conditions. In practice, for symmetrical considerations, only two motions of the chromophore, in-plane and out-of-plane, are contributing to the depolarizing process (Figure 1). The intrinsic anisotropy of the whole membrane architecture would exert differential hindrances on the two main rotational motions. This was first studied with the perylene probe in the steady-state mode by Shinitzky *et al.* (1971), in detergent micelles, and in the time-resolved mode in membranes (Barkley *et al.*, 1981; Salesse *et al.*, 1981; Brand *et al.*, 1985). Nevertheless, perylene location and orientation are not well defined.

A useful set of probes, derived from anthracene, was first synthesized by Waggoner and Stryer (1970): the *n*-(9-anthroyloxy) fatty acid derivatives of the stearic and palmitic series (*n*-AS, *n* being related to the carbon number of substitution on the holder fatty acid). They have been studied by many authors, including ourselves, in recent years both in model and biological membranes (Cadenhead *et al.*, 1977; Tilley *et al.*, 1979; De Paillerets *et al.*, 1981, 1984; Thulborn and Beddard, 1982; Vincent *et al.*, 1982a, 1982b; Kutchai *et al.*, 1983; Vincent and Gallay, 1984). These probes are likely to fit in the membrane with their acyl chains parallel to those of the phospholipids and allow a labeling at a graded series of depths in the bilayer, as demonstrated by fluorescence quenching and energy transfer experiments (Thulborn and Sawyer, 1978; Haigh *et al.*, 1979; Thulborn, 1981). NMR studies have shown that the anthracene ring and the carbon atom to which it

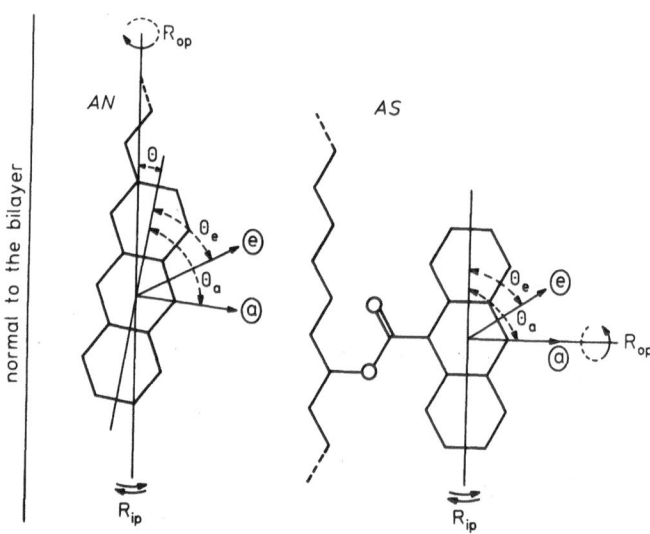

FIGURE 1. Schematic representation of the in-plane and out-of-plane modes of rotation of (*left*) 9-(2-anthryl)-nonanoic acid (AN) and (*right*) n-(9-anthroyloxy)-stearic acid (AS). (e), direction of the emission oscillator. (a), direction of the absorption oscillator at 380 nm. θ_a and θ_e are the angles between the long symmetry axis of the fluorophore and the absorption and emission oscillators, respectively. Reprinted with permission from Elsevier Science Publishers. From Vincent *et al.* (1985).

is attached are locked together in such a way that they move as a unit (Johns *et al.*, 1979) and experience types of motion similar to that of a lipid in bilayers (Kuroda *et al.*, 1986). Moreover, replacement of one phospholipid acyl chain in a phosphatidylcholine by n-AS derivatives and subsequent incorporation into model membrane systems evidence similar depolarization properties of the anthroyl ring as a function of its substitution position as compared to the ones of the related free anthroyl fatty acid (Figure 2). For each derivative, the anthracene rotation is more hindered when the holder fatty acid is bound to the glycerol moiety of a lysolecithin molecule. However, the hindrances appear to be reinforced in the 2-carbon region. The probe motion can be impeded by the glycerol backbone and in a general way by the adjacent palmitoyl chain, as seems indicated by the systematically higher limiting anisotropy values obtained with the phospholipid probes as compared to the fatty acid ones (Figure 2). Nevertheless, the shape of the profile remains identical for both kinds of probes, and artifacts such as transverse slipping of the fatty acid probes can be excluded.

More recently, Tocanne's group in Toulouse, France, synthesized a very promising series of fluorescent fatty acid analogues—9-(2-anthryl)-nonanoic acid—which, as demonstrated by monolayer, microcalorimetry, and fluorescence studies, are weakly perturbing even when present at relatively high concentrations in the membrane host system (De Bony and Tocanne, 1983, 1984; De Bony *et al.*, 1984); this is not the case for PC-DPH (Parente and Lentz, 1985). Moreover, these molecules can readily be incorporated into the membrane lipids of the bacterium *Micrococcus luteus*, mainly in the *sn-*2 glycerol position, through the regular metabolic pathway without degradation (Welby and Tocanne, 1982). Either in the fatty acid form or as an integral part of a phospholipid, an optimal alignment of the anthracene group and of the fatty acyl chains of the membrane phospholipids can be reasonably assumed (Figure 1). Thus, an evaluation of the order parameter from the limiting anisotropy value in the fluorescence anisotropy decay curve can be performed, following the published formalism (Lipari and Szabo, 1980; Zannoni *et al.*, 1983; Van der Meer *et al.*, 1984).

However, the low absorption coefficient of the 2-substituted anthracene fluorophore makes routine experiments with classical light sources, such as a thyratron gated flash lamp in membrane systems (Vincent *et al.*, 1985), difficult, but by using very fast repetition sources such as mode-locked laser or synchrotron radiation, these probes may be of great importance in future studies of the dynamic properties of the membrane structure.

2.2. Excited-State Characteristics

An important question regarding the interpretation of anisotropy decay data concerns the total intensity decay characteristics. Classically, fluorescence anisotropy is defined as

$$r(t) = \frac{D(t)}{S(t)} \quad \text{in time-resolved mode}$$

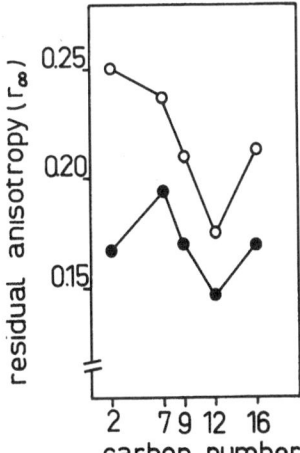

FIGURE 2. Comparison between the residual anisotropy values (r_α) of a set of *n*-(9-anthroyloxy) fatty acids (●) and the similar set of fatty acid probes as an integral part of a lecithin (○). The lipid matrix is POPC. Excitation wavelength was 380 nm. Temperature was −20°C (C. Wolf, M. Vincent, and J. Gallay, unpublished data).

where

$$D(t) = I_{vv}(t) - I_{vh}(t) \tag{1}$$

and

$$S(t) = I_{vv}(t) + 2I_{vh}(t) \tag{2}$$

$S(t)$ is the total fluorescence intensity decay necessary for the classical computation of the rotational parameters from the difference curve $D(t)$ (Vincent et al., 1982a).

In simple cases, $S(t)$ decays monoexponentially as

$$S(t) = I_o \exp(-t/\tau) \tag{3}$$

where τ is the excited-state lifetime. However, in the majority of instances reported so far concerning membrane probes such as DPH and derivatives, anthroyloxy fatty acids, or parinaric acids, the total intensity does not decay as a single exponential; rather, two or more time components are needed to improve the recovery of the experimental data. This must be taken into account to extract the rotational parameters from the anisotropy decay data, but until very recently this was not performed. Several decay times for the total intensity decay can be interpreted, for example, as arising from different ground or excited states of the probe corresponding to different environments within the membrane. The populations in these different environments must exchange slowly on the time scale of the fluorescence decay rate in order to give rise to definite excited-state lifetimes. Anisotropy, being additive, is the average over these environments. There can be a correlation between lifetimes (environments) and correlation times or order parameters. There can also be a complete independence between both characteristics. However, one should be aware that the order parameter, which is extracted from the long-time part of the anisotropy curve, is more influenced by the long-lifetime species than by the short-lifetime ones (Gallay et al., 1982; Rehorek et al., 1985; Lakowicz et al., 1987). Therefore, a better understanding of $S(t)$ must be achieved for refined interpretation of anisotropy decay data.

3. STEROL–PHOSPHOLIPID INTERACTIONS IN MODEL MEMBRANES

3.1. Cholesterol–Phospholipid Interactions: Lecithin as Bilayer Matrix

3.1.1. Rodlike Fluorescent Probes

Cholesterol is a major component of many biological membranes and is present at various lipid molar ratios as high as 1 : 1 in erythrocyte cellular membrane or myelin sheath. A 3-βOH, a flexible C17 sidechain, and a rigid nucleus are prerequisites for the maximum efficiency of phospholipid–sterol interactions (Demel and De Kruijff, 1976). In general, the cholesterol molecule acts by destabilizing the membrane structure in the gel phase (spacing and "fluidifying" effect) while the contrary is true in the liquid-crystalline phase [condensing and "rigidifying" effect (Demel and De Kruijff, 1976)]. As recognized by Demel *et al.* (1972b) in monolayer studies, the extent of unsaturation of the fatty acid constituents is preponderant for this effect, indicating the importance of Van der Waals interactions.

The first time-resolved fluorimetric evidences of the ordering effect of cholesterol of the liquid-crystalline phase in membrane model systems were performed almost simultaneously by Veatch and Stryer (1977) and Chen *et al.* (1977). Not long after, Kawato *et al.* (1978) were able, within their previously established wobbling-in-cone model (Kinosita *et al.*, 1977), to correlate the restricted rotation of the DPH probe with the increased hindrances exerted by the surrounding phospholipid acyl chains upon addition of cholesterol in the liquid-crystalline phase. The wobbling-in-cone model, which assumes a unique symmetry axis for the cylindrical probe that is allowed to diffuse in a restricted medium, was subsequently developed in terms of the orientational order parameter (Heyn, 1979; Jaehnig, 1979; Lipari and Szabo, 1980; Engel and Prendergast, 1981; Hare, 1983; Zannoni *et al.*, 1983; Van der Meer *et al.*, 1984; Szabo, 1984), making possible comparisons with other spectroscopic techniques (ESR or NMR). It is difficult, because of the size of the DPH molecule, to compare the order parameter derived from the limiting anisotropy to that obtained for one or two methylene segments of a fatty acyl chain (Engel and Prendergast, 1981); however, recent experiments (Kintanar *et al.*, 1986) have shown reasonably good agreement between the order parameter measured by ^{2}H-NMR and fluorescence anisotropy decay in pure phospholipids and in phospholipid–cholesterol mixtures.

In our laboratory, a comparative study of the effect of cholesterol in three lipidic systems—DPPC, SOPC, and OSPC—was performed at similar reduced temperatures both in the gel and in the liquid-crystalline phases (Z. Chraibi, J. Gallay, and M. Vincent, unpublished data). The results emphasize the effects of the presence and location of a *cis* double bond on the lipid ordering and its influence on the PC–cholesterol interaction, as visualized with DPH as probe.

In pure lecithin vesicles in the gel phase, at the same reduced temperature, the introduction of a *cis* double bond reduces significantly the DPH order parameter (Table I). This effect is more important for the *sn*-1 unsaturated isomer (OSPC): the DPH order parameter decreases by about 6.6 and 10.3% in SOPC and OSPC, respectively. A destabilization of the gel phase is thus observed due to the presence of a kink at the *cis* double bond level. The unsaturated fatty acid cannot pack with the other acyl chain as well as

Table I
Mean Excited-State Lifetime and Anisotropy Decay Parameters of DPH in Various Lecithins in the Gel Phase[a] as a Function of Cholesterol Mole Fraction (X)

	X	$\langle\tau\rangle$[b] (nsec)	ϕ (nsec)	r_∞	S[c]
DPPC	0.00	9.24	3.2	0.310	0.898
	0.10	9.25	3.7	0.320	0.913
	0.20	9.08	3.5	0.324	0.919
	0.33	8.66	12.7	0.292	0.872
	0.50	8.93	10.3	0.277	0.849
SOPC	0.00	8.90	5.6	0.270	0.839
	0.10	9.20	6.8	0.274	0.845
	0.20	9.10	5.3	0.283	0.858
	0.33	9.73	7.6	0.292	0.872
	0.50	9.96	6.5	0.305	0.891
OSPC	0.00	7.31	7.3	0.249	0.805
	0.10	8.74	6.4	0.266	0.832
	0.20	8.90	6.2	0.282	0.857
	0.33	8.40	7.8	0.282	0.857
	0.50	8.05	4.0	0.245	0.799

[a] 20, 0, and 1.5°C for DPPC, SOPC, and OSPC, respectively.
[b] $\langle\tau\rangle$ is the mean excited-state lifetime (Gallay et al., 1982).
[c] S is the orientational order parameter, where $S^2 = r_\infty/r_0$, r_∞ is obtained from $r(t) = (r_0 - r_\infty)\exp(-t/\phi) + r_\infty$, and $r_0 = 0.384$ (Gallay et al., 1982).

do fully saturated chains. Such a destabilization explains the large shift in transition temperature (from 54.5°C for DSPC to 6.3 and 8.9°C for SOPC and OSPC, respectively; Z. Chraibi, J. Gallay, and M. Vincent, unpublished data) (Davis et al., 1981). The packing defect is more important for OSPC, in which the double bond occupies a slightly deeper position than in SOPC as remarked by Davis et al. (1981) in their differential scanning calorimetry studies. Under such conditions, the effect of the double bond can be different (Barton and Gunstone, 1975).

Addition of cholesterol in the gel phase in the saturated system leads to the well-known "fluidizing" effect (Demel and De Kruijff, 1976), whereas in the unsaturated system a slight ordering is observed. Again, the sn-1 and sn-2 isomers do not behave identically; SOPC is more ordered than OSPC upon cholesterol addition. Such a difference in the cholesterol–lecithin interactions in the gel phase has also been observed by differential scanning calorimetry studies (Davis and Keough, 1983) and can be explained by steric hindrances.

In the liquid-crystalline phase, always at similar reduced temperatures, the order parameter of DPH is higher in the unsaturated systems than in the saturated ones (Table II). This was also observed in POPC (Vincent and Gallay, 1984). The effect is stronger for the *sn*-1 unsaturated isomer (11 and 24% increase in the DPH order parameter for SOPC and OSPC, respectively). Addition of cholesterol evokes a similar effect in saturated and unsaturated systems. However, when both acyl chains are monounsaturated (DOPC), the cholesterol effect is weaker (Gallay *et al.*, 1985). The rate of reorientation of the molecular axis is always higher in the saturated systems than in the unsaturated ones. Since the double bond is relatively rigid as compared to the rest of the acyl chain, it can slow down the rotation. In any case, cholesterol has no effect on this parameter.

It has to be remarked that the double-bond effect has been interpreted rather differently by Stubbs *et al.* (1981). At high temperatures relative to the melting transition temperature, they observed a decrease of the DPH order parameter upon introduction of a single *cis* double bond. However, one should stress that, as pointed out by Seelig and Seelig (1977), a membrane lipid system must be subject to the same average molecular forces. Therefore, it is observed by the fluorescence anisotropy technique, in agreement with ^{2}H-NMR, that the unsaturated systems may be more ordered than the saturated

Table II
Mean Excited-State Lifetime and Anisotropy Decay Parameters of DPH in Various Lecithins in the Liquid-Crystalline Phase[a] as a Function of Cholesterol Mole Fraction (X)

	X	$\langle\tau\rangle$[b] (nsec)	ϕ (nsec)	r_x	S[c]
DPPC	0.00	7.12	1.2	0.055	0.378
	0.10	7.80	1.0	0.109	0.533
	0.20	8.40	1.2	0.170	0.665
	0.33	8.83	1.9	0.214	0.747
	0.50	9.29	—	0.265	0.831
SOPC	0.00	7.70	3.5	0.070	0.427
	0.10	8.10	3.7	0.118	0.554
	0.20	8.37	2.9	0.157	0.639
	0.33	8.70	3.4	0.216	0.750
	0.50	9.47	3.7	0.269	0.837
OSPC	0.00	8.00	3.2	0.096	0.500
	0.10	8.11	3.4	0.142	0.608
	0.20	8.42	3.9	0.182	0.688
	0.33	8.70	3.5	0.243	0.796
	0.50	9.08	2.8	0.275	0.846

[a] 50, 19, and 18°C for DPPC, SOPC, and OSPC, respectively.
[b] $\langle\tau\rangle$ is the mean excited-state lifetime (Gallay *et al.*, 1982).
[c] S is the orientational order parameter, where $S^2 = r_x/r_0$, r_x is obtained from $r(t) = (r_0\text{-}r_x)\exp(\text{-}t/\phi) + r_x$, and $r_0 = 0.384$ (Gallay *et al.*, 1982).

ones depending on the reduced temperature at which the experiments are performed (Seelig and Seelig, 1977). The higher ordering can be explained by additional steric constraints exerted by the double bond on the rest of the acyl chains.

From the above data, we see that, despite a wide uncertainty in its orientation and localization, DPH can still provide specific information at the molecular level, for example, about the unequivalence of the fatty acid chains of a phospholipid.

By using other rodlike fluorescent probes—*trans*- and *cis*-parinaric acids and parinaroylphosphatidylcholines—similar effects of cholesterol on saturated phosphatidylcholine fatty acyl chains were found by Wolber and Hudson (1981). Cholesterol increased the order parameter measured in fluid phosphatidylcholine phase without affecting to a large extent the rate of angular relaxation. An important point is the virtual absence of difference when *trans*-parinaric acid or *trans*-parinaroylphosphatidylcholine is used as probe. Note also that in the same lipidic system (DPPC), DPH reported a very similar order parameter in the liquid-crystalline phase as *trans*-PNA-lecithin (Wolber and Hudson, 1981; Vincent and Gallay, 1984) at the same temperature of 50°C.

3.1.2. Disk-Shaped Fluorescent Probes

Since the above-mentioned conclusions are "vertically" averaged with regard to the uncertainty in fluorescent probe location, the use of well-located dyes is crucial to delineate differential effects of cholesterol and unsaturation at different depths through the bilayer. Different groups (Thulborn and Sawyer, 1978; Thulborn and Beddard, 1982; Vincent *et al.*, 1982b; Kutchai *et al.*, 1983) have tried to answer this question by working with the set of *n*-AS.

The above-mentioned dual effect of cholesterol on the thermotropic properties of saturated lecithin vesicles is well evidenced in steady-state fluorescence anisotropy measurements by using rodlike probes such as DPH or parinaric acid derivatives. On the other hand, disk-shaped molecules such as anthroyloxy fatty acid derivatives did not show such behavior, at least in the steady-state mode of illumination (Thulborn and Sawyer, 1978; Vincent *et al.*, 1982b). In particular, the melting appears to be less cooperative in pure lecithin systems, and upon cholesterol addition, the steady-state anisotropy values obtained with the *n*-AS probes are systematically higher than the ones with no cholesterol, whatever the lipid phase.

It has been shown that valuable correlations between steady-state and residual anisotropy values could be established in various membrane model systems for rodlike probes (Jaehnig, 1979; Van Blitterswijk *et al.*, 1981;

Hare, 1983; Van der Meer *et al.*, 1984). For such probes, the excitation and emission dipole orientations are quite collinear and, moreover, nearly parallel to both the probe axis and the transverse membrane axis, leading to the existence of only one depolarizing motion. This is not the case for planar molecules such as anthroyloxy fatty acids for which both in-plane and out-of-plane modes of rotation are contributing to the incident light depolarization process (Vincent *et al.*, 1982a). When selecting an adequate excitation wavelength (in the red part of the excitation spectrum), it is possible to partially separate the two intrinsic modes of rotation (Vincent *et al.*, 1982a). In the gel phase of lecithin liposomes, only the in-plane mode of rotation was shown to be hindered. By plotting the residual anisotropy values (r_x) versus the carbon number of substitution of the fluorophore (Figure 3), a graded hindering to the probe motion depending on the membrane depth is evidenced. When cholesterol is added to dipalmitoyllecithin vesicles (Figure 4), the major effect is observed in the C7 and C9 regions. A very interesting observation is that in the hydrophobic membrane core (C12 and C16), cholesterol minimizes the hindrances present in the gel phase (spacing effect). Thus, there is no more contradiction firstly with data from other spectroscopic techniques (Stoffel *et al.*, 1974; Oldfield *et al.*, 1978; Semer and Gelerinter, 1979; Dahl, 1981) and secondly with DPH or parinaric acid results. Both probes are assumed to be mainly localized and able to sense membrane fatty acyl chain packing close to the center of the bilayer (i.e., the C12 and C16 region). Moreover, cholesterol has very little effect on the membrane cohesion forces near the aqueous interface region labeled with 2-AS in agreement with results from other physical techniques (Stockton *et al.*, 1974; Bush *et al.*, 1980).

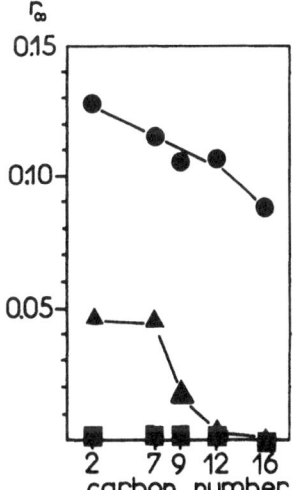

FIGURE 3. Residual anisotropy values (r_x) of anisotropy decays of *n*-(9-anthroyloxy) fatty acid probes in DPPC vesicles. The excitation wavelength was 381 nm and the temperature was 21 (●), 37 (▲), and 47°C (■). Reprinted with permission from the American Chemical Society. From Vincent *et al.* (1982a).

The introduction of a *cis* double bond in the fatty acid chain of lecithin and the subsequent analysis of its effect with the set of *n*-anthroyloxy probes underline specific repercussions on the membrane properties. Three main regions through the bilayer can be evidenced in the gel phase. Compared to the limiting anisotropy values profile of saturated lecithins (Figure 3), an increase in the acyl chain cohesion is observable near the polar interface. A sharp lowering of the hindrances is observed in the intermediate region of the bilayer, that is, where the *cis* double bond introduces a kink in the membrane (Figure 5). Finally, the *cis* double bond effect on the gel phase lecithin properties appears to be very localized since the r_α values increase again in the center of the bilayer. These strong disturbances in the gel phase lead to a drastic shift of the temperature of the main gel to liquid-crystalline phase transition and a correlated loss in cooperativity of the process (see Quinn, 1981).

Monitoring of differential effect as a function of the membrane depth has been tentatively performed in the past with the set of anthroyloxy probes but only in the steady-state mode of illumination. As a conclusion, the authors postulated a "microviscosity barrier in the lipid bilayer due to the presence of phospholipids containing unsaturated acyl chains" (Thulborn *et al.*, 1978). It is absolutely necessary to monitor time-resolved fluorescence measurements to obtain direct and unambiguous information on the fatty acyl chain degree of packing, otherwise the steady-state data provided by disk-shaped fluorophores can be misleading. The now well-described quality of information contained in steady-state fluorescence anisotropy measurements (order and dynamics are mixed) is more likely to be encountered with disk-shaped fluorophores than rod-shaped ones.

Addition of cholesterol on monounsaturated lecithin vesicles at temperatures where the system is in the gel state induces a strengthening of the lipid packing in the fatty acyl chain region C7 to C9 (Figure 5), corresponding to the known location of the sterol ring (Franks, 1976). In the double bond region, the lipid ordering is slightly enhanced while a sharp decrease of this

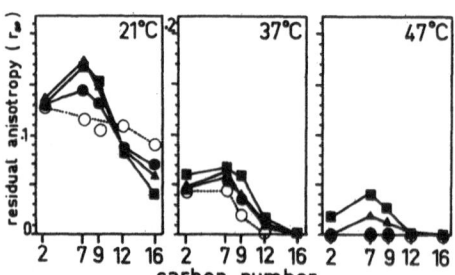

FIGURE 4. Residual anisotropy values (r_α) for the *n*-(9-anthroyloxy) fatty acid derivatives in mixed cholesterol–DPPC vesicles at three temperatures. Cholesterol–DPPC mole fractions are 0.0 (○), 0.1 (●), 0.2 (▲), and 0.33 (■). Reprinted with permission from Academic Press. From Vincent *et al.* (1982b).

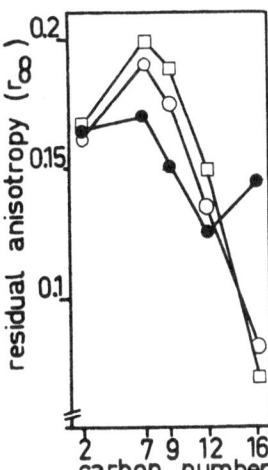

FIGURE 5. Residual anisotropy values (r_∞) of anisotropy decays of *n*-(9-anthroyloxy) fatty acid probes. The excitation wavelength was 381 nm and the temperature was $-10°C$. (●) POPC vesicles; (○) cholesterol–POPC vesicles (0.2 mole fraction); (□) cholesterol–POPC vesicles (0.33 mole fraction). Reprinted with permission from the American Chemical Society. From Vincent and Gallay (1984).

parameter is observed in the deepest hydrophobic part of the membrane (Figure 5) as it was found in saturated systems.

Not only similar but also extended conclusions are obtained with the anthryl-nonanoic acid derivatives (De Bony and Tocanne, 1983). For these probes, the fluorescent moiety is located in the deepest hydrophobic part of the bilayer. It extends from the C9 to the C18 and the information on the lipid acyl chains ordering is an average over this region. Of interest is the observation that behavior of this kind of probe in membrane model systems is about the same either as a pure fatty acid or as a part of a lecithin, although a slightly higher hindrance to the rotation is found when going from the pure fatty acid to the esterified one (Vincent *et al.*, 1985). From the geometrical properties of the excitation and emission dipoles, the out-of-plane rotational motion of the anthryl moiety can be monitored specifically and is an index of the cylindrical rotation of the fatty acid. This motion is undoubtedly hindered (Vincent *et al.*, 1985) in agreement with the data reported by Marsh (1980) with ESR techniques on similar systems. Cholesterol, even when present in lecithin bilayers at high molecular ratio (0.5), only slightly affects this motion. The effect is more important in unsaturated than in saturated lecithins, which can be explained by the rigidity introduced in the acyl chain at the double bond level. The oleic chain moves therefore as a rotating unit of a larger size than a stearic fatty acid chain. Its motion around the long axis is more sensitive to frictional forces or steric hindrances since the double bond kink is not fully aligned with this axis and does not fit well with the sterol ring system.

The wobbling motion of the anthryl moiety is by contrast strongly affected by cholesterol addition in both gel and liquid-crystalline phases. This motion is also hindered and cholesterol enhances the hindrances without affecting the reorientation rate. The above-mentioned averaging over about 10 Å does not allow one to observe any "spacing" effect of cholesterol in the gel phase (Vincent *et al.*, 1985), since it appears to be specifically located near the methyl end of the fatty acid chains.

3.2. Cholesterol–Phospholipid Interactions: Phospholipids Other Than Lecithin as Bilayer Matrix

Cellular membranes do not contain the same amount of cholesterol (Demel and De Kruijff, 1976). A negative gradient of cholesterol has been observed from the periphery to the interior of the cell. One of the factors that can dictate the preferential incorporation of cholesterol within plasma membranes versus inner mitochondrial membranes, for instance, may be the large difference in phospholipid compositions between these two membranes. Since cardiolipin is a characteristic phospholipid almost exclusively present in inner mitochondrial membranes, this prompted us to study its interactions with cholesterol in model systems (ethanol-injected vesicles) with DPH and parinaric acids as fluorescent probes (Gallay *et al.*, 1985; Gallay and Vincent, 1986). As reported by these probes, the fatty acid chain degree of packing is not significantly affected upon incorporation of cholesterol into cardiolipin vesicles up to 0.20 mole fraction (Figures 6 and 7). This is in strong contrast with unsaturated lecithin vesicles such as DOPC, for which one-third of the maximal effect is observed at the same molar ratio. More interesting is the fact that when the well-known maximum level mole ratio of 0.50 is reached in lecithin membrane systems, cardiolipin vesicles still keep on picking cholesterol up to 0.80 mole ratio (Figures 6 and 7), suggesting cholesterol–cardiolipin interactions with a stoichiometry of 1 : 1 between cholesterol molecule and cardiolipin fatty acid chain. While the cholesterol effect is biphasic in cardiolipin systems, a continuous effect is evidenced in lecithin systems (Figures 6 and 7).

One could argue that the high unsaturation degree of bovine heart cardiolipin (about 80% linoleic acid) as compared to DOPC could explain in part the above-mentioned phenomenon. However, Van Blitterswijck *et al.* (1987), in their systematic steady-state anisotropy study of phospholipid cholesterol interactions with DPH as fluorescent probe, have shown that the lipid ordering in dilinoleoylphosphatidylcholine (DLPC) was sensitive to cholesterol addition in a rather similar way as DOPC (Gallay *et al.* 1985). Similar values of steady-state anisotropy were obtained without cholesterol (0.085–0.090 at 20–25°C) and with 0.50 mole fraction of cholesterol (0.17 in DLPC, 0.19 in DOPC). The effect is less important than in saturated systems as already mentioned. Therefore, when forced, cardiolipin vesicles can accom-

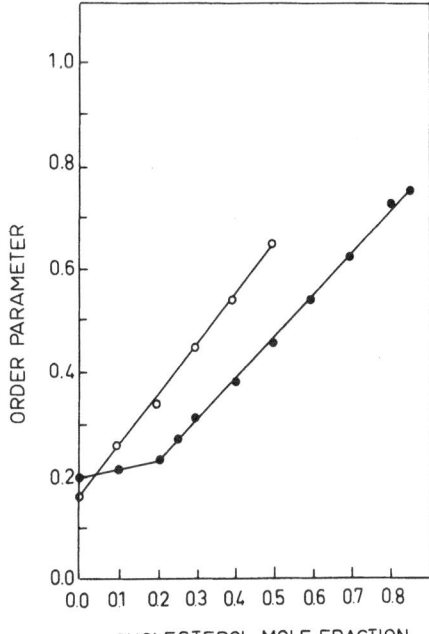

FIGURE 6. Variation of the average orientational order parameter of DPH in cardiolipin vesicles (●) and in DOPC (○) at 25°C as a function of the cholesterol mole fraction. Reprinted with permission from Elsevier Science Publishers. From Gallay *et al.* (1985).

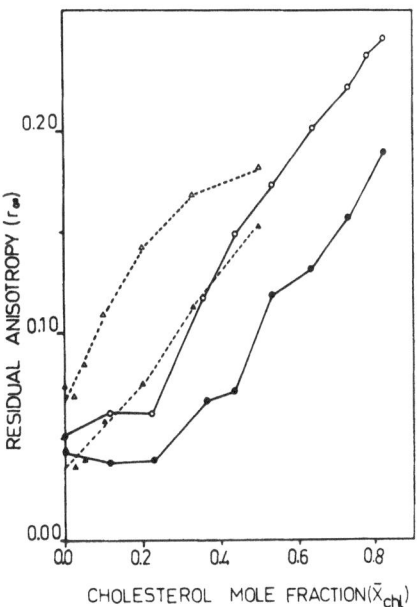

FIGURE 7. Variation of the residual fluorescence anisotropy (r_x) of *cis*- and *trans*-parinaric acids in cardiolipin and in DOPC vesicles as a function of the cholesterol mole fraction: *trans*-PnA in (○) cardiolipin and (△) DOPC vesicles; *cis*-PnA in (●) cardiolipin and (▲) DOPC vesicles. Reprinted with permission from the American Chemical Society. From Gallay and Vincent (1986).

modate a large amount of cholesterol without any change in acyl chain ordering, which could indicate that head group repulsions can manage enough space between cardiolipin molecules to allow insertion of cholesterol. Conversely, this suggests that cardiolipin–cholesterol interactions are rather weak. This would result in partial cholesterol clustering in this phospholipid as seems to be revealed by the differential behavior of *cis*- and *trans*-PNA, which are known to segregate differently into liquid and solid lipid phases (Sklar *et al.*, 1979). As suggested (Gallay and Vincent, 1986), this would in turn explain the relative low level of cholesterol in cardiolipin–rich membranes as well as a ready exchange of this sterol from these membranes to sterol transport proteins, for instance.

3.3. Cholesterol Chemical Modification: Effect on Phospholipid Fatty Acyl Chains Order and Dynamics

The structural requirements for the sterol interactions with membrane phospholipids in model systems are well documented (Demel and De Kruijff, 1976; Bloch, 1983; Yeagle, 1985). However, relatively few attempts have been performed to study the influence of chemical modifications of the cholesterol molecule on lipid ordering by time-resolved fluorescence measurements.

The DPH order parameter calculated either by steady-state (Pottel *et al.*, 1983; Van der Meer *et al.*, 1986) or time-resolved anisotropy measurements is very sensitive to the modification of the sterol chemical structure (Luken *et al.*, 1980; Vincent and Gallay, 1983). For instance, isomerization of the 3β-OH into 3α-OH leads to a lower ordering effect in the liquid-crystalline phase and a smaller spacing effect in the gel phase (Figure 8). This has been attributed to the different orientation of the 3α-OH isomer with respect to cholesterol (Dufourc *et al.*, 1984). Suppression of the C5–C6 double bond increases the spacing effect efficiency in the gel phase, whereas the ordering effect in the liquid-crystalline phase remains similar to the one displayed by cholesterol. Such an effect of the C5–C6 double bond has been observed by Ranadive and Lala (1987) and may also be related to the orientation of the sterol. In this respect, cholestanol has been shown to exhibit a larger molecular area than cholesterol in monolayer films (Demel *et al.*, 1972a); this can be interpreted as reflecting a tilted orientation.

The oxidation of the 3β-OH group also displays a strong effect on the DPH order parameter (Vincent and Gallay, 1983) both in the gel and in the liquid-crystalline phase (Figure 8). This effect is amplified by the presence of a conjugated C4–C5 double bond. The influence of the side chain is evidenced by the androstanol curve. In the gel phase, a strong spacing effect occurs due to the insertion of the ring system between lipid molecules. In

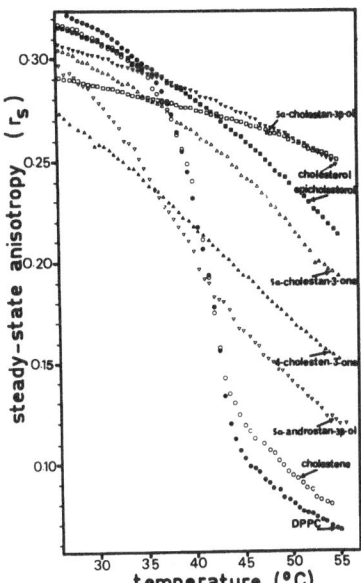

FIGURE 8. Thermal dependence of the steady-state anisotropy of DPH in DPPC sonicated vesicles without steroid (●) and with cholesterol (▼), 5α-cholestan-3β-ol (□), 4-cholesten-3-one (▲), 5α-androstan-3β-ol (▽), 5α-cholestan-3-one (△), epicholesterol (■), and cholestene (○). Sterol mole fraction was 0.5. Reprinted with permission of Academic Press. From Vincent and Gallay (1983).

the liquid-crystalline phase, the absence of acyl chain–side chain interaction allows only a small ordering effect.

Even more interesting is the last set of membrane structure effectors: the cholesterol derivatives with a polar group on the side chain. Oxysterols are involved in the regulation of cholesterol synthesis, in echinocyte formation, and in inhibition of leukocyte chemotaxis (Kandutsch *et al.*, 1978; Yachnin *et al.*, 1979; Richert *et al.*, 1984). Oxygenated sterols, not cholesterol, may play the primary role in arterial cell injury and lesion development (Imai *et al.*, 1980). They are also cytotoxic in cultured cells (Hietter *et al.*, 1984). Membrane effects have been invoked to explain these phenomena (Yachnin *et al.*, 1979) and some experimental clues have recently been published (Hagiwara *et al.*, 1982; Egli *et al.*, 1984; Theunissen *et al.*, 1986). In our laboratory, we have focused on sterols oxygenated on the side chain by comparing the effect of 22,*R*- and 22,*S*-hydroxycholesterol with 22-keto- and 20,*S*-hydroxycholesterol. It was shown that a single stereospecific substitution by a hydroxyl group on C22 dramatically depressed the cholesterol efficiency of membrane lipid order regulation. The 22,*R*-hydroxycholesterol, the first intermediate in the conversion of cholesterol into pregnenolone, exhibits a very poor effect on membrane structure compared to the 20,*S*-hydroxy- or 22-ketocholesterol, for which the effect resembles that of cholesterol (Figure 9). 22,*S*-hydroxycholesterol also exerts a strong effect. Differential scanning calorimetry as well as permeability measurements correlate with the membrane fluidity efficiency (Hagiwara *et al.*, 1982). These authors

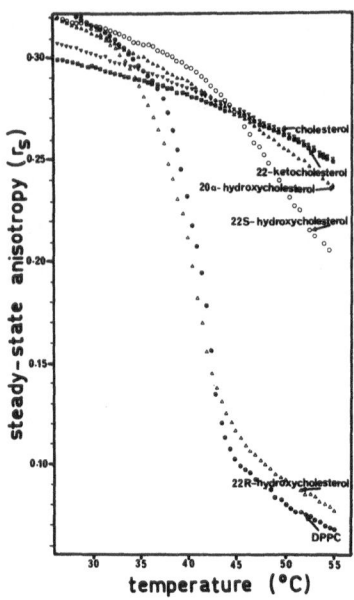

FIGURE 9. Thermal dependence of the steady-state anisotropy of DPH in DPPC sonicated vesicles without steroid (●) and with 22,R-hydroxycholesterol (△), 22,S-hydroxycholesterol (○), 20,S-hydroxycholesterol (▲), 22-ketocholesterol (■), and cholesterol (▼). Sterol mole fraction was 0.5. Reprinted with permission of Academic Press. From Vincent and Gallay (1983).

have also claimed that 23,R- and 23,S-hydroxycholesterol behave in the same manner as cholesterol itself.

The particular behavior of 22,R-hydroxycholesterol must rest on specific molecular characteristics making it a unique sterol derivative with, in principle, the full chemical elements needed to display membrane lipid ordering regulation. This is suggested by monomolecular film studies (Gallay *et al.*, 1984). The compression isotherm of either the pure sterol or the DOPC–sterol 1 : 1 monomolecular film exhibited a transition at about 103 Å², corresponding to the area of revolution of the sterol ring system. This suggests that the 22,R-hydroxycholesterol seems to be oriented with its sterol nucleus lying flat at the air–water interface, in contrast to all the other sterol derivatives known so far with a side chain on C17. The side chain must present a kinked conformation, shortening to a large extent the sterol long molecular axis (Nakane and Ikekawa, 1977).

At this stage, it could be of value if the above-mentioned data could be unified under a single concept. As stressed in the literature (Israelashvili and Mitchell, 1975; Israelashvili *et al.*, 1977; Cullis and de Kruijff, 1979; O'Leary, 1983), dynamic geometrical factors do make an appreciable contribution to interactions of phospholipids with other amphipathic molecules. This has led to the "shape concept," which includes the dynamical properties of the lipid molecules (fast rotation around the long axis, wobbling motion of the long axis), the inter- and intramolecular interactions (hydrogen bonding, electrostatic forces), and the hydration properties of the polar groups.

Although the exact dynamical shapes of lecithins and sterols are not precisely known, from molecular models and monolayer studies some insight into the shapes can be obtained. Saturated lecithins are like cylinders with an average cross section of the hydrophobic region similar to that of the polar head group (De Kruijff *et al.*, 1985). Cholesterol is more like a cone in this respect but hydration of the polar group may increase its size and make some kind of cylinder form at the end. The monomolecular film data (Demel *et al.*, 1972b) evidence a molecular area for cholesterol at moderate compression of about 39 Å^2, which corresponds approximately to the cross section of the cylinder of revolution around the long molecular axis. Other sterols, such as cholestenone, display a higher value (about 50 Å^2), which indicates that the ring system is no longer perpendicularly oriented to the air–water interface and that, due to fast axial rotation, the angle of the dynamic cone is larger than for cholesterol. The shape concept has been applied successfully to the bilayer–hexagonal H_{II} phase-inducing effect by sterols (Gallay and de Kruijff, 1982) and a direct relation has been found between the sterol molecular area and the lamellar–hexagonal H_{II} phase transition temperature. Such a relation seems to hold for the sterol ordering effect. A direct relation between the sterol molecular area and DPH order parameter in the liquid-crystalline phase reveals the contribution of the geometrical factors for maximizing the lecithin–cholesterol interactions (Figure 10). All the studied C27–sterols fit on the plot except the 22,*R*-hydroxycholesterol, which appears again as an exception because of its unique molecular features (Nakane and Ikekawa, 1977).

By using the anthroyloxy derivatives, it has been possible to locate the effects of such cholesterol derivatives at different depths through the bilayer of saturated lecithins in the gel phase. A disordering effect with regard to the degree of packing of fatty acyl chains of saturated lecithins is observed with the 22,*R*-hydroxycholesterol derivative. This is especially true in the polar phospholipid head group region as it would be expected from its orientation (Figure 11). On the contrary, the 22,*S*-hydroxy-, 20,*S*-hydroxy-, and 22-ketocholesterol derivatives behave as ordering effectors, even more than cholesterol (Figures 11 and 12). However, these oxysterols do not induce any spacing effect in the deepest hydrophobic region of the bilayer (Figures 11 and 12). This seems to indicate a stronger interaction with the C17 side chain due to a possible lower flexibility in these last three derivatives as compared to cholesterol.

Finally, we have to mention the recent use of fluorescent sterol derivatives as probes. Dehydroergosterol, when incorporated into lecithin vesicles with or without cholesterol, reports both sterol–lipid and sterol–sterol interactions (Schroeder *et al.*, 1987). Although indirect estimations of the limiting anisotropy were obtained by differential phase fluorimetry methods [devel-

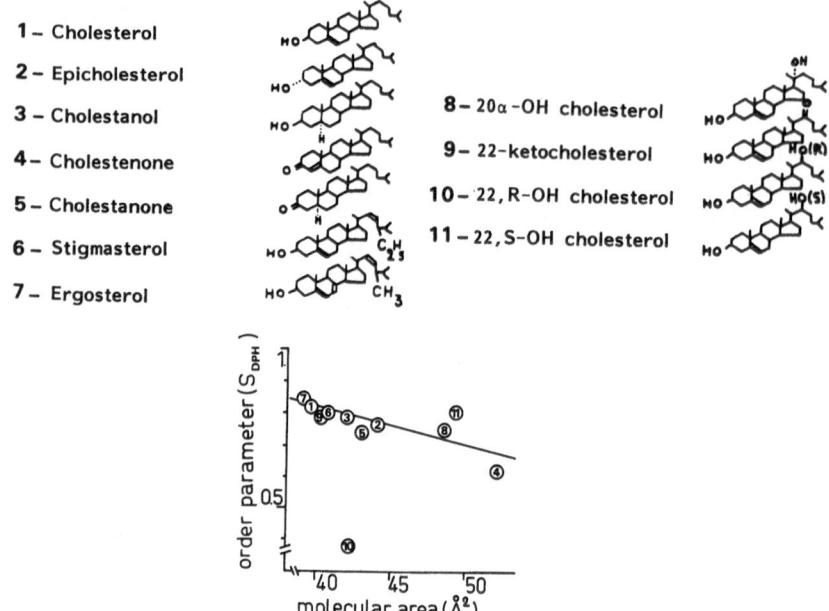

1 – Cholesterol

2 – Epicholesterol

3 – Cholestanol

4 – Cholestenone

5 – Cholestanone

6 – Stigmasterol

7 – Ergosterol

8 – 20α-OH cholesterol

9 – 22-ketocholesterol

10 – 22,R-OH cholesterol

11 – 22,S-OH cholesterol

FIGURE 10. Plot of DPH order parameter (S) computed from residual anisotropy values (Vincent and Gallay, 1983) versus molecular area (from Demel *et al.*, 1972a, and Gallay *et al.*, 1984) of some cholesterol derivatives embedded in DPPC vesicles in the liquid-crystalline phase.

oped by Lakowicz *et al.* (1984) from the theory of Weber (1978)], these sterol probes appear, at low mole percent (<5%), similar in motional properties to the cholesterol molecule (Schroeder *et al.*, 1987). At higher con-

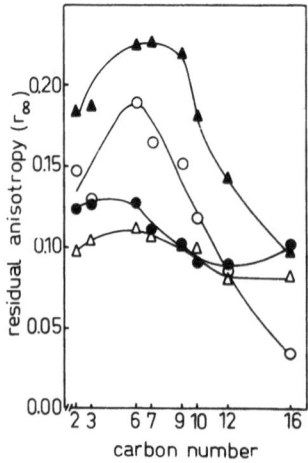

FIGURE 11. Residual anisotropy values (r_∞) of the *n*-(anthroyloxy) fatty acid in DSPC vesicles (●) at room temperature (gel phase) and in DSPC–sterol vesicles: cholesterol (○), 22,*R*-hydroxycholesterol (△), and 22,*S*-hydroxycholesterol (▲). Sterol mole fraction was 0.33 (M. Vincent and J. Gallay, unpublished results).

centrations, these sterol probes laterally segregate, as reported by the enhancement of the excited-state lifetime and limiting anisotropy values.

4. CONCLUDING REMARKS

If changes in membrane lipid composition alter the functional properties of biological membranes, these changes should correlate with modifications of the physical properties of the membrane lipids. In view of the problem of interpreting the signals arising from probe molecules inserted into membranes—due to the disturbing effect of the probe molecule itself on membrane lipids, as well as the partitioning of the probes into different lipid domains in the membrane—the use of various well-located dyes is a desirable feature allowing a more decisive interpretation of the results. Fluorescence methods, in contrast to other spectroscopic ones, allow one to work with very small amounts of biological material and with very low probe concentration, thus avoiding the disturbance of the whole membrane architecture. Fluorescence lifetimes are extremely sensitive to the environment but until very recently they were very difficult to interpret with certainty. Current analysis methods (maximum entropy, Livesey and Brochon, 1987; Livesey *et al.*, 1987) are expected to allow a precise description of the multiexponential behavior of the total fluorescence intensity decay in terms of a continuous distribution of excited-state lifetimes (in this case 150 free parameters). It is worthwhile to note that, with the maximum entropy method, no *a priori* assumption on the nature of the distribution is needed. This contrasts with the recently published work of Fiorini *et al.* (1987), where fits are obtained with symmetric distri-

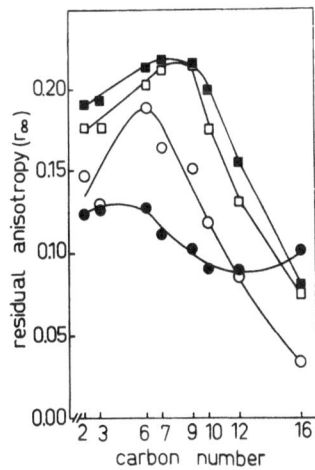

FIGURE 12. Residual anisotropy values (r_∞) of the *n*-(anthroyloxy) fatty acid in DSPC vesicles (●) at room temperature (gel phase) and in DSPC–sterol vesicles: cholesterol (○), 20,*S*-hydroxycholesterol (■), and 22-ketocholesterol (□). Sterol mole fraction was 0.33 (M. Vincent and J. Gallay, unpublished results).

bution functions: uniform, Gaussian, or Lorentzian. An illustration of the
method is provided in Figure 13 for *trans*-PnA incorporated into cardiolipin
vesicles in the absence and presence of cholesterol. When analyzed with the
maximum entropy method, the induction of a short-time component in the
intensity decay of this probe is thus clearly evidenced upon cholesterol addition as is an enhancement of the long-time component (Figure 13). This
would indicate two effects: (1) the short-time component can be linked to a
direct sterol–probe interaction (within cholesterol clusters?) and (2) the increase in proportion to the longest one and the enhancement of the lifetime
values can be related to the sterol ordering effect. The existence of a lifetime
distribution with three maxima in a pure lipid vesicle system remains to be
clarified. This analysis is quite impossible by using mathematical treatments
such as traditional deconvolution methods or nonlinear least squares regres-

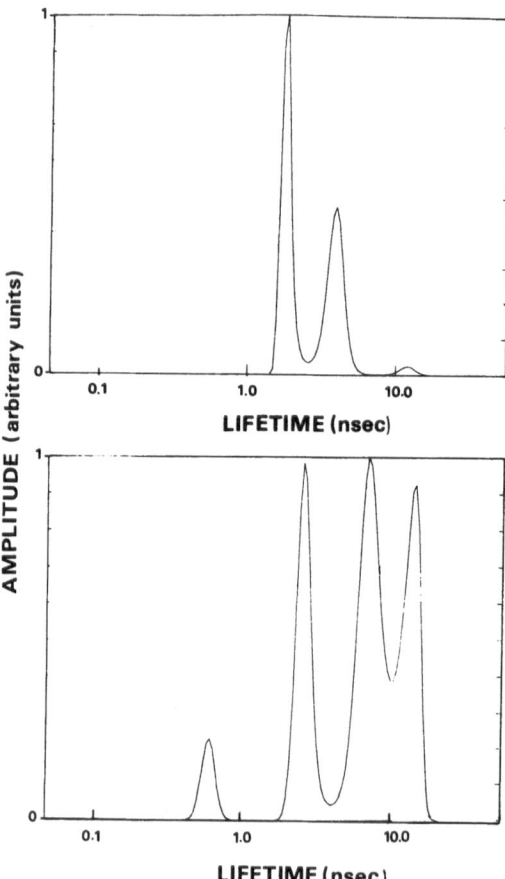

FIGURE 13. Reconstituted maximum entropy spectra (150 equally
spaced vectors) (Livesey *et al.*, 1987;
Livesey and Brochon, 1987) of *trans*-
parinaric acid in cardiolipin (upper
part) and in cardiolipin cholesterol
(lower part) vesicles in the liquid-
crystalline phase (M. Vincent and J.
Gallay, unpublished results).

sion analysis. The advent of such a powerful tool, soon available for time-resolved anisotropy data (A. K. Livesey and J. C. Brochon, personal communication), needs a systematical reevaluation of probe behavior in membranes and other systems.

The degree of fluorescence polarization, with regard to lipid "fluidity" and its main modulating factors—proteins, cholesterol, phospholipid chemical structure, and unsaturation degree—appears to have a formidable complexity. This approach, even in the time-resolved mode, implicitly assumes depolarizing motion processes arising from a simple emitting species, which is clearly a rough approximation. Correlation of the rotational relaxation times with total intensity decay times has to be evaluated in order to get a more accurate description of the membrane entity. This could be accomplished by working with highly sophisticated means, including both signal collection (high-frequency excitation sources) and analysis and also specifically located probes well-designed as membrane lipid labels.

ACKNOWLEDGMENTS. We wish to thank Dr. J. C. Brochon and Dr. A. K. Livesey for providing us with the Maximum Entropy Program for the analysis of total fluorescence intensity decays of *trans*-PnA in cardiolipin vesicles.

5. REFERENCES

Bangham, A. D., 1981, Introduction, in *Liposomes: From Physical Structure to Therapeutic Applications* (C. G. Knight, ed.), pp. 1–17, Elsevier/North-Holland Biomedical Press, Amsterdam.

Barenholz, Y., Suurkuusk, J., Mountcastle, D., Thompson, T.E., and Biltonen, R. L., 1976, A calorimetric study of the thermotropic behaviour of aqueous dispersions of natural and synthetic sphingomyelins, *Biochemistry* **15** : 2441–2447.

Barkley, M. D., Kowalczky, A. A., and Brand, L., 1981, Fluorescence decay studies of anisotropic rotations of small molecules, *J. Chem. Phys.* **75** : 3581–3593.

Barton, P. G., and Gunstone, F. O., 1975, Hydrocarbon chain packing and molecular motion in phospholipid bilayer formed from unsaturated lecithins, *J. Biol. Chem.* **250** : 4470–4476.

Berkhout, T. A., Visser, A. J. W. G., and Wirtz, K. W. A., 1984, Static and time-resolved fluorescence studies of fluorescent phosphatidylcholine transfer protein of bovine liver, *Biochemistry* **23** : 1505–1513.

Bloch, K. E., 1983, Sterol structure and membrane function, *CRC Crit. Rev. Biochem.* 47–92.

Brand, L., Knutson, J. R., Davenport, L., Beechem, J. M., Dale, R. E., Waldbridge, D. G., and Kowalczyk, A. A., 1985, Time-resolved fluorescence spectroscopy: Some applications of associative behaviour to studies of proteins and membranes, in *Spectroscopy and the Dynamics of Molecular Biological Systems* (P. M. Bayley and R. E. Dale, eds.), pp. 259–305, Academic Press, London.

Bush, S. F., Adams, R. G., and Levin, I. W., 1980, Structural reorganization in lipid bilayer systems: Effect of hydration and sterol addition on Raman spectra of dipalmitoylphosphatidylcholine multilayers, *Biochemistry* **19** : 4429–4436.

Cadenhead, D., Kellner, B. M. J., Jacobson, K., and Papahadjopoulos, D., 1977, Fluorescent probes in model membranes. 1. Anthroyl fatty acid derivatives in monolayers and liposomes of dipalmitoylphosphatidyl choline, *Biochemistry* **16** : 5386–5392.

Chen, L. A., Dale, R. E., Roth, S., and Brand, L., 1977, Nanosecond time-dependent fluorescence depolarization of diphenylhexatriene in dimyristoyl lecithin vesicles and the determination of "microviscosity," *J. Biol. Chem.* **252** : 2163–2169.

Cranney, M., Cundall, R. B., Jones, G. R., Richards, J. T., and Thomas, E. W., 1983, Fluorescence lifetime and quenching studies on some interesting diphenylhexatriene membrane probes, *Biochim. Biophys. Acta* **735** : 418–425.

Cullis, P. R., and De Kruijff, B., 1979, Lipid polymorphism and the functional roles of lipids in biological membranes, *Biochim. Biophys. Acta* **559** : 399–420.

Dahl, C. E., 1981, Effect of sterol on acyl chain ordering in phosphatidylcholine vesicles: A deuterium nuclear magnetic resonance and electron spin resonance study, *Biochemistry* **20**: 7158–7161.

Davis, P. J., and Keough, K. M. W., 1983, Differential scanning calorimetry studies of aqueous dispersions of mixtures of cholesterol with some mixed-acid and single-acid phosphatidylcholines, *Biochemistry* **22** : 6334–6340.

Davis, P. J., Fleming, B. D., Coolbear, K. P., and Keough, K. M. W., 1981, Gel to liquid-crystalline transition temperature of water dispersions of two pairs of positional isomers of unsaturated mixed-acid phosphatidylcholine, *Biochemistry* **20** : 3633–3636.

De Bony, J., and Tocanne, J. F., 1983, Synthesis and physical properties of phosphatidylcholine labeled with 9-(2-anthryl) nonanoic acid, a new fluorescent probe, *Chem. Phys. Lipids* **32**: 105–121.

De Bony, J., and Tocanne, J. F., 1984, Photo-induced dimerization of anthracene phospholipids for the study of the lateral distribution of lipids in membrane, *Eur. J. Biochem.* **143** : 373–379.

De Bony, J., Martin, G., Welby, M., and Tocanne, J. F., 1984, Evidence for a homogeneous lateral distribution of lipids in a bacterial membrane, *FEBS Lett.* **174** : 1–6.

De Foresta, B., Rogard, M., and Gallay, J., 1986, Temperature effects on adenylate cyclase activity and fluidity of bovine adrenal cortex plasma membranes, *Biochem. Soc. Trans.* **14**: 1011–1012.

De Kruijff, B., Cullis, P. R., Verkleij, A. J., Hope, M. J., Van Echteld, C. J. A., and Tarashi, T., 1985, Lipid polymorphism and membrane function, in *The Enzymes of Biological Membranes* (E. W. Martonosi, ed.), pp. 131–204, Plenum Press, New York.

Demel, R. A., and De Kruijff, B., 1976, The function of sterols in membranes, *Biochim. Biophys. Acta* **457** : 109–132.

Demel, R. A., Bruckdorfer, K. R., and Van Deenen, L. L. M., 1972a, Structural requirements of sterols for the interactions with lecithins at the air–water interface, *Biochim. Biophys. Acta* **255** : 311–320.

Demel, R. A., Geurts Van Kessel, W. S. M., and Van Deenen, L. L. M., 1972b, The properties of polyunsaturated lecithins in monolayers and liposomes and the interactions of these lecithins with cholesterol, *Biochim. Biophys. Acta* **266** : 26–40.

De Paillerets, C., Gallay, J., Vincent, M., Rogard, M., and Alfsen, A., 1981, Membrane lipid dynamics and enzymic activity in bovine adrenal cortex microsomes, *Biochim. Biophys. Acta* **664** : 134–142.

De Paillerets, C., Gallay, J., and Alfsen, A., 1984, Effect of cholesterol and protein content on membrane fluidity and 3β-hydroxysteroid dehydrogenase activity in mitochondrial inner membranes of bovine adrenal cortex, *Biochim. Biophys. Acta* **772** : 183–191.

Devaux, P. F., and Seigneuret, M., 1985, Specificity of lipid–protein interactions as determined by spectroscopic techniques, *Biochim. Biophys. Acta* **822** : 63–125.

Dufourc, E. J., Parish, E. J., Chotrakorn, S., and Smith, I. C. P., 1984, Structural and dynamic details of cholesterol–lipid interactions as revealed by deuterium NMR, *Biochemistry* 23 : 6062–6071.

Egli, U. H., Streuli, R. A., and Dubler, E., 1984, Influence of oxygenated sterol compounds on phase transitions in model membranes. A study with differential scanning calorimetry, *Biochemistry* 23 : 148–152.

Engel, L. W., and Prendergast, F. G., 1981, Values for and significance of order parameters and "cone angles" of fluorophore rotation in lipid bilayers, *Biochemistry* 20 : 7338–7345.

Fiorini, R., Valentino, M., Wang, S., Glaser, M., and Gratton, E., 1987, Fluorescence lifetime distributions of 1,6-diphenyl-1,3,5-hexatriene in phospholipid vesicles, *Biochemistry* 26 : 3864–3870.

Franks, N. P., 1976, Structural analysis of hydrated egg lecithin and cholesterol bilayers, *J. Mol. Biol.* 100 : 345–358.

Gallay, J., and De Kruijff, B., 1982, Correlation between molecular shape and hexagonal H promoting ability of sterols, *FEBS Lett.* 143 : 133–136.

Gallay, J., and Vincent, M., 1986, Cardiolipin–cholesterol interactions in the liquid-crystalline phase: A steady state and time-resolved fluorescence anisotropy study with *cis*- and *trans*-parinaric acids as probes, *Biochemistry* 25 : 2650–2656.

Gallay, J., Vincent, M., and Alfsen, A., 1982, Dynamic structure of bovine adrenal cortex microsomal membranes studied by time-resolved fluorescence anisotropy of DPH, *J. Biol. Chem.* 257 : 4038–4041.

Gallay, J., De Kruijff, B., and Demel, R., 1984, Sterol–phospholipid interactions in model membranes. Effect of polar group substitution in the cholesterol side-chain at C-20 and C-22, *Biochim. Biophys. Acta* 789 : 96–104.

Gallay, J., De Foresta, B., and Vincent, M., 1985, Cardiolipin vesicles can accommodate cholesterol up to 0.80 mole fraction, i.e. one molecule per cardiolipin fatty acid chain, *FEBS Lett.* 191 : 13–16.

Gallay, J., Vincent, M., Nicot, C., and Waks, M., 1987, Conformational aspects and rotational dynamics of ACTH 1-24 and glucagon in reverse micelles, *Biochemistry* 26 : 5738–5747.

Hagiwara, H., Nagasaki, T., Inada, Y., Saito, Y., Yasuda, T., Kojima, H., Morisaki, M., and Ikekawa, N., 1982, The interaction of cholesterol analogues with phospholipid — the effect of length of side-chain and configuration of a hydroxyl group introduced in the side chain, *Biochem. Int.* 5 : 329–338.

Haigh, E. A., Thulborn, K. R., and Sawyer, W. H., 1979, Comparison of fluorescence energy transfer and quenching methods to establish the position and orientation of components within the transverse plane of the lipid bilayer. Application to the gramicidin A–bilayer interaction, *Biochemistry* 18 : 3525–3532.

Hare, F., 1983, Simplified derivation of angular order and dynamics of rodlike fluorophores in models and membranes, *Biophys. J.* 42 : 205–218.

Hare, F., Amiell, J., and Lussan, C., 1979, Is an average viscosity tenable in lipid bilayers and membranes? A comparison of semi-empirical equivalent viscosities given by unbound probes: A nitroxide and a fluorophore, *Biochim. Biophys. Acta* 555 : 388–408.

Heyn, M. P., 1979, Determination of lipid order parameters and rotational correlation times from fluorescence depolarization experiments, *FEBS Lett.* 108 : 359–364.

Hieter, H., Trifilieff, E., Richert, L., Beck, J. P., Luu, B., and Ourisson, G., 1984, Antagonistic action of cholesterol towards the toxicity of cholesterol on cultured hepatoma cells, *Biochem. Biophys. Res. Commun.* 120 : 657–664.

Hope, M. J., Bally, M. B., Mayer, L. D., Janoff, A. S., and Cullis, P. R., 1986, Generation of multilamellar and unilamellar phospholipid vesicles, *Chem. Phys. Lipids* 40 : 89–107.

Illsley, N. P., Liu, H. Y., and Verkman, A. S., 1987, Lipid domain structure correlated with membrane protein function in placental microvillus vesicles, *Biochemistry* **26** : 446–454.

Imai, H., Werthessen, N. T., Subramanyam, V., LeQuesne, P. W., Soloway, A. H., and Kanisawa, M., 1980, Angiotoxicity of oxygenated sterols and possible precursors, *Science* **207** : 651–653.

Israelashvili, J. N., and Mitchell, D. J., 1975, A model for the packing of lipids in bilayer membranes, *Biochim. Biophys. Acta* **389** : 13–19.

Israelashvili, J. N., Mitchell, D. J., and Ninnam, B. W., 1977, Theory of self-assembly of lipid bilayers and vesicles, *Biochim. Biophys. Acta* **470** : 185–201.

Jaehnig, F., 1979, Molecular theory of lipid membrane order, *J. Chem. Phys.* **70** : 3279–3290.

Jaehnig, F., Vogel, H., and Best, L., 1982, Unifying description of the effect of membrane proteins on lipid order. Verification for melittin/dimyristoyl phosphatidylcholine systems, *Biochemistry* **21** : 6790–6798.

Johns, S., Willing, R., Thulborn, K., and Sawyer, W. H., 1979, ^{13}C-NMR studies on fluorescent probes: ^{13}C chemical shifts and longitudinal relaxation times of n-hydroxy-fatty (n = 2,6,9, and 12) acids and n-(9-anthroyloxy)-stearic (n = 6,12) acids, *Chem. Phys. Lipids* **24** : 11–16.

Kandutsch, A. A., Chen, H. W., and Heiniger, H. J., 1978, Biological activity of some oxygenated sterols, *Science* **201** : 498–501.

Kawato, S., Kinosita, K., and Ikegami, A., 1978, Effect of cholesterol on the molecular motion in the hydrocarbon region of lecithin bilayers studied by nanosecond fluorescence techniques, *Biochemistry* **17** : 5026–5031.

Keh, E., and Valeur, B., 1981, Investigation of water-containing inverted micelles by fluorescence polarization. Determination of size and internal fluidity, *J. Colloid Interface Sci.* **79**: 465–478.

Kinosita, K., Kawato, S., and Ikegami, A., 1977, A theory of fluorescence polarization decay in membranes, *Biophys. J.* **20** : 289–305.

Kintanar, A., Kunwar, A. C., and Oldfield, E., 1986, Deuterium nuclear magnetic resonance spectroscopic study of the fluorescent probe diphenylhexatriene in model membrane systems, *Biochemistry* **25** : 6517–6524.

Kleinfeld, A. M., Dragsten, P., Klausner, R. D., Pjura, W. D., and Mayatoshi, E. D., 1981, The lack of relationship between fluorescence polarization and lateral diffusion in biological membranes, *Biochim. Biophys. Acta* **649** : 471–480.

Kuroda, Y., Matsuzaki, K., Hanada, T., and Nakagaki, M., 1986, Mobility of fluorescent probe molecules in lipid bilayer vesicles as studied by steady-state and time-dependent nuclear Overhauser effect measurements in ^{1}H-nuclear magnetic resonance spectroscopy, *Biochim. Biophys. Acta* **859** : 171–179.

Kutchai, H., Chandler, L. H., and Zavoico, G. B., 1983, Effects of cholesterol on acyl chain dynamics in multilamellar vesicles of various phosphatidylcholines, *Biochim. Biophys. Acta* **736** : 137–149.

Lakowicz, J. R., Gratton, E., Cherek, H., Maliwal, B. P., and Laczko, G., 1984, Determination of time-resolved fluorescence emission spectra and anisotropies of a fluorophore–protein complex using frequency-domain phase-modulation fluorometry, *J. Biol. Chem.* **259** : 10967–10972.

Lakowicz, J. R., Cherek, H., Gryczynski, I., Joshi, N., and Johnson, M. L., 1987, Enhanced resolution of fluorescence anisotropy decays by simultaneous analysis of progressively quenched samples, *Biophys. J.* **51** : 755–768.

Lentz, B. R., Barenholz, Y., and Thompson, T. E., 1976, Fluorescence depolarization studies of phase transitions and fluidity in phospholipid bilayers: 1. Single component phosphatidylcholine liposomes, *Biochemistry* **15** : 4521–4528.

Lipari, G., and Szabo, A., 1980, Effect of librational motion on fluorescence depolarization and nuclear magnetic resonance relaxation in macromolecules and membranes, *Biophys. J.* **30** : 489–506.

Liu, B. M., Cheug, H. C., Chen, K. H., and Habercom, M. S., 1980, Fluorescence decay kinetics of pyrene in membrane vesicles, *Biophys. Chem.* **12** : 341–355.

Livesey, A. K., and Brochon, J. C., 1987, Analysing the distribution of decay constants in pulse fluorometry using the maximum entropy method, *Biophys. J.* **52** : 693–706.

Livesey, A. K., Delaye, M., Licinio, P., and Brochon, J. C., 1987, Maximum entropy analysis of dynamic parameters via the Laplace transform, *Faraday Discuss. Chem. Soc.* **83** : paper 14.

Luken, D. W., Esfahani, M., and Devlin, T. M., 1980, Effect of sterols on diphenylhexatriene fluorescence in lecithin vesicles, *FEBS Lett.* **114** : 48–50.

Marsh, D., 1980, Molecular motion in phospholipid bilayers in the gel phase: Long axis rotation, *Biochemistry* **19** : 1632–1637.

Mulders, F., Van Langen, H., Van Ginkel, G., and Levine, Y. K., 1986, The static and dynamic behaviour of fluorescent probe molecules in lipid bilayers, *Biochim. Biophys. Acta* **859** : 209–218.

Nakane, M., and Ikekawa, N., 1977, Studies on steroids. Part 41. Conformational analysis of steroidal side chains by proton magnetic resonance spectroscopy, *J. Chem. Soc. Perkin I*, 1426–1428.

Nicot, C., Vacher, M., Vincent, M., Gallay, J., and Waks, M., 1985, Membrane proteins in reverse micelles: Myelin basic protein in a membrane-mimetic environment, *Biochemistry* **24** : 7024–7032.

Oldfield, E., Meadows, M., Rice, D., and Jacobs, R., 1978, Spectroscopic studies of specifically deuterium labeled membrane systems. Nuclear magnetic resonance investigation of the effects of cholesterol in model systems, *Biochemistry* **17** : 2727–2740.

O'Leary, T., 1983, A simple theoretical model for the effect of cholesterol and polypeptides on lipid membranes, *Biochim. Biophys. Acta* **731** : 47–53.

Op den Kamp, J. A. F., 1979, Lipid asymmetry in membranes, *Annu. Rev. Biochem.* **48** : 47–71.

Owicki, J. C., and McConnell, H. M., 1980, Lateral diffusion in inhomogeneous membranes. Model membranes containing cholesterol, *Biophys. J.* **30** : 383–398.

Parente, R. A., and Lentz, B. R., 1985, Advantages and limitations of 1-palmitoyl-2-((-2(4-(6-phenyl-*trans*-1,3,5-hexatrienyl)phenyl)ethyl)carbonyl)-3-phosphatidylcholine as a fluorescent membrane probe, *Biochemistry* **24** : 6178–6185.

Pottel, H., Van der Meer, W., and Herreman, W., 1983, Correlation between the order parameter and the steady-state fluorescence anisotropy of 1,6-DPH and an evaluation of membrane fluidity, *Biochim. Biophys. Acta* **730** : 181–186.

Prendergast, F. G., Haughland, R. P., and Callahan, P. J., 1981, 1-(4-Trimethylamino)phenyl-6-phenylhexa-1,3,5-triene: Synthesis, fluorescence properties and use as a fluorescent probe of lipid bilayers, *Biochemistry* **20** : 7333–7338.

Quinn, P. J., 1981, The fluidity of cell membranes and its regulation, *Prog. Biophys. Mol. Biol.* **38** : 1–104.

Ranadive, G. N., and Lala, A. K., 1987, Sterol–phospholipid interaction in model membrane: Role of C5–C6 double bond in cholesterol, *Biochemistry* **26** : 2426–2431.

Rehorek, M., Dencher, N. A., and Heyn, M. P., 1985, Long-range lipid–protein interactions. Evidence from time-resolved fluorescence depolarization and energy transfer experiments with bacteriorhodopsin-DML vesicles, *Biochemistry* **24** : 5980–5988.

Richert, L., Castagna, M., Beck, J. P., Rong, S., Luu, B., and Ourisson, G., 1984, Growth-rate-related and hydroxysterol-induced changes in membrane fluidity of culture hepatoma cells: Correlation with 3-OH-3-CH3 glutaryl Co-A reductase activity, *Biochem. Biophys. Res. Commun.* **120** : 192–198.

Salesse, R., Brochon, J. C., and Garnier, J., 1981, Mesure des déclins de l'intensité et de l'anisotropie de fluorescence du pérylène en présence de membranes d'érythrocytes de pigeon, *Biochimie* 63 : 915–920.

Schroeder, F., Barenholz, Y., Gratton, E., and Thompson, T. E., 1987, A fluorescence study of dehydroergosterol in phosphatidylcholine bilayer vesicles, *Biochemistry* 26 : 2441–2448.

Seelig, A., and Seelig, J., 1977, Effect of a cis-double bond on the structure of a phospholipid bilayer, *Biochemistry* 16 : 45–50.

Seelig, J., and Seelig, A., 1980, Lipid conformation in model membranes and biological membranes, *Q. Rev. Biophys.* 13 : 19–61.

Seigneuret, M., and Devaux, P. F., 1984, ATP-dependent asymmetric distribution of spin-labeled phospholipids in the erythrocyte membrane: Relation to shape changes, *Proc. Natl. Acad. Sci. U.S.A.* 81 : 3751–3755.

Semer, R., and Gelerinter, E., 1979, A spin label study of the effects of sterols on egg lecithin bilayers, *Chem. Phys. Lipids* 23 : 201–211.

Shimshick, E. J., and McConnell, H. M., 1973, Lateral phase separation in binary mixture of cholesterol and phospholipids, *Biochemistry* 12 : 2351–2360.

Shinitzky, M., and Barenholz, Y., 1974, Dynamics of the hydrocarbon layer in liposomes of lecithin and sphingomyelin containing dicetylphosphate, *J. Biol. Chem.* 249 : 2652–2657.

Shinitzky, M., and Barenholz, Y., 1978, Fluidity parameters of lipid region determined by fluorescence polarization, *Biochim. Biophys. Acta* 515 : 367–394.

Shinitzky, M., and Yuli, I., 1982, Lipid fluidity at the submicroscopic level: Determination by fluorescence polarization, *Chem. Phys. Lipids* 30 : 261–282.

Shinitzky, M., Dianoux, A. C., Gitler, C., and Weber, G., 1971, Microviscosity and order in the hydrocarbon region of micelles and membranes determined with fluorescent probes. 1. Synthetic micelles, *Biochemistry* 10 : 2106–2113.

Sklar, L. A., Hudson, B. S., and Simoni, R. D., 1977, Conjugated polyene fatty acids as fluorescent probes: Synthetic phospholipid membrane studies, *Biochemistry* 16 : 819–828.

Sklar, L. A., Milijanich, G. P., and Dratz, E. A., 1979, Phospholipid lateral phase separation and the partition of cis-parinaric acid and trans-parinaric acid among aqueous, solid lipid and fluid lipid phases, *Biochemistry* 18 : 1707–1716.

Stockton, G. W., Polnaszek, C. F., Leitch, L. G., Tulloch, A. P., and Smith, I. P. C., 1974, A study of mobility and order in model membranes using ^{2}H-NMR relaxation rates and quadrupole splittings of specifically deuterated lipids, *Biochem. Biophys. Res. Commun.* 60 : 844–850.

Stoffel, W., Tunggall, B. D., Zierenberg, O., Schreiber, E., and Binczek, E., 1974, ^{13}C nuclear magnetic resonance studies of lipid interaction in single and multicomponent lipid vesicles, *Hoppe-Seyler's Z. Physiol. Chem.* 355 : 1367–1380.

Stubbs, C. D., Kouyama, T., Kinosita, K., and Ikegami, A., 1981, Effect of double bond on the dynamic properties of hydrocarbon region of lecithin bilayers, *Biochemistry* 20 : 4257–4262.

Stubbs, C. D., Kinosita, K. J., Munkonge, F., Quinn, P. J., and Ikegami, A., 1984, The dynamics of lipid motion in sarcoplasmic reticulum membranes determined by steady-state and time-resolved fluorescence measurements on DPH and related molecules, *Biochim. Biophys. Acta* 775 : 374–380.

Sune, A., Bette-Bobillo, P., Bienvenüe, A., Fellman, P., and Devaux, P. F., 1987, Selective outside–inside translocation of aminophospholipids in human platelets, *Biochemistry* 26 : 2972–2978.

Szabo, A., 1984, Theory of fluorescence depolarization in macromolecules and membranes, *J. Chem. Phys.* 81 : 160–167.

Teissié, J., 1987, Polar head molecular packing of dipalmitoylglycerophosphocholine in the gel state: A fluorescence investigation, *Biochemistry* 26 : 840–846.

Theunissen, J. J. H., Jackson, R. L., Kempen, H. J. M., and Demel, R. A., 1986, Membrane properties of oxysterols, influence of water permeability and redistributions between membranes, *Biochim. Biophys. Acta* **860** : 66–74.

Thulborn, K. R., 1981, The use of n-(9-anthroyloxy) fatty acids as fluorescent probes for biomembranes, in *Fluorescent Probes* (G. S. Beddard and M. A. West, eds.), pp. 113–139, Academic Press, London.

Thulborn, K. R., and Beddard, G., 1982, The effects of cholesterol on the time-resolved emission anisotropy of 12-(9-anthroyloxy) stearic acid in dipalmitoylphosphatidylcholine bilayers, *Biochim. Biophys. Acta* **693** : 246–252.

Thulborn, K. R., and Sawyer, W. R., 1978, Properties and location of a set of fluorescent probes sensitive to the fluidity gradient of the lipid bilayer, *Biochim. Biophys. Acta* **511** : 125–140.

Thulborn, K. R., Treloar, F. E., and Sawyer, W. H., 1978, A microviscosity barrier in the lipid bilayer due to the presence of phospholipid containing unsaturated acyl chains, *Biochem. Biophys. Res. Commun.* **81** : 42–49.

Tilley, L., Thulborn, K. R., and Sawyer, W. H., 1979, An assessment of the fluidity gradient of the lipid bilayer as determined by a set of n-(9-anthroyloxy) fatty acids (n = 2,6,9,12, and 16), *J. Biol. Chem.* **254** : 2592–2594.

Van Blitterswijk, W. J., Van Hoeven, R. P., and Van der Meer, B. W., 1981, Lipid structural order parameters (reciprocal of fluidity) in biomembranes derived from steady-state fluorescence polarization measurements, *Biochim. Biophys. Acta* **644** : 323–332.

Van Blitterswijk, W. J., Van der Meer, W., and Hilkmann, H., 1987, Quantitative contributions of cholesterol and the individual classes of phospholipids and their degree of fatty acyl (un)saturation to membrane fluidity measured by fluorescence polarization, *Biochemistry* **26** : 1746–1756.

Van der Meer, W., Pottel, H., Herreman, W., Amelot, M., Hendrickx, H., and Schroeder, H., 1984, Effect of orientational order on the decay of the fluorescence anisotropy in membrane suspensions, *Biophys. J.* **46** : 512–523.

Van der Meer, W., Van Hoeven, R. P., and Blitterswijk, W. J., 1986, Steady-state fluorescence polarization data in membranes. Resolution into physical parameters by an extended equation for restricted rotation of fluorophores, *Biochim. Biophys. Acta* **854** : 38–44.

Veatch, W. R., and Stryer, L., 1977, Effect of cholesterol on the rotational mobility of diphenylhexatriene in liposomes: A nanosecond fluorescence anisotropy study, *J. Mol. Biol.* **117** : 1109–1113.

Vincent, M., and Gallay, J., 1983, Steroid–lipid interactions in sonicated dipalmitoylphosphatidylcholine vesicles: A steady-state and time-resolved fluorescence anisotropy study with all-*trans*-1,6-diphenyl-1,3,5-hexatriene as probe, *Biochem. Biophys. Res. Commun.* **113** : 799–810.

Vincent, M., and Gallay, J., 1984, Time-resolved fluorescence anisotropy study of effect of a cis double bond on structure of lecithin and cholesterol–lecithin bilayers using n-(9-anthroyloxy) fatty acids as probes, *Biochemistry* **23** : 6514–6522.

Vincent, M., De Foresta, B., Gallay, J., and Alfsen, A., 1982a, Nanosecond fluorescence anisotropy decays of n-(9-anthroyloxy) fatty acids in dipalmitoylphosphatidyl vesicles with regard to isotropic solvents, *Biochemistry* **21** : 708–716.

Vincent, M., De Foresta, B., Gallay, J., and Alfsen, A., 1982b, Fluorescence anisotropy decays of n-(9-anthroyloxy) fatty acids in dipalmitoylphosphatidylcholine vesicles. Localization of the effects of cholesterol addition, *Biochem. Biophys. Res. Commun.* **107** : 914–921.

Vincent, M., Gallay, J., De Bony, J., and Tocanne, J. F., 1985, Steady-state and time-resolved fluorescence anisotropy study of phospholipid molecular motion in the gel phase using 1-palmitoyl-2-(9-(2-anthryl)-nonanoyl)-sn-glycero-3-phosphocholine as probe, *Eur. J. Biochem.* **150** : 341–347.

Visser, A. J. W. G., Santema, J. S., and Van Hoek, A., 1984, Spectroscopic and dynamics characterization of FMN in reversed micelles entrapped water pools, *Photochem. Photobiol.* **39** : 11–16.

Vos, M. H., Kooyman, R. P. H., and Levine, Y. K., 1983, Angle-resolved fluorescence depolarization experiments on oriented membrane lipid system, *Biochem. Biophys. Res. Commun.* **116** : 462–468.

Vos, K., Laane, C., and Visser, A. J. W. G., 1987, Spectroscopy of reversed micelles, *Photochem. Photobiol.* **45** : 863–878.

Waggoner, A. S., and Stryer, L., 1970, Fluorescent probes of biological membranes, *Proc. Natl. Acad. Sci. U.S.A.* **67** : 579–589.

Wahl, Ph., Kasai, M., and Changeux, J. P., 1971, A study on the motion of proteins in excitable membrane fragments by nanosecond fluorescence polarization spectroscopy, *Eur. J. Biochem.* **18** : 332–341.

Weber, G., 1971, Theory of fluorescence depolarization by anisotropic Brownian rotations. Discontinuous distribution approach, *J. Chem. Phys.* **55** : 2399–2407.

Weber, G., 1978, Limited rotational motion: Recognition by differential phase fluorimetry, *Acta Phys. Polon. A* **54** : 859–865.

Welby, M., and Tocanne, J. F., 1982, Evidence for the incorporation of a fluorescent anthracene fatty acid into the membrane lipids of *Micrococcus luteus, Biochim. Biophys. Acta* **689** : 173–176.

Wolber, P. K., and Hudson, B. S., 1981, Fluorescence lifetime and time-resolved polarization anisotropy studies of acyl chain order and dynamics in lipid bilayers, *Biochemistry* **20** : 2800–2810.

Yachnin, S., Streuli, R. A., Gordon, L. J., and Hsu, R. C., 1979, Alterations of peripheral blood cell membrane function and morphology by oxygenated sterols: A membrane insertion hypothesis, *Curr. Top. Hematol.* **2** : 245–271.

Yeagle, P. L., 1985, Cholesterol and cell membranes, *Biochim. Biophys. Acta* **822** : 267–287.

Zachowski, A., Fabvre, E., Cribier, S., Herve, P., and Devaux, P. F., 1986, Outside–inside translocation of amino phospholipids in the human erythrocyte membranes is mediated by a specific enzyme, *Biochemistry* **25** : 2585–2590.

Zannoni, C., Arcioni, A., and Cavatorta, P., 1983, Fluorescence depolarization in liquid crystals and membrane bilayers, *Chem. Phys. Lipids* **32** : 179–250.

Chapter 5

Fluorescence Polarization to Evaluate the Fluidity of Natural and Reconstituted Membranes

Elsie M. B. Sorensen

1. INTRODUCTION

1.1. Aims and Scope of This Chapter

Fluorescence polarization measurements assess movement of membrane constituents under near-normal conditions, as those membranes are challenged with toxic and/or nontoxic ions. Data interpretation of fluorescence polarization measurements are contingent upon an understanding of the interactions between the plasma membrane and other cellular constituents. Therefore, a brief summarization of a number of known interactions between the plasma membrane and other cellular components forms the basis of evaluations of membrane fluidity using fluorescence polarization.

Plasma membranes act as complex, dynamic, fluid barriers for individual cells, without which life as we know it would be impossible. These highly selective filters consist of continuous double layers of mobile phospholipid

Elsie M. B. Sorensen   Department of Pharmacology and Toxicology, College of Pharmacy, University of Texas, Austin, Texas 78712.

molecules in which membrane proteins are embedded. Since the 1970s, when researchers first recognized that membrane constituents are in constant motion, numerous technological advancements have shown that phospholipids move laterally, rotate, flex acyl chains, and migrate from the internal leaflet to the external leaflet (i.e., flip-flop) (Chapman, 1975; Kimelberg, 1977; Quinn and Chapman, 1980). Neighboring phospholipids exchange places with their neighbors by lateral diffusion at about 10^7 times per second, moving a total distance of about 2 μm per second. Flip-flop motion, in contrast, occurs less than once in 2 weeks (Kornberg and McConnell, 1971; Bangham, 1975; Quinn, 1976; Alberts et al., 1983).

At one time, eukaryotic plasma membranes were depicted as simple fluid bilayers containing suspended globular protein units (Singer, 1971; Singer and Nicolson, 1972). This simplistic approach often hinders present assessments of the effects of ions on membranes and ion transport through membranes. Now numerous studies reveal the complexity of the plasma membrane and the diversity of associations among the constituents of plasma membranes and adjacent molecules (Figure 1). Composed of about 40–50% lipid and 50–60% protein in noncovalent union, plasma membranes exhibit considerable variability in composition for different cell types (Quinn, 1976; Harrison and Lunt, 1980). Extrinsic proteins are suspended in the external and internal faces of the phospholipid bilayer. Intrinsic proteins are organized perpendicular to phospholipid acyl chains and transport ions through the bilayer. Collars of phospholipid molecules, organized around ion channel intrinsic proteins, are thought to bind ions present in the extracellular milieu and alter carrier function (Jacobson and Papahadjopoulos, 1975).

In addition, peripheral membrane proteins embedded in this fluid, heterogeneous layer are linked to cytoskeletal proteins. Large glycoprotein and protein molecules such as fibronectin, collagen, and glycosaminoglycans are connected to cytoskeletal proteins via integral proteins. Therefore, technological advancements since the 1970s show that the plasma membrane is a complex, dynamic partition separating living organisms from the external milieu. But this is only the beginning of a complete understanding of fluidity measurements of plasma membranes.

Individual animal cells are covered with plasma membranes having variable lipid content and correspondingly variable fluidity. Whetton et al. (1983) isolated three of the major subfractions of hepatocyte plasma membranes (i.e., sinusoidal, contiguous, and canalicular) for assessment of inherent membrane fluidity and alterations in fluidity following challenge with benzyl alcohol. The contiguous or lateral plasma membranes were most rigid and had the highest cholesterol and sphingomyelin content. Canalicular plasma membranes were more fluid than contiguous membranes and sinusoidal (i.e., blood facing) plasma membranes were the most fluid. Contiguous and sin-

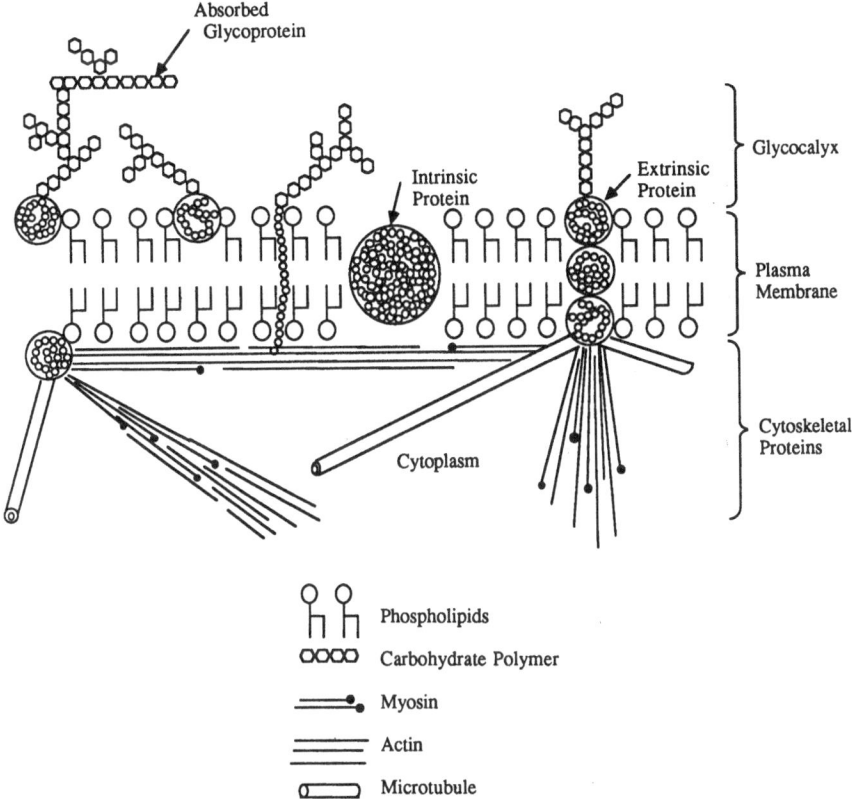

FIGURE 1. Diagrammatic representation of the complex relation existing between the protein and lipid molecules of the plasma membrane, cytoskeletal components, and glycocalyx. Size relations approximate those of natural membranes magnified 8,000,000×.

usoidal membranes contained cholesterol and sphingomyelin concentrations paralleling fluidity alterations. In response to the fluidizing agent benzyl alcohol, the more fluid membranes (i.e., sinusoidal) were less fluidized by the alcohol than were the contiguous or canalicular plasma membranes. Therefore, the lipid composition of different membranes on the same cell is capable of modulating membrane fluidity.

Other scientists have shown similar differences in the responsiveness of plasma membranes to perturbating agents. Canalicular membranes did not respond to calcium-induced membrane stabilization, whereas the sinusoidal fraction and the sinusoidal plus contiguous membrane fractions became less fluid with the addition of calcium (Storch *et al.*, 1983). However, the electron spin resonance studies of Storch *et al.* (1983) could not be compared with the fluorescence polarization studies of Whetton *et al.* (1983) because the former group evaluated mixtures of membranes. Despite these differences, both studies illustrate the variation in the composition of the plasma membrane of individual cells, as well as differences in the responsiveness of different isolates of plasma membranes to benzyl alcohol (Whetton *et al.*, 1983) or calcium (Storch *et al.*, 1983). Therefore, differences in lipid composition serve a modulatory function in cellular responsiveness.

The elevated cholesterol level of hepatocyte canalicular membranes is thought to protect hepatocytes from the detergent action of bile salts (Storch *et al.*, 1983). The study by Lowe and Coleman (1981) corroborates the study by Storch *et al.* (1983) and indicates that erythrocyte membranes of low fluidity are more resistant to destabilization by detergents. Therefore, an individual hepatocyte is covered with plasma membranes of variable lipid composition, correspondingly different membrane fluidity, and variable responsiveness to agents capable of perturbating cells.

Documentation of membrane asymmetries in the phospholipids present in the inner and outer phospholipid leaflets would suggest that alterations in the concentrations of intracellular ions might impact markedly on steady-state fluorescence polarization. Release of intracellular stores of calcium from mitrochondria, for example, might alter the steady-state fluorescence polarization of probes more so than extracellular changes in the levels of calcium. Verkleij *et al.* (1973) documented the predominance of phosphatidylserine on the inner leaflet of the phospholipid bilayer. Therefore, if intracellular calcium concentrations were elevated, chelation of adjacent internal phosphatidylserines might alter fluorescence polarization in the absence of alterations in external ionic concentrations.

Normally, ions bound to the plasma membrane are in a constant state of flux. Alterbach and Seelig (1984) found that calcium binds to individual lipid head groups of the membrane surface for less than 10^{-5} sec and that individual calcium ions complex with two membrane lipid head groups simultaneously. Moreover, alterations in membrane fluidity can occur as a result of changes in the composition of extracellular fluid. Because biological systems are constant thermally, conformational changes such as lipid phase transitions (i.e., localized islands of solid aggregates of phospholipids in a sea of fluid phospholipids) must be caused by variables other than temperature. For example, an increase in pH from 7 to 9 increases the charge on phosphatidic acid from one to two per polar head group (Trauble and Eibl, 1974). In addition, lateral expansion, decreased density, and decreased thickness result from the gel to liquid phase transition (Oldfield and Chapman, 1972; Nagle, 1973a,b). Therefore, as a result of rapid ion–membrane interactions and alterations in the ionic milieu, fluidity changes are altered on a moment-by-moment basis *in vivo*.

Therefore, despite the level of sophistication of a specific procedure such as steady-state fluorescence polarization, alterations in plasma membranes depend on intricate and dynamic interactions among diverse intracellular and extracellular molecules. These interactions may or may not affect the fluidity of membranes directly. The complexity of these interactions limits the value of any method of measuring membrane motion when that method is used alone. However, despite the number of ramifications, fluidity measurements of ion–membrane interactions are consistent with published studies using alternate methods such as electron spin resonance and phase-transition temperature measurements. Fluorescence polarization results provide valuable insights into ion–membrane interactions. Therefore, based on recognition of the inherent limitations of fluorescence polarization procedures, this chapter summarizes a portion of the scientific data base elucidating membrane–ion interactions. Emphasis is placed on those intermolecular interactions in natural and reconstituted eukaryotic plasma membranes measurable using fluorescence polarization and comparable techniques. The physiological significance of these interactions is stressed throughout. Moreover, the prevalence of reported studies on calcium-induced alterations of membrane fluidity results in a greater data base for calcium than that for other ions.

1.2. Mechanism of Action and Biological Significance of Fluorescence Polarization Measurements of Membrane Fluidity

Although alterations in membrane fluidity can occur as a result of changes in both the external and internal bilayer phospholipid leaflet, those occurring as a result of external perturbation of the membrane occur initially and are therefore emphasized in this chapter. Both toxicological and natural interactions are considered, with emphasis on the mechanism of action occurring when ions contact plasma membranes because the plasma membrane represents the first line of defense in the protection of hepatocytes and other cells from xenobiotics.

In this regard, cadmium is one toxic ion of special interest because the interaction of cadmium with plasma membranes exemplifies the significance of fluorescence polarization in assessment of ion–membrane interactions. Direct, membrane-active effects of cadmium reduce membrane fluidity (Bevan *et al.*, 1983; Nealon *et al.*, 1984; Sorensen *et al.*, 1985). Measurements of phase transition temperatures and steady-state fluorescence polarization provide the same results—fluidity is reduced when cadmium levels are elevated. The phase transition temperature (T_m) of artificial membranes is reduced (an indication of reduced fluidity) when the membranes are exposed to low concentrations of cadmium (Bevan *et al.*, 1983). Reduced T_m is thought to result from the chelation of the single anionic groups on the hydrophilic portion of two adjacent phospholipids, such as phosphatidic acid or phosphatidylserine (Figure 2).

The results of some laboratories suggest that calcium may mediate the irreversible cell death produced by a number of toxic agents (Schanne *et al.*, 1979; Casini and Farber, 1981; Chenery *et al.*, 1981). Other studies show that the presence of calcium results in beneficial effects (Smith *et al.*, 1981; Edmondson and Bang, 1981) or no effect (Stacey and Klaassen, 1982). Differences in the composition of tissue-culture media are believed to be involved in these controversial results. Calcium augmented detrimental effects of the hepatotoxins tested when complete tissue medium was used but coun-

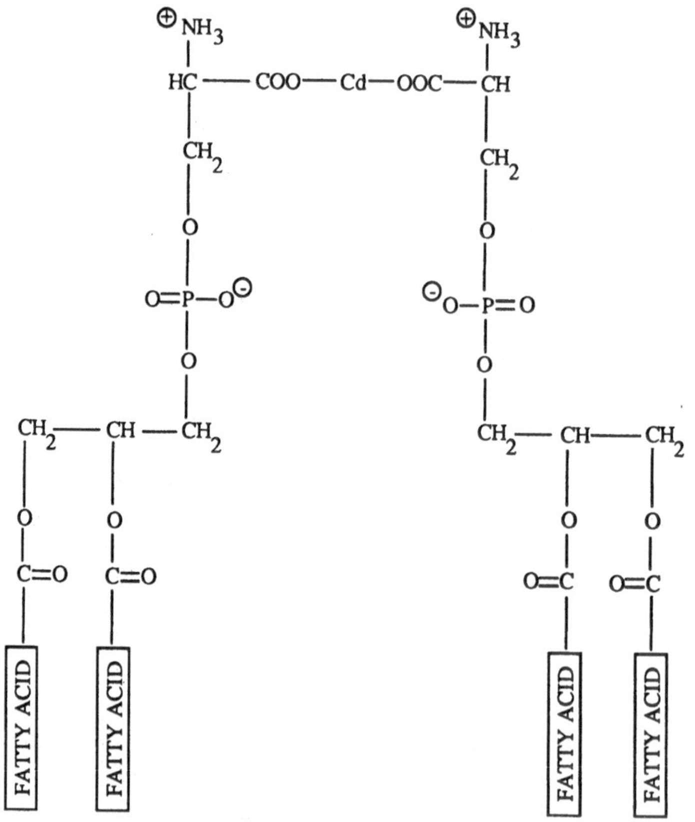

FIGURE 2. Representation of the possible interaction of cadmium with two adjacent phospholipids (i.e., phosphatidylserine) in the plasma membrane. Phase separations result as a number of these chelates form solid phase islands in a sea of fluid phospholipids.

tered detrimental effects or showed no change when buffers or salt solutions were used. Complete tissue-culture media were used in several studies that showed that calcium augmented the detrimental effects of a number of hepatotoxins (Schanne et al., 1979; Casini and Farber, 1981), whereas several studies that showed a beneficial or nonexistent effect utilized buffers or salt solutions as the exposure solution (Smith et al., 1981; Stacey and Klaassen, 1982). Other possibilities include differences in hepatocyte-isolation procedures and/or differences in inherent mitochondrial calcium loads for various species of rats due to variable calcium levels in water or food supplies. Although the complete explanation of these reported discrepancies remains to be determined, primary cultures of murine parenchymal hepatocytes were used to show that the presence of extracellular calcium lessened, but did not entirely prevent, cadmium-induced hepatotoxicity when buffered solutions were used for exposures of cells.

Evaluation of ion–membrane effects using steady-state fluorescence polarization has been used to address the controversial explanations of the role of calcium in irreversible cell death induced by structurally unrelated xenobiotics. Intracellular calcium loads increase following exposure of liver cells to hepatotoxins (Reynolds, 1963; Judah et al., 1964; Farber and El-Mofty, 1975). Because extracellular calcium levels are 1000-fold greater than intracellular levels, even slight xenobiotic-induced alterations in the integrity of plasma membranes could result in passive equilibrium of calcium as calcium rushes across the damaged plasma membranes. Therefore, elevation of intracellular calcium levels would result in, rather than initiate, cellular injury.

Steady-state fluorescence polarization studies were conducted using the same conditions used for functional and morphological evaluations. Cultured hepatocytes exposed to cadmium in buffered, balanced salt solutions containing calcium had increased survival, higher levels of total urea, less cellular rounding, reduced lactate dehydrogenase release, less pronounced membrane fluidity changes, and fewer superficial blebs than cells exposed to cadmium in the absence of calcium (Sorensen and Acosta, 1982, 1983, 1984a,b,c, 1985; Sorensen et al., 1985).

Calcium and cadmium would not be expected to bind as strongly to zwitterionic head groups, such as those on phosphatidylcholine, as to single anionic binding sites on the head group of phosphatidylserine. As a consequence, phosphatidylserine has been shown to be involved in the binding of calcium to trigger phase separations resulting in domain formation (Ohnishi and Ito, 1974; Jacobson and Papahadjopoulos, 1975). Phosphatidic acid caused a similar effect (Trauble and Eibl, 1974; Hartmann et al., 1977). Chelation of a number of these phospholipids would result in circular islands of less fluid phospholipids, surrounding intrinsic proteins in a sea of more fluid phospholipids (Figure 3).

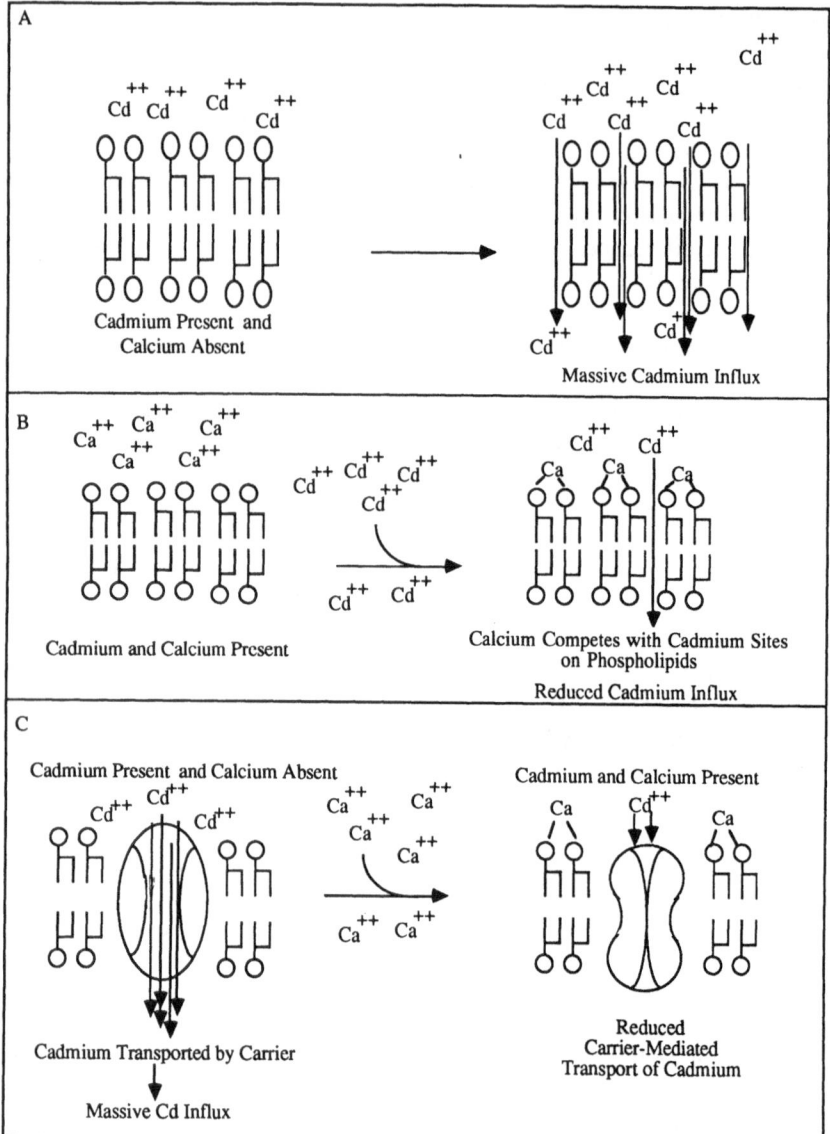

FIGURE 3. Diagrams of three possible mechanisms that could explain interactions between cadmium and calcium at the plasma membrane to cause reported fluidity changes. These are not considered to be mutually exclusive mechanisms (see text).

Mechanistically, cations such as calcium could compete with cadmium for anionic binding sites in the lipid bilayer since these cations have similar ionic radii (0.97 versus 0.99 Å) (*CRC Handbook of Chemistry and Physics,*

1981). Following binding to phospholipid head groups, therefore, either cation could effectively trigger phase separations (Schlatz and Marinetti, 1972; Ohnishi and Ito, 1974; Jacobson and Papahadjopoulos, 1975; Hartmann *et al.*, 1977) in proximity to the protein carriers to shut off carrier function (Ohnishi and Ito, 1974). In this respect, we should note that the presence of increasing concentrations of calcium stabilized isolated plasma membranes (Gordon *et al.*, 1978; Livingstone and Schachter, 1980; Storch *et al.*, 1983), as well as artificial phospholipid liposomes (Trauble and Eibl, 1974; Ohnishi and Ito, 1974; Jacobson and Papahadjopoulos, 1975; Hartmann *et al.*, 1977), and significantly reduced the cadmium-induced release of LDH from cultured hepatocytes (Sorensen and Acosta, 1982).

At least three mechanisms could reasonably explain the ability of calcium ions to diminish the cadmium-induced hepatotoxicity observed in this study: (1) calcium could stabilize membranes to effectively impede cadmium influx into hepatocytes (Schlatz and Marinetti, 1972; Ohnishi and Ito, 1974; Trauble and Eibl, 1974; Jacobson and Papahadjopoulos, 1975; Hartmann *et al.*, 1977), (2) calcium may occupy anionic sites on the plasma membranes which would otherwise be occupied by cadmium (Ohnishi and Ito, 1974; Trauble and Eibl, 1974; Jacobson and Papahadjopoulos, 1975; Hartmann *et al.*, 1977), and/or (3) calcium might bind to phospholipid head groups in proximity to cadmium–zinc carrier protein molecules to form lipid domains circumscribing the carrier to block carrier function (Ohnishi and Ito, 1974; Trauble and Eibl, 1974; Jacobson and Papahadjopoulos, 1975; Hartmann *et al.*, 1977) (Figure 3). Yousef (1983), for example, demonstrated that plasma membranes exposed to 4-mM (or higher) concentrations of calcium had reduced loss of low molecular weight polypeptides; this supported the second mechanism, which could explain the protective role of calcium on cadmium-challenged hepatocytes. Moreover, Farris *et al.* (1984), in addition, showed that omission of calcium from the complete culture medium resulted in increased lipid peroxidation, depletion of glutathione, glutathione disulfide formation, the efflux of reduced glutathione, and increased vitamin E uptake in isolated rat hepatocytes (Farriss *et al.*, 1984).

Potassium permeability adenylate cyclase activity and Na^+–K^+-dependent adenosine triphosphatase activity are inhibited by the presence of ambient calcium levels and altered by fluidity changes (Hepp *et al.*, 1970; Boyer and Reno, 1975; Wisher and Evans, 1975; Houslay *et al.*, 1976; Kolb and Adam, 1976; Blitzer and Boyer, 1978; Dipple and Houslay, 1978; Keefe *et al.*, 1979; Latham and Kashgarian, 1979; Giraud *et al.*, 1981). Fluidity changes can alter the activity of enzymes embedded in the plasma membrane (see review by Kimelberg, 1977). Therefore, phase separations (manifested by alterations in fluidity) could play an important role in modulating enzyme activity and other functions in biomembranes.

Jacobson and Paphadjopoulos (1975) speculate on the involvement of calcium in the regulation of carrier transport. Acidic lipid domains circumscribe transport peptides and would be chelated by calcium to inactivate carriers such as those for valinomycin (Krasne et al., 1971). Release of calcium might fluidize the carrier to allow valinomycin conductance. The second possibility involves the existence of asymmetrical carriers with acidic lipids on one side of the carrier (Bretcher, 1973; Zwaal et al., 1973). Calcium chelation to the acidic lipids might create further carrier asymmetry to impede conductance in the presence of sufficient extracellular calcium. Transmembrane conductance proceeds only when calcium is released from the carrier. A third possibility involves membrane disordering preventing or reducing carrier-mediated transport prior to the addition of calcium (Figure 4).

As indicated previously, the plasma membrane of the typical animal cell is the first line of defense in the protection of those cells from toxic ions, while allowing entry of beneficial ions. The capacity of any one ion to interact with plasma membranes is contingent upon the chemical and physical relation between that toxic ion and beneficial ions required by the body. Evaluation of the direct, intermolecular events that occur in ion-challenged membranes using fluorescence polarization can provide an understanding of the manner by which the plasma membrane withstands that impact of toxic ions. The steady-state fluorescence polarization of probes such as 1,6-diphenyl-1,3,5-hexatriene (DPH), incorporated into the isolated plasma membranes, can be used to quantify membrane fluidity changes induced by toxic ions.

The motive force or mechanism of action for phase separation is the formation of energy-gaining aggregates of calcium and phosphatidylserine or phosphatidic acid in the two-dimensional bilayer. In 1968, Papahadjopoulos proposed a monomolecular model for phosphatidylserine–calcium and phosphatidic acid–calcium complexes consisting of a linear polymeric arrangement. The enthalpy (chemical energy of a physical system) decrease due to the aggregate formation of calcium chelates and the Van der Waals interaction among closely packed alkyl chains would overcome the entropy (i.e., energy unavailable for the performance of useful work) decrease resulting from phase separation. Based on electron spin resonance spectra, Ohnishi and Ito (1974) propose the necessity of at least two chelating sites in the adjacent phospholipid head groups to develop the chelated structure.

In contrast, phosphatidylinositol has only one anionic site and would not form aggregates (Ohnishi and Ito, 1974). Despite the presence of two anionic sites in cardiolipin, steric hindrance was thought to prevent aggregation because the lipid has two anionic sites per four alkyl chains. Biologically, the chelation of phospholipids adjacent to integral proteins could shut off carrier function. Because steady-state fluorescence polarization methods average changes over the entire membrane, localized fluidity changes adjacent to

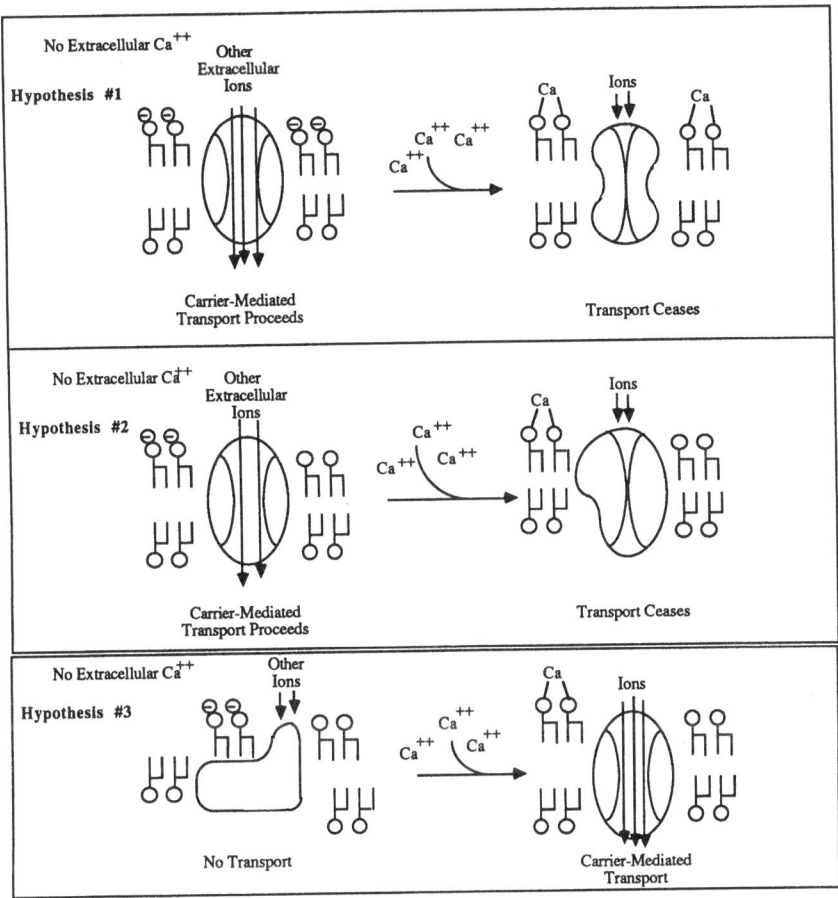

FIGURE 4. Three possible mechanisms involving extracellular calcium in carrier transport. Transport (e.g., valinomycin) could be inhibited in the presence of calcium (Hypothesis #1); calcium could be bound to carriers having an asymmetrical arrangement of acidic lipids, which chelate calcium and prevent carrier-mediated transport (Hypothesis #2); and/or carriers could be distorted in the absence of extracellular calcium (Hypothesis #3) (see text).

carrier proteins are not thought to be measured at the present time with this method.

Interaction of the two divalent cations (i.e., cadmium and calcium) with the plasma membrane probably accounts for the reduced cellular binding and influx of cadmium found in primary cultures of rat hepatocytes in the presence of extracellular calcium (Sorensen, 1988). These results show what appears to represent competition between the two cations for sites on the plasma membrane. Therefore, fluorescence polarization was used to assess further the intermolecular alterations at the plasma membrane during simultaneous exposure of isolated membranes to the two cations (Nealon *et al.*, 1984; Sorensen *et al.*, 1984).

Changes in plasma membrane fluidity were evaluated using steady-state fluorescence polarization of 1,6-diphenyl-1,3,5-hexatriene (DPH) incorporated into isolated plasma membranes from human erythrocytes or rat hepatocytes. The presence of increasing concentrations of calcium (0.5–4 mM), cadmium (50–500 μM), or both decreased the motional freedom of the fluorescent probe molecules in plasma membranes derived from both types of plasma membranes (Table I). Cadmium-induced effects were from three to ten times greater than those of calcium. Higher concentrations of cadmium in the presence of calcium increased the anisotropy parameter, which leveled off at lower cadmium concentrations. The presence of calcium diminished the overall effects of cadmium on these membranes.

Addition of increasing concentrations of calcium to the plasma membranes of erythrocytes resulted in an increased anisotropy parameter (a measure of increased stability in the membranes). This difference was statistically

Table 1

Comparison of Membrane Fluidity and Morphological Changes Characteristic of Cytotoxicity[a]

Calcium[b]	Cadmium concentration[c]	Anisotropy parameter[d]	RVP of blebs[e]	RVP of flattened cells[f]
Absent	0	1.168 ± 0.008	3.20 ± 0.88	87.8 ± 1.4
	50	1.191 ± 0.005	16.13 ± 2.09	50.2 ± 9.6
	100	1.184 ± 0.014	23.81 ± 1.70	9.5 ± 2.3
	200	1.226[g] ± 0.007	36.57 ± 4.85	7.1 ± 2.7
	500	1.232[h] ± 0.016	—	—
Present	0	1.216 ± 0.011	3.89 ± 1.74	90.1 ± 1.9
	50	1.230 ± 0.015	5.47[i] ± 1.17	84.0[k] ± 3.5
	100	1.240 ± 0.020	16.40[j] ± 2.62	74.4[i] ± 3.3
	200	1.244 ± 0.010	18.03[i] ± 1.66	53.3[i] ± 6.3
	500	1.252 ± 0.013	—	—

[a]Reprinted by permission from Nealon *et al.* (1984).
[b]A positive symbol indicates the presence of 1.8 mM calcium in the medium, negative symbol indicates the absence of calcium.
[c]In μM.
[d]Mean (± 1 standard error of the mean, S.E.M.).
[e]Mean (± 1 S.E.M.) relative volume percentage (RVP) for the presence of blebs on the surface of hepatocytes. Increases in RVP of blebs are indicative of cellular damage.
[f]Mean (± 1 S.E.M.) RVP for the more normal, flattened cells. Decreases in the RVP of flattened cells are indicative of increasing cellular damage.
[g]Differs from the respective control, $p \leq 0.01$.
[h]Differs from the respective control, $p \leq 0.05$.
[i]Differs from the paired control, $p \leq 0.001$.
[j]Differs from the paired control, $p \leq 0.05$.
[k]Differs from the paired control, $p \leq 0.01$.

significant following addition of 0.5 mM calcium to membranes ($p < 0.05$ compared to that value obtained from the 0-μM control plasma membranes; however, the anisotropy parameter did not level off until the calcium concentration reached 2.0 mM ($p < 0.01$). Erythrocyte plasma membranes exposed in the absence of calcium to increasing concentrations of cadmium had anisotropy parameter values that were significantly different from the 0-μM control values following the addition of 50 μM cadmium ($p < 0.01$). These cadmium concentrations were 10 times less than the calcium concentration required to produce the same relative change. The presence of 1.8 or 3.6 mM calcium in the membrane suspension during the addition of increasing levels of cadmium resulted in a significant increase in the anisotropy parameter following the addition of only 50 μM cadmium ($p < 0.001$).

Addition of increasing levels of cadmium thereafter resulted in a minimal change in the anisotropy parameter. No significant differences were observed between data obtained using 1.8 or 3.6 mM extracellular calcium. The same pattern was observed for plasma membranes derived from hepatocytes, although the anisotropy parameter values for those membranes exposed in the presence of calcium to increasing concentrations of cadmium were not significantly different from 0-μM control values. Moreover, the anisotropy parameter value was not significantly different from the 0-μM control value until 200 μM cadmium had been added to hepatocyte plasma membranes that had not been exposed to calcium simultaneously.

Therefore, the presence of 10-fold lower concentrations of cadmium (in the absence of calcium) decreased the fluidity of plasma membrane to a degree similar to that obtained using calcium alone. This result is obtained regardless of the source of the plasma membranes. The presence of 1.8 or 3.6 mM calcium in plasma membrane suspensions—which were simultaneously exposed to increasing levels of cadmium (in the 0–500-μM concentration angle)—resulted in no significant change in fluidity for the plasma membranes beyond that obtained using the 50-μM cadmium concentration. This effect was considered due to a higher binding affinity of cadmium for the phosphatidic acid and phosphatidylserine head groups than for calcium, as previously reported by Feigenson (1982). Alternatively, differences in the binding affinity of cadmium and calcium for sites in the plasma membrane could result in the displacement of calcium by cadmium and could account for the small elevation of the anisotropy parameter with the addition of small concentrations of cadmium in the presence of calcium.

These data provide evidence for the concept that calcium ions can effectively compete with cadmium ions for binding sites on the phospholipid head groups. A reduction in the binding of cadmium ions might result in decreased entrance of those ions into the intracellular region via active or passive transport mechanisms. Binding of cadmium ions to the plasma membrane might represent the first step in expression of cadmium-induced cyto-toxic effects providing an explanation for the diminished hepatotoxicity of cadmium in the presence of calcium.

Other studies provide an understanding of the mechanisms that may govern cadmium influx into parenchymal hepatocytes. Interactions between cadmium and zinc have been documented. Isolated rat hepatocytes, maintained in medium 60–75 min prior to exposure to 3 μM cadmium, for example, accumulated twice as much cadmium (3.8 nmol/mg protein) as zinc (2 nmol/mg protein) during a 20-hr incubation period (Failla et al., 1979). In this study, it was observed that, once accumulated by parenchymal hepatocytes, cadmium was less mobile during exit–exchange processes than was its biologically related congener zinc; moreover, cadmium influx enhanced zinc influx. Since zinc and cadmium share a common transport mechanism (Huan et al., 1980) and since zinc can effectively prevent a number of toxic manifestations of cadmium (Flick et al., 1971), zinc and perhaps other divalent cations such as calcium might compete with intracellular enzyme or metallothionein-binding sites.

Alternatively, and probably concurrently, divalent cations such as calcium might decrease the lipid fluidity of plasma membrane (Gordon et al., 1978) by binding directly to anionic sites of the lipid bilayer, probably to phospholipid polar head groups and to sialic acid residues (Schlatz and Marinetti, 1972), and triggering phase separations (Ohnishi and Ito, 1974; Trauble and Eibl, 1974; Jacobson and Papahadjopoulos, 1975; Hartmann et al., 1977) to form lipid domains that circumscribe protein carrier molecules to prevent carrier function (Ohnishi and Ito, 1974).

2. METHODOLOGY

2.1. Theory of Fluorescence Polarization for Ion–Membrane Measurements

Immediate interactions of divalent cations with plasma membranes are monitored using steady-state fluorescence polarization of fluorescent probes incorporated into plasma membranes. One or more fluorescent probes are dissolved in a solvent, which is then volatilized. Freshly isolated plasma membranes are then incubated with the probe to allow incorporation into the bilayer. Although a variety of fluorescent probes are now available, 1,6-diphenyl-1,3,5-hexatriene (DPH) is frequently used for this purpose. Therefore, it serves as the basis of reference for this discussion.

Briefly, the fluorescent probe is excited with polarized light (Figure 5).

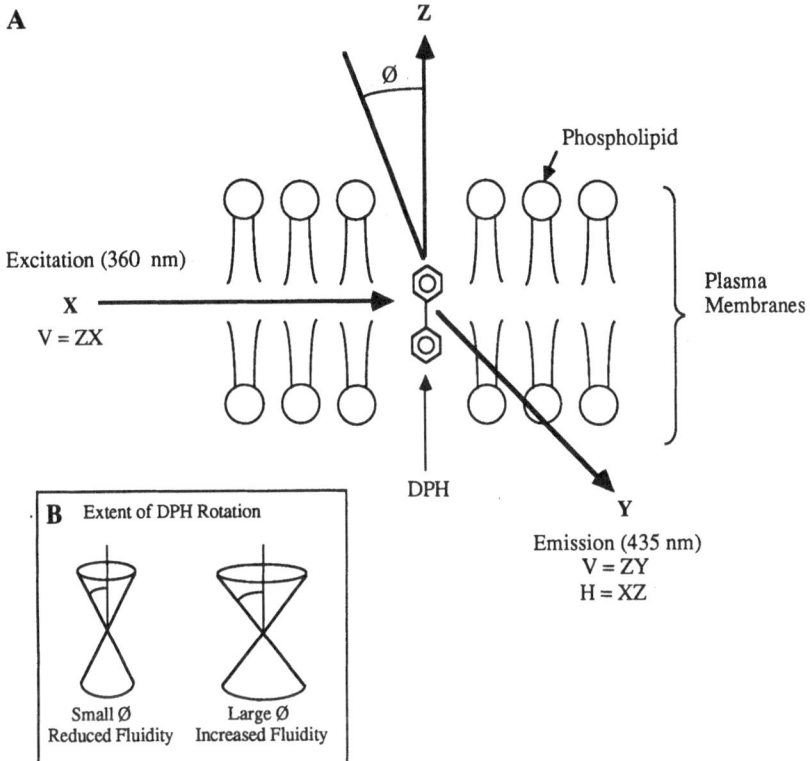

FIGURE 5. (A) Diagrammatic representation of DPH (1,6-diphenyl-1,3,5-hexatriene) molecule integrated in plasma membranes. Following excitation of DPH in the ZX plane with 360-nm light, the excited DPH molecule rotates as it emits light at 435 nm. Intensities of emitted light in the ZY and XZ planes are measured for computation of the measured anisotropy and the anisotropy parameter. The rotation of excited DPH molecules can be computed from the emission intensities in the ZY and XZ planes. (B) This method is used to determine the extent of DPH rotation. The larger the angle ϕ, the more mobile the phospholipids adjacent to the probe and the greater the membrane fluidity.

During the excited-state lifetime of the probe (11.2 nsec), light is emitted from the probe. Vertical and horizontal components of the emitted light are monitored during the excited-state lifetime of the probe to provide a measure of DPH rotation. The greater the rotation, the greater the fluidity of phospholipids adjacent to the probe. To make direct measurements of intermolecular events within plasma membranes, ions or combinations thereof are added to labeled membranes. This technique has been useful in monitoring the interactions of increasing concentrations of divalent metal cations, such as cadmium or calcium, with plasma membranes. This interaction probably represents the first step in the series of cytotoxic or nontoxic physiologic reactions at the plasma membranes of parenchymal hepatocytes.

Determination of the fluorescence polarization of DPH incorporated into membranes is based on straightforward theoretical considerations. As indicated labeled membrane suspensions are irradiated with polarized light at 360 nm with a vector orientation in the vertical (i.e., parallel) plane (Figure 5). Only DPH molecules oriented in the vertical plane become excited. The electron charge distribution on the molecule is altered; this can be envisioned as a change in the electron cloud shape from that of a cigar to that of a dumbbell. During the excited-state lifetime, the excited DPH molecule rotates, resulting in an increased angle ϕ. The DPH molecules are constantly excited with 360-nm polarized light, hence the use of the term "steady-state" fluorescent polarization. During constant excitation, the intensities of the vertical (F_{VV}) and horizontal (F_{VH}) components of the emitted light are measured with photomultiplier tubes at 435 nm. When membrane fluidity increases, a greater angle of rotation (ϕ) results. The steady-state anisotropy value, r, of the fluorescence signal can be evaluated as follows:

$$r = \frac{F_{VV} - F_{VH}\, g}{F_{VV} + 2F_{VH}\, g} \tag{1}$$

where F_{VV} and F_{VH} are the vertical and horizontal components of the fluorescence following excitation with vertically polarized light (Eftink, 1983). Measurements can be collected using a SPEX Spectrofluorometer in an L format with a 360-nm excitation (5-nm bandpass) and a 435-nm emission (20-nm bandpass). The factor g, which serves to correct for emission monochromator polarization effects, is equal to the ratio I_{HV}/I_{HH}, where I_{HV} and I_{HH} are the measured vertical and horizontal emission intensities, respectively, resulting from a horizontally polarized excitation source. The degree of fluorescence depolarization can be expressed as the fluorescence anisotropy parameter, $(r_0/r - 1)^{-1}$, which is proportional to the apparent rotational relaxation time of the probe (Brasitus and Schachter, 1980). Maximal limiting anisotropy, r_0, of DPH is obtained from suspension of DPH in glycerin at subzero temperatures to immobilize the probe; this value has been taken as 0.365 (Shinitzky and Barenholz, 1974). Calculation of the contribution of scattered light (Lakowicz, 1983) is required for these studies.

Procedures for the isolation of plasma membranes of erythrocytes of hepatocytes have been provided (Figures 6 and 7) to illustrate the relative simplicity of some available procedures. Plasma membranes were isolated from the erythrocytes of human blood using a procedure modified from Hanahan and Ekholm (1974). Erythrocytes were separated from fresh whole blood,

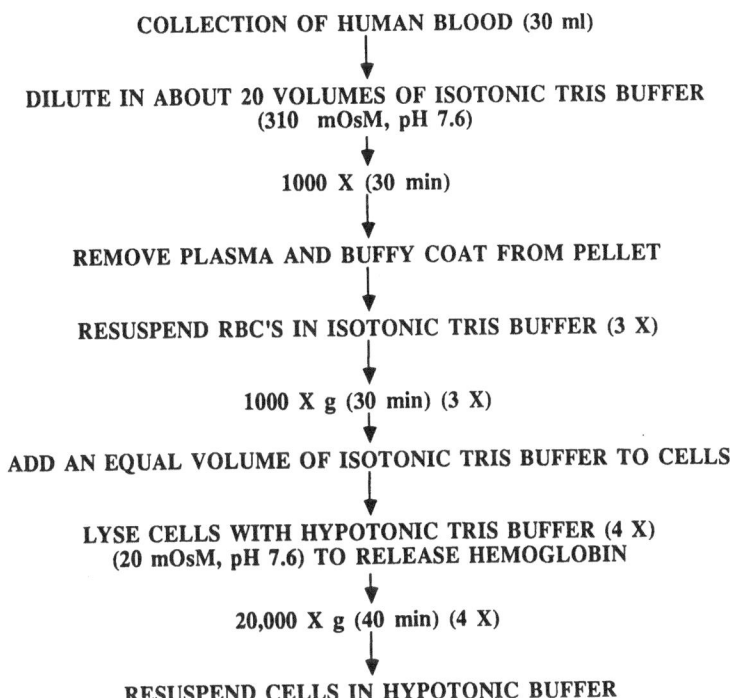

COLLECTION OF HUMAN BLOOD (30 ml)

DILUTE IN ABOUT 20 VOLUMES OF ISOTONIC TRIS BUFFER
(310 mOsM, pH 7.6)

1000 X (30 min)

REMOVE PLASMA AND BUFFY COAT FROM PELLET

RESUSPEND RBC'S IN ISOTONIC TRIS BUFFER (3 X)

1000 X g (30 min) (3 X)

ADD AN EQUAL VOLUME OF ISOTONIC TRIS BUFFER TO CELLS

LYSE CELLS WITH HYPOTONIC TRIS BUFFER (4 X)
(20 mOsM, pH 7.6) TO RELEASE HEMOGLOBIN

20,000 X g (40 min) (4 X)

RESUSPEND CELLS IN HYPOTONIC BUFFER

FIGURE 6. Isolation of erythrocyte plasma membranes modified from Hanahan and Ekholm (1974).

washed in an isotonic Tris buffer (310 mOsM, pH 7.6), and subsequently lysed followed by repeated washings in a hypotonic Tris buffer (20 mOsM, pH 7.6). Repeated lysis removed the contained hemoglobin. Plasma membranes were resuspended in an isotonic Tris-saline buffer (0.15 M NaCl, 10 mM Tris, pH 7.6) prior to incorporation of the fluorescent probe. Examination of plasma membranes prepared by this procedure revealed the purity and relative homogeneity of the sample.

A simplified procedure used to isolate plasma membranes from rat hepatocytes was modified from those developed by Ray (1970), Brunette and Till (1971), and Lesko et al. (1973). The procedure utilizes the surface-active properties of a two-polymer system of dextran and polyethylene glycol (MW 6000) to separate plasma membranes at the interface of the two polymers. Successive centrifugations provide a plasma membrane isolate of high purity (Nealon et al., 1984; Sorensen et al., 1985). Once separated, membranes were washed and resuspended in Tris buffer. Protein determination was conducted using a microburet procedure (Itzhaki and Gill, 1964). Lipid/protein ratios can be based on measurements such as those reported by Jain (1972).

In discussion of probe–membrane interactions, the purity and compo-

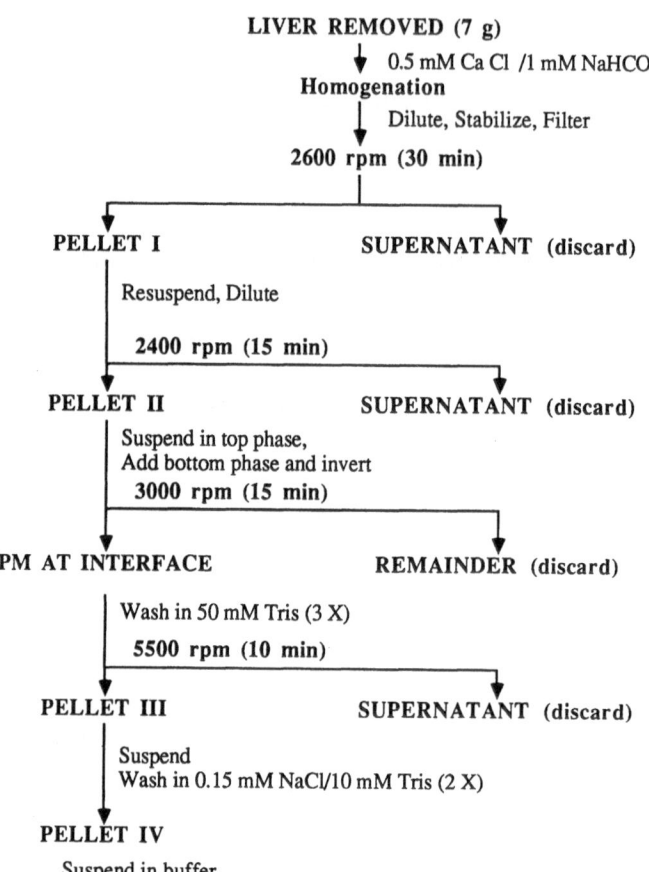

LIVER REMOVED (7 g)

0.5 mM Ca Cl /1 mM NaHCO

Homogenation

Dilute, Stabilize, Filter

2600 rpm (30 min)

PELLET I   SUPERNATANT (discard)

Resuspend, Dilute

2400 rpm (15 min)

PELLET II   SUPERNATANT (discard)

Suspend in top phase,
Add bottom phase and invert

3000 rpm (15 min)

PM AT INTERFACE   REMAINDER (discard)

Wash in 50 mM Tris (3 X)

5500 rpm (10 min)

PELLET III   SUPERNATANT (discard)

Suspend
Wash in 0.15 mM NaCl/10 mM Tris (2 X)

PELLET IV

Suspend in buffer

FIGURE 7. Hepatocyte plasma membrane isolation procedure modified from Ray (1970), Brunette and Till (1971), and Lesko *et al.* (1973).

sition of plasma membranes is of importance. Presumably, the two-polymer isolation procedure presented here yields a mixture of canalicular, contiguous, and sinusoidal hepatocyte plasma membranes from the entire cell. Results of fluidity measurements can vary in accordance with membrane composition, as previously mentioned. Membranes isolated from each of the three major areas (i.e., sinusoidal or blood facing, canalicular, or contiguous or lateral) respond differently to calcium because of differences in lipid composition (Storch *et al.*, 1983; Whetton *et al.*, 1983). Therefore, isolation procedures resulting in a greater composition of one than that present in another would result in differences in fluidity measurements.

Membranes can be loaded at 4–5°C overnight with DPH as previously described (Schachter and Shinitzky, 1977) such that the lipid/probe ratio is approximately 500 : 1. In any given experiment, mixtures of ions and buffers are combined with DPH-loaded plasma membranes and monitored at a specific temperature within about 10 min. Five or more measurements can be made on each sample monitored to verify the accuracy of first samples measured in any sequence of experiments. Matrix diagrams of concentrations can be prepared to obtain a specific concentration range when several ions are involved.

The possibility of interaction between ions and the fluorophore can be eliminated by examination of both the fluorescence intensity values and the excited-state lifetime of the fluorescent probe following addition of increasing concentrations of the specific ion of interest. Also, each of the suspensions of plasma membranes (which have not been labeled with the fluorescent probe) must be evaluated to verify the absence of scattered light contributions to F_{VV}. This eliminates the possibility that excess particulate in each isolate will increase the turbidity of the aqueous suspension of membranes such that light scattering results, as previously discussed in detail by Lakowicz (1983).

An example of the stepwise procedure involved in measurements of membrane fluidity has been published (Nealon et al., 1984). This procedure details a rapid, efficient, and inexpensive method of monitoring intermolecular membrane events (i.e., motional changes in the phospholipid molecules adjacent to the fluorescent probe). Two assumptions are necessary prerequisites. First, relatively pure membranes must be isolated from cells using one of a number of published procedures (Ray, 1970; Brunette and Till, 1971; Lesko et al., 1973; Hanahan and Ekholm 1974). Second, polarizers must be properly aligned (Jameson, 1984). As a final check, calculate the total fluorescence intensity, F, for each sample as follows:

$$F = (I_{VV} - I_{SV}) + 2 (I_{VH} - I_{SH})g \qquad (2)$$

Unidirectional changes in the total fluorescent intensity upon the addition of an ion to membranes could indicate changes in the excited-state lifetime of DPH, leading to erroneous fluorescence anisotropy values. These changes appear not to be problematic (Nealon et al., 1984; Sorensen et al., 1985).

This technique provides a sensitive means of direct detection of potential toxic interactions at the plasma membrane, well in advance of manifestations of the functional impairment of cellular integrity (i.e., enzyme leakage, morphological changes, and functional changes).

2.2. Probe–Membrane Interactions

The fluorescent probe DPH has a rodlike configuration and partitions in the hydrocarbon core of the membrane since it is hydrophobic in character.

Therefore, it is distant from ions present in the aqueous phase (Figure 5A). Probe molecules failing to partition into the membrane would not contribute to the observed fluorescence signal by virtue of the low quantum yield in water (Shinitzky and Barenholz, 1974). Moreover, no apparent rationale exists to support the possibility for formation of a complex between an aromatic molecule such as DPH and ions. Moreover, quenching of the probe has not been shown to occur because the excited-state lifetime for DPH (11.2 nsec) was not changed by the presence of ions, nor was the total fluorescence intensity of DPH in membranes altered by addition of these ions (Sorensen *et al.*, 1985; Nealon *et al.*, 1984). Whether this lack of interaction exists between every probe molecule and every ion has yet to be determined.

Different fluorescent probes are available to monitor specific membrane regions. Diphenylhexatriene (DPH) is the most widely used because it is a small, uncharged, rod-shaped molecule that distributes throughout hydrophobic regions deep within membrane, as mentioned (Shinitzky and Barenholz, 1974). Distribution within the membrane core and an 11.2-nsec fluorescence lifetime make this probe ideal for indirect measurement of membrane motion. Probe rotation provides a measure of membrane viscosity (dynamic component), when probe wobble reflects changes in lipid order (static component) (Kinosita *et al.*, 1981; Van Blitterswijk *et al.*, 1981).

Use of alternate fluorescence probes can provide conflicting results, which suggests that a battery of probes might be better suited for studies attempting to relate fluorescence polarization changes to ion flux. Membrane surface changes can be measured using TMA-DPA (1- [4-(trimethylammonium) phenyl] -6-phenyl-1,3,5-hexatriene) (Harris *et al.*, 1984; Harris and Bruno, 1985) or 1-AP (1-aminopyrine) (Harris and Schroeder, 1981). Ashley and Brammer (1984), for example, used steady-state fluorescence polarization of two fluorescent probes in synaptosomal lipids in an effort to evaluate the disposition of synaptosomal membrane lipids during the process of neurotransmitter release and during membrane interactions with calcium. Calcium promoted phase transitions in negatively charged phospholipids (Cullis and De Kruijff, 1979) and reduced the mobility of one fluorescent probe used in the Ashley and Brammer study (i.e., dns-DPPE, dansylated dipalmitoylphosphatidylethanolamine). Fluorescence polarization measurements of this large asymmetric probe (i.e., dns-DPPE) required evaluation of rotation around three major axes (Weber, 1971) to derive an average rate of rotation (Faucon and Lussan, 1973). Mobility of dns-DPPE increased with thermal increase; however, the presence of 2 mM calcium progressively decreased probe mobility. Therefore, the Ashley and Brammer (1984) study suggested that the association of calcium with synaptosomal lipids resulted in aggregation of acidic phospholipids, causing phase separations that may underlie a channel-forming mechanism for neurotransmitter release. Physiologically, this explanation is intuitively satisfying because fluidity changes adjacent to carriers would almost certainly precede the opening or closing of intrinsic protein channels.

Whereas the dns-DPPE probe showed calcium-induced changes in fluidity, the use of 1,6-diphenyl-1,3,5-hexatriene (DPH) as a fluorescent probe for synaptosomal membrane preparations did not result in a change in lipid order after depolarization of synaptosomes with 150 μM veratridine, even in the presence of 2 mM calcium (Ashley and Brammer, 1984). This unexpected result might be due to the distribution of the probe into additional synaptosomal membrane systems (Grunberger *et al.*, 1982) or to averaging effects of DPH fluorescence polarization measurements—either of which could swamp localized changes in phospholipids adjacent to calcium channels. Nonetheless, use of more than one probe could facilitate a better understanding of the effects of perturbating agents, especially when the first probes fail to show anticipated results.

Differences in the ratio of protein to lipid (as well as differences in the lipids present) apparently play a role in slight differences for membranes derived from different sources. Plasma membranes from rat hepatocytes are reported to be comprised of 85% protein and 15% lipid, compared to those from erythrocytes, which are reported to have approximately equal portions of each (Jain, 1972). Other studies indicate that the protein/lipid ratio is approximately the same for plasma membranes derived from rat hepatocytes (Harrison and Lunt, 1980). Published differences in anisotropy parameters for different plasma membranes could therefore be a consequence. of the difference in lipid/protein ratios for membranes derived from different sources. A reasonable explanation for higher anisotropy parameter in erythrocyte plasma membranes might be the higher cholesterol/phospholipid ratio for the erythrocyte compared to that of rat hepatocytes (i.e., 0.90 : 0.95 for the erythrocyte and 0.25 : 0.45 for the rat hepatocytes) (Nealon *et al.*, 1984; Sorensen *et al.*, 1985). Regardless of the exact mechanism for the differences, the plasma membranes from erythrocytes have consistently higher anisotropy values (i.e., were less fluid).

In addition to differences in the procedures used to isolate parenchymal hepatocytes, instrumental differences can result in variations in the absolute values of anisotropy parameters reported in various studies. Livingstone and Schachter (1980), for example, reported anisotropy parameter values of 1.480 (± 0.020, S.E.) for hepatocyte membranes derived from isolated rat hepatocytes using a discontinuous sucrose density gradient and 1.680 (± 0.010, S.E.) for membranes that had been exposed to 0 or 4 mM calcium for 2 hr at 37°C. These values contrasted with values of 1.397 (± 0.006, S.E.) and 1.483 (± 0.008, S.E.) for exposures to the respective concentrations of calcium (Nealon *et al.*, 1984; Sorensen *et al.*, 1985). Storch and her colleagues (1983), moreover, showed that the absolute anisotropy parameter for various subfractions of rat hepatocyte plasma membranes also depended on the preparation procedure for the subfraction. Use of the Wisher and Evans (Harrison and Lunt, 1980) isolation procedure resulted in absolute values of 4.280 ($\pm$ 0.660, S.E.) for the canalicular subfraction; whereas use of the Inoue et al. (1982) procedure gave anisotropy values of 2.480 ($\pm$ 0.120, S.E.) for this subfraction.

2.3. Probe–Ion Interactions

The possibility of direct interaction between the probe and the toxic agent must be considered to eliminate false positives. If such an interaction existed between the DPH molecule and the specific ion under test, changes in membrane fluidity would be artifactual. Because of its hydrophobic character, the DPH probe is located in the hydrocarbon core of the membrane and is therefore distant from ions that are present in the aqueous phases on either side of the membrane. In the event that some of the probe molecules fail to partition into the membrane, these would remain in solution and not contribute to the observed fluorescence signal because they do not capture sufficient energy to fluoresce (Shinitzky and Barenholz, 1974). In addition, no apparent rationale would support the possibility of the existence of a complex between an aromatic molecule (such as DPH) and an ion, if ions do penetrate membranes in a particular experiment. Measurements of the excited-state lifetime can be made to preclude the possibility of a change in the excited-state lifetime of DPH in membranes as a result of the addition of ions to the plasma membrane. Both values were found to be 11.2 nsec (Nealon et al., 1984; Sorensen et al., 1985). The observations were consistent with those of others: no changes occur in the excited-state lifetime of membrane-associated DPH when cations such as calcium are added (Livingstone and Schachter, 1980). In addition, the presence of cadmium did not change the total fluorescence intensity of DPH in the membranes. Therefore, these two evaluations of fluorescence intensity and excited-state lifetime can be used to eliminate the possibility of fluorescence quenching, which would suggest the existence of interactions between the probe and the ions being added.

3. CURRENT ADVANCEMENTS IN THE MEASUREMENT OF ION–MEMBRANE INTERACTIONS USING FLUORESCENCE POLARIZATION

3.1. Natural Membranes

Changes in fluorescence polarization of DPH can be measured in whole brain synaptic plasma membranes (Harris et al., 1984; Harris and Bruno,

1985). Sodium and calcium uptake are inhibited by membrane disordering compounds such as ethanol, chloroform, pentobarbital, and others. All compounds inhibit sodium influx in a fashion that is linearly correlated with the fluorescence polarization of DPH and TMA-DPH ($r = 0.90$ and 0.66, respectively). Inhibition of calcium uptake, however, is not correlated with changes in fluorescence polarization using either membrane probe ($r = 0.36$ and 0.25, respectively). Although these results suggest that inhibition of calcium flux does not parallel fluorescence polarization changes, as does inhibition of sodium flux, the use of fluorescent polarization of whole brain preparations averages differences in sodium and calcium flux which are known to occur in different brain regions (Elrod and Leslie, 1980; Harris and Stokes, 1982; Harris and Bruno, 1985). Moreover, the use of DPH or TMA-DPH fluorescence polarization equalizes effects over the entire plasma membrane and reduces the possibility of detecting minute fluidity changes hypothesized to occur in critical regions near ion channels—particularly calcium channels.

The controversial nature of these studies is evidenced by other research involving synaptic plasma membranes. Following administration of 5–6% (v/v) ethanol for 7 days, seizure-prone and seizure-resistant mice were assessed to determine physical properties of whole brain preparations of synaptic plasma membranes (Harris *et al.*, 1984). Electrical and pentylenetetrazole methods for seizure induction were used to evaluate central nervous system excitability. Excitability of the central nervous system was the same in both genetic strains. Despite the similarity of neuronal excitability and definitive elevations in withdrawal seizures, no differences were observed in fluorescence polarization of DPH or TMA-DPH. These probes partition equally over the hydrophobic region of lipid bilayers and are used to measure the stability of the lower and upper regions of phospholipid acyl chains, respectively. Again, differences in brain regions were not considered in this study, nor were measurements of alternate probes, such as dns-DPPE, which could detect differences in the fluidity of phospholipids adjacent to ionic channels.

Isolated plasma membranes from hepatocytes also have been evaluated for ion-induced effects. Livingstone and Schachter (1980) compared effects of 4-mM concentrations of five divalent cations on the DPH anisotropy parameters of isolated hepatocyte plasma membranes. Calcium increased the anisotropy parameter (i.e., stabilized the plasma membranes) by 13.5% ($p < 0.001$) and strontium increased the value by 6.1% ($p < 0.05$), but barium, magnesium, and zinc had no significant effect. A brief incubation in EDTA reversed strontium-induced stabilization but did not reverse the calcium-induced stabilization. Lipid composition of hepatocyte plasma membranes was altered by incubation of membranes in 4 mM calcium and was believed due to modulation of membrane-bound enzymes. Calcium also bound directly to plasma membranes with a half-time of 10–15 min at 37°C. It was proportional to calcium concentrations in the 0–4-mM range and was readily reversed on addition of excess EDTA. Therefore, calcium interacted both directly and indirectly with plasma membranes in the Livingstone and Schachter (1980) study. These results agree with those of Ohnishi and Ito (1974), who observed that phase separation in phosphatidylserine–phosphatidylcholine membranes decreased in the order calcium > barium > strontium. As with the Livingstone and Schachter (1980) study, magnesium did not affect phase separation in the Ohnishi and Ito (1974) membranes. Explanation of the difference in results noted in the two studies for barium has not been elucidated, although it was suggested that differences in ionic radius of cations may be responsible for the selectivity.

In conclusion, membrane perturbating agents such as cadmium (Sorensen *et al.*, 1985; Nealon *et al.*, 1984) and calcium (Livingstone and Schachter, 1980) interact with membranes in a concentration-dependent way. Moreover, these agents may interfere with either lipids or proteins or both (Livingstone and Schachter, 1980).

3.2. Reconstituted Membranes

The use of completely or partially reconstituted plasma membranes is useful in determining the extent of interaction between various membrane components. Livingstone and Schachter (1980), for example, prepared liposomes of membrane lipids to determine the effects of increasing concentrations of calcium on changes in the anisotropy parameter. Although calcium increased the anisotropy parameter in both natural membranes and liposomes, the magnitude and pattern of the responses differed considerably. Values for the natural membranes leveled off between 2.0 and 4.0 mM calcium, whereas values for liposomes rose linearly in the same concentration range. Moreover, the anisotropy values of natural membranes were two to four times that of liposomes in the range 0.5–2.0 mM calcium. A modulatory role of membrane-bound enzymes is therefore probably involved in altering lipid composition in plasma membranes exposed to increasing concentrations of calcium.

Alternatively, and probably concomitantly, the presence of membrane proteins is a recognized enhancer of the lipid order parameter of reconstituted membranes. Shinitzky (1984) reviewed the effect of protein content on lipid dynamics. Because of low compressibility (Brandts *et al.*, 1970; Li *et al.*, 1976) and large volume relative to that of phospholipids (i.e., about 50–100 times that of phospholipids), membrane proteins represent rigid domains in a fluid lipid matrix. Increases in protein content have been shown to stabilize fluid lipid bilayers (Shinitzky and Inbar, 1976; Cooper, 1977; Fraley *et al.*, 1978; Herreman *et al.*, 1981). Proper physiological membrane function apparently is controlled within a narrow range due to the presence of cholesterol and proteins, which stabilize membranes and increase the lipid order parameter (Shinitzky and Inbar, 1976).

To evaluate the effects of divalent and monovalent cations on lipid bilayers, Trauble and Eibl (1974) investigated phase transition temperatures with the addition of divalent cations (calcium and magnesium) and monovalent cations (lithium, sodium, and potassium) to suspensions of membranes. The monovalent cations fluidized bilayers whereas divalent cations stabilized bilayers. Similar results are obtained with nerve excitation and sensory transduction where cation-induced structural changes in biomembranes are evoked. Axons, for example, are stabilized into a resting state by perfusion of the outer surface with divalent cations but are induced into an excitatory state by perfusion with monovalent cations (Tasaki, 1968).

4. CRITICAL EVALUATION OF THE SIGNIFICANCE OF ION–MEMBRANE MEASUREMENTS

4.1. Advantages of Fluorescence Polarization for Evaluation of Ion–Membrane Interactions

Although it is difficult to demonstrate a direct relation between fluidity measurements and the toxic response of cells, several published studies correlate fluidity changes with alterations of physiological significance in the toxicity response. In Duchenne-type muscular dystrophy, for example, significant changes in membrane viscosity (Sha'afi et al., 1975) were associated with changes in membrane-bound enzymes (Rodan et al., 1974). Storch and Schachter (1984), moreover, have shown that a dietary starve–refeed regimen can alter plasma membrane fatty acid composition (i.e., increasing the double-bond index and decreasing the proportion of saturated fatty acids). These changes were observed with concurrent increases in membrane fluidity and the specific activity of $(Na^+ + K^+)$-dependent adenosine triphosphatase (Storch and Schachter, 1984). Changes in membrane fluidity could therefore ultimately lead to secondary or tertiary effects, which in turn could result in changes in membrane fluidity.

Applications of the fluorescence polarization method for evaluation of fluidity changes, which eventually result in cellular injury or death, are numerous. Membrane fluidity changes are believed to occur well within 1 sec and probably represent the earliest of all changes following interaction of ions with cells. Moreover, the presence of measurable changes in the anisotropy parameter (and therefore membrane fluidity) represents a direct interaction between ions and the plasma membrane. Subsequent membrane changes can result in loss of membrane integrity (indicated by enzyme release), in reduced synthesis of metabolic waste products (such as urea), and in the detrimental morphological changes (Sorensen and Acosta, 1983, 1984a,b,c, 1985; Sorensen et al., 1984; Sorensen, 1988). To illustrate the correspondence that exists between fluidity changes and cytotoxic effects, Table I provides results of a morphometric analysis of cytotoxic changes that result 30 min (i.e., column 4) and 60 min (column 5) following incubation of cells in the same concentrations of cadmium and calcium under the same conditions as those used for the fluidity measurements. These data show that parallel effects were observed for the measurements. Although it is presently impossible to prove that fluidity changes lead to cellular injury, results of fluorescence polarization measurements do parallel those currently available using other parameters.

4.2. Limitations of Fluorescence Polarization for Measurement of Ion–Membrane Interactions

Theoretical considerations for fluorescence polarization are straightforward and the technique is simple; however, a number of precautions should be considered for meaningful data interpretation. Sample preparation techniques, culture procedures, and membrane source must be evaluated and standardized to avoid complications. Storch and her colleagues (Storch et al., 1983), for example, showed that the anisotropy values for various subfractions of rat hepatocyte plasma membranes depended on the preparation procedure for the subfraction. The use of rate zonal-centrifugation isolation procedures (Harrison and Lunt, 1980) resulted in absolute values significantly greater than those obtained using nitrogen cavitation and calcium precipitation (Inoue et al., 1982).

A number of other variables can affect membrane fluidity and could interfere with the interpretation of ion–membrane interactions. Examples include differences in temperature, pH, atmospheric pressure, and media composition (i.e., calcium concentration, lipid/protein ratio, and cholesterol/phospholipid ratio), as well as others (Shinitzky and Yuli, 1982). For example, Thompson and his colleagues (Fukushima et al., 1976; Thompson and Nozawa, 1977; Martin and Thompson, 1978) have demonstrated the marked change in phospholipid composition of membranes and lipid fluidity when growth temperatures were varied for the unicellular, ciliated protozoan Tetrahymena pyriformis. Therefore, measurement of membrane fluidity of cells incubated in the same media could result in different values than those published because of differences in isolation and incubation procedures. Because of the dependence of fluidity measurements upon calcium concentrations, lipid/protein ratios, and phospholipid/cholesterol ratios, the same cell type (incubated or grown in different media) could produce markedly different absolute and relative anisotropy values. From these considerations, one must observe strict adherence to consistency in experimental conditions prior to measurement of membrane fluidity. Because differences in anisotropy parameters are minute even with the most pronounced ionic challenge, procedural consistency increases the chance of obtaining meaningful data that can be statistically analyzed.

4.3. Substantiation of the Fluorescence Polarization Measurements of Ion–Membrane Interactions

The results of fluorescence polarization data can be compared with those of other procedures such as phase transition temperature measurements. For example, fluorescence polarization studies indicated that cadmium exerts about the same effect on anisotropy parameters at concentrations 10 times lower than those used for calcium. Using phospholipid bilayer vesicles, Bevan and his colleagues (1983) found that a 0.1-mM cadmium concentration elevates the phase transition temperatures (T_m) of the artificial lipid vesicles to the same degree as 1.0 mM calcium. The elevation of T_m is believed to represent chelation of two phosphatidylserine or phosphatidic acid head groups and subsequent defluidization of the membrane. Feigenson (1982), moreover, found that cadmium had a higher binding affinity for phosphatidic acid than did calcium; therefore, cadmium had a greater ability to induce the gel phase (or to reduce the fluidity of plasma membranes) in this model membrane system than did calcium. Lis *et al.* (1980), however, observed that no difference existed in the adsorption of these two divalent cations with respect to binding to phosphatidylcholine bilayers.

5. CONCLUDING REMARKS

In conclusion, fluorescence polarization studies provide a valuable tool in the assessment of the interaction between ions and plasma membranes. Published research in this area involves exposure of isolated membranes to one ion (Livingstone and Schachter, 1980; Storch *et al.*, 1983) or to more than one ion simultaneously (Nealon *et al.*, 1984; Sorensen *et al.*, 1985). From these studies, as well as others, ratios of ions appear to be important, as does ionic charge, ionic radius, and binding affinity to the components of the plasma membrane. Much research remains to elucidate interactions of a complex mixture of extracellular ions with plasma membranes. Moreover, ionic interactions with the internal phospholipid bilayer might prove as important as those with the external phospholipid bilayer.

Therefore, with the proper understanding of the potential hazards that might arise when conducting fluorescence polarization measurements, useful information has been derived from evaluations of fluidity changes in the plasma membranes. This information can lead to the evaluation of direct intermolecular changes, which occur as toxicants interact with the plasma membrane, the first barrier in cellular defense.

6. REFERENCES

Alberts, B., Bray, D., Lewis, J., Raff, M., Roberts, K., and Watson, J. D., 1983, *Molecular Biology of the Cell*, pp. 255–317, Garland Publishing, New York.

Alterbach, C., and Seelig, J., 1984, Ca^{2+} binding to phosphatidylcholine bilayers as studied by deuterium magnetic resonance. Evidence for the formation of a Ca^{2+} complex with two phospholipid molecules, *Biochemistry* **23** : 3913–3920.

Ashley, R. H., and Brammer, M. J., 1984, A fluorescence polarization study of calcium and phase behaviour in synaptosomal lipids, *Biochim. Biophys. Acta* **769** : 363–369.

Bangham, A. D., 1975, Models of cell membranes, in *Cell Membranes: Biochemistry, Cell Biology and Pathology* (G. Weissmann and R. Claiborne, eds.), pp. 24–34, Hospital Practice, New York.

Bevan, D. R., Worrell, W. J., and Barfield, K. D., 1983, The interaction of Ca^{2+}, Mg^{2+}, Zn^{2+}, Cd^{2+}, and Hg^{2+} with phospholipid bilayer vesicles, *Colloids and Surfaces* **6** : 365–376.

Boyer, J. L., and Reno, D., 1975, Properties of (Na^+ and K^+)-activated ATPase in rat liver plasma membranes enriched with bile canaliculi, *Biochim. Biophys. Acta* **401** : 59–72.

Blitzer, B. L., and Boyer, J. L., 1978, Cytochemical localization of Na^+, K^+-ATPase in the rat hepatocyte, *J. Clin. Invest.* **62** : 1104–1108.

Brandts, J. F., Oliveira, R. J., and Westort, C., 1970, Thermodynamics of protein denaturation. Effect of pressure on the denaturation of ribonuclease A, *Biochemistry* **9** : 1039–1047.

Brasitus, T. A., and Schachter, D., 1980, Lipid dynamics and lipid–protein interactions in rat enterocyte basolateral and microvillus membranes, *Biochemistry* **19** : 2763–2769.

Brétcher, M. S., 1973, Membrane structure: Some general principles, *Science* **181** : 622–629.

Brunette, D. M., and Till, J. E., 1971, A rapid method for the isolation of L-cell surface membranes using an aqueous two-phase polymer system, *J. Membr. Biol.* **5** : 215–224.

Casini, A. F., and Farber, J. L., 1981, Dependence of the carbon-tetrachloride-induced death of cultured hepatocytes on the extracellular calcium concentration, *Am. J. Pathol.* **105** : 138–148.

Chapman, D., 1975, Lipid dynamics in cell membranes, in *Cell Membranes: Biochemistry, Cell Biology and Pathology* (G. Weissman and R. Claiborne, eds.), pp. 13–22, Hospital Practice, New York.

Chenery, R., George M., and Krishna, G., 1981, The effect of ionophore A23187 and calcium on carbon tetrachloride-induced toxicity in cultured rat hepatocytes, *Toxicol. Appl. Pharmacol.* **60** : 241–252.

Cooper, R. A., 1977, Abnormalities of cell-membrane fluidity in the pathogenesis of disease, *N. Engl. J. Med.* **297** : 371–377.

CRC Handbook of Chemistry and Physics, Ed. 62, 1981, CRC Press, Boca Raton, FL.

Cullis, P. R., and De Kruijff, B., 1979, Lipid polymorphism and the functional roles of lipids in biological membranes, *Biochim. Biophys. Acta* **559** : 399–420.

Dipple, I., and Houslay, M. D., 1978, The activity of glucagon-stimulated adenylate cyclase from rat liver plasma membranes is modulated by the fluidity of its lipid environment, *Biochem J.* **174** : 179–190.

Edmondson, J. W., and Bang, N. U., 1981, Deleterious effects of calcium deprivation on freshly isolated hepatocytes, *Am. J. Physiol.* **241** : C3–C8.

Eftink, M., 1983, Quenching-resolved emission anisotropy studies with single and multitryptophan-containing proteins, *Biophys. J.* **43** : 323–326.

Elrod, S. V., and Leslie, S. W., 1980, Acute and chronic effects of barbiturates on depolarization-induced calcium influx into synaptosomes from rat brain regions, *J. Pharmacol. Exp. Ther.* **212** : 131–136.

Failla, M. L., Cousins, R. J., and Mascenik, M. J., 1979, Cadmium accumulation and metabolism by rat liver parenchymal cells in primary monolayer culture, *Biochim. Biophys. Acta* **583** : 63–72.

Farber, J. L., and El-Mofty, S. K., 1975, The biochemical pathology of liver cell necrosis, *Am. J. Pathol.* **81** : 237–250.

Farriss, M. W., Olafsdottir, K., and Reed, D. J., 1984, Extracellular calcium protects isolated hepatocytes from injury, *Biochem. Biophys. Res. Commun.* **121** : 102–110.

Faucon, J., and Lussan, C., 1973, Aliphatic chain transitions of phospholipid vesicles and phospholipid dispersions determined by polarization of fluorescence, *Biochim. Biophys. Acta* **307** : 459–466.

Feigenson, G. W., 1982, Fluorescence quenching in model membranes, *J. Biophys. Soc.* **37** : 165.

Flick, D. F., Kraybill, H. F., and Dimitroff, J. M., 1971, Toxic effects of cadmium: A review, *Environ. Res.* **4** : 71–91.

Fraley, R. T., Jameson, D. M., and Kaplan, S., 1978, The use of the fluorescent probe α-parinaric acid to determine the physical state of the intracytoplasmic membranes of the photosynthetic bacterium, *Rhodopseudomonas sphaeroides*, *Biochim. Biophys. Acta* **511** : 52–69.

Fukushima, H., Martin, C. E., Iida, H., Kitajima, Y., Thomson, G. A., Jr., and Nozawa, Y., 1976, Changes in membrane lipid composition during temperature adaptation by a thermotolerant strain of *Tetrahymena pyriformis*, *Biochim. Biophys. Acta* **431** : 165–179.

Giraud, F., Claret, M., Bruckdor, K. R., and Chailley, B., 1981, The effects of membrane lipid order and cholesterol on the internal and external cationic sites of the Na^+-K^+ pump in erythrocytes, *Biochim. Biophys. Acta* **647** : 249–258.

Gordon, L. M., Sauerheber, R. D., and Esgate, J. A., 1978, Spin label studies on rat liver and heart plasma membranes: Effects of temperature, calcium, and lanthanum on membrane fluidity, *J. Supramol. Struct.* **9** : 229–310.

Grunberger, D., Haimovitz, R., and Shinitzky, M., 1982, Resolution of plasma membrane lipid fluidity in intact cells labelled with diphenylhexatriene, *Biochim. Biophys. Acta* **688** : 764–774.

Hanahan, D. J., and Ekholm, J. E., 1974, The preparation of red cell ghosts (membranes), in *Methods in Enzymology* (S. Fleischer and L. Packer, eds.), Vol. 31, pp. 168–172, Academic Press, New York.

Harris, R. A., and Bruno, P., 1985, Membrane disordering by anesthetic drugs: Relationship to synaptosomal sodium and calcium fluxes, *J. Neurochem.* **44** : 1274–1281.

Harris, R. A., and Schroeder, F., 1981, Effects of ethanol and related drugs on the physical and functional properties of brain membranes, *Curr. Alcohol.* **8** : 461–468.

Harris, R. A., and Stokes, J. A., 1982, Effects of a sedative and a convulsant barbiturate on synaptosomal calcium transport, *Brain Res.* **242** : 157–163.

Harris, R. A., Baxter, D. M., Mitchell, M. A., and Hitzemann, R. J., 1984, Physical properties and lipid composition of brain membranes from ethanol tolerant-dependent mice, *Mol. Pharmacol.* **25** : 401–409.

Harrison, R., and Lunt, G. G., 1980, Membrane components, in *Biological Membranes: Their Structure and Function*, pp. 62–101, Wiley, New York.

Hartmann, W., Galla, H. J., and Sackman, E., 1977, Direct evidence of charge-induced lipid domain structure in model membranes, *FEBS Lett.* **78** : 169–172.

Hepp, K. D., Edel, R., and Wieland, O., 1970, Hormone action on liver adenyl cyclase activity, *Eur. J. Biochem.* **17** : 171–177.

Herreman, W., Van Tornout, P., Van Cauwelaert, F. H., and Hanssens, I., 1981, Interaction of α-lactalbumin with dimyristoyl phosphatidylcholine vesicles. II. A fluorescence polarization study, *Biochim. Biophys. Acta* **640** : 419–429.

Houslay, M. D., Hesketh, T. R., Smith, G. A., Warren, G. B., and Metcalfe, J. C., 1976, The lipid environment of the glucagon receptor regulates adenylate cyclase activity, *Biochim. Biophys. Acta* **436** : 495–504.

Huan, P. C., Smith, B., Bohdan, P., and Corrigan, A., 1980, Effect of zinc on cadmium influx and toxicity in cultured CHO cells, *Biol. Trace Element Res.* **2** : 211–220.

Inoue, M., Kinne, R., Tran, T., and Arias, I. M., 1982, Rat liver canalicular plasma membrane vesicles: Isolation and topological characterization, *Fed. Proc.* **41** : 916.

Itzhaki, R. F., and Gill, D. M., 1964, A micro-buret method for estimating proteins, *Anal. Biochem.* **9** : 101–410.

Jacobson, K., and Papahadjopoulos, D., 1975, Phase transitions and phase separations in phospholipid membranes induced by changes in temperature, pH, and concentration of bivalent cations, *Biochemistry* **14** : 152–161.

Jain, M. K., 1972, *The Biomolecular Lipid Membrane: A System*, pp. 383–409, Van Nostrand Reinhold, New York.

Jameson, D. M., 1984, Fluorescein hapten: An immunological probe, in *Fluorescence: Principles, Methodologies, and Applications* (E. W. Voss, Jr., ed.), pp. 23–48, CRC Press, Boca Raton, FL.

Judah, J. D., Ahmed, K., and McLean, A. E. M., 1964, Possible role of ion shifts in liver injury, in *Ciba Foundation Symposium on Cellular Injury* (A. V. S. de Reuck and J. Knight, eds.), pp. 187–205, Churchill, London.

Keefe, E. B., Scharschmidt, B. F., Blankenship, N. M., and Okner, R. K., 1979, Studies of relationships among bile flow, liver plasma membrane Na^+-K^+-ATPase, and membrane microviscosity in the rat. *J. Clin. Invest.* **64** : 1590–1598.

Kimelberg, H. K., 1977, The influence of membrane fluidity on the activity of membrane-bound enzymes, in *Dynamic Aspects of Cell Surface Organization. Cell Surface Reviews* (G. Poste and G. L. Nicolson, eds.), Vol. 3, pp. 205–293, Elsevier, Amsterdam.

Kinosita, K., Jr., Kawato, S., Ikegami, A., Yoshida, S., and Orii, Y., 1981, The effect of cytochome oxidase on lipid chain dynamics: A nanosecond depolarization study, *Biochim. Biophys. Acta* **647** : 7–17.

Kolb, H. A., and Adam, G., 1976, Regulation of ion permeabilities of isolated rat liver cells by external calcium concentration and temperature, *J. Membr. Biol.* **26** : 121–151.

Kornberg, R. D., and McConnell, H. M., 1971, Lateral diffusion of phospholipids in a vesicle membrane, *Proc. Natl. Acad. Sci. U.S.A.* **68** : 2564–2568.

Kasne, S., Eisenmann, G., and Szabo, G., 1971, Freezing and melting of lipid bilayers and the mode of action of nonactin, valinomycin and gramicidin, *Science* **174** : 412–415.

Lakowicz, J. R., 1983, *Principles of Fluorescence Spectroscopy*, pp. 111–153, Plenum Press, New York.

Latham, P. S., and Kashgarian, M., 1979, The ultrastructural localization of transport ATPase in the rat liver at nonbile canalicular plasma membranes, *Gastroenterology* **76** : 988–996.

Lesko, L., Donlon, M., Marinetti, G. V., and Hare, J. D., 1973, A rapid method for the isolation of rat liver plasma membranes using an aqueous two-phase polymer system, *Biochim. Biophys. Acta* **311** : 173–179.

Li, T. M., Hook, J. W., III, Drickamer, H. G., and Weber, G., 1976, Plurality of pressure-denatured forms in chymotrypsinogen and lysozyme, *Biochemistry* **15** : 5571–5580.

Lis, L. J., Lis, W. T., Parsegian, V. A., and Rand, R. P., 1980, Adsorption of divalent cations to a variety of phosphatidylcholine bilayers, *Biochemistry* **20** : 1771–1778.

Livingstone, C. J., and Schachter, D., 1980, Calcium modulates the lipid dynamics of rat hepatocyte plasma membranes by direct and indirect mechanisms, *Biochemistry* **19** : 4823–4827.

Lowe, P. J., and Coleman, R., 1981, Membrane fluidity and bile salt damage, *Biochim. Biophys. Acta* **640** : 55–65.

Martin, C. E., and Thompson, G. A., Jr., 1978, Use of fluorescence polarization to monitor intracellular membrane changes during temperature acclimation. Correlation with lipid compositional and ultrastructural changes, *Biochemistry* **17** : 3581–3591.

Nagle, J. F., 1973a, Theory of biomembrane phase transitions, *J. Chem. Phys.* **58** : 252–264.

Nagle, J. F., 1973b, Lipid bilayer phase transition: Density measurements and theory, *Proc. Natl. Acad. Sci. U.S.A.* **70** : 3443–3444.

Nealon, D. G., Sorensen, E. M. B., and Acosta, D., 1984, A fluorescence polarization procedure for the evaluation of the effects of cadmium and calcium on plasma membrane fluidity, *J. Tissue Culture Methods* **9** : 11–17.

Ohnishi, S., and Ito, T., 1974, Calcium-induced phase separations in phosphatidylserine–phosphatidylcholine membranes, *Biochemistry* **13** : 881–887.

Oldfield, E., and Chapman, D., 1972, Dynamics of lipids in membranes: Heterogeneity and the role of cholesterol, *FEBS Lett.* **23** : 285–297.

Papahadjopoulos, D., 1968, Surface properties of acidic phospholipids: Interaction of monolayers and hydrated liquid crystals with uni- and bi-valent metal ions, *Biochim. Biophys. Acta* **163** : 204–254.

Quinn, A. J., 1976, *The Molecular Biology of Cell Membranes* (A. J. Quinn, ed.), pp. 47–75, University Park Press, Baltimore.

Quinn, A. J., and Chapman, D., 1980, The dynamics of membrane structure, *CRC Crit. Rev. Biochem.* **8** : 1–117.

Ray, T. K., 1970, A modified method for the isolation of the plasma membrane from rat liver, *Biochim. Biophys. Acta* **196** : 1–9.

Reynolds, E. S., 1963, Liver parenchymal cell injury. I. Initial alterations of the cell following poisoning with carbon tetrachloride, *J. Cell Biol.* **19** : 139–157.

Rodan, S. B., Hintz, R. L., Sha'afi, R. I., and Rodan, G. A., 1974, The activity of membrane bound enzymes in muscular dystrophic chicks, *Nature* **252** : 589–591.

Schachter, D., and Shinitzky, M., 1977, Fluorescence polarization studies of rat intestinal microvillus membranes, *J. Clin. Invest.* **59** : 536–548.

Schanne, F. A. X., Kane, A. B., Young, E. E., and Farber, J. L., 1979, Calcium dependence of toxic cell death: A final common pathway, *Science* **206** : 700–702.

Schlatz, L., and Marinetti, G. V., 1972, Calcium binding to the rat liver plasma membrane, *Biochim. Biophys. Acta* **290** : 70–83.

Sha'afi, R. I., Rodan, S. B., Hintz, R. L., Fernandez, S. M., and Rodan, G. A., 1975, Abnormalities in membrane microviscosity and ion transport in genetic muscular dystrophy, *Nature* **254** : 525–526.

Shinitzky, M., 1984, Membrane fluidity and cellular functions, in *Physiology of Membrane Fluidity* (M. Shinitzky, ed.), pp. 1—51, CRC Press, Boca Raton, FL.

Shinitzky, M., and Barenholz, Y., 1974, Dynamics of the hydrocarbon layer in liposomes of lecithin and sphingomyelin containing dicertylphosphate, *J. Biol. Chem.* **249** : 2652–2657.

Shinitzky, M., and Inbar, M., 1976, Microviscosity parameters and protein mobility in biological membranes, *Biochim. Biophys. Acta* **433** : 133–149.

Shinitzky, M., and Yuli, I., 1982, Lipid fluidity at the submacroscopic level; determination by fluorescence polarization, *Chem. Phys. Lipids* **30** : 261–282.

Singer, S. J., 1971, The molecular organization of biological membranes, in *Structure and Function of Biological Membranes* (L. I. Rothfield, ed.), pp. 145–222, Academic Press, New York.

Singer, S. J., and Nicolson, G. L., 1972, The fluid mosaic model of the structure of cell membranes, *Science* **175** : 720–731.

Smith, M. T., Thor, H., and Orrenius, S., 1981, Toxic injury to isolated hepatocytes is not dependent on extracellular calcium, *Science* **213** : 1257–1259.

Sorensen, E. M. B., and Acosta, D., 1982, Protective effect of calcium on cadmium-induced cytotoxicity in cultured rat hepatocyte, *In Vitro* **18** : 288.

Sorensen, E. M. B., and Acosta, D., 1983, Semiquantitative morphologic analysis to quantify the cytotoxicity of cadmium to cultured hepatocyte, *In Vitro* **19** : 287.

Sorensen, E. M. B., and Acosta, D., 1984a, Morphometric analysis of the protective effect of calcium on cadmium-challenged cultured rat hepatocytes, *In Vitro* **20** : 284.

Sorensen, E. M. B., and Acosta, D., 1984b, Morphometric analysis of cadmium-induced cytotoxicity in cultured hepatocyte, *Toxicologist* **4** : 163.

Sorensen, E. M. B., and Acosta, D., 1984c, Cadmium-induced hepatotoxicity as evaluated by morphometric analysis, *In Vitro* **20** : 763–770.

Sorensen, E. M. B., and Acosta, D., 1985, Protective effects of calcium in the amelioration of cadmium-induced cytotoxicity in cultured murine parenchymal hepatocytes, in *Alternative Methods in Toxicology* (A. M. Goldberg, ed.), Vol. 3., pp. 101–139, Mary Ann Liebert, Inc., Publishers, New York.

Sorensen, E. M. B., Smith, N. K. R., Boecker, C. S., and Acosta, D., 1984, Calcium amelioration of cadmium-induced cytotoxicity in cultured rat hepatocytes, *In Vitro* **20** : 771–779.

Sorensen, E. M. B., Nealon, D. G., and Acosta, D., 1985, Effects of cadmium and calcium on the fluidity of plasma membranes, *Toxicol. Lett.* **25** : 319–326.

Sorensen, E. M. B., 1988, Modulatory effect of calcium on the influx and binding of cadmium in primary cultures of neonatal rat hepatocytes, *Toxicol. Lett.* **41** : 39–48.

Stacey, N. H., and Klaassen, C. D., 1982, Lack of protection against chemically induced injury to isolated hepatocytes by omission of calcium from the incubation medium, *J. Toxicol. Environ. Health* **9** : 267–276.

Storch, J., Schachter, D., Inoue, M., and Wolkoff, A. W., 1983, Lipid fluidity of hepatocyte plasma membrane subfractions and their differential regulation by calcium, *Biochim. Biophys. Acta* **727** : 209–212.

Storch, J., and Schachter, D., 1984, Dietary induction of acyl chain desaturases alters the lipid composition and fluidity of rat hepatocyte plasma membranes, *Biochemistry* **23** : 1165–1170.

Tasaki, I., 1968, *Nerve Excitation: A Macromolecular Approach*, Charles C. Thomas, Springfield, IL.

Thompson, G. A., Jr., and Nozawa, Y., 1977, *Tetrahymena*: A system for studying dynamic membrane alterations within the eukaryotic cell, *Biochim. Biophys. Acta* **472** : 55–92.

Trauble, H., and Eibl, H., 1974, Electrostatic effects on lipid phase transitions; membrane structure and ionic environment, *Proc. Natl. Acad. Sci. U.S.A.* **71** : 214–219.

Van Blitterswijk, W. J., Van Hoeven, R. P., and Van der Meer, B. W., 1981, Lipid structural order parameters (reciprocal of fluidity) in biomembranes derived from steady-state fluorescence polarization measurements, *Biochim. Biophys. Acta* **644** : 323–332.

Verkleij, A. J., Zwaal, R. R. A., Roelofsen, B., Comfurius, P., Kastelign, D., and Van Deenen, L. L. M., 1973, The asymmetric distribution of phospholipids in the human red cell membranes, *Biochim. Biophys. Acta* **323** : 178–193.

Weber, G., 1971, Theory of fluorescence depolarization by anisotropic Brownian rotations. Discontinuous distribution approach, *J. Chem. Phys.* **55** : 2399–2407.

Whetton, A. D., Houslay, M. D., Dodd, N. J. F., and Evans, W. H., 1983, The lipid fluidity of rat liver membrane subfractions, *Biochem. J.* **214** : 851–854.

Wisher, M. H., and Evans, W. H., 1975, Functional polarity of the rat hepatocyte surface membrane, *Biochem. J.* **146** : 375–388.

Yousef, I. M., 1983, Effect of Ca^{2+} ions on liver cell plasma membrane polypeptides, *Can. J. Biochem. Cell Biol.* **61** : 293–300.

Zwaal, R. F. A., Reolofsen, B., and Colley, C. M., 1973, Localization of red cell membrane constituents, *Biochim. Biophys. Acta* **300** : 159–182.

Chapter 6

Fluidity of Thyroid Plasma Membranes

Hugo Depauw, Marc De Wolf, Guido Van Dessel,
Herwig Hilderson, Albert Lagrou, and Wilfried Dierick

1. INTRODUCTION

The main function of the thyroid cell is the secretion of thyroid hormones
(T_4, thyroxine; T_3, triiodothyronine). Thyroid hormones are first synthesized
as part of thyroglobulin. Newly synthesized thyroglobulin is iodinated and
vectorially transported to the apical region of the cell and released for storage
in the lumen by exocytosis. When hormones are needed, thyroglobulin is
removed from the luminal content by endocytosis and transferred to the lysosomes, where it is hydrolysed by cathepsin D liberating the hormones.
Nonhormonal active iodotyrosines (3-monoiodotyrosine; 3,5-diiodotyrosine)
are also formed. The hormones are released in the bloodstream and the io-
dotyrosine deiodinated. The iodide formed is partly reutilized by the cell
(Taurog, 1978). This bidirectional transport of thyroglobulin is compart-
mented and implies concomitant transfer of membrane (Herzog, 1981). Every
step in thyroid metabolism is activated by TSH (thyroid stimulating hormone,

**Hugo Depauw, Marc De Wolf, Herwig Hilderson, Albert Lagrou, and Wilfried
Dierick**   RUCA–Laboratory for Human Biochemistry, University of Antwerp, B2020
Antwerp, Belgium.   **Guido Van Dessel**   UIA–Laboratory for Pathological Biochem-
istry, University of Antwerp, B2020 Antwerp, Belgium.

thyrotropin). It evokes intracellular metabolic responses by (1) recognition and binding to the cell surface, (2) transmission of information across the cell membrane, and (3) activation of internal metabolic pathways. Apart from the glycoprotein portion of the TSH receptor, other membrane constituents such as phospholipids and glycolipids are involved in the mechanism of TSH action (Macchia *et al.*, 1970; Yamashita *et al.*, 1970; Haye and Jacquemin, 1971; Yamashita and Field, 1973; Moore and Wolff, 1974; Amir *et al.*, 1976). From these studies it has been suggested that certain phospholipids (PC and PS) are involved in the coupling mechanism between the occupied receptor and the catalytic component of the adenylate cyclase system. It has also been shown that the enzymic activity might be modulated by changes in membrane lipid composition and fluidity (Klein *et al.*, 1978; Sandermann, 1978; Bakardjieva *et al.*, 1979; Houslay *et al.*, 1981; Salesse *et al.*, 1982a; Houslay and Gordon, 1983; McOsker *et al.*, 1983). As a consequence, lipid microviscosity has gained attention with regard to its involvement in the regulation of plasma membrane functions.

2. THYROID PLASMA MEMBRANES

Plasma membranes from thyroid are characterized by the presence of specific TSH and thyroglobulin receptors in different domains and by the occurrence of an active transport system for iodide. They capture and process signals from the environment (e.g., TSH, thyroid stimulating immunoglob-

Abbreviations used in this chapter: AC, adenylate cyclase; ANS, 8-anilino-1-napthalenesulfonate; 12-AS, 12-(9-anthroyloxy)stearic acid; BA, benzyl alcohol; CH, cholesterol; Con A, concanavalin A; CPZ, chlorpromazine; D_{diff}, lateral diffusion coefficient of pyrene; DPH, 1,6-diphenyl-1,3,5-hexatriene; DPPC, dipalmitoylphosphatidylcholine; Dol, dolichol; DMP, dolichylmonophosphate; GD_3, N-acetylneuraminyl-N-acetylneuraminylgalactosylglucosylceramide; GM_1, galactosyl-N-acetylgalactosaminyl(N-acetylneuraminyl)galactosylglucosylceramide; GT_{1b}, N-acetylneuraminylgalactosyl-N-acetylgalactosaminyl(N-acetylneuraminyl-N-acetylneuraminyl)galactosylglucosylceramide; G_s-protein, stimulatory GTP-binding regulatory protein; FITC, fluorescein-isothiocyanate; ns-TP, nonspecific lipid transfer protein; L, light mitochondrial fraction; P, microsomal fraction; L + P, combined light mitochondrial and microsomal fractions; P_2 (P_3) fraction, enriched plasma membrane fraction (Section 2.1.1); PB, phenobarbital; PL, phospholipid; PA, phosphatidic acid; PC, phosphatidylcholine; PE, phosphatidylethanolamine; PI, phosphatidylinositol; PS, phosphatidylserine; PM, plasma membranes; PM_1, nonbinding plasma membrane fraction; PM_2, binding plasma membrane fraction; PM_3, second binding plasma membrane fraction; RCA, *Ricinus communis* agglutinin; ϕ, rotational correlation time; r_s, steady-state fluorescence anisotropy; τ, fluorescence excited-state lifetime; RSA, relative specific activity; S_{DPH}, lipid structural order parameter as measured with DPH; SUV, small unilamellar vesicles; TMA-DPH, 1-(4-trimethylammoniumphenyl)-6-phenyl-1,3,5-hexatriene; TSH, thyroid-stimulating hormone (thyrotropin); T_3, triiodothyronine; T_4, thyroxine; WGA, wheat germ agglutinin; w/w, weight to weight ratio.

ulins) and control the nature and extent of material entering and leaving the cell (thyroglobulin, iodide, T_4, and T_3). The thyroidal cell surface membrane is highly differentiated and constructed of geographically distinct areas facing different environments (stroma, follicular lumen, cell–cell) and specialized for different physiological functions. From a methodological point of view, the plasma membrane fraction is the *in vitro* model enabling one to study its structure and function at a molecular level.

2.1. Enriched Plasma Membrane Fractions

No single procedure is equally effective for the purification of plasma membranes of different cell types. For each tissue, cell disruption and membrane purification techniques have to be adapted (Hilderson *et al.*, 1980). Because thyroids consist of follicles enmeshed in thick, tough connective tissue, rather drastic procedures are required for homogenizing resulting in damage to some subcellular organelles, disruption of membranes into small vesicles, and polydispersity of plasma membrane fragments. Two methods are described here for bovine thyroid tissue, both starting with a differential pelleting step. In the first method this step is followed by flotation through a discontinuous sucrose gradient, in the second by isopycnic continuous sucrose gradient centrifugation in a zonal rotor.

The first method is an adaptation of the procedure of Yamashita and Field (1973) as modified by Kidroni *et al.* (1980). Thyroid tissue is homogenized in a Waring blender and the ensuing homogenate filtered through cheesecloth and subjected to differential pelleting. The combined light mitochondrial and particulate fractions $(L + P)$, containing the highest relative specific activity for plasma membrane markers, are further fractionated by discontinuous sucrose gradient centrifugation. Plasma membrane fractions with the highest relative specific activity for $5'$-nucleotidase are collected at the interfaces between 0.91 and 1.10 M sucrose (P_2 fraction) and between 1.10 and 1.25 M sucrose (P_3 fraction). In P_2 the yield of plasma membrane proteins amounts to 15–25 mg per 200 mg of anatomically prepared thyroid tissue with an average purification of 25.

In order to perform a continuous sucrose gradient centrifugation in a zonal rotor, thyroid tissue, cut into small blocks, is homogenized (Waring blender) in 60 mM $NaHCO_3$, pH 7.4, 3 mM EGTA. A 15 min $\times$ 73,300 g pellet resuspended in 0.25 M sucrose, 1 mM $NaHCO_3$, 5 mM imidazole buffer pH 7.4, 1 mM EGTA is applied on top of a linear sucrose gradient (20–50% w/w) in a HS-zonal rotor and spun (18 hr at 10,400 g_{max}). With this approach a 20-fold purification of plasma membranes can be obtained. As shown in Figure 1, the profiles of different plasma membrane markers do not coincide, reflecting the biochemical polarity of the cell surface mem-

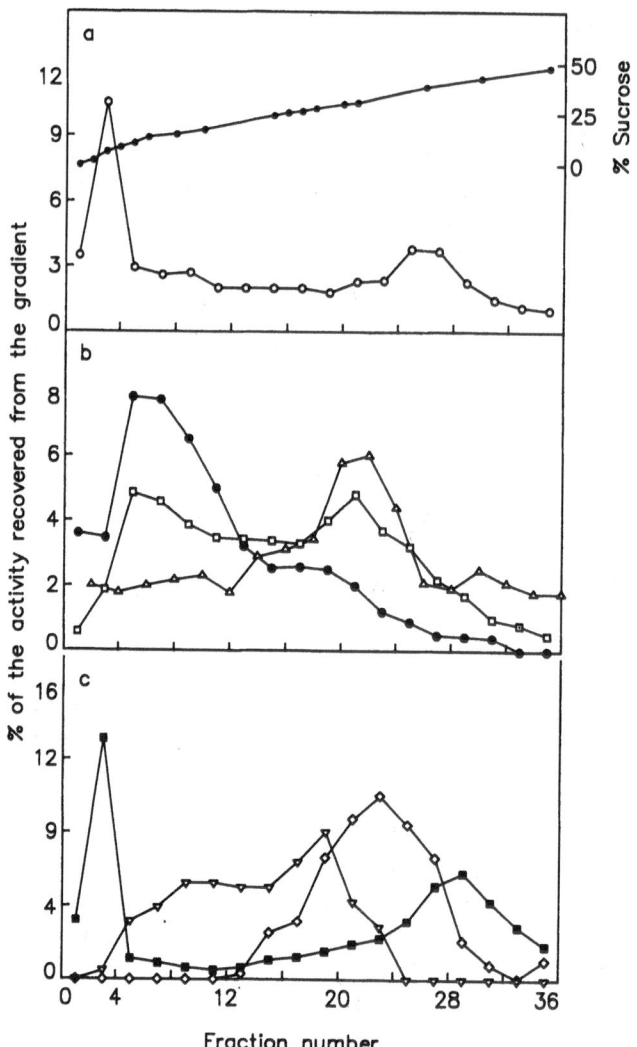

FIGURE 1. Zonal centrifugation of a 73,000 g pellet from a NaHCO₃ homogenate in a HS zonal rotor. (a) ●—●, slope of the gradient; ○—○, proteins; (b) plasma membrane markers: △—△, alkaline phosphatase; □—□, NAD-glycohydrolase; ●—●, 5′-nucleotidase; (c) assessment of contaminants: ■—■, β-D-N-acetylglucosaminidase; ▽—▽, guaiacol peroxidase; ◇—◇, cytochrome *c* oxidase.

brane. The bulk of 5′-nucleotidase activity equilibrates at low densities with a peak at density 1.08. Alkaline phosphatase, on the other hand, shows a rather broad distribution (1.09–1.18) with a peak at a density of 1.16. NAD-glycohydrolase, an integral protein (De Wolf *et al.*, 1985), is distributed over the entire gradient. In the lower density region it mainly follows $5'$-nucleotidase (peak around density 1.08); in the higher density region its distribution parallels that of alkaline phosphatase (peak around density 1.16). This enzyme is not associated with one particular region of the cell surface but seems rather uniformly distributed. NaF-stimulated adenylate cyclase displays maximal activity at somewhat higher density than $5'$-nucleotidase. Although registration of zonal profiles provides information with regard to plasma membrane heterogeneity, it is difficult to gain insight in the topological origin of the plasma membrane fragments. Experiments using radioactive or fluorescent-labeled specific ligands (TSH, thyroglobulin, lectins, or toxins) could probably provide this information.

2.2. Chemical Characterization of Purified Plasma Membranes

The plasma membrane fraction P_2 displays a protein/phospholipid ratio (w/w) of 1.67 and a cholesterol/phospholipid molar ratio of 0.55. The phospholipid composition of plasma membranes does not deviate appreciably from that of whole tissue except for the higher sphingomyelin level (22.5 versus 14.0%, Table I). This composition as well as the molar ratios of cholesterol/phospholipids and sphingomyelin/phospholipids and the protein/phospholipid ratio all closely resemble those reported for rat liver plasma membranes (Ansell *et al.*, 1973; Lowe and Coleman, 1982). The fatty acid composition of total phospholipids of plasma membranes differs from that of whole tissue in C16:1—fatty acids being present in much higher amounts—and in C20:0, C20:4, and C22:6—fatty acids being present in much smaller amounts (Table I). The ratio of unsaturated/saturated fatty acids is 1.50 for whole tissue and 1.37 for plasma membranes.

2.3. Enzymic Characterization of Purified Plasma Membranes

Current methods for the isolation of plasma membranes from thyroid tissue result in the formation of polydisperse vesicles and small fragments. Therefore, an assessment of purity cannot only be approached by morphological criteria. One has also to rely on the use of markers such as enzymes and other membrane constituents. Because of the heterogeneity of the thyroidal membrane, it is essential to verify the representativity of the isolated fractions by determining several plasma membrane markers. Furthermore, as biochemical heterogeneity parallels structural heterogeneity, a given plasma membrane marker is sometimes only valid for a particular region of the plasma membrane. Therefore, several enzymic markers must be used for monitoring the isolation and establishing the purity of the fractions. The relevant data are summarized in Table II. The enrichment factor for succinate dehydrogenase, a mitochondrial marker, is 0.7. There is still a significant

Table I
Chemical, Phospholipid, and Fatty Acid Composition of Bovine Thyroid

Compound[a]	Whole tissue[b]	P$_2$ fraction[b]
Total phospholipids	61 ± 8	592 ± 81
Cholesterol + CHE	12 ± 2	153 ± 22
Lipid-bound sialic acid	0.13 ± 0.04	5.2 ± 1.1
Total asialo glycolipids	0.16 ± 0.02	5.9 ± 0.5
Phospholipids[c]		
PI	6.5 ± 0.7	5.5 ± 0.6
PS	5.6 ± 0.8	6.1 ± 0.9
Sphingomyelin	14.0 ± 1.8	22.5 ± 2.5
LysoPC ⎫		1.2 ± 0.4
PC ⎭	43.0 ± 2.0	38.7 ± 3.4
Alkenyl-PE	18.3 ± 1.4	15.4 ± 1.2
PE	9.9 ± 0.7	8.5 ± 0.9
Cardiolipin	2.8 ± 0.5	—
Other phospholipids	< 1	2.0 ± 0.3
Fatty acids[d]		
14 : 0	—	3.3 ± 0.4
A	2.3 ± 0.4	2.8 ± 0.2
16 : 0	22.0 ± 0.5	23.3 ± 1.2
16 : 1	tr	7.9 ± 0.8
B	1.5 ± 0.3	1.4 ± 0.2
18 : 0	13.1 ± 1.1	10.9 ± 0.8
18 : 1	28.2 ± 0.9	28.9 ± 1.3
18 : 2	10.6 ± 0.4	11.2 ± 0.6
20 : 0	1.7 ± 0.3	1.0 ± 0.3
20 : 4	11.1 ± 0.7	3.4 ± 0.9
22 : 0	1.7 ± 0.3	1.9 ± 0.4
22 : 6	7.8 ± 0.9	4.1 ± 1.1

[a]Values are expressed as μg/mg protein. CHE, cholesteryl ester.
[b]All data are average values from three separate experiments.
[c]Values are expressed as percentage of total phospholipid phosphorus.
[d]Values are expressed as percentage of total fatty acids. A and B, aldehydes, hexadecanal and octodecanal; tr, trace.

contamination by β-hexosaminidase (lysosomal marker) and glucose-6-phosphatase (endoplasmic reticulum marker) with enrichment factors of 1.4 and 3.9, respectively.

2.4. Subfractionation of Thyroidal Plasma Membranes

Plasma membrane fractions (e.g., P$_2$) can be subfractionated by affinity chromatography on Con A–Sepharose (Resch *et al.*, 1981), yielding a non-binding fraction PM$_1$ and a binding fraction PM$_2$. Therefore, a suspension of thyroid plasma membranes in 0.02 M Hepes buffer pH 7.0 containing 0.14 M KCl is mixed with the Con A–Sepharose gel by stirring for 2 min. The

Table II
Characterization of Thyroid Plasma Membranes and Plasma Membrane Subfractions[a]

Enzyme or compound	Homogenate	P_2	PM_1	PM_2
Protein (%)		100	72 ± 1	14 ± 1
5'-Nucleotidase	7.7 ± 0.5	202.5 ± 14.9	235.7 ± 15.3	158.2 ± 6.4
		(26 ± 4)	(30 ± 2)	(20 ± 2)
Alkaline phosphatase	1.2 ± 0.4	35.8 ± 6.5	29.9 ± 6.1	53.7 ± 7.1
		(30 ± 6)	(25 ± 4)	(45 ± 3)
(Na^+, K^+)ATPase	2.4 ± 0.5	51.2 ± 7.8	59.4 ± 7.6	37.8 ± 5.4
		(21 ± 3)	(25 ± 3)	(16 ± 3)
Leucine aminopeptidase	0.6 ± 0.1	11.1 ± 0.9	10.8 ± 0.8	19.7 ± 2.1
		(19 ± 2)	(18 ± 2)	(33 ± 2)
NADase	0.7 ± 0.1	13.3 ± 0.2	16.8 ± 1.2	12.6 ± 1.2
		(19 ± 2)	(24 ± 2)	(18 ± 2)
Glucose-6-phosphatase	2.7 ± 0.4	10.5 ± 0.5	11.4 ± 0.6	9.7 ± 0.5
		(4 ± 0.5)	(4.2 ± 0.5)	(3.5 ± 0.4)
β-Hexosaminidase	29.0 ± 2.2	40.6 ± 2.9	45.2 ± 5.4	37.7 ± 3.8
		(1.4 ± 0.1)	(1.6 ± 0.2)	(1.1 ± 0.1)
Succinate dehydrogenase	0.38 ± 0.05	0.27 ± 0.08	—	—
		(0.7 ± 0.1)		
Cholesterol/phospholipids		0.55 ± 0.07	0.56 ± 0.06	0.66 ± 0.07
Protein/phospholipid		1.67 ± 0.11	1.80 ± 0.13	2.02 ± 0.15

[a]Specific activities are expressed as nmol/min per mg protein. Numbers in parentheses are RSA. Cholesterol/phospholipids is a molar ratio. The number of experiments is five; statistics refer to the average deviation. P_2, enriched plasma membrane fraction; PM_1, nonbinding plasma membrane fraction; PM_2, binding plasma membrane fraction.

mixture is poured into a column and allowed to stand for 20 min at 4°C. A nonbinding fraction (PM_1) is eluted from the column using the suspension buffer as eluent. A second fraction (PM_2) can be eluted with the same buffer solution supplemented with 0.1 M α-methylmannoside. To this end, the gel is transferred to a beaker, stirred (2 min), and transferred again to the column and eluted. A second binding plasma membrane fraction (PM_3) can eventually be obtained after stirring the gel for an additional period of 15 min.

2.5. Characterization of Thyroid Plasma Membrane Subfractions

Subfractions of plasma membranes that can be separated may reflect the existence of membrane heterogeneity. For thyroid, this view is substantiated by the different distribution patterns observed for several plasma-membrane-associated enzymes (Table II). The specific activities of 5′-nucleotidase and of ($Na^+ + K^+$)-ATPase are enhanced in PM_1, whereas those of leucine aminopeptidase and alkaline phosphatase are enriched in PM_2. In several epithelial systems, leucine aminopeptidase has been shown to be mainly located in the apical cell surface (Carlsen *et al.*, 1983). Recent electron microscopic studies using horseradish peroxidase-labeled antibodies directed against leucine aminopeptidase (V. Herzog, personal communication) strongly suggest that the thyroidal leucine aminopeptidase is also predominantly located at the apical cell surface. ($Na^+ + K^+$)-dependent ouabain-sensitive ATPase, at the other hand, appears to be mainly located in basolateral membranes (Biber *et al.*, 1983). Therefore, from the distribution of these enzymes over PM_1 and PM_2 it is suggested that the nonbinding fraction PM_1 is enriched in basolateral membranes and the binding fraction PM_2 in apical membranes. 5′-Nucleotidase seems to be concentrated in basolateral membranes, whereas alkaline phosphatase is more associated with luminal plasma membranes. NADase slightly prefers basolateral membranes, although from Figure 1 a homogeneous distribution is more likely to occur.

As shown in Table II, the cholesterol/phospholipid molar ratio and the protein/phospholipid ratio (w/w) of PM_1 are similar to those found in P_2. This is not surprising, as in terms of proteins, PM_1, represents about 75% of P_2. These ratios, however, are higher in PM_2.

3. FLUIDITY OF THYROID PLASMA MEMBRANES

By introducing their fluid mosaic model, Singer and Nicolson (1972) put great emphasis on the dynamical aspects of a membrane structure. In this model, the membrane constituents are subject to a number of motions and the resultant dynamic state is commonly described by the all-inclusive term "fluidity." In recent years, biomembrane fluidity has become the object of intense research. Many properties and functions of biological membranes are discussed in terms of changes in "membrane lipid fluidity" (Shinitzky, 1984). Maintenance of "membrane fluidity" within narrow limits is presumably a prerequisite for proper functioning of a cell (Sinensky, 1974). Modification of "membrane lipid fluidity" has been reported to alter important cellular processes or functions such as the regulation of membrane-bound enzymes (Kimelberg, 1977; Sandermann, 1978; Le Grimellec et al., 1982; Stubbs, 1983; Stubbs and Smith, 1984), ligand–receptor binding (Heron et al., 1980; Ginsberg et al., 1982), membrane fusion (Prives and Shinitzky, 1977; Cullis and De Kruijff, 1979), growth and vitality of cells (Shinitzky, 1984), and the transmembrane movement of water, ions, and nonelectrolytes (Jain, 1980).

3.1. Fluidity Measurements

The physical state of the thyroid membrane was investigated by following the motional characteristics of several fluorescent probes embedded in the membrane such as 1,6-diphenyl-1,3,5-hexatriene (DPH), 1-(4-trimethyl-ammoniumphenyl)-6-phenyl-1,3,5-hexatriene (TMA-DPH), 12-(9-anthroyl-oxy)stearic acid (12-AS), and 1-anilino-8-naphthalene sulfonate (ANS). The motional characteristics were monitored by measuring the steady-state anisotropy of these fluorescent probes and by calculation of the lipid structural order parameter (S_{DPH}) from steady-state anisotropy determinations of the hydrophobic probe DPH. Because these fluorescent probes are located at different levels of the lipid bilayer, simultaneous evaluation of the steady-state anisotropies of the fluorophores allows the estimation of changes in membrane fluidity in different regions of the bilayer.

Since fluorescence anisotropy parameters do not necessarily reflect lateral translational diffusion rates of lipids or proteins in cell membranes (Kleinfeld et al., 1981), the lateral diffusion coefficient of the hydrophobic fluorescent probe pyrene (D_{diff}) was also measured by following its excimer formation.

Static and dynamic fluorescence parameters were recorded on a SLM 4800 spectrofluorometer equipped with a Hewlett-Packard 85 calculator and a 7225A plotter. Temperature was controlled by a Lauda thermostated waterbath and measured inside the fluorescent cuvettes with an AD 590 probe (Analog Devices).

3.2. Fluidity of Thyroid Subcellular Fractions

Upon DPH incorporation into aliquots of different subcellular fractions of bovine thyroid [obtained after different pelleting according to Hilderson et al., (1980)] at the same probe/lipid ratio, it was observed that the value of the steady-state anisotropy r_s was highest in the fractions most enriched in plasma membranes (L fraction, P_2 fraction) (Depauw et al., 1985), indicating that the structural order of thyroid plasma membranes is greater than that of the intracellular membranes. This is in agreement with their higher cholesterol/phospholipid and sphingomyelin/phospholipid molar ratios. The nuclear fraction showed the highest fluidity and the highly purified plasma membranes (P$_2$ fraction) were still more rigid than the membranes recovered from the L fraction (Figure 2).

3.3. Fluidity of a P$_2$ Fraction in Reconstituted Thyroid Plasma Membranes

In order to study the relation between thyroid plasma membrane composition and its structural or fluidity characteristics, reconstitution with the nonionic detergent octylglucoside was used as an approach. Large unilamellar vesicles were prepared by detergent dialysis from octylglucoside-solubilized membrane constituents. All native membrane constituents were incorporated in the large vesicles according to their ratios in a P$_2$ fraction. It has been shown that in these artificial vesicles glycoproteins and glycolipids have the same topological distribution as in the native membranes (Petri and Wagner, 1979).

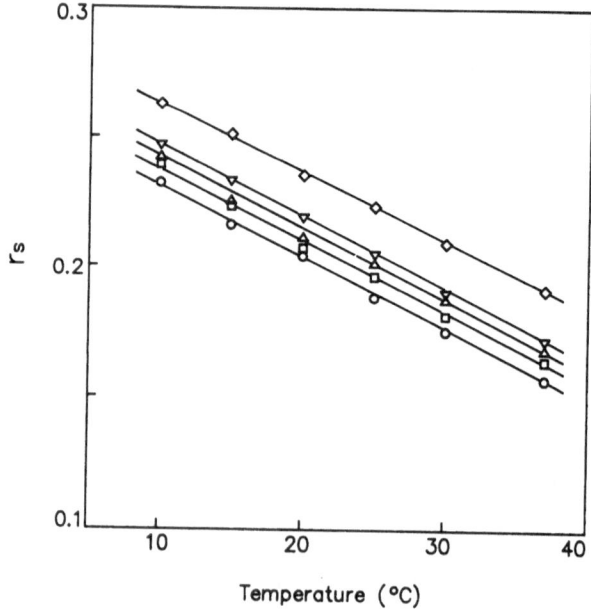

FIGURE 2. Temperature dependence of the fluorescence steady-state anisotropy (r_s) of diphenylhexatriene incorporated into aliquots of different subcellular fractions of bovine thyroid: O—O, nuclear fraction; □—□, mitochondrial fraction; △—△, microsomal fraction; ▽—▽, light mitochondrial fraction; ◇—◇, enriched plasma membrane fraction P$_2$. The diphenylhexatriene/lipid ratio was always 1 : 250.

From the temperature dependence of the fluorescence excited-state life-time (τ) of DPH, embedded in these large vesicles with a different degree of reconstitution, it is clear that τ varies considerably not only with temperature but also with membrane composition (Figure 3). Large lipid vesicles prepared from phospholipids alone display the lowest τ value. Incorporation of neutral lipids, acidic glycolipids, or membrane proteins leads to an increase of the τ value, whereas addition of neutral glycolipids leads to a decrease. The higher the degree of reconstitution the more the value of the lifetime found in the P_2 fraction is approached. The τ value for the total lipid extract of a P_2 fraction is intermediate between the values for vesicles prepared from phospholipids alone and from the P_2 fraction (Figure 3). Heterogeneity analysis of phase and modulation-average lifetimes for DPH in these different reconstituted bilayer systems and native membranes seems to be consistent with a monoexponential decay model. This is in contrast with several pulse

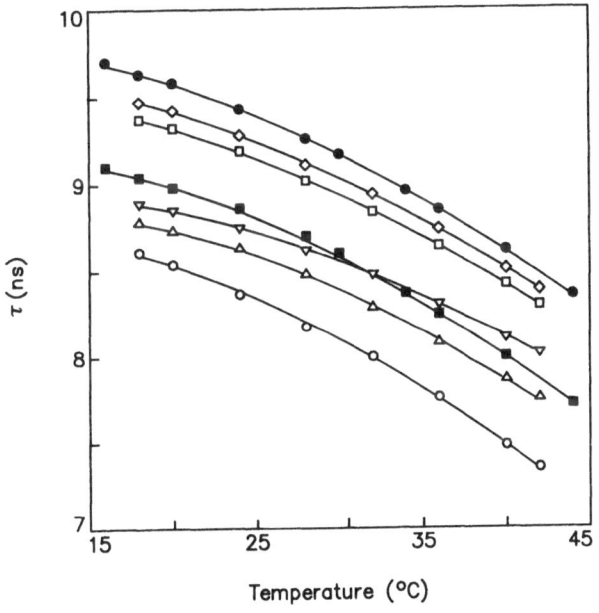

Temperature (°C)

FIGURE 3. Temperature dependence of the fluorescence excited-state lifetime (τ) of diphenylhexatriene embedded in large vesicles prepared by detergent dialysis from a P_2 fraction ($\bullet-\bullet$); a total lipid fraction of P_2 ($\blacksquare-\blacksquare$); a phospholipid extract of P_2 ($\bigcirc-\bigcirc$); a mixture of phospholipids and neutral lipids ($\square-\square$); a mixture of phospholipids, neutral lipids, and neutral glycolipids ($\triangle-\triangle$); a mixture of phospholipids, neutral lipids, neutral glycolipids, and acidic glycolipids ($\triangledown-\triangledown$); and a mixture of phospholipids, neutral lipids, neutral glycolipids, acidic glycolipids, and membrane proteins ($\diamond - \diamond$). All these native membrane constituents were mixed according to their ratio in the P_2 fraction. The diphenyl-hexatriene/lipid ratio was always kept at 1 : 250.

fluorimetric studies, which revealed biexponential decay kinetics for DPH in artificial vesicle preparations as well as native membranes (Cundall and Dale, 1983). It should, however, be noted that, using analogous systems with the phase and modulation technique, in most cases no similar biexponential decays were observed. High standard errors in lifetime determinations are thought to be partially responsible for this discrepancy (Cundall and Dale, 1983).

The temperature dependence of the structural order parameter S_{DPH} of DPH embedded in vesicles with a different degree of reconstitution does not reveal any significant breakpoint (Figure 4a). This indicates that in the temperature range studied no lipid phase transitions would occur. Similar results were already reported for other types of biological membranes (Van Blitterswijk *et al.*, 1981). Large lipid vesicles prepared from phospholipids alone display the lowest S_{DPH} value. Incorporation of neutral lipids, acidic glycolipids, or membrane proteins according to their native ratios provokes an increase in the S_{DPH} value, whereas neutral glycolipids have no distinct effect on this value. A similar phenomenon occurs when acidic glycolipids, neutral glycolipids, or membrane proteins are incorporated into vesicles containing both phospholipids and neutral lipids. After complete reconstitution of all plasma membrane constituents, a S_{DPH} value is obtained corresponding nicely to this in native plasma membranes (P_2 fraction). This is also the case when S_{DPH} of vesicles prepared from total lipids of plasma membranes is compared to S_{DPH} found in a reconstituted system containing phospholipids, neutral lipids, and neutral glycolipids. As a consequence, large vesicles prepared by detergent dialysis of octylglucoside are suitable systems for studying thyroid plasma membrane fluidity and the contribution of plasma membrane constituents to this fluidity.

The lipid order parameter of S_{DPH} of bovine thyroid plasma membranes is less than average but similar to that observed for rat liver plasma membranes and mouse thymocyte plasma membranes (S_{DPH} bovine thyroid plasma membranes: 0.71; S_{DPH} rat liver plasma membranes: 0.74; S_{DPH} mouse thymocyte plasma membranes: 0.71). As for other membrane systems (Van Blitterswijk *et al.*, 1981), the lipid order parameter S_{DPH} of bovine thyroid plasma membrane is mainly determined by the neutral lipids. The latter, however, does not seem to increase the viscous drag imposed on the rotation of DPH as evidenced by the decrease in rotational correlation time ϕ upon addition of neutral lipids (Table III). A further decrease of ϕ is observed upon incorporation of plasma membrane proteins in artificial vesicles with a different degree of reconstitution in contrast with the data reported by Heyn (1979).

When following the lateral diffusion coefficient D_{diff} of pyrene as a function of temperature, similar fluidity trends are observed as with S_{DPH} or τ, apart from the fact that the incorporation of acidic glycolipids produces

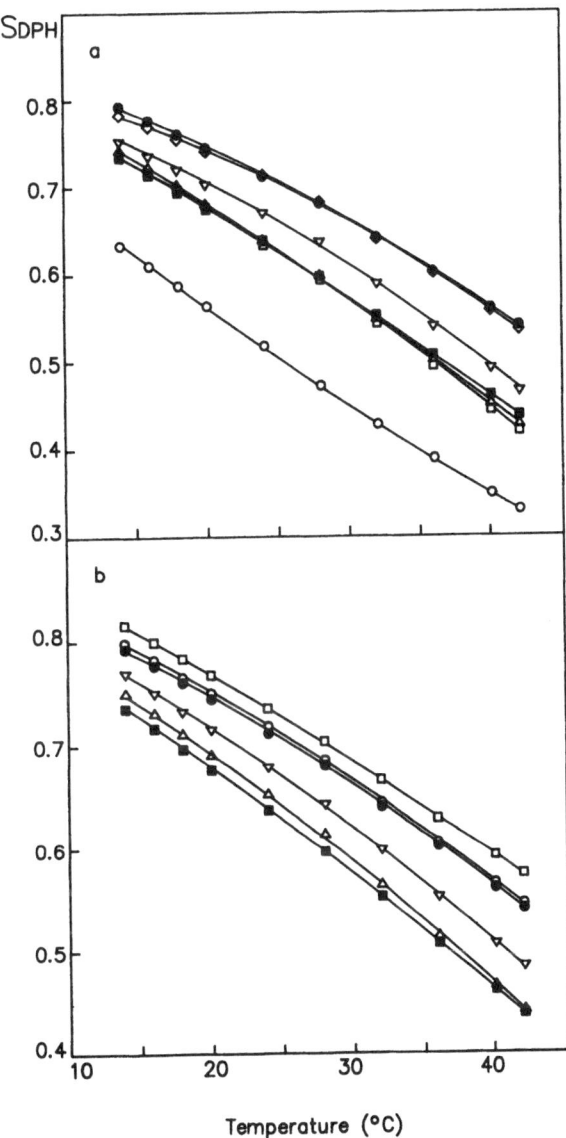

FIGURE 4. Temperature dependence of the structural order parameter (S_{DPH}) of diphenylhexatriene embedded in large vesicles prepared by detergent dialysis from: (a) the same reconstituted systems as in Figure 3. For explanation of symbols see legend to Figure 3; (b) ●—●, P_2 fraction; ○—○, PM_1 fraction; □—□, PM_2 fraction; ■—■, total lipids P_2 fraction; △—△, total lipids PM_1 fraction; ▽—▽, total lipids PM_2 fraction. All these native membrane constituents were mixed according to their ratio in a P_2 fraction. The diphenylhexatriene/lipid molar ratio was always 1 : 250.

Table III
Fluorescence Parameters for DPH and Pyrene in Bovine Thyroid Plasma Membranes and in Reconstituted Large Unilamellar Vesicles at 24°C[a]

	r_s	r_x	r_f	S_{DPH}	D_{diff} (cm^2/sec)($\times 10^8$)	τ (nsec)	ϕ (nsec, $r_x \neq 0$)
P$_2$	0.225 ± 0.003	0.200 ± 0.004	0.025 ± 0.005	0.712 ± 0.007	0.98 ± 0.13	9.43 ± 0.21	1.39 ± 0.28
PM$_1$	0.229 ± 0.003	0.205 ± 0.004	0.024 ± 0.005	0.720 ± 0.007	1.01 ± 0.12	9.22 ± 0.25	1.33 ± 0.28
PM$_2$	0.234 ± 0.004	0.212 ± 0.005	0.022 ± 0.006	0.733 ± 0.009	0.82 ± 0.11	9.34 ± 0.23	1.28 ± 0.34
tlip P$_2$	0.196 ± 0.007	0.161 ± 0.009	0.035 ± 0.011	0.638 ± 0.018	1.78 ± 0.26	8.84 ± 0.36	1.55 ± 0.42
tlip PM$_1$	0.202 ± 0.006	0.169 ± 0.008	0.033 ± 0.010	0.654 ± 0.015	1.76 ± 0.22	8.29 ± 0.35	1.41 ± 0.42
tlip PM$_2$	0.214 ± 0.007	0.185 ± 0.009	0.029 ± 0.011	0.684 ± 0.017	1.72 ± 0.24	8.03 ± 0.37	1.29 ± 0.48
PL	0.155 ± 0.005	0.107 ± 0.007	0.048 ± 0.009	0.520 ± 0.017	1.51 ± 0.18	8.37 ± 0.11	1.67 ± 0.32
PL + NL	0.194 ± 0.003	0.159 ± 0.004	0.035 ± 0.005	0.634 ± 0.008	1.12 ± 0.16	9.20 ± 0.41	1.60 ± 0.23
PL + NL + NGL	0.196 ± 0.004	0.161 ± 0.005	0.035 ± 0.006	0.638 ± 0.010	1.82 ± 0.22	8.65 ± 0.21	1.52 ± 0.26
PL + NL + NGL + AGL	0.208 ± 0.003	0.177 ± 0.004	0.031 ± 0.005	0.669 ± 0.008	2.15 ± 0.32	8.75 ± 0.22	1.45 ± 0.23
PL + NL + NGL + AGL + MP	0.226 ± 0.007	0.201 ± 0.009	0.025 ± 0.011	0.713 ± 0.016	1.04 ± 0.20	9.27 ± 0.38	1.37 ± 0.59

[a]Results are means of three separate experiments. PL, phospholipids; NL, neutral lipids; AGL, acidic glycolipids; NGL, neutral glycolipids; MP, membrane proteins; tlip, total lipids.

an increase in D_{diff} and that the effects of neutral glycolipids and membrane proteins are more pronounced than the effect of neutral lipids (Figure 5a). Complete reconstitution again gives rise to a D_{diff} value very close to that found in native membranes. This is also the case for large lipid vesicles prepared from the total lipid extract of the P_2 fraction and a reconstituted system containing phospholipids, neutral lipids, and neutral glycolipids (Figure 5a). The excited-state lifetime of the excimer of pyrene has a more pronounced temperature dependency than that of DPH and its increase also parallels the decrease in fluidity.

From Table III it is clear that the fluorescence anisotropy r_s is mainly determined by the neutral lipids. The increase in steady-state anisotropy with a higher degree of reconstitution is due to an increase in the static part r_∞ and hence to S_{DPH}. The presence of neutral lipids or membrane proteins actually leads to a decrease in the rotational correlation time ϕ (dynamic effect). The lipid order parameter (S_{DPH}) is mainly determined by the neutral lipids, whereas D_{diff} is more sensitive to membrane proteins. Incorporation of membrane proteins markedly decreases D_{diff} (approximately 50%) but provokes a minor increase in S_{DPH} (6%). Neutral glycolipids do not display any significant effect on S_{DPH}. However, they increase D_{diff}. Gangliosides enhance S_{DPH} as well as D_{diff}. Thus, an increase in S_{DPH} is not always paralleled by a decrease in D_{diff} (see also B. Wieb Van der Meer, Chapter 1 in this volume).

Table III clearly demonstrates that both neutral lipids and proteins have a major effect on thyroid membrane fluidity. However, estimation of their relative contribution to membrane fluidity is dependent on the type of fluorescence probe technique applied—S_{DPH} being mainly affected by the presence of neutral lipid, and D_{diff} being more sensitive to plasma membrane proteins.

In contrast to the rat liver systems where it has been shown that insulin produces a rapid and marked increase in membrane fluidity (Luly and Shinitzky, 1979), addition of thyroid effectors such as thyrotropin and cholera toxin does not result in a significant effect on bovine thyroid plasma membrane fluidity. These data are in contrast to those reported by Beguinot *et al.* (1983), where it was shown that TSH caused a significant increase in the fluorescence polarization of DPH when incorporated into a strain of functioning rat thyroid cells. However, one has to consider that the authors were using viable cells where during the 30-min preincubation at 37°C a progressive incorporation of this fluorescent probe in intracellular membranes (e.g., endocytosis, membrane fusion, membrane recycling) might have occurred. Thyrotropin, in turn, by affecting all these membrane phenomena, might change the fluidity values, which are in fact a weighted average of all labeled lipid domains (Van Hoeven *et al.*, 1979).

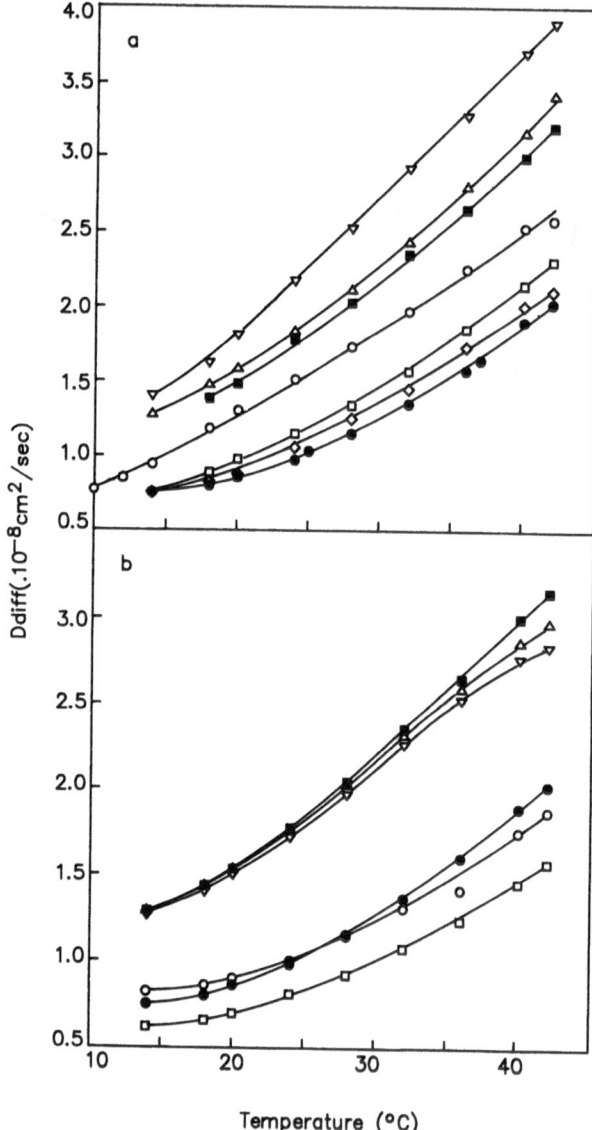

FIGURE 5. Temperature dependence of the lateral diffusion coefficient (D_{diff}) of pyrene embedded in large unilamellar vesicles prepared by detergent dialysis from: (a) the same reconstituted systems as in Figure 3 (for explanation of symbols see legend to Figure 3); (b) ●—●, P_2 fraction; ○—○, PM_1 fraction; □—□, PM_2 fraction; ■—■, total lipids P_2 fraction; △—△, total lipids PM_1 fraction; ▽—▽, total lipids PM_2 fraction. All the native membrane constituents were mixed according to their ratio in a P_2 fraction. The pyrene/lipid molar ratio was always 0.01.

3.4. Fluidity Characteristics of Plasma Membrane Subfractions

From studies on the fluidity characteristics of plasma membrane subfractions (PM_1 and PM_2) and unilamellar vesicles prepared from their total lipid extracts (Depauw et al., 1985), it became clear that the temperature dependence of S_{DPH} for the PM_1 and PM_2 fractions again does not reveal any significant breakpoint. From the temperature dependence of the lateral diffusion coefficient of pyrene (Figure 5b), it is clear that the apical membrane fraction (PM_2) is less fluid than its basolateral counterpart (PM_1). This difference in fluidity is probably due to a different distribution of membrane proteins and/or acidic glycolipids, since nearly identical D_{diff} values are found in large unilamellar vesicles prepared from the total lipid extracts of both PM_1 and PM_2 (Figure 5b). However, the temperature dependence of the structural order parameter of DPH shows that the difference in S_{DPH} values obtained in PM_1 and PM_2 is also found in vesicles prepared from their total lipid extracts (Figure 4b). Therefore, it is suggested that both lipid as well as protein are at the basis of the observed differences in fluidity. This view is supported by the different protein/phospholipid (w/w) and cholesterol/phospholipid molar ratios in PM_1 and PM_2 (Table II). Thus, subfractions of thyroid plasma membranes enriched in luminal membranes (PM_2) display a somewhat lower fluidity than their basolateral counterpart (enriched in PM_1) as evidenced by all fluorescence parameters studied (Table III). It is rather unlikely that this difference in fluidity is the result of a preferential retention of contaminating membrane fragments in PM_1 and PM_2 because of the lack of any significant differential enrichment of contaminating membrane markers in each subfraction (Table II). The difference in fluidity, however, can be a consequence of differences in protein composition of PM_1 and PM_2, which are probably maintained in vivo by their links to elements of the cytoskeleton as well as of differences in lipid composition. These results point to the presence of a lateral fluidity gradient in the plasma membranes of thyroid follicular cells. The existence of a similar fluidity gradient has already been proposed for rat erythrocytes (Brasitius and Schachter, 1980) and dog kidney epithelial cells (Le Grimellec et al., 1982). The physiological relevance of such a fluidity gradient remains to be established.

4. MODULATION OF THE ADENYLATE CYCLASE ACTIVITY BY MANIPULATING THE PLASMA MEMBRANE COMPOSITION

The membrane-bound enzyme adenylate cyclase forms a transmembrane complex composed of at least three distinct classes of protein components—hormone receptors, the guanine-nucleotide-binding regulatory protein(s) (G-proteins), and the catalyst—and is regulated by interaction of these components in a relatively unperturbed membrane (Rodbell, 1980; Ross and Gilman, 1980; Gilman, 1984; Levitzki, 1984, 1987).

4.1. Incorporation of Phospholipids

Membrane lipids play an important role in the regulation of the activity of membrane-bound enzymes (Kimelberg, 1977; Sandermann, 1978) and adenylate cyclase appears to be no exception (Houslay and Gordon, 1983). Hormone stimulation of adenylate cyclase is strongly influenced by the membrane lipid environment (Ross and Gilman, 1980; Houslay and Gordon, 1983). Optimal hormonal control can be significantly affected by manipulations changing the physicochemical state and/or composition of the lipid environment of the adenylate cyclase system (Houslay *et al.*, 1976; Dipple and Houslay, 1978; Engelhard *et al.*, 1978; Klein *et al.*, 1978; Bakardjieva *et al.*, 1979; Sinensky *et al.*, 1979; Briggs and Lefkowitz, 1980; Rimon *et al.*, 1980; Houslay *et al.*, 1980, 1981; McOsker *et al.*, 1983). More specifically, an important role for lipids has been implicated in the TSH-responsive adenylate cyclase system (Yamashita and Field, 1973; Moore and Wolff, 1974; Amir *et al.*, 1976; Dacremont *et al.*, 1984); certain lipids would be involved in the coupling mechanism between the occupied TSH-receptor and the catalytic component. Several experimental approaches have been employed to evaluate the role of phospholipids in the functioning of the adenylate cyclase system.

Attempts have been made to modify the phospholipid composition and then to determine the effect of such alterations on the adenylate cyclase activity. In the literature, membrane phospholipid composition has been altered by treatment of membranes with various phospholipases (Rubalcava and Rodbell, 1973; Limbird and Lefkowitz, 1976; Low and Finean, 1977; Lad *et al.*, 1979), organic solvents (Rottem *et al.*, 1973), reconstitution of detergent-solubilized adenylate cyclase preparations with specific phospholipids (Levey, 1971; Hebdon *et al.*, 1981; Ross, 1982), fusion of phospholipid vesicles with membranes (Houslay *et al.*, 1976), and modification *in vivo* (Engelhard *et al.*, 1976, 1978). All these methods, however, suffer from certain drawbacks and it has not been possible to draw a general conclusion from these studies. In recent years the use of lipid transfer proteins (both specific and nonspecific) to modify membrane lipid composition has obtained increasing attention (Barsukov *et al.*, 1978; Dyatlovitskaya *et al.*, 1979; McOsker *et al.*, 1983). Such proteins promote exchanges of lipid molecules between lipid vesicles and biological membranes (Bloj and Zilversmit, 1981) under relatively nonperturbing conditions.

Recently, it was found that a nonspecific lipid transfer protein (ns-TP) purified from beef liver (Crain and Zilversmit, 1980) is able to modify the lipid composition of bovine thyroid plasma membranes (H. Depauw, M. De Wolf, G. Van Dessel, H.J. Hilderson, A. Lagrou, and W. Dierick, in preparation). Upon incubation of bovine thyroid plasma membranes with ns-TP and SUV containing a specific phospholipid and cholesterol to a cholesterol/phospholipid molar ratio equal to that of the plasma membranes (no net exchange of cholesterol), the plasma membranes were enriched in that particular phospholipid. This was of great value in studying the influence of the environmental lipid molecules on the TSH signal expression.

Because the adenylate cyclase system is inactivated during the purification procedure of bovine thyroid PM, this enzyme was determined on a fraction obtained after differential pelleting (Dierick and Hilderson, 1967) of a thyroid homogenate. The L fraction was used for this purpose because it displays the highest enrichment of plasma membranes.

From Table IV it is clear that the nonspecific lipid transfer protein of beef liver is able to promote exchange of anionic (PS, PI, PA) as well as zwitterionic (PC, PE) phospholipids. The transfer of anionic phospholipids in the presence of 13–14 units of ns-TP is approximately two times higher than that of PE. The addition of an equimolar amount of PI to PC vesicles provokes a twofold increase in PC transfer activity. A similar enhancement of phospholipid exchange by ns-TP when acidic phospholipids are present in donor or acceptor vesicles has been reported previously (Dicorleto and Zilversmit, 1977; Crain and Zilversmit, 1980). The spontaneous, nonprotein-mediated transfer of the acidic phospholipids (PS, PI, PA) is, respectively, two and ten times higher than that of PE and PC (Table IV).

Incubation of bovine thyroid plasma membranes with lipid-exchange proteins and PI-containing liposomes results in a marked decrease in TSH-stimulated adenylate cyclase activity (Table IV). Incubation of PM with 13–14 units of ns-TP and PS- or PA-containing SUV results in approximately the same percentage lipid transfer as with PI-containing SUV (~41%) but provokes a much smaller decrease in TSH-stimulated AC activity (~27% versus 66%) (Table IV). The percentage inhibition of TSH-stimulated AC activity varies linearly with the amount of each phospholipid incorporated (Figure 6). Calculation of the slopes of the regression lines reveals that PI is two to four times more effective as an inhibitor of TSH stimulation than the other phospholipids tested. A similar specificity for PI has been reported previously for the isoproterenol-stimulated AC of turkey erythrocytes (McOsker et al., 1983). In an attempt to restore the enzyme activity, the PI-modified PM are treated with PI-specific phospholipase C or are incubated with ns-TP and PC/CH-SUV. Incubation with PC/CH-SUV results in a removal of 60% of the incorporated PI, while treatment with PI-specific phospholipase C results in hydrolysis of 86% of the incorporated PI. However, neither method

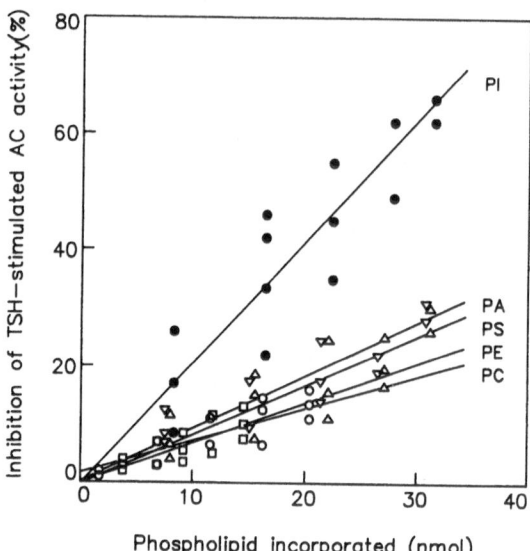

FIGURE 6. Inhibition of TSH-stimulated adenylate cyclase activity as a function of the amount of each phospholipid incorporated: phosphatidylcholine (○), phosphatidylethanolamine (□), phosphatidylserine (△), phosphatidic acid (▽), and phosphatidylinositol (●). Bovine thyroid PM (0.6 μmol phospholipid) were incubated with SUV (0.15 μmol phospholipid) containing the indicated phospholipid and increasing amounts of ns-TP. Data were fit to a straight line by the method of linear regression.

used reverses the observed inhibition of TSH-stimulated adenylate cyclase activity. Assuming that phospholipid "flip-flop" in bovine thyroid PM is similar to that reported for rat erythrocytes (Bloj and Zilversmit, 1981), it is reasonable to suggest that the incorporated PI is chiefly located in the outer monolayer of the PM vesicles. These data suggest that inhibition of TSH stimulation is due to a nonremovable pool of PI located at the outer leaflet of the PM phospholipid bilayer. In this respect it has been shown that treating PM with PI-specific phospholipase C can lead to a selective release of some membrane-bound enzymes in soluble form (Low and Finean, 1978; Low and Zilversmit, 1980; Shukla *et al.*, 1980), pointing out that PI may be involved in the attachment of these enzymes to the external surface of the PM. Therefore, it is not excluded that by association with some PM proteins (such as a component of the adenylate cyclase system) part of the incorporated PI becomes no longer removable.

Incorporated PI may interfere with one or more of the sequential events between initial TSH binding and ultimate activation of the catalyst of adenylate cyclase. Binding experiments demonstrate that TSH binding to bovine thyroid PM is not affected by PI incorporation. Thus, PI appears to act as an inhibitor of TSH-stimulated cyclase at a step beyond TSH binding. Basal activity and stimulation of AC by forskolin, which is reported to act at the level of the catalyst itself (Seamon and Daly, 1981), as well as activation of adenylate cyclase by cholera toxin, which is known to act at the level of the G_s-protein (Sahyoun and Cuatrecasas, 1975; Cassel and Pfeuffer, 1978), were inhibited by approximately 10% in PI-modified PM, under conditions where TSH stimulation was decreased by 66% (Table IV). These results indicate that the PI effect does not interfere with the activating coupling between the α subunit of the G_s-protein and the catalyst nor does it act at the level of the catalyst itself. Thus, by elimination, PI probably acts as a negative modulator of TSH receptor expression by interfering with the efficiency of the activating coupling between the occupied TSH receptor and the G_s-protein. This can be due to changes in fluidity or to a direct interaction of the acidic phospholipid PI with the TSH receptor. However, our system, incorporation of PI in bovine thyroid PM does not affect the steady-state anisotropy* of TMA-DPH, while the anisotropy of DPH has only a slight tendency to increase. This suggests that the PI effect is not due to changes in membrane fluidity but rather to a specific and direct interaction of this phospholipid with the TSH receptor.

According to Gilman (1984) the hormone–receptor complex acts catalytically to activate G_s and presumably facilitates a conformational change of G_s. It is tempting to suggest that PI may be associated to the occupied TSH receptor inducing a conformational change that leads to less affinity for the interaction with the G_s-protein.

In our system, incorporation of PI, in the absence or presence of ns-TP, has no effect on TSH binding although other investigators found that addition of acidic phospholipids to thyroid PM (PI ≈ PA > PS) diminished TSH receptor binding (Omodeo-Sale et al., 1978; Aloj et al., 1979b). An explanation to be considered is that incomplete removal of PI-containing liposomes from PM after the preincubation period partially blocks TSH binding due to a direct interaction with the hormone. In the turkey erythrocyte system, PI incorporation has also no effect on β-adrenergic receptor binding (McOsker et al., 1983).

Several studies have shown that in thyroid PM, phosphatidylcholine is

---

*Throughout this and the following sections of this chapter, the steady-state fluorescence anisotropy (r_s) is used as a measure of the membrane lipid fluidity without further resolution of the components that determine r_s. For many current biological applications, changes in r_s, whether owing to changes in the rotational correlation time (φ), the limiting anisotropy $(r_∞)$, or both, are of significance and are often designated as changes in membrane fluidity. The term ''membrane fluidity'' is used here in a general sense to express the overall motional freedom of the lipid molecules. So one must bear in mind that the anisotropy parameter so used reflects the overall motional freedom of the lipid molecules without distinguishing the specific mechanisms affecting their motions.

Table IV
Inhibition of TSH-Stimulated Adenylate Cyclase Activity by Incorporation of Various Phospholipids in Bovine Thyroid PM[a]

SUV	ns-TP	Lipid transfer (%)	cAMP	Inhibition (%)
—	—	—	0.60	—
—	+	—	0.60	<1
PC/CH (1 : 0.4)	—	1.1	0.59	2
	+	13.5	0.51	14
PC/PI/CH (1 : 1 : 0.8)	—	11.0	0.50	17
	+	42.4	0.20	66
PC/PE/CH (1 : 1 : 0.8)	—	5.1	0.58	3
	+	19.3	0.54	10
PC/PS/CH (1 : 1 : 0.8)	—	10.7	0.56	7
	+	41.6	0.44	26
PC/PA/CH (1 : 1 : 0.8)	—	10.0	0.56	7
	+	41.0	0.43	28

Influence of Some Effectors on Adenylate Cyclase Activity in Unmodified and PI-Modified Bovine Thyroid PM

Agonist	cAMP			Inhibition (%)	
	Control	PC/CH-SUV	PC/PI/CH-SUV	PC/CH-SUV	PC/PI/CH-SUV
—	0.29	0.28	0.27	3	7
TSH (50 mU/ml)	0.59	0.51	0.20	14	66
Cholera toxin (5 μM)	1.80	1.57	1.58	13	12
Forskolin (10 μM)	1.30	1.14	1.16	12	11

[a]Bovine thyroid PM (0.6 μmol phospholipid) were incubated with SUV (0.15 μmol phospholipid) and 13–14 units of ns-TP for 1 hr at 37°C in a total volume of 0.5 ml. Control incubations contained bovine thyroid PM with or without ns-TP only. Lipid transfer is expressed as percentage of the indicated phospholipid. Percentage of inhibition was determined relative to control incubations. SUV compositions are expressed as molar ratios. cAMP is expressed as pmol/mg protein per 30 min.

probably a critical phospholipid for the hormone responsiveness of adenylate cyclase (Yamashita and Field, 1973; Moore and Wolff, 1974; Amir *et al.*, 1976). These observations, taken together with our results, therefore suggest that phospholipids may play a dual role in the regulation of AC activity— PC being probably an essential component for TSH stimulation and PI, if present in the outer half of the bilayer, acting as a negative modulator of TSH receptor expression. A similar modulatory action of PI has been suggested previously (McOsker *et al.*, 1983) for the β-adrenergic receptor-mediated regulation of AC in turkey erythrocytes.

4.2. Incorporation of Gangliosides

Gangliosides are important components of plasma membranes and have been shown to affect adenylate cyclase activity (Parington and Daly, 1979;

Dacremont *et al.*, 1984). Studies on TSH binding to thyroid membranes have implicated higher-order gangliosides as a potential component of the thyrotropin receptor (Kohn and Shifrin, 1982). In contrast, other investigators have reported that under *physiological conditions* gangliosides do not block TSH binding and consequently are not a part of the TSH receptor (Beckner *et al.*, 1981). However, apart from binding TSH, gangliosides might have a more important role in the transmission of the TSH signal through the membrane (Mullin *et al.*, 1976, 1978; Kohn, 1978; Aloj *et al.*, 1979a; Valente *et al.*, 1982). A specific inhibitory effect of gangliosides on basal and stimulated adenylate cyclase activity in human thyroid membranes has been reported (Dacremont *et al.*, 1984). A possible role of gangliosides as cofactors of membrane adenylate cyclase has already been suggested by Parington and Daly (1979).

Spontaneous incorporation of gangliosides in membranes is a well-established phenomenon (Kanda *et al.*, 1982; Felgner *et al.*, 1983; Brown *et al.*, 1985). However, less information exists on the protein-mediated transfer of gangliosides between membranes. Bloj and Zilversmit (1981) reported that GM_1 exchange between membranes is catalyzed by ns-TP from beef liver. We were able to demonstrate that the incorporation of other gangliosides such as GD_3 and GT_{1b} in bovine thyroid plasma membranes is also catalyzed by ns-TP. The protein-mediated exchange of GT_{1b} or GD_3 is twice as high as that of GM_1 (H. Depauw, M. De Wolf, G. Van Dessel, H. J. Hilderson, A. Lagrou, W. Dierick, in preparation). These data are in agreement with those of Felgner *et al.* (1983), who observed that the transfer rate for GT_{1b} is substantially larger than that for GM_1. However, taking into account the spontaneous exchange (approximately 8% for GT_{1b} and GD_3 and 1% for GM_1), ns-TP seems to be more effective for GM_1 (enhancement of the exchange activity is 18 times versus 4.5 times).

Gangliosides are incorporated into biological membranes with the lipophilic ceramide portion embedded in the membrane and the hydrophilic carbohydrate portion protruding into the surrounding medium (Irwin, 1974). This structural feature provides the potential for selective control of membrane characteristics both through fluidity changes and through specific interaction of the carbohydrate moiety with membrane-surface-associated components such as hormone receptors. Thus, gangliosides could be involved in the organization of the components of the membrane structure and hence in the regulation of adenylate cyclase.

Incorporation of gangliosides by ns-TP in bovine thyroid PM is found to have a significant inhibitory effect on TSH-stimulated adenylate cyclase activity with the following order of efficacy: $GT_{1b} > GD_3 > GM_1$ (Table V). Gangliosides carrying a disialosyl group are stronger inhibitors. These data may suggest that there exists a correlation between the extent of inhi-

Table V

Stimulated Adenylate Cyclase Activity in Native and Ganglioside-Modified Bovine Thyroid PM[a]

Agonist	cAMP (pmol/mg protein per 30 min)				Inhibition (%)		
	Control	$+GM_1$	$+GD_3$	$+GT_{1b}$	$+GM_1$	$+GD_3$	$+GT_{1b}$
TSH (50 mU/ml)	0.60	0.47	0.38	0.32	23	37	47
Cholera toxin (5 μM)	1.80	1.62	1.28	1.22	10	29	32
Forskolin (10 μM)	1.20	1.16	1.15	1.14	3	4	5
NaF (10 mM)	3.00	2.79	2.52	2.46	7	16	18

Effect of Ganglioside Incorporation in Bovine Thyroid PM by ns-TP on Membrane Fluidity

SUV	r_{sDPH}	r_{s12-AS}	$r_{sTMA-DPH}$
Control	0.172	0.132	0.225
PC/CH/GM₁ (1 : 0.4 : 0.1)	0.173	0.133	0.228
PC/CH/GT₁b (1 : 0.4 : 0.1)	0.195	0.136	0.239
PC/CH/GD₃ (1 : 0.4 : 0.1)	0.205	0.138	0.226

[a]Bovine thyroid PM (0.6 μmol phospholipid) were incubated with ganglioside-containing SUV and ns-TP for 1 hr at 37°C in a total volume of 0.5 ml. This incubation resulted in an incorporation of 11 nmol of each ganglioside into bovine thyroid PM. Percentage of inhibition of adenylate cyclase activity was determined relative to control incubations, which contained bovine thyroid PM with ns-TP alone. SUV compositions are expressed as molar ratios. Anisotropy measurements were performed at 37°C. The fluorescent probe/phospholipid molar ratio was always 1 : 500.

bition and the structure of the gangliosides. Our results are in agreement with those of Dacremont *et al.* (1984) except that in human thyroid membranes GD_3 is more inhibitory than GT_{1b}. However, comparison with these results is difficult because they incubated gangliosides with the PM where we incorporated gangliosides into the PM using ns-TP. Binding experiments demonstrate that TSH binding to bovine thyroid PM is inhibited by ganglioside incorporation with the same order of efficacy ($GT_{1b} > GD_3 > GM_1$) and to the same extent as their inhibitory effect on TSH stimulation. Therefore, it is suggested that the decrease in TSH-stimulated AC activity upon ganglioside incorporation is chiefly the result of a drop in TSH binding. Mullin *et al.* (1976) also observed that inhibition of thyrotropin binding is critically altered by the number and location of the sialic acid residues within the ganglioside molecule.

Although it is shown that incorporation of gangliosides into bovine thyroid PM inhibits the TSH-stimulated state of AC and TSH binding, it is possible that they also exert an inhibitory effect at another point subsequent to hormone binding. As shown in Table V, forskolin stimulation is not significantly affected by ganglioside modification of the plasma membranes, indicating that the gangliosides do not act at the level of the catalyst of adenylate cyclase. Incorporation of GT_{1b} or GD_3 in bovine thyroid plasma membranes gives rise to a substantial decrease in cholera-toxin-stimulated adenylate cyclase activity ($\sim 30\%$) and to a lesser degree to a decrease in NaF-stimulated activity ($\sim 17\%$), whereas GM_1 incorporation does not significantly affect these stimulated activities (Table V). These results indicate that gangliosides might also interfere with the efficiency of the activating coupling between the G_s-protein and the catalyst of adenylate cyclase.

This can be due to changes in membrane fluidity or to a direct interaction of the gangliosides with the G_s-protein. The inhibitory effects on cholera toxin and NaF stimulation are paralleled by changes in fluorescence steady-state anisotropy: GT_{1b} modification of the plasma membranes provokes a slight increase in TMA-DPH anisotropy, whereas the anisotropy of DPH is substantially enhanced after incorporation of GD_3 or GT_{1b} (Table V). The fluorescence anisotropy of 12-AS is unaffected, GM_1 modification of the plasma membranes has no effect on the fluorescence anisotropy of any of the probes tested. These results are in agreement with other physicochemical studies in which it is shown that gangliosides may have significant influence on molecular motion in phosphatidylcholine bilayers. Their presence generally results in a rigidification of the bilayer or an increase in lipid order (Sharom and Grant, 1978; Bertoli et al., 1981; Uchida et al., 1981; Hitzemann et al., 1984). In Section 3.3, we have shown that incorporation of bovine thyroid membrane gangliosides in reconstituted vesicles provokes an increase in both the lipid order parameter of DPH and the lateral diffusion coefficient of pyrene. This means that the ganglioside-expanded membranes might display at once higher values of both the lipid order parameter and the lateral diffusion coefficient resulting in an easier removal of lipid molecules by ns-TP.

Gangliosides incorporated by ns-TP are believed to be located almost entirely in the outer leaflet of the membrane bilayer. Nevertheless, they are able to inhibit the enzymic activity of cholera-toxin- and NaF-stimulated cyclase, both activities residing in the cytoplasmic side of the lipid bilayer. This can be due to the presence of inside-out vesicles or open structures in the plasma membrane preparations. Electron micrographs reveal that thyroid plasma membranes are almost exclusively present under the form of closed vesicular structures (H. Depauw, unpublished observations). Upon treatment of bovine thyroid plasma membrane preparations with the diazonium salt of sulfanilic acid, a maximum 20% of the activity of the ecto-enzyme 5'-nucleotidase escapes from inhibition. This indicates that there is only a minor proportion of inside-out vesicles and/or open structures present. Therefore, gangliosides incorporated at the outer side of the bilayer would exert an influence on enzymic activities residing in the cytoplasmic side of the bilayer. The observed changes in membrane fluidity after ganglioside modification of the plasma membranes may be responsible for the decrease in cholera-toxin- and NaF-stimulated cyclase activities. The greater inhibition of cholera-toxin-stimulated enzyme activity could be due to the fact that this activity is much more sensitive to changes in membrane fluidity than the NaF-stimulated one. In Section 3.3, it is shown that changes in temperature provoke significant changes in membrane fluidity. The data of Van Sande *et al.* (1979), taken together with our own results, show that stimulation of adenylate cyclase by cholera toxin decreases significantly when lowering the temperature of the adenylate cyclase incubation, whereas NaF stimulation was more or less conserved. These observations indeed suggest that the cholera-toxin-stimulated state of the enzyme is much more sensitive to changes in membrane fluidity than the NaF-stimulated one.

The extent of all these effects can be correlated with the structure of the ganglioside molecules: gangliosides with a disialosyl group display the largest effects and the extent of the effects, especially in the case of inhibition of TSH stimulation, seems to be dependent on the number of sialic acid residues.

4.3. Incorporation of Dolichol and Dolichyl Derivatives

Some mammalian organs, in particular the pituitary and thyroid gland, contain relatively large amounts of free and esterified dolichol (Van Dessel *et al.*, 1979; Eggens *et al.*, 1983). It is still a matter of debate whether or not dolichols provide one of the structural elements of the membrane bilayer (McCloskey and Troy, 1980; Vigo *et al.*, 1984; Valtersson *et al.*, 1985). Recent studies have shown that dolichol and its derivatives are present under several cytoplasmic- and membrane-associated forms (Appelkvist *et al.*, 1985; Steen *et al.*, 1986). Because of the rather large variations in dolichol content of thyroid tissue (Hilderson *et al.*, 1984), the question can be addressed whether these variations affect the physicochemical state of thyroid membranes. Indeed, the special poly-cis geometry and the unusual length (about 100 Å) of these very hydrophobic molecules should endow them with some unique physical properties when they interact with phospholipid bilayers. Dolichylphosphate functions as a chemical carrier of saccharide units during the membrane assembly of mammalian glycoproteins (Waechter and Lennarz, 1976). Although a substantial literature concerning dolichylphosphate has accumulated (Hemming, 1974; Lennarz, 1975; Parodi and Leloir, 1979), little is known about its effects on membrane motional freedom.

In respect to the influence of dolichol on the membrane fluidity of PC bilayers, discrepancies in results and interpretations are found in the literature (Lai and Schutzbach, 1984; Vigo *et al.*, 1984; Valtersson *et al.*, 1985) probably due to differences in experimental conditions. Valtersson and collaborators studied membrane fluidity using ^{31}P-NMR, small angle x-ray scattering, and differential scanning calorimetry. They observed that the influence of dolichol on bilayer fluidity is dependent on the number of isoprene units present in the dolichol molecule. They state that dolichol does not affect *membrane fluidity* in PC bilayers (as measured by their techniques). On the other hand, Vigo and collaborators using the techniques of differential scanning calorimetry and fluorescence depolarization reported that incorporation of dolichol in multilamellar liposomes composed only of PC increases the motional freedom of the bilayer. Finally, Lai and Schutzbach (1984) demonstrated that the presence of dolichol in PE/PC vesicles induces membrane leakage whereas the presence of dolichol in pure PC membranes inhibits membrane leakage. The dolichol-induced inhibition of membrane leakage in pure PC vesicles was explained by suggesting a decrease in PC membrane fluidity by dolichol. The latter statement was confirmed in our laboratory, as incorporation of dolichol in DPPC-SUV results in an overall decrease in *membrane fluidity* (H. Depauw, M. DeWolf, G. Van Dessel, H. J. Hilderson, A. Lagrou, and W. Dierick, in preparation 1986). In DPPC/PE-SUV, however, dolichol does not significantly affect the lipid order at the hydrophilic–hydrophobic interface of the membrane at all temperatures below the transition temperature but provokes a large increase in the lipid motional freedom in the hydrophobic membrane core in the same temperature range (Table VI). These results may explain the different behavior of PC and PC/PE vesicles vis-à-vis membrane leakage after incorporation of dolichol.

From Table VI it is clear that in DPPC vesicles low concentrations of dolichylmonophosphate decrease the bilayer motional freedom of the hydrophobic core below the transition temperature, while high concentrations increase the motional freedom in the same membrane region. Above the transition temperature DMP decreases the bilayer fluidity of the hydrophobic core at all concentrations, as evidenced by an increase in DPH anisotropy. These results are in agreement with those of Vigo *et al.* (1984). At the hydrophobic–hydrophilic interface of the DPPC bilayer DMP acts to "stiffen" the membrane at all concentrations, as indicated by an increase in TMA-DPH anisotropy. The membrane fluidity near the bilayer center is not significantly affected by dolichylmonophosphate except below the transition temperature at low concentrations where it acts to rigidify the membrane, as evidenced by an increase in 12-AS anisotropy. In contrast to the effects on DPPC vesicles, incorporation of dolichylmonophosphate in PE-containing SUV provokes an overall increase in the bilayer motional freedom below the transition temperature.

Apparently, the effect of dolichols and dolichyl derivatives on the physicochemical state of a bilayer depends on their chain length, their concentration, the temperature, and the lipid composition of the membrane model.

The different effects of dolichol or dolichylmonophosphate on membrane

Table VI
Effects of Dolichol and Dolichylmonophosphate on the Steady-State Fluorescence Anisotropies of Different Fluorescent Probes Embedded in Small Unilamellar Vesicles[a]

SUV T(°C)	$r_{s\text{DPH}}$				$r_{s\text{TMA-DPH}}$				$r_{s\text{12-AS}}$			
	Control	+Dol (10:1)	+DMP (10:1)	+DMP (1000:1)	Control	+Dol (10:1)	+DMP (10:1)	+DMP (1000:1)	Control	+Dol (10:1)	+DMP (10:1)	+DMP (1000:1)
PC-SUV												
20	0.337	0.355	0.322	0.352	0.340	0.390	0.368	0.373	0.160	0.175	0.166	0.178
25	0.332	0.352	0.317	0.347	0.335	0.385	0.363	0.370	0.158	0.173	0.164	0.176
48	0.082	0.094	0.092	0.093	0.190	0.227	0.197	0.205	0.076	0.086	0.083	0.081
52	0.073	0.085	0.083	0.083	0.183	0.220	0.194	0.201	0.074	0.084	0.081	0.077
PC/PE-SUV												
20	0.295	0.265	0.285	0.280	0.300	0.308	0.288	0.298	0.180	0.170	0.165	0.163
25	0.285	0.255	0.275	0.270	0.295	0.303	0.282	0.293	0.177	0.167	0.164	0.161
48	0.122	0.114	0.120	0.120	0.222	0.237	0.218	0.224	0.107	0.102	0.105	0.099
52	0.114	0.106	0.112	0.112	0.220	0.235	0.216	0.222	0.104	0.100	0.102	0.096

[a] Control SUV are composed of PC or PC/PE alone. PC/PE-SUV contained equimolar amounts of PC and PE. Dolichol or dolichylmonophosphate was incorporated into the control SUV at phospholipid/dolichol or DMP molar ratios as indicated in the table. The fluorescent probe/phospholipid molar ratio was always 1 : 500. The results are the means of three separate experiments.

fluidity between PC and PE/PC vesicles suggest that the interactions of these compounds with the two model membranes occur via different mechanisms. It is known that PC is a typical bilayer-forming lipid (Small, 1970), whereas PE prefers to adopt the hexagonal (HII) phase (Cullis and De Kruijff, 1978; Tilcock *et al.*, 1982; Seddon *et al.*, 1983). It has been proposed that dolichol and dolichylmonophosphate might promote hexagonal II phase formation (Lai and Schutzbach, 1984; Valtersson *et al.*, 1985) in PE-containing membranes, causing an increase in membrane permeability and fluidity. In PC membranes, on the other hand, dolichol should have a much lower solubility than in PE membranes and as a consequence might be present in clusters (Valtersson *et al.*, 1985). McCloskey and Troy (1980) noted the self-aggregation of neutral polyisoprenoids at molar ratios of polyisoprenoid/phospholipid greater than 1 : 200. Formation of such dolichol clusters could account for the rigidifying effect as observed in Table VI. The results of Valtersson and collaborators are consistent with a localization of dolichylmonophosphate in which the phosphate group is oriented to the water interface. Upon incorporation of dolichylmonophosphate there is a considerable increase in membrane rigidity in PC-SUV at the lipid–water interface (Table VI). This might be due to the molecule acting as a rigid "anchor" in this membrane region, consequently restricting the movement of DPPC molecules. Below the transition temperature a rigidifying effect is also observed at the hydrophobic core of the DPPC bilayer at low DMP concentrations, while at high DMP concentrations the motional freedom is increased. The latter phenomenon might be explained by segregation of dolichylmonophosphate from the rest of the lipids as the concentration of dolichylmonophosphate increases. Separate domains of DPPC and dolichylmonophosphate might occur at a molar ratio of 1 : 10, creating irregularities in the packing at the border of the domains where DPPC molecules can move more freely (Vigo *et al.*, 1984).

In order to investigate a potential relation between the effects of dolichol or DMP on the lipid order in artificial membranes and their effects on biomembranes, dolichol and DMP were incorporated in bovine thyroid membranes using the nonspecific lipid transfer protein from beef liver. It is noteworthy that thus far no studies have been published indicating that incorporation of dolichol or DMP in biomembranes is catalyzed by ns-TP from beef liver. It is observed that the spontaneous exchange activity, as well as the protein-mediated exchange activity, of DMP is higher than that of dolichol (18.1 and 6.4% versus 11.2 and 1.4%). This was evidently brought about by the presence of the phosphate group in dolichylmonophosphate. In Sections 4.1 and 4.2, we already mentioned similar events in respect to acidic phospholipids and gangliosides. Apparently the presence of a negatively charged group on a lipid molecule causes them to be removed more easily. In the case of acidic phospholipids and gangliosides, this effect is accompanied by a tendency to increase the fluorescence anisotropy of DPH or an enhancement in the lipid order. It is not excluded that this is also occurring upon incorporation of dolichol or dolichylmonophosphate, although this is not immediately apparent from Table VII. Maybe other phenomena are superimposing.

In thyroid gland, high levels of dolichol have been reported (Van Dessel *et al.*, 1979; Eggens *et al.*, 1983). Therefore, the effects of dolichol or dolichylmonophosphate on bovine thyroid membrane physical properties were investigated. Subcellular fractions enriched in plasma membranes or endoplasmic reticulum membranes display a dolichol/phospholipid molar ratio of approximately 0.032. The dolichylmonophosphate/phospholipid molar ratio is approximately 1% of that of dolichol. Incubation of bovine thyroid plasma membranes with ns-TP and dolichol-containing SUV results in an incorporation of one molecule of dolichol for every 100 phospholipid molecules. This incorporation gives rise to an overall increase in membrane lipid order as indicated by an increase in the anisotropies of DPH, TMA-DPH, and 12-AS (Table VII). Incorporation of five molecules of dolichylmonophosphate for every 1000 phospholipid molecules does not affect the membrane motional freedom.

Incorporation of dolichol into endoplasmic reticulum membranes (Table VIII) also results in an overall increase in membrane lipid order. In contrast to plasma membranes, dolichylmonophosphate provokes an increase in DPH and TMA-DPH anisotropies. PE enrichment of membranes does not affect the effects of dolichol or DMP on the fluidity of endoplasmic reticulum membranes, except that the rigidifying effects of dolichol near the bilayer center and of DMP at the hydrophobic core are abolished (Table VIII). Thus, PE modification seems to neutralize partially the increase in lipid order brought about by dolichol or DMP. However, PE modification does not result in the

Table VII
Effects of Dolichol and Dolichylmonophosphate on the Fluorescence Anisotropy Parameter of Different Fluorophores Embedded in Bovine Thyroid PM[a]

T (°C)	r_{sDPH}			$r_{sTMA-DPH}$			r_{s12-AS}		
	Control	+ DMP	+ Dol	Control	+ DMP	+ Dol	Control	+ DMP	+ Dol
12	0.240	0.243	0.257	0.294	0.297	0.306	0.200	0.205	0.210
18	0.223	0.227	0.239	0.277	0.276	0.288	0.183	0.186	0.193
32	0.184	0.186	0.198	0.238	0.235	0.249	0.144	0.142	0.152
37	0.170	0.173	0.183	0.224	0.221	0.235	0.130	0.126	0.138

[a]Bovine thyroid PM (0.6 µmol phospholipid) were incubated with 16 units of ns-TP and either PC/Dol/CH (1 : 0.1 : 0.4 mol/mol) or PC/DMP/CH (1 : 0.03 : 0.4 mol/mol) SUV (0.6 µmol phospholipid) for 1 hr at 37°C in a total volume of 0.5 ml. Incubation resulted in an incorporation of 10 molecules dolichol or 5 molecules DMP for every 1000 phospholipid molecules. Control incubations contained bovine thyroid PM with ns-TP and PC/CH-SUV.

Table VIII
Effects of Dolichol and Dolichylmonophosphate on the Fluorescence Anisotropy Parameter of Different Fluorophores Embedded in Native and PE-Modified Membranes of the Endoplasmic Reticulum[a]

Membrane T (°C)	r_{sDPH}			$r_{sTMA\text{-}DPH}$			$r_{s12\text{-}AS}$		
	Control	+ DMP	+ Dol	Control	+ DMP	+ Dol	Control	+ DMP	+ Dol
Native membranes									
12	0.230	0.240	0.243	0.283	0.302	0.312	0.189	0.191	0.199
18	0.213	0.223	0.226	0.267	0.286	0.296	0.173	0.175	0.184
32	0.174	0.186	0.189	0.228	0.247	0.257	0.134	0.134	0.144
37	0.161	0.172	0.175	0.215	0.234	0.244	0.121	0.121	0.130
PE-modified membranes									
12	0.240	0.239	0.250	0.294	0.304	0.309	0.199	0.200	0.204
18	0.223	0.223	0.233	0.277	0.287	0.292	0.184	0.183	0.188
32	0.184	0.185	0.194	0.237	0.248	0.253	0.144	0.146	0.146
37	0.171	0.172	0.181	0.224	0.234	0.240	0.130	0.131	0.132

[a]PE-modified membranes were obtained by incubating membranes (0.6 μmol phospholipid) with SUV (0.3 μmol phospholipid, PC/PE/CH 1 : 1 : 0.3 mol/mol) and 16 units of ns-TP for 1 hr at 37°C in a total volume of 0.5 ml. Dolichol or DMP was incorporated in native or PE-modified membranes as described in Table VII, except that the cholesterol/phospholipid molar ratio of the SUV is 0.3. PE is incorporated at a ratio of 1 molecule for every 20 phospholipid molecules. Dolichol and DMP are incorporated at respective ratios of 10 and 5 molecules for every 1000 phospholipid molecules. Control incubations contained native or PE-modified membranes with ns-TP and PC/CH-SUV.

conversion of an increase in lipid order caused by dolichol or DMP into an increase in lipid motional freedom as was the case with PC and PC/PE vesicles.

Obviously there is no simple relation between the effects of dolichol or DMP on the fluidity of PC and PC/PE model membranes on one hand and bovine thyroid membranes on the other hand. Other membrane components present in bovine thyroid membranes might modify or neutralize the influence of dolichol or DMP on membrane fluidity. Therefore, it is premature to conclude that these isoprenoids play a role as modulators of the physical properties of thyroid membranes.

4.4. Addition of Membrane-Perturbing Drugs

Local anesthetics have been thought to expand the lipid bilayer (Seeman, 1972), to cause it to thicken (Ashcroft *et al.*, 1977; Haydon *et al.*, 1977), to increase the "molten" portion of the bilayer (Trudell, 1977; Vanderkooi *et al.*, 1977), or to melt a lipid annulus that surrounds the protein (Lee, 1976a). These effects are taken to be consequences of the drug-induced "fluidization" of the membrane by anesthetics (Seeman, 1972; Roth, 1979; Miller, 1981). Spin-labeling experiments have provided evidence that they may either increase or decrease the degree of molecular order in phospholipid membranes, depending on lipid composition, anesthetic concentration, and solution pH (Butler *et al.*, 1973; Neal *et al.*, 1976; Pang and Miller, 1978). In a more recent theory it is argued that local anesthetics affect the membrane not by an amorphous fluidization of the hydrophobic core of the lipid bilayer but by a restructuring of its hydrogen belts, that is, the regions occupied by the CO and OH groups of the membrane lipids (Brockerhoff, 1982).

Several investigators have studied the effects of membrane-perturbing drugs on adenylate cyclase activity (Schroeder, 1985). The neutral local anesthetic benzyl alcohol, a well-known membrane fluidizer, was used to investigate the relation between changes in fluidity and changes in adenylate cyclase activity (Dipple and Houslay, 1978; Gordon *et al.*, 1980b). Similarly, cationic and anionic drugs may also affect the membrane physical state and even might have a parallel effect on both membrane fluidity and adenylate cyclase activity (Houslay *et al.*, 1981; Salesse *et al.*, 1982a, 1982b). It has also been suggested that owing to the asymmetric distribution of charged phospholipids in the membrane, bilayer-perturbing drugs of opposite charge may act preferentially at one or the other side of the bilayer (Sheetz and Singer, 1974). In this manner they might selectively modulate the activity of the adenylate cyclase complex (Houslay *et al.*, 1981; Salesse *et al.*, 1982a, 1982b; Houslay and Gordon, 1983). In the absence of hormone, adenylate cyclase should be sensitive to the lipid environment of the cytosol side of the bilayer only (Houslay and Palmer, 1978; Houslay, 1979). However, in the presence of hormone, the occupied receptor and the catalyst of adenylate cyclase interact to form a transmembrane complex spanning the lipid bilayer (Houslay *et al.*, 1977; Houslay, 1981) and as a consequence adenylate cyclase should be sensitive to the lipid environment of both halves of the bilayer (Houslay, 1979, 1981). Chlorpromazine (cationic drug) in conjunction with phenobarbital (anionic drug) and benzyl alcohol (neutral drug) were used in order to modulate selectively the different components of the adenylate cyclase system in bovine thyroid plasma membranes (H. Depauw, M. De Wolf, G. Van Dessel, H. J. Hilderson, A. Lagrou, and W. Dierick, in preparation). The membranes were incubated with the drugs prior to the addition of coupling factors.

As previously reported for liver plasma membranes (Gordon *et al.*, 1980b; Houslay and Gordon, 1983), adipocytes (Sauerheber *et al.*, 1982), intestinal brush border (Brasitius and Schachter, 1980), and kidney basolateral as well as brush-border membranes (Martin *et al.*, 1985; Carrière and Le Grimellec, 1986), benzyl alcohol decreases the anisotropy of DPH in bovine thyroid plasma membranes. A significant decrease in DPH anisotropy commences at 10^{-2} M BA (Figure 7b). Experiments using TMA-DPH and ANS show a progressive fluidization at the bilayer–water interface of the membrane at concentrations above 2×10^{-2} M BA. The fluidity near the lipid bilayer center is only slightly increased at the highest concentration tested (0.1 M

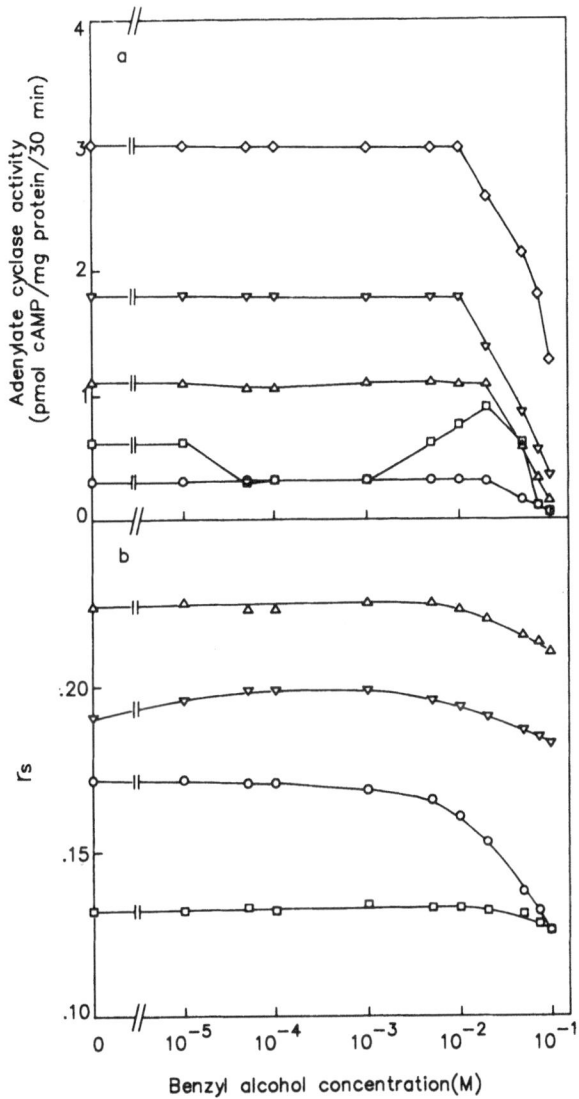

FIGURE 7. Effect of benzyl alcohol on (a) the adenylate cyclase activity and (b) the physical state of bovine thyroid plasma membranes. The change in adenylate cyclase activity (a) is shown for basal (O—O), TSH-stimulated (□—□), forskolin-stimulated (△—△), cholera-toxin-stimulated (▽—▽), and fluoride-stimulated (◇—◇) enzyme activity. The membranes were incubated with the drugs prior to the addition of coupling factors. The physical state of the membrane was evaluated by following the steady-state fluorescence anisotropy (b) of several fluorophores embedded in the membranes: DPH (O—O), 12-AS (□—□), TMA-DPH (△—△), and ANS (▽—▽). Measurements were performed at 37°C. The fluorophore/lipid molar ratio was always 1 : 500.

BA). These data are in agreement with the localization of BA in membranes, that is, adsorbed at the hydrophobic–polar interface of the lipid bilayer (Ashcroft *et al.*, 1977). The acyl chains of the phospholipids would become more disordered as they fold around to fill the space created by the insertion of the small benzyl alcohol molecule into the bilayer–water interface (Coster and Laver, 1986).

Benzyl alcohol at concentrations below 10^{-2} M does not affect basal and stimulated AC activities except for the TSH-stimulated one. The TSH-stimulated adenylate cyclase activity appears to be initially inhibited at low concentrations of BA (5×10^{-5}–10^{-3} M) and then reincreased up to or higher than (hyperactivation) its original level (5×10^{-3}–5×10^{-2} M) (Figure 7a). The initial inhibition of TSH stimulation is paralleled by a slight increase in ANS anisotropy (Figure 7b) and by an inhibition of TSH binding (Table IX). The latter inhibition may be due to a direct interaction of the alcohol with the receptor protein or to changes in the membrane physicochemical state as detected by the ANS fluorophore. This question is not yet elucidated. Restoration of ANS anisotropy to its original level is concomitant with restoration (or even hyperactivation) of the TSH-stimulated state of adenylate cyclase and of the TSH binding to their original levels. These results may point to the presence of two different binding sites for BA on the TSH receptor: one displaying both high affinity for the receptor and inhibitory effect on TSH binding and the other displaying low affinity for the receptor but restoring the TSH binding and the TSH-stimulated activity of the adenylate cyclase

Table IX
Effect of Drug Treatment on Binding of TSH to Bovine Thyroid Membranes[a]

Drug	Concentration (mM)	TSH binding vs. control (%)
—	—	100
Chlorpromazine	0.1	106
	0.5	75
	1.0	55
Phenobarbital	0.1	104
	1.0	98
	10.0	97
Benzyl alcohol	0.01	105
	0.5	28
	10.0	98
	50.0	92

[a]The membranes were incubated with the drugs prior to the binding assay. Binding of TSH to membranes not exposed to drugs was taken as 100%.

system. The hyperactivation of TSH-stimulated adenylate cyclase coincides with a progressive fluidization of the membrane as evidenced by the decrease in ANS, DPH, and TMA-DPH anisotropies (Figure 7b). Fluidization may lead to hyperactivity by relieving an inhibitory control exerted on the TSH receptor expression. At benzyl alcohol concentrations beyond 2×10^{-2} M, a progressive inhibition of basal as well as all stimulated adenylate cyclase activities is observed (Figure 7a). This inhibition is paralleled by an overall decrease in membrane rigidity, except near the center of the bilayer (Figure 7b). A possible explanation for these phenomena may be that a minimum membrane rigidity is required for the activity of the enzyme. Inhibition would be the result of a "too fluid" lipid bilayer. The overall decrease in membrane rigidity and adenylate cyclase activity may also point to a collapse of the membrane structure.

The anionic local anesthetic phenobarbital has been demonstrated to enhance the bilayer fluidity in neutral or positively charged liposomes, but not in negatively charged ones (Papahadjopoulos et al., 1975; Lee, 1976b). Houslay et al. (1981) observed that phenobarbital interacts preferentially with the external half of the bilayer of liver plasma membranes. The latter observation was to be expected as the negatively charged phospholipids are found almost exclusively at the inner surface of the liver plasma membranes (Higgins and Evans, 1978). Houslay et al. (1981) state that charged drugs exhibit "a greater tendency" to interact with one or the other half of the bilayer, but there is no evidence that any of these charged drugs should exclusively reside in only one half of the bilayer. This statement seems to be very applicable in the case of bovine thyroid plasma membranes, where all stimulated activities of adenylate cyclase are affected by phenobarbital (Figure 8a). Phenobarbital allows a hyperstimulation of fluoride-, cholera-toxin-, and forskolin-stimulated adenylate cyclase activities in the range 5×10^{-5}–10^{-3} M. At 10^{-4} M PB, however, a depression of cholera-toxin-stimulated activity is observed (Figure 8a). This depression is paralleled by a significant rise in ANS anisotropy (Figure 8b). In Section 4.2, it has already been mentioned that changes in fluidity may have a large effect on cholera toxin stimulation. This may explain the observation that only cholera toxin stimulation is affected by a rise in ANS anisotropy. In contrast to benzyl alcohol, phenobarbital is nearly without effect on the anisotropies of DPH, TMA—DPH, and 12-AS (Figure 8b). Other investigators have shown that PB preferentially fluidizes the external half of the membrane bilayer (Houslay et al., 1981; Sweet and Schroeder, 1986). However, the ability of PB to increase the fluidity of bilayers has been shown to be dependent on their cholesterol content (Pang and Miller, 1978), and moreover it has been demonstrated that PB has no effect on or even slightly decreased the fluidity at temperatures above 30°C or below 10°C (Houslay et al., 1981). At PB concentrations

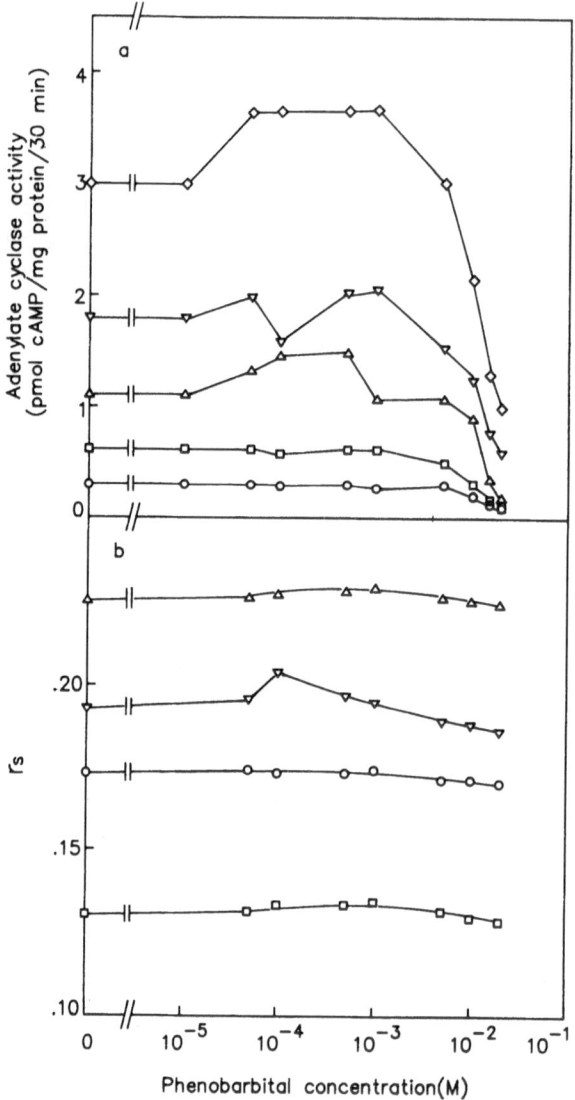

FIGURE 8. Effect of phenobarbital on (a) the adenylate cyclase activity and (b) the physical state of bovine thyroid plasma membranes. The change in adenylate cyclase activity (a) is shown for basal (O—O), TSH-stimulated (□—□), forskolin-stimulated (△—△), choleratoxin-stimulated (∇—∇), and fluoride-stimulated (◇—◇) enzyme activity. The membranes were incubated with the drugs prior to the addition of coupling factors. The physical state of the membrane was evaluated by following the steady-state fluorescence anisotropy (b) of several fluorophores embedded in the membranes: DPH (O—O), 12-AS (□—□), TMA-DPH (△—△), and ANS (∇—∇). Measurements were performed at 37°C. The fluorophore/lipid molar ratio was always 1 : 500.

beyond 5×10^{-3} M a progressive inhibition of basal as well as stimulated cyclase activities is observed, which is paralleled by a slight decrease in membrane rigidity (Figure 8). The latter phenomena might be due to the displacement of annular lipids from around the enzyme (Gordon *et al.*, 1980a, 1980b).

Chlorpromazine, being a cationic drug, is expected to act preferentially on the cytoplasmic half of the lipid bilayer (Houslay *et al.*, 1981). In bovine thyroid plasma membranes, the psychoactive drug CPZ elicits a hyperactivating effect on cholera-toxin- and NaF-stimulated adenylate cyclase activities in the 5×10^{-5}–3×10^{-4} M concentration range (Figure 9a). A further increase in CPZ concentration leads to a progressive inhibition of these stimulated activities. Forskolin- and TSH-stimulated cyclase activities are progressively inhibited at CPZ concentrations beyond 2×10^{-4} M (Figure 9a). This is in agreement with the results of other investigators who observed a substantial inhibition of thyroidal thyrotropin-sensitive adenylate cyclase by CPZ (Wolff and Jones, 1970; Yamashita *et al.*, 1970; Moore and Wolff, 1974). At CPZ concentrations higher than 5×10^{-5} M, the fluidity of bovine thyroid plasma membranes in the vicinity of the upper portions of the acyl chains is monotonously increased, as indicated by a drop in TMA-DPH anisotropy (Figure 9b). In contrast, CPZ exhibits a rigidifying effect near the bilayer center, as shown by a steep increase in 12-AS anisotropy. A significant drop in DPH anisotropy is observed at CPZ concentrations above 5×10^{-5} M (Figure 9b). Cholera toxin and NaF stimulation pass through the maximum of their activity curves in the same concentration range, where a substantial decrease in TMA-DPH and DPH steady-state anisotropies is observed (Figure 9). This suggests that the maxima in the activity curves may be the result of an increase in fluidity. A possible explanation for this phenomenon could be that a minimum membrane rigidity is required for maximal activity of the enzyme. Inhibition would be the result of a "too-fluid" lipid bilayer. These results may support the view of a fluidity-controlled coupling between the G_s-protein and the catalyst of adenylate cyclase. It is very unlikely that changes in the fluidity of the acyl chains near the center of the lipid bilayer are involved in activity changes because the increase in 12-AS anisotropy is never paralleled by any change in adenylate cyclase activity. At chlorpromazine concentrations higher than 5×10^{-4} M, Breton *et al.* (1977) observed membrane denaturation due to a partial solubilization of the proteins. At this concentration we also observed an overall decrease in adenylate cyclase activities as well as in steady-state anisotropies. The progressive inhibition of TSH-stimulated cyclase activity at CPZ concentrations above 10^{-4} M is paralleled by a monotonous decrease in DPH and TMA-DPH anisotropy values (Figure 9) and by a progressive loss of TSH binding (as indicated in Table IX). The decrease in TSH binding is probably due to a parallel loss of TSH

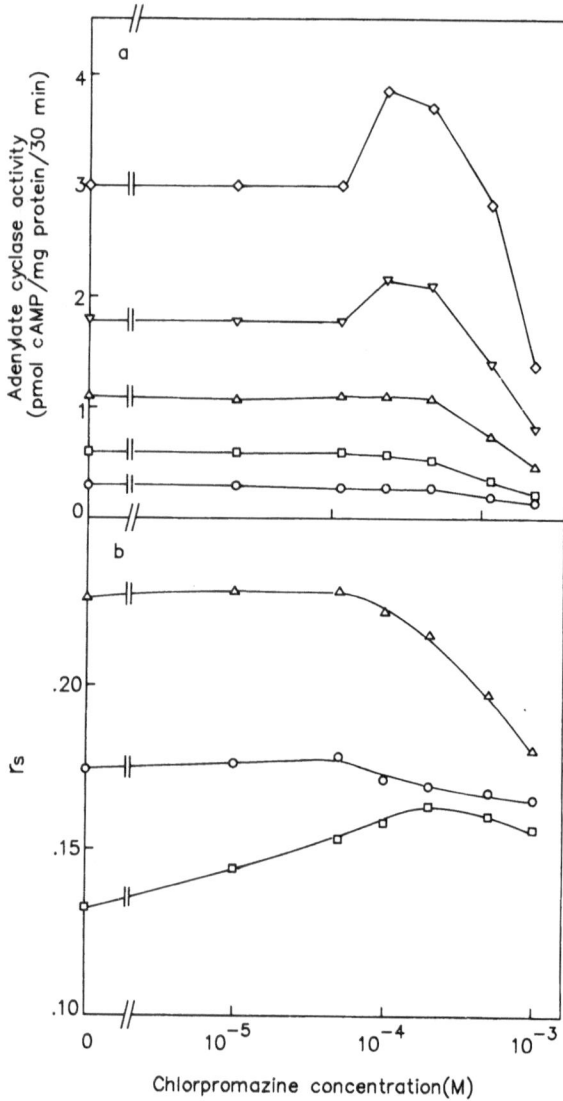

FIGURE 9. Effect of chlorpromazine on (a) the adenylate cyclase activity and (b) the physical state of bovine thyroid plasma membranes. The change in adenylate cyclase activity (a) is shown for basal (O—O), TSH-stimulated (□—□), forskolin-stimulated (△—△), choleratoxin-stimulated (▽—▽), and fluoride-stimulated (◇ — ◇) enzyme activity. The membranes were incubated with the drugs prior to the addition of coupling factors. The physical state of the membrane was evaluated by following the steady-state fluorescence anisotropy (b) of several fluorophores embedded in the membranes: DPH (O—O), 12-AS (□—□), and TMA-DPH (△—△). Measurements were performed at 37°C. The fluorophore/lipid molar ratio was always 1 : 500.

receptors (by partial solubilization at high CPZ concentrations) as previously shown for other receptor systems (Salesse et al., 1982b). Apparently the drug influences the lipid mobility at each level of the lipid bilayer in a different way (Figure 9b). It is not excluded that different cholesterol/phospholipid molar ratios at each level as well as membrane proteins play an important role in this phenomenon (Breton et al., 1977; Salesse et al., 1982a). This is indeed the case as ESR studies by Pang and Miller (1978), using 5-doxyl stearic acid as a probe, showed the existence of a crossover from an ordering to a disordering effect at a certain cholesterol composition, which varies with the perturber, being 26 mol % for CPZ. These data are in agreement with the observed drop in TMA-DPH anisotropy; as in our system, the cholesterol content is certainly higher than 26 mol %. In synaptic membranes Breton et al. (1977) were able to demonstrate a predominant role of proteins in the effects induced by CPZ.

From this discussion it is obvious that the selectivity of any of these drugs for one or the other membrane leaflet is certainly not absolute and that fluidity changes in the nonpreferred leaflet can also lead to activity changes. The above results demonstrate that there is not always a parallel effect of drugs on bilayer fluidity and adenylate cyclase stimulation. Moreover, fluidization of the bilayer may have opposite effects on the adenylate cyclase activity depending on the stimulating agent. One must also consider the possibility that other effects induced by the drugs (e.g., modification of membrane proteins by direct interaction) may be superimposed on fluidization effects.

5. INVOLVEMENT OF MEMBRANE FLUIDITY ON HUMAN NORMAL AND PATHOLOGICAL THYROID GLANDS

Changes in membrane composition as well as in fluidity may modulate the activity of the adenylate cyclase system (Houslay et al., 1976; Dipple and Houslay, 1978; Bakardjieva et al., 1979; Salesse et al., 1982a; Houslay and Gordon, 1983; McOsker et al., 1983). Differences in both composition and physicochemical state of membranes as well as in adenylate cyclase activities could be demonstrated when comparing thyroid plasma membranes obtained from patients with normal thyroid and patients with multinodular goiter euthyroid gland (Depauw et al., 1987). Normal human thyroid membranes display higher protein/phospholipid w/w ratios (8.59 versus 4.26), higher cholesterol/phospholipid molar ratios (0.69 versus 0.38), and a lower 5'-nucleotidase specific activity (8.3 versus 18.4 nmol/min per mg protein). Both normal and multinodular goiter euthyroid plasma membranes show similar basal (0.34 versus 0.35 pmol cAMP/mg protein per 30 min), TSH-stim-

Table X
**Fluidity Characteristics of Human Normal and Multinodular Goiter
Euthyroid Plasma Membranes at 37°C[a]**

	$r_{s12\text{-AS}}$	$r_{s\text{TMA-DPH}}$	$r_{s\text{DPH}}$	S_{DPH}	D_{diff} (cm^2/sec) $(\times 10^8)$
Normal	0.165 ± 0.003	0.258 ± 0.004	0.228 ± 0.004	0.719 ± 0.008	0.82 ± 0.12
Goiter	0.132 ± 0.003	0.169 ± 0.003	0.151 ± 0.003	0.506 ± 0.008	1.22 ± 0.18

[a]All data are average values from three separate experiments.

ulated (0.95 versus 0.98), and forskolin-stimulated adenylate cyclase activities (1.47 versus 1.52), whereas this enzyme is less stimulated by NaF and cholera toxin in normal thyroid membranes (6.46 versus 9.80 and 3.26 versus 4.80).

After detergent solubilization of proteins (1.5% DOC, pH 7.5), SDS-gel electrophoresis, nitrocellulose transfer (25 mM Tris–192 mM glycine, 20% methanol, pH 8.3), and FITC-lectin visualization, more proteins are manifest as an intense peak in the high molecular weight region than with Coomassie blue staining. The densitometric profiles of membrane proteins of normal thyroid, using different fluorochrome-labeled lectins, are distinct from those of membrane proteins of multinodular goiter euthyroid gland.

From Table X it is obvious that membranes derived from the multinodular goiter euthyroid gland display a higher fluidity at any depth of the bilayer than those prepared from normal tissue. In thyroid a higher value of r_s or S_{DPH} can be correlated with more abundant neutral lipids, whereas D_{diff} can be lowered by the presence of more plasma membrane proteins (Section 3.3). Hence, it is not surprising that human multinodular goiter euthyroid membranes, displaying lower protein/phospholipid and cholesterol/phospholipid ratios, show lower anisotropy values, a lower order parameter, and a higher D_{diff} value. The observed increase in NaF and cholera toxin stimulation in the goiter membranes with respect to normal thyroid membranes is probably related to the higher fluidity in the former membranes.

These data support the view that the coupling between the G_s-protein and the catalyst of the adenylate cyclase system is probably dependent on membrane fluidity. The TSH-stimulated cyclase activity being obviously unaffected by changes in the physicochemical state of the membrane suggests that the activating coupling between the occupied TSH receptor and the G_s-protein is the rate-limiting step in the TSH activation of adenylate cyclase. The observation that TSH stimulation is unaffected by changes in membrane fluidity is in agreement with the slight temperature dependency of the activation of adenylate cyclase activity by TSH (Van Sande *et al.*, 1979).

6. REFERENCES

Aloj, S. M., Lee, G., Consiglio, E., Formisano, S., Minton, A. P., and Kohn, L. D., 1979a, Dansylated thyrotropin as a probe of hormone–receptor interactions, *J. Biol. Chem.* **254** : 9030–9039.

Aloj, S. M., Lee, G., Grollman, E. F., Beguinot, F., Consiglio, E., and Kohn, L. D., 1979b, Role of phospholipids in the structure and function of the thyrotropin receptor, *J. Biol. Chem.* **254** : 9040–9049.

Amir, S. M., Goldfine, J., and Ingbar, S. H., 1976, Properties of the interaction between bovine thyrotropin and bovine thyroid plasma membranes, *J. Biol. Chem.* **251** : 4693–4699.

Ansell, G. B., Hawthorne, J. N., and Dawson, R. M. C., 1973. The phospholipid composition of mammalian tissues, in *Form and Function of Phospholipids*, 2nd ed. (G. B. Ansell, R. M. C. Dawson, and J. N. Hawthorne, eds.), BBA Library Vol. 3, pp. 441–482, Elsevier, Amsterdam.

Appelkvist, E. L., Chojnacki, T., and Dallner, G., 1985, The presence of dolichol in liver supernatant, *Acta Chem. Scand. B* **39** : 72–74.

Ashcroft, R. G., Coster, H. G. L., and Smith, J. R., 1977, Local anesthetic benzyl alcohol increases membrane thickness, *Nature* **269** : 819–820.

Bakardjieva, A., Galla, H. J., and Helmreich, E. J. M., 1979, Modulation of the β-receptor adenylate cyclase interactions in cultured Chang liver cells by phospholipid enrichment, *Biochemistry* **18** : 3016–3023.

Barsukov, L. I., Kulikov, V. I., Simakova, I. M., Tikhonova, G. V., Ostrovskii, D. N., and Bergelson, L. D., 1978, Manipulation of phospholipid composition of membranes with the aid of lipid exchange proteins. Incorporation of phosphatidylcholine into protoplasts of *Micrococcus lysodeikticus*, *Eur. J. Biochem.* **90** : 331–336.

Beckner, S. K., Brady, R. O., and Fishman, P. H., 1981, Reevaluation of the role of gangliosides in the binding and action of thyrotropin, *Proc. Natl. Acad. Sci. U.S.A.* **78** : 4848–4852.

Beguinot, F., Formisano, S., Rotella, C. M., Kohn, L. D., and Aloj, S. M., 1983, Structural changes caused by thyrotropin in thyroid cells and in liposomes containing reconstituted thyrotropin receptor, *Biochem. Biophys. Res. Commun.* **110** : 48–54.

Bertoli, E., Masserini, M., Sonnino, S., Ghidoni, R., Cestaro, B., and Tettamanti, G., 1981, Electron paramagnetic resonance studies on the fluidity and surface dynamics of egg phosphatidylcholine vesicles containing gangliosides, *Biochim. Biophys. Acta* **467** : 196–202.

Biber, J., Rechkemmer, G. Bodmer, M., Schröder, P., Haaze, W., and Murer, H., 1983, Isolation of basolateral membranes from columnar cells of the proximal colon of the guinea pig, *Biochim. Biophys. Acta* **735** : 1–11.

Bloj, B., and Zilversmit, D. B., 1981, Lipid transfer proteins in the study of artificial and natural membranes, *Mol. Cell. Biochem.* **40** : 163–172.

Brasitius, T. A., and Schachter, D., 1980, Lipid dynamics and lipid–protein interactions in rat enterocyte basolateral and microvillus membranes, *Biochemistry* **19** : 2763–2769.

Breton, J., Viret, J., and Leterrier, F., 1977, Calcium and chlorpromazine interactions in rat synaptic plasma membranes. A spin-label and fluorescence probe study, *Arch. Biochem. Biophys.* **179** : 625–633.

Briggs, M. M., and Lefkowitz, R. J., 1980, Parallel modulation of catecholamine activation of adenylate cyclase and formation of the high-affinity agonist–receptor complex in turkey erythrocyte membranes by temperature and *cis*-vaccenic acid, *Biochemistry* **19** : 4461–4466.

Brockerhoff, H., 1982, Anesthetics may restructure the hydrogen belts of membranes, *Lipids* **17** : 1001–1003.

Brown, R. E., Sugar, I., and Thompson, T. E., 1985, Spontaneous transfer of gangliotetra-osylceramide between phospholipid vesicles, *Biochemistry* **24** : 4082–4091.

Butler, K. W., Schneider, H., and Smith, I. C. P., 1973, The effects of local anesthetics on lipid multilayers. A spin probe study, *Arch. Biochem. Biophys.* **154** : 548–554.

Carlsen, J., Christiansen, K., and Bro, B., 1983, Purification of microvillus membrane vesicles from pig small intestine by adsorption chromatography on Sepharose, *Biochim. Biophys. Acta* **727** : 412–415.

Carrière, B., and Le Grimellec, C., 1986, Effects of benzyl alcohol on enzyme activities and D-glucose transport in kidney brush-border membranes, *Biochim. Biophys. Acta* **857** : 131–138.

Cassel, D., and Pfeuffer, T., 1978, Mechanism of cholera toxin action: Covalent modification of the guanyl nucleotide-binding protein of the adenylate cyclase system, *Proc. Natl. Acad. Sci. U.S.A.* **75** : 2669–2673.

Coster, H. G. L., and Laver, D. R., 1986, The effect of benzyl alcohol and cholesterol on the acyl chain order and alkane solubility of bimolecular phosphatidylcholine membranes, *Biochim. Biophys. Acta* **861** : 406–412.

Crain, R. C., and Zilversmit, D. B., 1980, Two nonspecific phospholipid exchange proteins from beef liver. 1. Purification and characterization, *Biochemistry* **19** : 1433–1439.

Cullis, P. R., and De Kruijff, B., 1978, The polymorphic phase behaviour of phosphatidyl-ethanolamines of natural and synthetic origin: A ^{31}p-NMR study, *Biochim. Biophys. Acta* **513** : 31–42.

Cullis, P. R., and De Kruijff, B., 1979, Lipid polymorphism and the functional roles of lipids in biological membranes, *Biochim. Biophys. Acta* **559** : 399–420.

Cundall, R. B., and Dale, R. E., 1983, Membrane structure and dynamics by fluorescence probe depolarization kinetics, in *Time-Resolved Fluorescence Spectroscopy in Biochemistry and Biology*, Vol. 69, pp. 555–612, Plenum Press, New York.

Dacremont, G., De Baets, M., Kaufman, J. M., Elewaut, A., and Vermeulen, A., 1984, Inhibition of adenylate cyclase activity of human thyroid membranes by gangliosides, *Biochim. Biophys. Acta* **770** : 142–147.

Depauw, H., De Wolf, M., Van Dessel, G., Hilderson, H. J., Lagrou, A., and Dierick, W., 1985, Fluidity characteristics of bovine thyroid plasma membranes, *Biochim. Biophys. Acta* **814** : 57–67.

Depauw, H., Peeters, C., Hilderson, H. J., De Wolf, M., Van Dessel, G., Lagrou, A., and Dierick, W., 1987, A note on the glycoproteins and fluidity of multinodular goiter euthyroid plasma membranes, in *Methodological Surveys in Biochemistry and Analysis* (E. Reid, G. Cook, and J. Lusio, eds.), Vol. 17, pp. 235–238. Plenum Press, New York.

De Wolf, M. J. S., Van Dessel, G. A. F., Lagrou, A. R., Hilderson, H. J. J., and Dierick, W. S. H., 1985, Topography, purification and characterization of thyroidal NAD$^+$ glycoh-ydrolase, *Biochem. J.* **226** : 415–417.

Dicorleto, P. E., and Zilversmit, D. B., 1977, Protein-catalyzed exchange of phosphatidylcho-line between sonicated liposomes and multilamellar vesicles, *Biochemistry* **16** : 2145–2150.

Dierick, W., and Hilderson, H. J. 1967, Subcellular structure of bovine thyroid. I. β-Glucu-ronidase: A marker enzyme for lysosomal particles, *Arch. Int. Physiol. Biochim.* **75** : 1–11.

Dipple, I., and Houslay, M. D., 1978, The activity of glucagon-stimulated adenylate cyclase from rat liver plasma membranes is modulated by the fluidity of its lipid environment, *Biochem. J.* **174** : 179–190.

Dyatlovitskaya, E. V., Lemenovskaya, A. F., and Bergelson, L. D., 1979, Use of protein-mediated lipid exchange in the study of membrane-bound enzymes. The lipid dependence of glucose-6-phosphatase, *Eur. J. Biochem.* **99** : 605–612.

Eggens, I., Chojnacki, T., Kenne, L., and Dallner, G., 1983, Separation, quantitation and distribution of dolichol and dolichylphosphate in rat and human tissues, *Biochim. Biophys. Acta* **751** : 355–368.

Engelhard, V. H., Esko, J. D., Storm, D. R., and Glaser, M., 1976, Modification of adenylate cyclase activity in LM cells by manipulation of the membrane phospholipid composition *in vivo*, *Proc. Natl. Acad. Sci. U.S.A.* **73** : 4482–4486.

Engelhard, V. H., Glaser, M., and Storm, D. R., 1978, Effect of membrane phospholipid compositional changes on adenylate cyclase in LM cells, *Biochemistry* **17** : 3191–3200.

Felgner, P. L., Thompson, T. E., Barenholz, Y., and Lichtenberg, D., 1983, Kinetics of transfer of gangliosides from their micelles to dipalmitoylphosphatidylcholine vesicles, *Biochemistry* **22** : 1670–1674.

Gilman, A. G., 1984, G proteins and dual control of adenylate cyclase, *Cell* **36** : 577–579.

Ginsberg, B. H., Jabour, J., and Spector, A. A., 1982, Effects of alterations in membrane lipid unsaturation on the properties of the insulin receptor of Ehrlich ascites cells, *Biochim. Biophys. Acta* **690** : 157–164.

Gordon, L. M., Dipple, I., Sauerheber, R. D., Esgate, J. A., and Houslay, M. D., 1980a, The selective effects of charged local anaesthetics on the glucagon- and fluoride-stimulated adenylate cyclase activity of rat liver plasma membranes, *J. Supramol. Struct.* **14** : 21–32.

Gordon, L. M., Sauerheber, R. D., Esgate, J. A., Dipple, I., Marchmont, R. J., and Houslay, M. D., 1980b, The increase in bilayer fluidity of rat liver plasma membranes achieved by the local anesthetic benzyl alcohol affects the activity of intrinsic membrane enzymes, *J. Biol. Chem.* **255** : 4519–4527.

Haydon, D. A., Hendry, B. M., Levinson, S. R., and Requena, J., 1977, The molecular mechanisms of anesthesia, *Nature* **268** : 356–358.

Haye, B., and Jacquemin, C., 1971, Interaction de la thyreostimuline avec ses recepteurs cellulaires: Effet de la phosphoplipase C sur la fixation et l'activité biologique, *FEBS Lett.* **18** : 47–52.

Hebdon, G. M., Levine, H., Sahyoun, N. E., Schmitges, C. J., and Cuatrecasas, P., 1981, Specific phospholipids are required to reconstitute adenylate cyclase solubilized from rat brain, *Proc. Natl. Acad. Sci. U.S.A.* **78** : 120–123.

Hemming, F. W., 1974, Lipids in glycan biosynthesis, in *Biochemistry of Lipids, Biochemistry Series One* (T. W. Goodwin, ed.), Vol. 4, pp. 39–98, University Park Press, Baltimore.

Heron, D. S., Shinitzky, M., Hershkowitz, M., and Samuel, D., 1980, Lipid fluidity markedly modulates the binding of serotonin to mouse brain membranes, *Proc. Natl. Acad. Sci. U.S.A.* **77** : 7463–7467.

Herzog, V., 1981, Pathways of endocytosis in secretory cells, *Trends Biochem. Sci.* **6** : 319–322.

Heyn, M. P., 1979, Determination of lipid order parameters and rotational correlation times from fluorescence depolarization experiments, *FEBS Lett.* **108** : 359–364.

Higgins, J. A., and Evans, W. H., 1978, Transverse organization of phospholipids across the bilayer of plasma-membrane subfractions of rat hepatocytes, *Biochem. J.* **174** : 563–567.

Hilderson, H. J., Van Dessel, G., Lagrou, A., and Dierick, W., 1980, The subcellular biochemistry of thyroid, *Subcell. Biochem.* **7** : 213–265.

Hilderson, H. E. L., Steen, L., De Wolf, M., Lagrou, A., Hilderson, H. J. J., Van Dessel, G., and Dierick, W., 1984, Spectrophotometric determination of dolichol and dolichyl derivatives using the Chugaev color reaction, *Anal. Biochem.* **141** : 116–120.

Hitzemann, R. J., Harris, R. A., and Loh, H. H., 1984, Synaptic membrane fluidity and function, in *Physiology of Membrane Fluidity* (M. Shinitzky, ed.),Vol. 2, pp. 109–126, CRC Press, Boca Raton, FL.

Houslay, M. D., 1979, Coupling of the glucagon receptor to adenylate cyclase, *Biochem. Soc. Trans.* **7** : 843–846.

Houslay, M. D., 1981, Mobile receptor and collision coupling mechanisms for the activation of adenylate cyclase by glucagon, *Adv. Cyclic Nucleotide Res.* **14** : 111–119.

Houslay, M. D., and Gordon, L. M., 1983, The activity of adenylate cyclase is regulated by the nature of its lipid environment, in *Membrane Receptors, Current Topics in Membranes and Transport* (B. R. Martin and A. Kleinzeller, eds.), Vol. 18, pp. 179–231, Academic Press, New York.

Houslay, M. D., and Palmer, R. W., 1978, Changes in the form of Arrhenius plots of the activity of glucagon-stimulated adenylate cyclase and other hamster liver plasma membrane enzymes occurring on hibernation, *Biochem. J.* **174** : 909–919.

Houslay, M. D., Hesketh, T. R., Smith, G. A., Warren, G. B., and Metcalfe, J. C., 1976, The lipid environment of the glucagon receptor regulates adenylate cyclase activity, *Biochim. Biophys. Acta* **436** : 495–504.

Houslay, M. D., Ellory, J. C., Smith, G. A., Hesketh, T. R., Stein, J. M., Warren, G. B., and Metcalfe, J. C., 1977, Exchange of partners in glucagon receptor–adenylate cyclase complexes. Physical evidence for the independent, mobile receptor model, *Biochim. Biophys. Acta* **467** : 208–219.

Houslay, M. D., Dipple, I., Rawal, S., Sauerheber, R. D., Esgate, J. A., and Gordon, L. M., 1980, Glucagon-stimulated adenylate cyclase detects a selective perturbation of the inner half of the liver plasma-membrane bilayer achieved by the local anaesthetic prilocaine, *Biochem. J.* **190** : 131–137.

Houslay, M. D., Dipple, I., and Gordon, L. M., 1981, Phenobarbital selectively modulates the glucagon-stimulated activity of adenylate cyclase by depressing the lipid phase separation occurring in the outer half of the bilayer of liver plasma membranes, *Biochem. J.* **197** : 675–681.

Irwin, L. N., 1974, Glycolipids and glycoproteins in brain function, *Rev. Neurosci.* **1** : 137–180.

Jain, M. K., 1980, Phase properties of bilayers, in *Introduction to Biological Membranes* (M. K. Jain and R. C. Wagner, eds.), pp. 53–175, Wiley, New York.

Kanda, S., Inoue, K., Nojima, S., Utsumi, H., and Wiegandt, H., 1982, Incorporation of spin-labeled ganglioside analogues into cell and liposomal membranes, *J. Biochem.* **91** : 1707–1718.

Kidroni, G., Spiro, M. J., and Spiro, R. G., 1980, Studies on thyroid cell surface glycoproteins: Isolation of plasma membranes and characterization of carbohydrate units, *Arch. Biochem. Biophys.* **203** : 151–160.

Kimelberg, H. K., 1977, The influence of membrane fluidity on the activity of membrane-bound enzymes, in *Dynamic Aspects of Cell Surface Organization* (G. Poste and G. L. Nicolson, eds.), *Cell Surface Reviews*, Vol. 3, pp. 205–293, Elsevier Biomedical, Amsterdam.

Klein, I., Moore, L., and Pastan, I., 1978, Effect of liposomes containing cholesterol on adenylate cyclase activity of cultured mammalian fibroblasts, *Biochim. Biophys. Acta* **506** : 42–53.

Kleinfeld, A. M., Dragsten, P., Klausner, R. D., Pjura, W. J., and Matayoshi, E. D., 1981, The lack of relationship between fluorescence polarization and lateral diffusion in biological membranes, *Biochim. Biophys. Acta* **649** : 471–480.

Kohn, L. D., 1978, Relationships in the structure and function of receptors for glycoprotein hormones, bacterial toxins and interferon, in *Receptors and Recognition* (P. Cuatrecasas and M. F. Greaves, eds.), Ser. A, Vol. 5, pp. 135–212, Chapman and Hall, London.

Kohn, L. D., and Shifrin, S., 1982, Receptor structure and function: An exploratory approach using the thyrotropin receptor as a vehicle, in *Horizons in Biochemistry and Biophysics. Hormone Receptors* (L. D. Kohn, ed.), Vol. 6, pp. 1–42, Wiley, New York.

Lad, P. M., Preston, M. S., Welton, A. F., Nielsen, T. B., and Rodbell, M., 1979, Effects of phospholipase A_2 and filipin on the activation of adenylate cyclase, *Biochim. Biophys. Acta* **551** : 368–381.

Lai, C.-S., and Schutzbach, J. S., 1984, Dolichol induces membrane leakage of liposomes composed of phosphatidylethanolamine and phosphatidylcholine, *FEBS Lett.* **169** : 279–282.

Lee, A. G., 1976a, Interactions between phospholipids and barbiturates, *Biochim. Biophys. Acta* **455** : 102–108.

Lee, A. G., 1976b, Interactions between amine anaesthetics and lipid mixtures, *Biochim. Biophys. Acta* **448** : 34–44.

Le Grimellec, C., Giocondi, M. C., Carrière, B., Carrière, S., and Cardinal, J., 1982, Membrane fluidity and enzyme activities in brush border and basolateral membranes of the dog kidney, *Am. J. Physiol.* **242** : F246–F253.

Lennarz, W. J., 1975, Lipid linked sugars in glycoprotein synthesis, *Science* **188** : 986–991.

Levey, G. S., 1971, Restoration of norepinephrine responsiveness of solubilized myocardial adenylate cyclase by phosphatidylinositol, *J. Biol. Chem.* **246** : 7404–7410.

Levitzki, A., 1984, Receptor to effector coupling in the receptor-dependent adenylate cyclase system, *J. Receptor Res.* **4** : 399–409.

Levitzki, A., 1987, Hypothesis. Regulation of adenylate cyclase by hormones and G-proteins, *FEBS Lett.* **211** : 113–118.

Limbird, L. E., and Lefkowitz, R. J., 1976, Adenylate cyclase-coupled β-adrenergic receptors: Effect of membrane lipid-perturbing agents on receptor binding and enzyme stimulation by catecholamines, *Mol. Pharmacol.* **12** : 559–567.

Low, M. G., and Finean, J. B., 1977, Effects of phosphatidylinositol-specific phospholipase C from *Staphylococcus aureus* on plasma membrane enzymes, *Biochem. Soc. Trans.* **5** : 1131–1132.

Low, M. G., and Finean, J. B., 1978, Specific release of plasma membrane enzymes by a phosphatidylinositol-specific phospholipase C, *Biochim. Biophys. Acta* **508** : 565–570.

Low, M. G., and Zilversmit, D. B., 1980, Role of phosphatidylinositol in attachment of alkaline phosphatase to membranes, *Biochemistry* **19** : 3913–3918.

Lowe, P. J., and Coleman, R., 1982, Fluorescence anisotropy from diphenylhexatriene in rat liver plasma membranes, *Biochim. Biophys. Acta* **689** : 403–409.

Luly, P., and Shinitzky, M., 1979, Gross structural changes in isolated liver cell plasma membranes upon binding insulin, *Biochemistry* **18** : 445–450.

Macchia, V., Tamburrini, O., and Pastan, I., 1970, Role of lecithin in the mechanism of TSH action, *Endocrinology* **86** : 787–792.

Martin, K. J., McConkey, C. L., Jr., and Stokes, T. J., Jr., 1985, Effects of benzyl alcohol on PTH receptor–adenylate cyclase system of canine kidney, *Am. J. Physiol.* **248** : E31–E35.

McCloskey, M. A., and Troy, F. A., 1980, Paramagnetic isoprenoid carrier lipids. 2. Dispersion and dynamics in lipid membranes, *Biochemistry* **19** : 2061–2066.

McOsker, C. C., Weiland, G. A., and Zilversmit, D. B., 1983, Inhibition of hormone-stimulated adenylate cyclase activity after altering turkey erythrocyte phospholipid composition with a nonspecific lipid transfer protein, *J. Biol. Chem.* **258** : 13017–13026.

Miller, K. W., 1981, in *Burger's Medical Chemistry*, 4th ed. (M. E. Wolf, ed.), pp. 623–644, Wiley, New York.

Moore, W. V., and Wolff, J., 1974, Thyroid-stimulating hormone binding to beef thyroid membranes. Relation to adenylate cyclase activity, *J. Biol. Chem.* **249** : 6255–6263.

Mullin, B. R., Fishman, P. H., Lee, G., Aloj, S. M., Ledley, F. D., Winand, R. J., Kohn, L. D., and Brady, R. O., 1976, Thyrotropin–ganglioside interactions and their relationship to the structure and function of thyrotropin receptors, *Proc. Natl. Acad. Sci. U.S.A.* **73** : 842–846.

Mullin, B. R., Pacuszka, T., Lee, G., Kohn, L. D., Brady, R. O., and Fishman, P. H., 1978,

Thyroid gangliosides with high affinity for thyrotropin: Potential role in thyroid regulation, *Science* **199** : 77–79.

Neal, M. J., Butler, K. W., Polnaszek, C. F., and Smith, I. C. P., 1976, The influence of anesthetics and cholesterol on the degree of molecular organization and mobility of ox brain white matter. Lipids in multibilayer membranes: A spin probe study using spectral simulation by the stochastic method, *Mol. Pharmacol.* **12** : 144–155.

Omodeo-Sale, F., Brady, R. O., and Fishman, P. H., 1978, Effect of thyroid phospholipids on the interaction of thyrotropin with thyroid membranes, *Proc. Natl. Acad. Sci. U.S.A.* **75** : 5301–5305.

Pang, K.-Y. Y., and Miller, K. W., 1978, Cholesterol modulates the effects of membrane perturbers in phospholipid vesicles and biomembranes, *Biochim. Biophys. Acta* **511** : 1–9.

Papahadjopoulos, D., Jacobson, K., Poste, G., and Shepherd, G., 1975, Effects of local anesthetics on membrane properties. I. Changes in the fluidity of phospholipid bilayers, *Biochim. Biophys. Acta* **394** : 504–519.

Parington, C. R., and Daly, J. W., 1979, Effect of gangliosides on adenylate cyclase activity in rat cerebral cortical membranes, *Mol. Pharmacol.* **15** : 484–491.

Parodi, A. J., and Leloir, L. R., 1979, The role of lipid intermediates in the glycosylation of proteins in the eucaryotic cell, *Biochim. Biophys. Acta* **559** : 1–37.

Petri, W. A., Jr., and Wagner, R. R., 1979, Reconstitution into liposomes of the glycoprotein of vesicular stomatitis virus by detergent dialysis, *J. Biol. Chem.* **254** : 4313–4316.

Prives, J., and Shinitzky, M., 1977, Increased membrane fluidity precedes fusion of muscle cells, *Nature* **268** : 761–763.

Resch, K., Schneider, S., and Szamel, M., 1981, Separation of right-side-out-oriented subfractions from purified thymocyte plasma membranes by affinity chromatography on concanavalin A–Sepharose, *Anal. Biochem.* **117** : 282–292.

Rimon, G., Hanski, E., and Levitski, A., 1980, Temperature dependence of β-receptor, adenosine receptor and sodium fluoride-stimulated adenylate cyclase from turkey erythrocytes, *Biochemistry* **19** : 4451–4460.

Rodbell, M., 1980, The role of hormone receptors and GTP-regulatory proteins in membrane transduction, *Nature* **284** : 17–22.

Ross, E. M., 1982, Phosphatidylcholine-promoted interaction of the catalytic and regulatory proteins of adenylate cyclase, *J. Biol. Chem.* **257** : 10751–10758.

Ross, E. M., and Gilman, A. G., 1980, Biochemical properties of hormone-sensitive adenylate cyclase, *Annu. Rev. Biochem.* **49** : 533–564.

Roth, S. H., 1979, Physical mechanisms of anesthesia, *Annu. Rev. Pharmacol. Toxicol.* **19**: 159–178.

Rottem, S., Cirillo, V. P., De Kruiff, B., Shinitzky, M., and Razin, S., 1973, Cholesterol in *Mycoplasma* membranes. Correlation of enzymic and transport activities with physical state of lipids in membranes of *Mycoplasma mycoides* var. *capri* adapted to grow with cholesterol concentrations, *Biochim. Biophys. Acta* **323** : 509–519.

Rubalcava, B., and Rodbell, M., 1973, The role of acidic phospholipids in glucagon action on rat liver adenylate cyclase, *J. Biol. Chem.* **248** : 3831–3837.

Sahyoun, N., and Cuatrecasas, P., 1975, Mechanism of activation of adenylate cyclase by cholera toxin, *Proc. Natl. Acad. Sci. U.S.A.* **72** : 3438–3442.

Salesse, R., Garnier, J., Leterrier, F., Daveloose, D., and Viret, J., 1982a, Modulation of adenylate cyclase activity by the physical state of pigeon erythrocyte membrane. 1. Parallel drug-induced changes in the bilayer fluidity and adenylate cyclase activity, *Biochemistry* **21** : 1581–1586.

Salesse, R., Garnier, J., and Daveloose, D., 1982b, Modulation of adenylate cyclase activity by the physical state of pigeon erythrocyte membrane. 2. Fluidity-controlled coupling between the subunits of the adenylate cyclase system, *Biochemistry* **21** : 1587–1590.

Sandermann, H., Jr., 1978, Regulation of membrane enzymes by lipids, *Biochim. Biophys. Acta* **515** : 209–237.

Sauerheber, R. D., Esgate, J. A., and Kuhn, C. E., 1982, Alcohols inhibit adipocyte basal and insulin-stimulated glucose uptake and increase the membrane lipid fluidity, *Biochim. Biophys. Acta* **691** : 115–124.

Schroeder, F., 1985, Fluorescence probes unravel asymmetric structure of membranes, *Subcell. Biochem.* **11** : 51–101.

Seamon, K. B., and Daly, J. W., 1981, Activation of adenylate cyclase by the diterpene forskolin does not require the guanine nucleotide regulatory protein, *J. Biol. Chem.* **256** : 9799–9801.

Seddon, J. M., Cevc, C., and Maesh, D., 1983, Colorimetric studies of the gel-fluid (L_β-L_α) and lamellar-inverted hexagonal (L_α-H_{II}) phase transitions in dialkyl- and diacylphosphatidylethanolamines, *Biochemistry* **22** : 1280–1289.

Seeman, P., 1972, The membrane actions of anesthetics and tranquilizers, *Pharmacol. Rev.* **24** : 583–655.

Sharom, F. J., and Grant, C. W. M., 1978, A model for ganglioside behaviour in cell membranes, *Biochim. Biophys. Acta* **507** : 280–293.

Sheetz, M. P., and Singer, S. J., 1974, Biological membranes as bilayer couples. A molecular mechanism of drug–erythrocyte interactions, *Proc. Natl. Acad. Sci. U.S.A.* **71** : 4457–4461.

Shinitzky, M., 1984, Membrane fluidity and cellular functions, in *Physiology of Membrane Fluidity* (M. Shinitzky, ed.), Vol. 1, pp. 1–51, CRC Press, Boca Raton, FL.

Shukla, S. D., Coleman, R., Finean, J. B., and Michell, R. H., 1980, Selective release of plasma membrane enzymes from rat hepatocytes by a phosphatidylinositol-specific phospholipase C, *Biochem. J.* **187** : 277–280.

Sinensky, M., 1974, Homeoviscous adaptation—A homeostatic process that regulates the viscosity of membrane lipids in *Escherichia coli, Proc. Natl. Acad. Sci. U.S.A.* **71** : 522–525.

Sinensky, M., Minneman, K. P., and Molinoff, P. B., 1979, Increased membrane acyl chain ordering activates adenylate cyclase, *J. Biol. Chem.* **254** : 9135–9141.

Singer, S. J., and Nicolson, G. L., 1972, The fluid mosaic model of the structure of cell membranes. Cell membranes are viewed as two dimensional solutions of oriented globular proteins and lipids, *Science* **175** : 720–731.

Small, D. M., 1970, Surface and bulk interactions of lipids and water with a classification of biologically active lipids based on these interactions, *Fed. Proc.* **29** : 1320–1326.

Steen, L., Van Dessel, G., De Wolf, M., Hilderson, H. J., Lagrou, A., and Dierick, W., 1986, On the subcellular localization of dolichol, dolicholkinase and dolicholmonophosphate phosphohydrolase in bovine thyroid. The occurrence of non membrane associated dolichol, *Arch. Int. Physiol. Biochim.* **94** : 893.

Stubbs, C. D., 1983, Membrane fluidity: Structure and dynamics of membrane lipids, *Essays Biochem.* **19** : 1–39.

Stubbs, C. D., and Smith, A. D., 1984, The modification of mammalian membrane polyunsaturated fatty acid composition in relation to membrane fluidity and function, *Biochim. Biophys. Acta* **779** : 89–137.

Sweet, W. D., and Schroeder, F., 1986, Charged anesthetics alter LM-fibroblast plasma membrane enzymes by selective fluidization of inner or outer membrane leaflets, *Biochem. J.* **239** : 301–310.

Taurog, A., 1978, Hormone synthesis, thyroid iodine metabolism, in *The Thyroid* (S. C. Werner and S. H. Ingbar, eds.), pp. 31–61, Harper & Row, New York.

240 Hugo Depauw *et al.*

Tilcock, C. P. S., Bally, M. B., Farren, S. B., and Cullis, P. R., 1982, Influence of cholesterol on the structural preferences of dioleoylphosphatidylethanolamine–dioleoylphosphatidylcholine systems: A phosphorus 31 and deuterium nuclear magnetic resonance study, *Biochemistry* **21** : 4596–4601.

Trudell, J. R., 1977, A unitary theory of anesthesia based on lateral phase separations in nerve membranes, *Anesthesiology* **46** : 5–10.

Uchida, T., Nagai, Y., Kawasaki, Y., and Wakayama, N., 1981, Fluorospectroscopic studies of various ganglioside and ganglioside–lecithin dispersions. Steady-state and time-resolved fluorescence measurements with 1,6-diphenyl-1,3,5-hexatriene, *Biochemistry* **20** : 162–169.

Valente, W. A., Vitti, P., Yavin, Z., Yavin, E., Rotella, C. M., Grollman, E. F., Taccafondi, R., and Kohn, L. D., 1982, Monoclonal antibodies to the thyrotropin receptor: Stimulating and blocking antibodies derived from the lymphocytes of patients with Graves disease, *Proc. Natl. Acad. Sci. U.S.A.* **79** : 6680–6684.

Valtersson, C., Van Duyn, G., Verkleij, A. J., Chojnacki, T., De Kruijff, B., and Dallner, G., 1985, The influence of dolichol, dolichylesters and dolichylphosphate on phospholipid polymorphism and fluidity in model membranes, *J. Biol. Chem.* **260** : 2742–2751.

Van Blitterswijk, W. J., Van Hoeven, R. P., and Van der Meer, B. W., 1981, Lipid structural order parameters (reciprocal of fluidity) in biomembranes derived from steady-state fluorescence polarization measurements, *Biochim. Biophys. Acta* **644** : 323–332.

Vanderkooi, J. M., Landesberg, R., Selick, H., and McDonald, G. G., 1977, Interaction of general anesthetics with phospholipid vesicles and biological membranes, *Biochim. Biophys. Acta* **464** : 1–16.

Van Dessel, G., Lagrou, A., Hilderson, H. J., Dommisse, R., Esmans, E., and Dierick, W., 1979, Isolation and identification of polyprenols from bovine thyroid gland, *Biochim. Biophys. Acta* **573** : 296–300.

Van Hoeven, R. P., Van Blitterswijk, W. J., and Emmelot, P., 1979, Fluorescence polarization measurements on normal and tumour cells and their corresponding plasma membranes, *Biochim. Biophys. Acta* **551** : 44–54.

Van Sande, J., Pochet, R., and Dumont, J. E., 1979, Dissociation by cooling of hormone and cholera toxin activation of adenylate cyclase in intact cells, *Biochim. Biophys. Acta* **585** : 282–292.

Vigo, C., Grossman, S. H., and Drost-Hansen, W., 1984, Interaction of dolichol and dolichylphosphate with phospholipid bilayers, *Biochim. Biophys. Acta* **774** : 221–226.

Waechter, C. J., and Lennarz, W. J., 1976, The role of polyprenol-linked sugars in glycoprotein synthesis, *Annu. Rev. Biochem.* **45** : 95–112.

Wolff, J., and Jones, A. B., 1970, Inhibition of hormone-sensitive adenyl cyclase by phenothiazines, *Proc. Natl. Acad. Sci. U.S.A.* **65** : 454–459.

Yamashita, K., and Field, J. B., 1973, The role of phospholipids in TSH stimulation of adenylate cyclase in thyroid plasma membranes, *Biochim. Biophys. Acta* **304** : 686–692.

Yamashita, K., Bloom, G., Rainard, B., Zor, U., and Field, J. B., 1970, Effects of chlorpromazine, propranolol and phospholipase C on thyrotropin and prostaglandin stimulation of adenyl cyclase–cyclic AMP system in dog thyroid slices, *Metabolism* **19** : 1109–1118.

Chapter 7

Spectroscopic Analysis of the Structure of Bacteriorhodopsin

Akira Ikegami, Tsutomu Kouyama, Hisako Urabe, and Kazuhiko Kinosita, Jr.

1. INTRODUCTION

Bacteriorhodopsin (bR) is the only protein found in the purple membrane, a specific patch in the plasma membrane of *Halobacterium halobium*. Bacteriorhodopsin accounts for 75% of the mass of the purple membrane and forms a two-dimensional crystalline lattice. It has retinal as chromophore with an absorption maximum around 570 nm, hence the name purple membrane. Light energy absorbed by the retinal chromophore in bR drives the protons actively across the purple membrane from the cytoplasmic to extracellular side. Although the mechanism of this light-driven proton pump has been the subject of a considerable number of investigations, the relation between the mechanism and tertiary structure of bR is still obscure [for reviews see Henderson (1977), Stoeckenius *et al.* (1979), Ottolenghi (1980), and Stoeckenius and Bogomolni (1982)].

A bacteriorhodopsin molecule consists of 248 amino acids. The amino acid sequence was determined (Khorana *et al.*, 1979; Ovchinnikov *et al.*, 1979); the retinal bound at Lys-216 via a Schiff base linkage (Bayley *et al.*,

Akira Ikegami, Tsutomu Kouyama, Hisako Urabe, and Kazuhiko Kinosita, Jr.   Institute of Physical and Chemical Research, Wako-shi, Saitama 351-01, Japan.

1981; Katre *et al.*, 1981; Lemke and Oesterhelt, 1981; Mullen *et al.*, 1981). The three-dimensional structure model (Henderson and Unwin, 1975; Unwin and Henderson, 1975) studied by electron diffraction analysis shows seven α-helixes that cross the membrane almost perpendicularly. The resolutions of the present diffraction analyses are too low, however, to indicate the disposition of each amino acid, or even retinal in bR (Hayward and Stroud, 1981; Agard and Stroud, 1981).

X-ray crystallographic analysis is a very powerful method for the detailed structural determination of proteins. The application of the method for the membrane proteins, however, is substantially restricted, for the crystallizations of most membrane proteins are prevented by their hydrophobic nature. Detergents used for the extraction of the proteins from membranes disturb the crystallization.

In considering the biologically important functions of membrane proteins, we need novel methods for the structural analysis of intact membrane proteins. As an attempt to find new methods, we tried to determine the three-dimensional location and orientation of the retinal chromophore in the purple membrane by the analyses of fluorescence energy transfer and polarized resonance Raman scattering. We have also estimated the location of Lys-41 using the fluorescence energy transfer techniques combined with fluorescence depolarization techniques.

Most analyses of fluorescence energy transfer in biological systems are qualitative rather than quantitative since the systems are too complicated and flexible. Purple membrane is a very suitable sample to test the quantitative analysis of fluorescence energy transfer for the following reasons: (1) the membrane is composed of only one protein, which forms a rigid two-dimensional crystal structure. (2) Retinal chromophore is bound stoichiometrically to bR and can be converted to fluorescent form *in situ*. (3) Information about the structure and orientation of retinal in bR can be specifically detected by the resonance Raman spectroscopy.

We predicted from the free-energy calculation seven amino acid sequences that should correspond to seven transmembrane α-helixes. By comparing the disposition of retinal and Lys-41 in the electron density map, together with seven predicted amino acid sequences, we estimated possible folding of the polypeptide chain of bR in the purple membrane.

2. PRINCIPLE OF THE FLUORESCENCE ENERGY TRANSFER TECHNIQUE

Consider a dilute system of fluorescent chromophore D. When the system is illuminated by a pulsed light at time $t = 0$, some fluorescent molecules are excited by absorbing photons. An excited molecule returns to the ground state either by emitting a photon as fluorescence (with a rate constant k_f) or via nonradiative processes (the sum of rate constants k_{nr}). The intensity of fluorescence $F(t)$ from the system and the quantum yield Q (the ratio of the number of emitted photons to the number of absorbed photons) are given by

$$F(t) = F_o \exp[-(k_f + k_{nr})\, t] = F_o \exp(-t/\tau) \qquad (1)$$

$$Q = k_f/(k_f + k_{nr}) \qquad (2)$$

where $\tau = 1/(k_f + k_{nr})$ is the lifetime of the excited state (also called the fluorescence lifetime).

When the system contains chromophores A around the chromophores D, and the absorption spectrum of A overlaps with the emission spectrum of D, the fluorescence energy of an excited singlet of D (donor) is transferred to an A molecule (acceptor) by the dipole–dipole coupling between them without emitting a photon. The rate of the fluorescence energy transfer calculated by Förster (1965) is

$$k_{tr} = (R_0/R)^6 \kappa^2 \tau^{-1} \qquad (3)$$

where R is the distance between the donor (D) and the acceptor (A), and τ is the lifetime of the donor in the absence of the acceptor. The efficiency factor R_0 is given from the spectral overlap between donor emission and acceptor absorption. The orientation factor κ can be calculated from the emission transition moment of the donor D and the absorption transition moment of the acceptor A. Although the value of κ^2 ranges between 0 and 4, the average value is calculated to be $\frac{2}{3}$ if the directions of emission or absorption moments are random.

When the excitation energy of a donor D is transferred to acceptors A, Eqs. (1) and (2) should be replaced by the following forms:

$$F(t) = F_0 \exp[-(k_f + k_{nr} + \Sigma k_{tr})t] \qquad (4)$$

$$Q = \frac{k_f}{k_f + k_{nr} + \Sigma k_{tr}} \qquad (5)$$

where Σk_{tr} is the sum of the rate of energy transfer from the donor to all acceptors. Thus, the intensity of donor fluorescence decays faster, and the quantum yield becomes smaller, than the corresponding quantities in the absence of acceptors. From these changes in $F(t)$ or Q, the value of k_{tr} can be estimated.

Since k_{tr} depends on the sixth power of R_0/R, and R_0 values range between 1 and 10 nm for most donor–acceptor systems, the value of k_{tr} gives the interchromophore distance of the order of several nanometers.

Applying the characteristics of the fluorescence energy transfer to properly selected sets of donor and acceptor in the membrane, we determined the three-dimensional location and orientation of the retinal chromophore in bacteriorhodopsin.

3. THREE-DIMENSIONAL DISPOSITION OF THE RETINAL CHROMOPHORE IN THE PURPLE MEMBRANE

3.1. In-Plane Location

We determined the most probable location and orientation of the retinal chromophore in the plane of the purple membrane by the crystallographic analysis of fluorescence energy transfer (Kouyama *et al.*, 1981a).

In the study, a fraction of the retinal chromophores in the purple membrane were converted to fluorescent derivatives—retroretinyllysine—according to Peters *et al.* (1976). The double bond in the Schiff base portion of retinal chromophores was first reduced by $NaBH_4$ under yellow light. Then, by further irradiation with near-ultraviolet light, the reduced chromophore was converted into a fluorescent derivative that exhibits a broad emission band in the visible region ($\lambda_{max} = 500$ nm) and a long fluorescence lifetime of about 20 nsec. In the first procedure of reduction, we applied different reaction times, so we obtained several products corresponding to different degrees of conversion p, in which native retinal and the fluorescent derivative coexist. The hexagonal crystal structure of the membrane was not affected by the conversion because the x-ray diffraction pattern did not change. Fluorescent derivative in the reduced purple membrane cannot be replaced by added free retinal, indicating that the fluorescent derivative occupies the same site as the native chromophore. Furthermore, both retinal and its derivative are immobile in the partially converted purple membrane, because the fluorescence anisotropy of the derivative and the flash-induced absorption dichroism of the acceptor were very close to the values theoretically expected for fixed chromophores (Kouyama *et al.*, 1981b). These facts support the validity of the following analysis.

As the emission spectrum of the fluorescent derivative nicely overlaps with the absorption spectrum of the native retinal chromophore, the fluorescence energy transfer from the derivatives to the native retinal is expected. In fact, the observed fluorescence decay $F(t)$ of the fluorescent derivative (the donor) was slower in membranes with a higher degree of conversion p, because the number of native chromophore (acceptor) neighbors to any donor decreases with the increase of p (Figure 1).

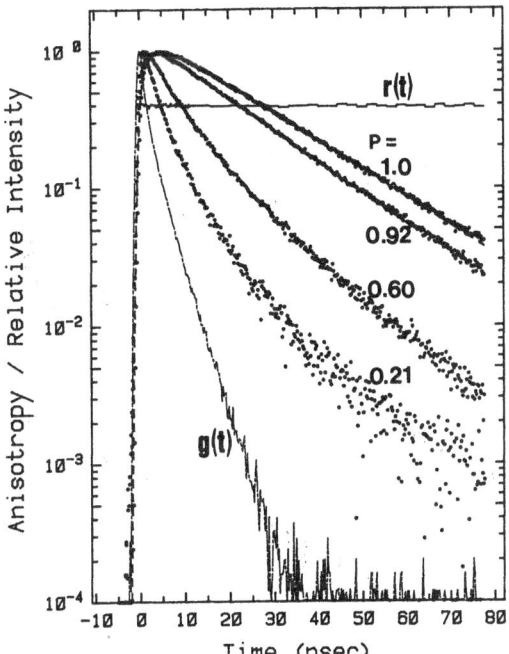

FIGURE 1. Fluorescence intensity decay $F(t)$ of the fluorescent retinal chromophore (converted from native retinal by reduction and ultraviolet irradiation) in the purple membranes at the indicated degree of conversion p. The horizontal curve $r(t)$ shows fluorescence anisotropy at $p = 0$ (Kouyama et al., 1981a).

Since bR molecules form the two-dimensional crystal structure in the purple membrane, the disposition of all retinal chromophore can be uniquely determined when the location and orientation of the chromophore in a unit cell are assigned. Thus, we can calculate the rate of energy transfer between any pair of donor and acceptor in the membrane by Eq. (3). If the conversion occurs in a random way, we can taken the probability of finding a donor or an acceptor at any chromophore site as p or $1 - p$, respectively. Then, fluorescence decay $F(t)$ of the donor and the fluorescence quantum yield Q of the donor at any value of p are given by Eqs. (4) and (5), as a function of three coordinates that assign the position and orientation of the retinal chromophore in the unit cell. The random conversion of the chromophore was confirmed from the fact that the initial rate of fluorescence decay at times much shorter than the lifetime was proportional to $1 - p$, because the same relation is theoretically predicted for the random mixture of donors and acceptors in the membrane. By the least square comparison between the calculated and experimentally obtained quantum yields at various p values, the most probable position and orientation of the retinal chromophore were obtained.

Because of the symmetry of the hexagonal lattice, there are 12 dispositions mathematically equivalent for the most probable disposition (Figure

2). Six dispositions could be excluded, because these dispositions are located outside the bacteriorhodopsin molecules. Among the remaining six dispositions, we proposed that the position and orientation shown by the thick arrow in Figure 2 were the most likely, because they did not overlap with the electron density map of the seven α-helixes. Recently, our proposed disposition has been supported by the neutron diffraction studies; Jubb *et al.* (1984) proposed almost the identical disposition shown by us, and Seiff *et al.* (1985) proposed almost the same position as ours but with a somewhat different direction.

3.2. Transmembrane Location

To determine the transmembrane location of the retinal chromophore, we (Kouyama *et al.*, 1983; Kometani *et al.*, 1987) again applied the energy transfer technique to the five systems of different combinations of donors and acceptors (Figure 3).

In systems (i)–(iii), the energy donor was the same fluorescent derivative, retroretinyllysine, used in the previous analysis of the in-plane location. In system (i), fluorescent membranes were dispersed in the aqueous solution of the energy acceptor, cobalt–ethylenediaminetetraacetate (Co-EDTA), but in system (ii) the membranes were embedded in solid Co-EDTA. To eliminate the surface charge effects on acceptors, we chose dried and solid forms in systems (ii) and (iii).

System (iii) was similar to system (ii) except that the acceptor was a positively charged tris(2,2'-bipyridyl) ruthenium(II) complex with a large radius of 0.7 nm instead of the negatively charged cobalt complex with a radius of 0.4 nm.

For the fluorescence energy transfer from a donor within a membrane to

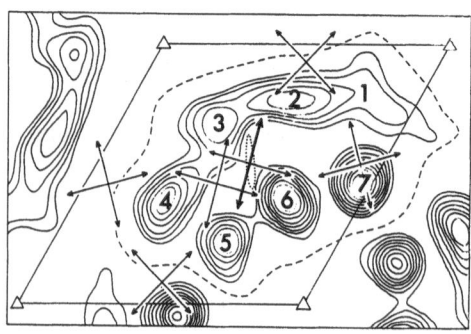

retinal

FIGURE 2. The most probable location and orientation of the retinal chromophore in the plane of purple membrane. The arrows indicate the polyene chain portion of retinal (see bottom). The dotted line associated with the thick arrow shows the uncertainty in the location; the error in the orientation was estimated to be about 20° (Kouyama *et al.*, 1981a). The numbering of seven helixes is cited from Engelman *et al.* (1980).

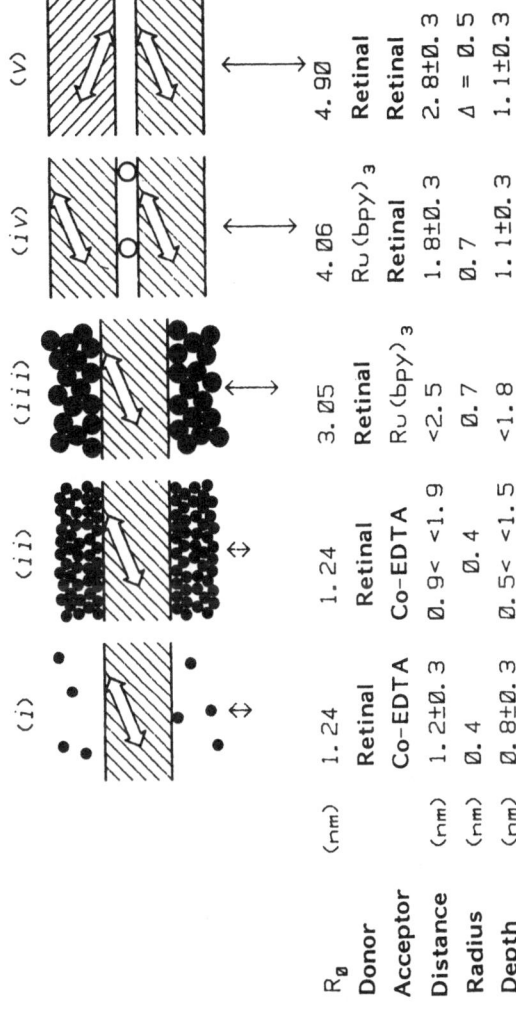

		(i)	(ii)	(iii)	(iv)	(v)
R_0	(nm)	1.24	1.24	3.05	4.06	4.90
Donor		Retinal	Retinal	Retinal	Ru(bpy)$_3$	Retinal
Acceptor		Co-EDTA	Co-EDTA	Ru(bpy)$_3$	Retinal	Retinal
Distance	(nm)	1.2±0.3	0.9< <1.9	<2.5	1.8±0.3	2.8±0.3
Radius	(nm)	0.4	0.4	0.7	0.7	Δ = 0.5
Depth	(nm)	0.8±0.3	0.5< <1.5	<1.8	1.1±0.3	1.1±0.3

FIGURE 3. Summary of the energy transfer experiments for the determination of the transmembrane location of retinal chromophore. Vertical arrows indicate the size of R_0 compared to the thickness of the membrane.

acceptors randomly distributed in the external space, the initial slope of the decay of donor fluorescence is given by

$$\left[\frac{dF\,(t)}{dt}\right]_{t=0} = \frac{1}{\tau}\left[1 + \left(\frac{\pi}{12}\right)\left(\frac{R_0^6}{z^3}\right)(1 + \cos^2\theta)C\right] \qquad (6)$$

where z is the distance of closest approach between the donor and acceptor, θ ($\sim 67°$) is the angle between the transition moment of retinal and the membrane normal, and C is the number of acceptor molecules in a unit volume. Determination of the initial slope with a reasonable accuracy is sufficient for the estimation of z, although the entire $F(t)$ is generally nonexponential. In system (i), in particular, we assume a rapid diffusion limit, because $F(t)$ decayed exponentially, suggesting that the diffusion of acceptor molecules (Co-EDTA) is much faster than the rate of energy transfer.

In system (iv), energy transfer from the luminescent ruthenium complex interspersed between purple membrane sheets to the native retinal chromophore was observed. In system (v) we observed the rate of fluorescence energy transfer from the reduced purple membrane to the native membranes stacked together in parallel. In the theoretical analysis of the rate of these energy transfers, we assume random stacking of membrane sheets irrespective of the asymmetric charge distribution of the native and reduced sheets.

The results for these five systems were all consistent with a location of the retinal chromophore at a depth of 1.0 ± 0.3 nm from a surface of the purple membrane with a thickness of 4.5 nm (Figure 3).

The remaining problem is to determine in which side of purple membrane, either the cytoplasmic or extracellular side, the retinal chromophore is located. We studied the problem also by the analysis of fluorescence energy transfer in oriented membranes. Oriented single sheets of purple membrane with the fluorescent derivative of retinal were prepared on a coverglass coated with polylysine according to Fisher's method (1984). If we spread the purple membranes on the coverglass at acidic pH, the upper surfaces of most membranes adsorbed were the cytoplasmic surface, whereas both cytoplasmic and extra surfaces appeared in almost even ratio at neutral pH. The difference was detected by the molecular weight measurements after the cleavage of the C-terminal part of bR by papain treatments. When the upper exposed surface is the cytoplasmic surface, the molecular weight of bR should be reduced because the C-terminal part extrudes from the cytoplasmic surface. We observed the fluorescence lifetime of the retinal derivatives in the oriented membrane, after spreading various energy acceptors over the membranes. As the acceptors, we used cytochrome c, hemoglobin, and ferritin molecules, which cannot penetrate into the space between the membrane and the coverglass because of their large size. In all cases, the lifetime of the retinal derivative was longer at neutral pH than at acidic pH. Thus, the results indicate that the retinal chromophore is located at 1.0 ± 0.3 nm below the cytoplasmic surface of the purple membrane.

3.3. Orientation of the Molecular Plane

To detect the molecular direction of the retinal chromophore in the purple membrane, we studied polarized resonance Raman scattering due to $C=C$ stretching vibrational modes (Ikegami *et al.*, 1987b). By resonance Raman scattering, the signal of vibrational modes of the retinal chromophore can be specifically detected; otherwise, the signal hides among the other signals from whole protein. The purple membranes were stacked on a conductive transparent glass by electrophoresis according to Varo (1983). The orientation of the membranes within the stacked film was checked by the measurement of small-angle x-ray scattering. The root mean square deviation of the orientation angle was less than $5°$.

Polarized resonance Raman scattering of the oriented purple membrane was observed for three scattering geometries relative to the incident radiation (Ar laser, 488 nm). Typical polarized Raman spectra in the 1500–1600-cm^{-1} region are shown in Figure 4. Here we assume that the X, Y, and Z axes are the membrane-fixed coordinate axes, with the Z axis parallel to the membrane normal. Then I_{ZZ} and I_{ZX} are scattered Raman intensities polarized in the Z and X directions, respectively, when the incident radiation is polarized in the Z direction. Two strong lines at 1530 and 1570 cm^{-1} indicate $C=C$ stretching vibrational bands of bR and M intermediate, respectively. The observed depolarization ratios in the three sample orientations were used for analysis after the correction for the effect of reflection and scattering.

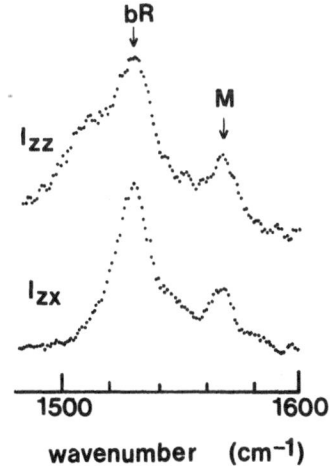

FIGURE 4. Polarized Raman spectra of the oriented purple membrane.

If we assume that the principal z axis of the Raman tensor of the $C=C$ stretching mode is parallel to the long axis of the chromophore and the y axis is included in the molecular plane of retinal, the Raman tensor should be given by

$$\begin{pmatrix} 0 & 0 & 0 \\ 0 & a & 0 \\ 0 & 0 & 1 \end{pmatrix}$$

The value of a was estimated from the depolarization ratio for the purple membrane suspension. Then the Raman depolarization ratios for the three geometries can be calculated as functions of Euler angles, which connect the membrane and molecular coordinates. The most probable values of the Euler angles were obtained from the comparison between the observed and calculated values of the depolarization ratios. The value of θ, the angle between Z and z axes, obtained for bR and M is $65° \pm 2°$, which agrees with reported values. From the polarization ratios of M intermediate, the molecular plane of the retinal chromophore was estimated to be almost perpendicular to the membrane surface. Recently, the same result was reported by Earnest *et al.* (1986) using polarized infrared spectroscopy.

The final disposition of retinal in bR estimated by the spectroscopic methods is shown in Figure 5.

4. IN-PLANE LOCATION OF NBD (7-CHLORO-4-NITROBENZO-2-OXA-1,3-DIAZOLE) BOUND TO Lys-41 IN THE PURPLE MEMBRANE

Fluorescence energy transfer techniques combined with fluorescence depolarization techniques were used for the determination of the location of NBD-Cl attached to Lys-41. As a fluorescent probe, 7-chloro-4-nitrobenzo-2-oxa-1,3-diazole (NBD-Cl) reacts specifically with Lys-41 of bR and induces a large shift in the absorption spectrum (Allegrini *et al.*, 1983), we can estimate its location from the fluorescence energy transfer without any disturbance from unreacted species. In Figure 6 the absorption and emission spectra of NBD bound to Lys-41 are compared with those of native and fluorescent retinal chromophore. From the overlapping of absorption and emission spectra of these chromophores, we studied fluorescence energy transfer from NBD to the native retinal chromophore, from the fluorescent derivative of retinal to NBD, and between NBD probes.

The fluorescence quantum yield of NBD labeled to native purple membrane

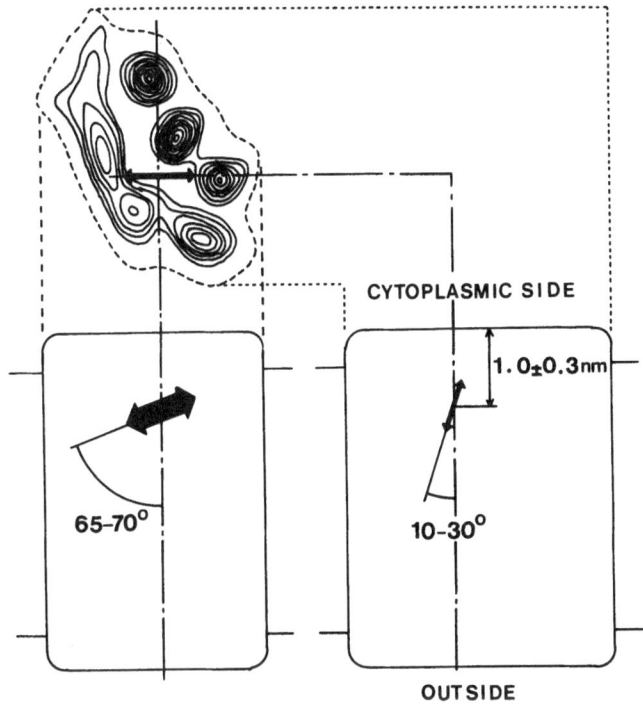

FIGURE 5. Three-dimensional location and orientation of retinal in bacteriorhodopsin.

was very low, indicating the efficient energy transfer to neighboring native retinal chromophores. The efficiency of this energy transfer was estimated to be close to 100% by monitoring the near-infrared fluorescence from native retinal. If the excitation energy of NBD were received almost exclusively by the nearest native retinal, and if the orientation of NBD were random, the distance between NBD and the nearest retinal would be 1.8 ± 0.3 nm.

A similar geometry was suggested from the analysis of the energy transfer from the fluorescent retinal derivative to NBD. The initial decay rate of fluorescence from the fluorescent retinal increases with the increase of the labeling ratio of NBD/protein. Then the difference between the initial decay rate of the fully labeled sample and that of unlabeled sample indicates the sum of transfer rates from a fluorescent retinal derivative to NBD at all binding sites. From this value, the distance between fluorescent retinal and the nearest NBD is calculated to be 2.1 nm when the orientation of NBD is random, that is, $\kappa^2 = \frac{2}{3}$.

Another restriction on the location of NBD–Lys-41 was obtained from experiments of the NBD-labeled reduced purple membrane solubilized with oc-tylglucoside. Upon solubilization, the rate of energy transfer from the fluorescent retinal derivative to NBD–Lys-41 became much slower; the distance between

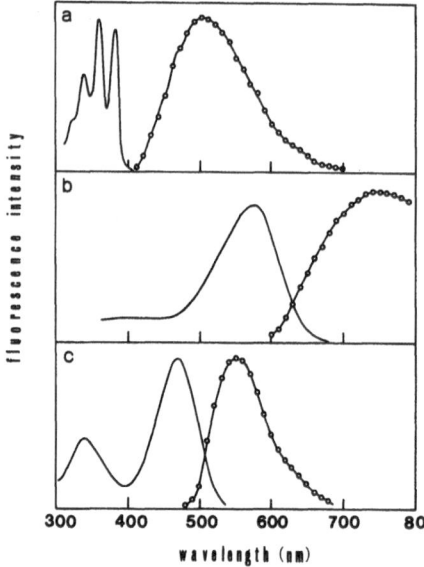

FIGURE 6. Absorption (—) and emission (○) spectra of (a) the fluorescent retinal chromophore converted by reduction and ultraviolet irradiation, (b) the native retinal, and (c) NBD–Lys-41 in the purple membrane.

the retinal and NBD–Lys-41 in monomeric bacteriorhodopsin was estimated to be 2.65 nm, if $\kappa^2 = \frac{2}{3}$.

The possible explanation of this result is that NBD–Lys-41 locates in a contacting area of two bacteriorhodopsin molecules and the distance to the retinal in a neighboring bacteriorhodopsin molecule is the shortest (Figure 7a).

The possible location of NBD–Lys-41 was further restricted by the following experiments, which indicate the energy migration between NBD probes. Purple membrane used for this experiment was fully reduced to eliminate the energy transfer from NBD–Lys-41 to native retinal. The fluorescence anisotropy of NBD–Lys-41 directly excited by polarized light became smaller at higher labeling ratio. This dependence disappeared when we dissolved the membrane with 40 mM octylglucoside. From the initial decay rate of fluorescence anisotropy of NBD fully labeled, the distance between the nearest neighboring NBD was estimated to be less than 20 Å. That is, NBD–Lys-41 must exist within 12 Å from a C3 symmetry axis in the purple membrane (Figure 7b).

In order to know the accurate location and orientation of a fluorescent probe in the membrane, we have to determine the five independent coordinates (x, y, z; θ, ψ) and thus at least five independent experimental quantities. To get a sufficient number of independent parameters from fluorescence energy transfer studies, we designed an energy transfer system in which the acceptor as well as the donor was highly fluorescent; that is, the fluorescent retinal derivative and NBD–Lys-41 were used as an energy donor and acceptor, respectively. When the fluorescence intensity and anisotropy of the donor and acceptor were ob-

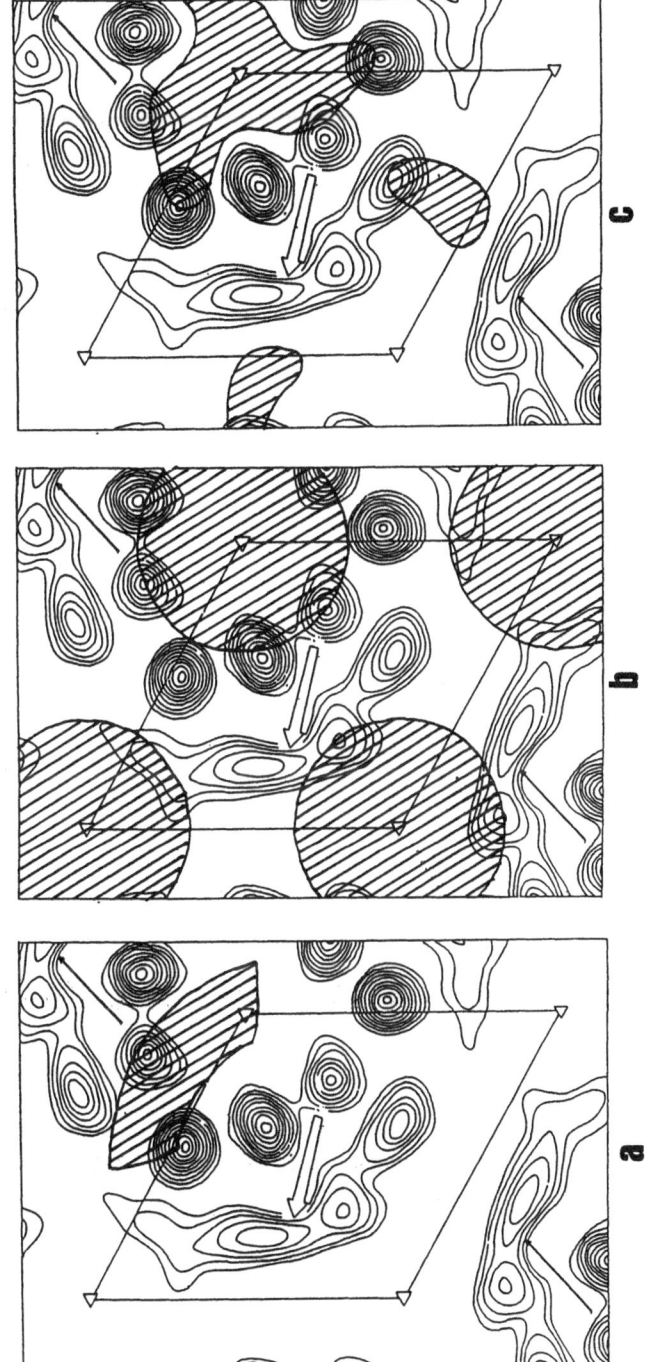

FIGURE 7. In-plane location of NBD bound to Lys-41 estimated by fluorescence energy transfer techniques.

served for samples with various labeling ratios of NBD/protein, four independent quantities were obtained. Another independent quantity was obtained by measuring the tilt angle of the chromophore with respect to the membrane normal. These data were combined with each other and the crystallographic analysis of energy transfer was carried out. The result of analysis suggested that NBD–Lys-41 exists in the hatched area shown in Figure 7c. At the same time, the retinal chromophore was suggested to exist within 16 Å vertically from NBD–Lys-41.

If all the results shown in Figure 7a–c are correct, Lys-41 must locate on the helix 7 in Figure 2.

5. CONFORMATIONAL PREDICTION OF BACTERIORHODOPSIN MOLECULE

To discuss the detailed structure of bacteriorhodopsin molecule in the membrane, we predicted amino acid sequences corresponding to the seven transmembrane α-helixes from the free-energy calculations. We calculated the free energies for all segments composed of 22 consecutive amino acid sequences of bR with three conformations, α-helix (α), antiparallel β-sheet of hairpin type (β), and random coil (γ) conformations in both hydrophobic ("membranelike") and hydrophilic ("waterlike") environments. We chose 22 amino acids because the length of the α or β conformation composed of 22 amino acids is almost the same as the thickness of the hydrophobic region of the membrane. By comparing the calculated free-energy values for six conformational states of segments composed of 22 amino acid sequences, we predicted amino acid sequences AB . . . G, denoted from the amino terminus, which correspond to seven α-helical segments (Ikegami *et al.*, 1987b). Similar predictions of the amino acid sequences were reported by several investigators (Engelman *et al.*, 1980; Agard and Stroud, 1982; Kimura *et al.*, 1982; Ovchinnikov, 1982; Trewhella *et al.*, 1983; Katre *et al.*, 1984). All predictions are very similar to each other; the amino acid numbers of the central position of predicted sequences are all confined within very narrow limits indicated on the center line of the membrane in Figure 8.

The transmembrane positions of the α-carbon of ionizable amino acids averaged over these predictions are shown in the figure. The transmembrane location of the retinal shown on the right-hand side of the figure suggests that Asp-96 is the most probable counterion of the Schiff base proton, although there remains the possibility of some deprotonated tyrosine. Furthermore, it suggests that the side chain of Lys-216 is extended to the cytoplasmic surface without large flexibility because Lys-216 locates near the center of the membrane. Therefore, the possible locations of the G helix in the membrane plane should be restricted to either helix 2 or 5 in Figure 2, because of the short distances from

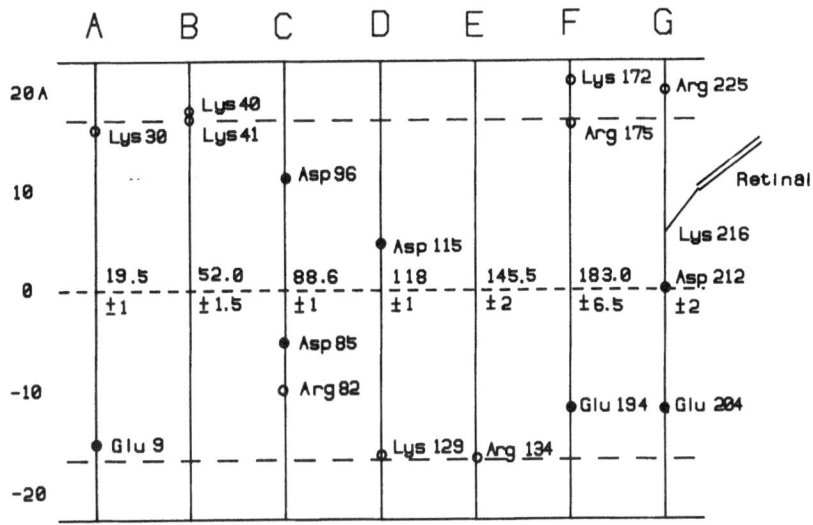

FIGURE 8. Transmembrane positions of the α-carbon of ionizable amino acids in bacteriorhodopsin averaged over seven predictions. The upper side is the cytoplasmic side of the membrane.

the in-plane location of the retinal ends. For similar reasons, the probable location of the C helix, which holds Asp-96, may be helix 6 or 3.

In-plane location of the B helix was estimated to be near helix 7 as described previously. Recently, Trewhella *et al.* (1986) reported that the most probable locations of A and B helixes are either helixes 1 or 7 or helixes 7 or 6 by neutron diffraction methods.

In light of these results, we tentatively assigned the seven sequences (A B. . .G) to the seven helixes 1-7-6-2-3-4-5 or 1-7-6-5-4-3-2 in the electron density map in Figure 2. Since the arrangement of the five sequences CDEFG in one model makes almost mirror symmetry for the retinal plane of the other model, the mutual distances between the retinal and these helixes are almost the same between two models. If we assume either of the two models, we can imagine the transmembrane distribution of ionizable groups around the chromophore. Based on this structural prediction, we recently proposed a mechanism of proton pumping (Ikegami *et al.*, 1987a).

6. REFERENCES

Allegrini, P. P., Sigrist, H., Schaller, J., and Zahler, P., 1983, Site-directed fluorogenic modification of bacteriorhodopsin by 7-chloro-4-nitrobenz-2-oxa-1,3-diazole, *Eur. J. Biochem.* **132** : 604–608.

Agard, D. A., and Stroud, R. M., 1982, Linking regions between helices in bacteriorhodopsin revealed, *Biophys. J.* **37** : 589–602.

Bayley, H., Huang, K. S., Radhakrishnan, R., Ross, A. H., Takagi, Y., and Khorana, H. G., 1981, Site of attachment of retinal in bacteriorhodopsin, *Proc. Natl. Acad. Sci. U.S.A.* **78**: 2225–2229.

Earnest, T. N., Roepe, P., Briaman, M. S., Gillespie, J., and Rothschild, K. J., 1986, Orientation of the bacteriorhodopsin chromophore probed by polarized Fourier transform infrared difference spectroscopy, *Biochemistry* **25** : 7793–7798.

Engelman, D. M., Henderson, R., McLachlan, A. D., and Wallace, B. A., 1980, Path of the polypeptide in bacteriorhodopsin, *Proc. Natl. Acad. Sci. U.S.A.* **77** : 2023–2027.

Fisher, K. A., 1984, Preparation of planar membrane monolayers for spectroscopy and electron microscopy, *Methods Enzymol.* **88** : 230–235.

Förster, T., 1965, Delocalized excitation and excitation transfer, in *Modern Quantum Chemistry* (O. Sinanoglu, ed.), pp. 93–137, Academic Press, New York.

Hayward, S. B., and Stroud, R. M., 1981, Projected structure of purple membrane determined to 3.7 Å resolution by low temperature electron microscopy, *J. Mol. Biol.* **151** : 491–517.

Henderson, R., 1977, The purple membrane from *Halobacterium halobium*, *Annu. Rev. Biophys. Bioeng.* **6** : 87–109.

Henderson, R., and Unwin, P. N. T., 1975, Three-dimensional model of purple membrane obtained by electron microscopy, *Nature* **257** : 28–32.

Ikegami, A., Kouyama, T., Kinosita, K., Jr., Urabe, H., and Otomo, J., 1987a, Structure change of bacteriorhodopsin and the mechanism of proton pump, in *Primary Processes in Photobiology*, Springer Proceedings of Physics, Vol. 20 (T. Kobayashi, ed.), pp. 173–182, Springer-Verlag, New York.

Ikegami, A., Kouyama, T., Kinosita, K., Jr., Otomo, J., Urabe, H., Fukuda, K., and Kataoka, R., 1987b, Conformational analysis of bacteriorhodopsin, in *Retinal Proteins* (Y. A. Ovchinnikov, ed.), pp. 307–315, VNU Science Press, Utrecht.

Jubb, J. S., Worceste, D. L., Crespi, H. J. L., and Zaccai, G., 1984, Retinal location in purple membrane of *Halobacterium halobium*: A neutron-diffraction study of membranes labeled *in vivo* with deuterated retinal, *EMBO J.* **3** : 1455–1461.

Katre, N. V., Wolber, P. K., Stoeckenius, W., and Stroud, R. M., 1981, Attachment site(s) of retinal in bacteriorhodopsin, *Proc. Natl. Acad. Sci. U.S.A.* **78** : 4068–4072.

Katre, N. V., Finer-Moore, J., Stroud, R. M., and Hayward, S. B., 1984, Location of an extrinsic label in the primary and tertiary structure of bacteriorhodopsin, *Biophys. J.* **46** : 195–204.

Khorana, H. G., Gerber, G. E., Herlihy, W. C., Gray, P. C., Anderegg, R. J., Nihei, K., and Biemann, K., 1979, Amino acid sequence of bacteriorhodopsin, *Proc. Natl. Acad. Sci. U.S.A.* **76** : 5046–5050.

Kimura, K., Mason, T. L., and Khorana, H. G., 1982, Immunological probes for bacteriorhodopsin, *J. Biol. Chem.* **257** : 2859–2867.

Kometani, T., Kinosita, K., Jr., Furuno, T., Kouyama, T., and Ikegami, A., 1987, Transmembrane location of retinal in purple membrane: Fluorescence energy transfer in maximally packed donor–acceptor systems, *Biophys. J.* **52** : 509–517.

Kouyama, T., Kimura, K., Kinosita, K., Jr., and Ikegami, A., 1981a, Location and orientation of the chromophore in bacteriorhodopsin: Analysis of fluorescence energy transfer, *J. Mol. Biol.* **153** : 337–359.

Kouyama, T., Kimura, K., Kinosita, K., Jr., and Ikegami, A., 1981b, Immobility of the chromophore in bacteriorhodopsin, *FEBS Lett.* **124** : 100–104.

Kouyama, T., Kinosita, K., Jr., and Ikegami, A., 1983, Fluorescence energy transfer studies of transmembrane location of retinal in purple membrane, *J. Mol. Biol.* **165** : 91–107.

Lemke, H. D., and Oesterhelt, D., 1981, Lysine 212 is a binding site of the retinal moiety in bacteriorhodopsin, *FEBS Lett.* **128** : 255–260.

Mullen, E., Johnson, A. H., and Akhtar, M., 1981, The identification of Lys-216 as the retinal binding residue in bacteriorhodopsin, *FEBS Lett.* **130** : 187–193.

Ottolenghi, M., 1980, The photochemistry of rhodopsins, *Adv. Photochem.* **12** : 97–200.

Ovchinnikov, Y. A., 1982, Rhodopsin and bacteriorhodopsin: Structure and function relationships, *FEBS Lett.* **148** : 179–191.

Ovchinnikov, Y. A., Abdulaev, N. Y., Feigina, M. Y., Kiselev, A. V., and Lobanov, N. A., 1979, The structural basis of the functioning of bacteriorhodopsin: An overview, *FEBS Lett.* **100** : 219–224.

Peters, J., Peters, R., and Stoeckenius, W., 1976, A photosensitive product of sodium borohydride reduction of bacteriorhodopsin, *FEBS Lett.* **61** : 128–134.

Seiff, F., Wallat, I., Ermann, P., and Heyn, P., 1985, Neutron-diffraction studies on the location of retinal in bacteriorhodopsin, *Proc. Natl. Acad. Sci. U.S.A.* **82** : 3227–3231.

Stoeckenius, W., and Bogomolni, R. A., 1982, Bacteriorhodopsin and pigments of halobacteria, *Annu. Rev. Biochem.* **52** : 587–616.

Stoeckenius, W., Lozier, R. H., and Bogomolni, R. A., 1979, Bacteriorhodopsin and the purple membrane of halobacteria, *Biochim. Biophys. Acta* **505** : 215–278.

Trewhella, J., Anderson, S., Fox, R., Gogol, E., Engelman, D., and Zaccai, G., 1983, Assignment of segments of the bacteriorhodopsin sequence to positions in the structural map, *Biophys. J.* **42** : 233–241.

Trewhella, J., Popot, J. L., Zaccai, G., and Engelman, D. M., 1986, Localization of two chymotryptic fragments in the structure of renatured bacteriorhodopsin by neutron diffraction, *EMBO J.* **5** : 3045–3049.

Unwin, P. N. T., and Henderson, R., 1975, Molecular structure determination by electron microscopy of unstained crystalline specimens, *J. Mol. Biol.* **94** : 425–440.

Varo, G., 1983, Dried oriented purple membrane samples, *Acta Biol. Acad. Sci. Hung.* **32** : 301–310.

Chapter 8

Structure and Dynamics of the Liver Microsomal Monoxygenase System

Christoph Richter, Josef Gut, and Barbara C. Kunz

1. INTRODUCTION

1.1. General Structure of Biological Membranes

Cell membranes regulate a variety of cellular processes ranging from permeability, transport, and excitability to intercellular interaction, morphological

---

Abbreviations used in this chapter: ANS, 1-anilino-8-naphthalene sulfonate; ANM, *N*-(1-anilinonaph-4-yl)maleimide; DPH, 1,6-diphenyl-1,3,5-hexatriene; DMPC, dimyristoylphosphatidylcholine; DOPC, dioleylphosphatidylcholine; DPPC, dipalmitoylphosphatidylcholine; E, eosin; EPR, electron paramagnetic resonance; FITC, fluorescein isothiocyanate; FRAP, fluorescence recovery after photobleaching; I-a, I-b, I-c, I-d, P-450 isozymes isolated from rat liver; LM-2, LM-4, P-450 isozymes isolated from rabbit liver; NMR, nuclear magnetic resonance; P-450, microsomal cytochrome P-450; P-450$_b$, P-450 isozyme isolated from rat liver; PA, dipalmitoylphosphatidic acid; PC, phosphatidylcholine; PE, phosphatidylethanolamine; PS, phosphatidylserine; PUFA, polyunsaturated fatty acid; SDS-PAGE, polyacrylamide gel electrophoresis; ST-EPR, saturation transfer EPR; T_c, temperature of the gel to liquid-crystalline phase transition; TRSP, time-resolved spatial photometry.

Christoph Richter, Josef Gut, and Barbara C. Kunz   Laboratory of Biochemistry, Swiss Federal Institute of Technology, Zürich CH-8092, Switzerland.   *Present address of J. G.:* Stanford University, Stanford, California 94025.   *Present address of B. C. K.:* National Institutes of Health, Bethesda, Maryland 20205.

differentiation, and fusion. Numerous models have been advanced to characterize the organization of lipids and proteins in cell membranes. Today, there is substantial agreement on the "fluid mosaic" model (Singer and Nicholson, 1972), which emphasizes the dynamic behavior of the membrane components. Both lipids and proteins can undergo a variety of motions: rotational motion around the axis perpendicular to the plane of the membrane; lateral diffusion in the plane of the membrane; in addition, lipids can "flip-flop" (exchange from one monolayer to the other) and undergo *trans–gauche* conformational changes in the phospholipid acyl chains, which give rise to an increased segmental mobility toward the center of the bilayer. Since the fluid mosaic model has been proposed, its rather crude and generalizing picture has been filled with some details. The refined picture shows a dynamic membrane in which molecular associations are tightly controlled, and in which long-range lateral motions are surprisingly restricted.

1.1.1. Lipid Mobility in Membranes

Three spectroscopic techniques, nuclear magnetic resonance (NMR), electron paramagnetic resonance (EPR), and fluorescence polarization, have been of great help in examining the motion of membrane lipids. NMR and EPR are particularly useful in defining fast motions of membrane lipids and the average degree of order in the packing of acyl chains in the bilayer.

As outlined in detail in other chapters, fluorescence polarization measurements can be done under steady-state or time-resolved conditions. Steady-state measurements are not unambiguous but are convenient and can be useful for comparative purposes. In contrast, time-resolved fluorescence measurements can give unambiguous information about lipid order and dynamics (Heyn, 1979). Fluorescence polarization has been used to determine "microviscosities" in membranes. However, it is now clear that with diphenyl-hexatriene (DPH), one of the most frequently used fluorescent probes (Shinitzky and Bahrenholz, 1978), the order of the acyl chains but not microviscosity is detected.

Membrane phospholipids show the following characteristics: their head groups and glycerol backbones are highly ordered; the acyl chains are tilted some 30° from the membrane normal for periods longer than 10^{-7} sec; oscillations and flexing of acyl chain carbons occur around the long axis of the chains and are greatest in the interior of the bilayer; the mobility of the acyl chains and their interaction with one another affect diffusional motion in the membrane; the lateral diffusion coefficient lies in the range of 10^{-8} cm²/sec.

1.1.2. Protein Mobility in Membranes

Many membrane proteins are mobile, being able to diffuse laterally and rotate around an axis perpendicular to the plane of the membrane. The Saffman–Delbrück model (Saffman and Delbrück, 1975) treats both the lateral and the rotational diffusion in terms of the Einstein–Stokes law for viscous friction adapted to two dimensions. The lateral diffusion coefficient D_L of a cylinder with radius a and height h spanning the membrane is

$$D_L = \frac{k_B T}{4\pi\eta h}\left[\ln\left(\frac{\eta h}{\eta_w a}\right) - \gamma\right] \qquad (1)$$

where k_B is Boltzmann's constant, T is the temperature, η is the viscosity of the fluid membrane, η_w is the viscosity of the surrounding aqueous phase, and γ is Euler's constant.

Fluorescence recovery after photobleaching (FRAP), a powerful technique for the measurement of lateral motion in membranes, was introduced more than 10 years ago (Peters et al., 1974). Extensions of this technique have been developed (Koppel and Sheetz, 1983). Laser photobleaching combined with time-resolved spatial photometry (TRSP) has recently been introduced to measure lateral motion (Kapitza and Jacobson, 1986). Values of the lateral diffusion coefficient obtained for various proteins in membranes range from 10^{-8} to 10^{-13} cm^2/sec.

The detection of rotational motion within membranes in the micro- to millisecond range was facilitated by the development of two areas of spectroscopic technology, one in EPR [saturation transfer EPR—ST-EPR, using nitroxide spin labels (Thomas, 1978)] and the other in optical spectroscopy (transient optical anisotropy), using long-lifetime chromophores (Cherry, 1978). In the optical technique, the principle is very similar to that used to study excited-state processes by fluorescence. The main difference is the use of probes with lifetimes in the micro- to millisecond range, in contrast to the nanosecond lifetimes of most fluorescent probes. The rate of rotational diffusion depends sensitively on the size of the diffusing molecule as well as on the viscosity of the lipid bilayer. The diffusion coefficient for rotation is

$$D_R = \frac{k_B T}{4\pi\eta h a^2} \qquad (2)$$

The measured rotational correlation times of membrane proteins are in the micro- to millisecond range, reflecting the various sizes and aggregation states of proteins. Because of the dependence on molecular volume, rotational diffusion provides a sensitive indicator of aggregation processes (Thomas, 1986).

The optical spectroscopy technique has been used extensively to investigate the mobility of microsomal cytochrome P-450 in various membranes.

A brief description of this technique is therefore given here. The method depends on the principles of (1) photoselection (orientation-sensitive excitation) and (2) orientation-dependent absorption measurements that monitor reorientation of the photoselected molecules. Photoselection is achieved by exciting a randomly distributed population of molecules with plane-polarized light, so that those molecules whose transition moment for absorption is parallel or at a small angle to the electric vector of the incident light are preferentially excited. Absorption signals arising from the excited molecules are dichroic. When excitation is by a brief pulse of light, the initial absorption dichroism decays as the molecules become randomized by Brownian motion. The transient absorption anisotropy $r(t)$ is defined as

$$r(t) = \frac{A_\parallel - A_\perp}{A_\parallel + 2A_\perp} \tag{3}$$

where $A_\parallel$ and $A_\perp$ are the changes in the absorbance of light polarized parallel and perpendicular to the plane of the exciting light. From the rate of decay, rotational relaxation times may be determined. For a membrane-bound mobile protein, $r(t)$ does not fall to zero but decays to a time-dependent finite value $r(\infty)$. From the ratio $r(0)/r(\infty)$ the coexistence of mobile and immobile proteins may be estimated.

Most membrane proteins lack an endogenous chromophore, which could be used to study rotational motion. They can be labeled with extrinsic probes such as eosin, erythrosin, or pyrene. Their use can be limited by problems generally encountered with extrinsic probes (e.g., lack of specificity, disturbance of the sample). However, some membrane proteins possess intrinsic chromophores that undergo reversible photochemical reactions that can be used to study rotational motion. Examples are the retinal chromophore rhodopsin (Cone, 1972) and bacteriorhodopsin (RaziNaqvi *et al.*, 1973). Flash photolysis of heme-CO has been used to study cytochrome oxidase (Kawato *et al.*, 1981, 1982a) and microsomal cytochrome P-450 (see below).

1.2. Peroxidation of Membrane Lipids

Lipid peroxidation in membranes is a complex process during which unsaturated acyl residues of phospholipids undergo reactions with molecular oxygen to yield lipid hydroperoxides. These then decompose to a variety of further products including alkanals, alkenals, hydroxyalkenals, and ketones (Esterbauer *et al.*, 1982). Lipid peroxidation comprises chain reactions initiated by the abstraction of a hydrogen atom from the unsaturated lipid by a reactive free radical. This reaction is followed by the addition of oxygen to give a lipid peroxy free radical and, in the case of polyunsaturated fatty acids

(PUFAs) containing isolated 1,4-dienes, by double-bond rearrangement to yield conjugated dienes with characteristic UV absorption around 233 nm. The lipid peroxy free radical can itself abstract a hydrogen atom from a neighboring molecule, which may be another PUFA substrate molecule to give the corresponding lipid hydroperoxide.

An important result of studies on lipid peroxidation has been the realization that many toxic agents can be metabolically activated, predominantly by the microsomal monoxygenase system, within cells to free radical intermediates that can initiate lipid peroxidation and result in cell injury. Irradiation of tissue, cells, or cell organelles may likewise produce reactive free radicals with similar consequences. Moreover, depletion of physiological protective compounds, such as vitamin E, ascorbate, urate, or glutathione, may also cause lipid peroxidation.

Studies on lipid peroxidation in biological systems have demonstrated the degradation of PUFAs, with a subsequent disorganization of membrane structure and disturbance of membrane function.

1.3. Microsomal Monoxygenase

1.3.1. Discovery of the Nature and Function of the Monoxygenase

Once it became possible to fractionate the particles of animal cells, the presence of NADH–cytochrome c reductase activity (Hogeboom, 1949) and NADPH–cytochrome c reductase activity (Hogeboom and Schneider, 1950) was discovered in submicroscopic particles prepared from liver homogenates. These particles, now called *microsomes*, were identified as vesicles derived from the endoplasmic reticulum (Palade and Siekevitz, 1956). Initially, the physiological significance of the cytochrome c reductase activities remained unclear, since cytochrome c and cytochrome oxidase were known to be present only in mitochondria and not in the endoplasmic reticulum. Then, Strittmatter and Ball (1952) discovered a new b-type cytochrome in the microsomal fraction prepared from rat liver. This cytochrome, today called *cytochrome b_5*, was solubilized and purified by Strittmatter and Velick (1956a). In addition, they isolated from microsomes a new flavoprotein that reduced cytochrome b_5 with NADH (Strittmatter and Velick, 1956b). The reduced form of cytochrome b_5 reacted rapidly with externally added cytochrome c, which explained the NADH–cytochrome c reductase activity in liver microsomes (Strittmatter and Velick, 1956b).

The isolation of the NADPH–cytochrome c reductase activity was first reported by Williams and Kamin (1962) and by Phillips and Langdon (1962). The properties of the purified enzyme suggested its identity with a flavoprotein previously isolated from an acetone powder of whole liver (Horecker,

1950). The physiological significance of NADPH–cytochrome c reductase in microsomes became clear following the discovery of another cytochrome, cytochrome P-450.

The occurrence in liver microsomes of the CO-binding pigment, which is now called *cytochrome P-450*, was first reported independently by two investigators (Garfinkel, 1958; Klingenberg, 1958). Today, it is well known that hemoproteins similar in spectral properties to hepatic microsomal cyto-chrome P-450 are widely distributed in the animal, plant, and microbial kingdoms. Indeed, an ever increasing number of P-450 isozymes is being characterized (Ortiz de Montellano, 1986). They participate as monoxygen-ases in various biochemical processes of metabolic importance. In animals, P-450s are also responsible for oxidative transformations of xenobiotics such as drugs, pesticides, carcinogens, and environmental pollutants. For this rea-son, P-450s are attracting broad attention among pharmacologists and on-cologists, as well as biochemists. In addition, many physical chemists have become interested in the anomalous physical properties of this class of hemo-proteins.

1.3.2. Topology of the Proteins

The reductase ($M_r \approx 78$ kDa) is anchored to the membrane of the en-doplasmic reticulum via a small ($M_r \approx 6$–10 kDa) hydrophobic segment (Black *et al.*, 1979; Gum and Strobel, 1979). The large hydrophilic part, which contains 1 molecule of FMN and FAD, protrudes from the membrane into the cytoplasmic space. The topology of the cytochrome P-450 isozymes ($M_r \approx 50$ kDa) in membranes, on the other hand, is currently not known in detail (De Pierre and Ernster, 1977). Limited proteolysis of microsomes does not release a soluble catalytic domain but results in conversion of P-450 to the catalytically inactive form called cytochrome P-420 (Sato *et al.*, 1969). It was therefore concluded that P-450 is not simply anchored to the membrane by a single hydrophobic domain as its reductase. Recent sequencing studies seemed to support this view. P-450 and its reductase can be isolated and reconstituted in an enzymatically active form. Successful reconstitution re-quires the presence of phospholipids (Lu *et al.*, 1969). The arrangement of the two enzymes in the membrane and their odd stoichiometry (there can be 20–30 cytochromes per reductase in the microsomal membrane of rat hepa-tocytes) have raised questions as to the mechanism of electron transfer from the reductase to the cytochrome and the functional interaction of the proteins.

P-450 is synthesized on membrane-bound polysomes and cotranslation-ally inserted into the membrane (Bar-Nun *et al.*, 1980). Experiments using specific antibodies (Thomas *et al.*, 1977; Matsuura *et al.*, 1981; Lemos-Chiarandini *et al.*, 1987) or proteases (Vlasuk *et al.*, 1982) as probes indicate that P-450 is exposed on the cytoplasmic face of the endoplasmic reticulum and suggest that no immunological determinants within the polypeptide are exposed on the luminal side of the membrane (Matsuura *et al.*, 1981). However, enhancement of the proteolytic susceptibility of P-450 in microsomes by detergent solubilization has been given as evidence for a transmembrane orientation (De Pierre and Dallner, 1975).

Recently, both protein and cDNA sequencing studies (Fujii-Kuriyama *et al.*, 1982; Heinemann and Ozols, 1983; Tarr *et al.*, 1983; Leighton *et al.*, 1984) have allowed new speculations concerning the topology of P-450 in membranes and the location of the heme-binding site in the cytochrome. The main phenobarbital-induced form of rabbit liver microsomal P-450, termed *P-450 LM-2*, is the first microsomal P-450 to be sequenced completely by protein chemical means (Heinemann and Ozols, 1983; Tarr *et al.*, 1983). The protein has eight hydrophobic regions long enough to span the membrane. They are separated by charged amino acid residues. The COOH terminus contains a large hydrophilic domain (about 115 amino acid residues). This domain contains regions of homology with several other P-450 isozymes and has been proposed by Heinemann and Ozols (1983) to contribute the thiolate heme ligand by a cysteinyl residue (Cys-436). The model advanced by Coon's group (Tarr *et al.*, 1983), however, erroneously predicted Cys-152, found within a long hydrophilic segment of α-helix close to the NH$_2$ terminus (about 51 amino acid residues), to be the only cysteinyl residue available to function as the fifth ligand to the heme iron atom. Both putative heme-binding domains are hydrophilic, suggesting that they are not located in the membrane's lipid bilayer. By assuming a globular shape for these hydrophilic segments, their diameters can be estimated to be 38 and 29 Å, respectively. It is now generally agreed that a cysteine residue corresponding to position 436 in LM-2 is indeed the fifth ligand of the heme iron.

2. MEMBRANE DYNAMICS AND ORDER STUDIED BY FLUORESCENCE

2.1. Biophysical Consequences of Lipid Peroxidation

Pyrene excimer formation, perylene and 3-methoxybenzanthracene rotational relaxation rates, and quenching of 4-dimethylaminochalcone fluorescence were used to investigate membrane viscosity in liposomes, liver microsomes, and liver mitochondria (Dobretsov *et al.*, 1977). Although these fluorescent probes sense different regions of the membrane, they all indicated a lowering of the membrane "fluidity" after lipid peroxidation. Increases in the fluorescence polarization of probe molecules intercalated in the hydrocarbon regions were also reported for aged chloroplasts (Ford and Barber,

1980). Treatment of human erythrocyte membranes with phenylhydrazine resulted in lipid peroxidation. Fluorescence polarization studies with DPH indicated a decrease in bulk lipid "fluidity" (Rice-Evans and Hochstein, 1981). In addition, the fluorescence intensity of 1-anilino-8-naphthalene sulfonate (ANS) was decreased and red-shifted in these membranes, indicating an increased solvent accessibility to the naphthalene ring system. DPH and ANS fluorescence polarization studies in membranes prepared from erythrocytes of increasing age also indicated a decreased "fluidity" possibly due to lipid peroxidation (Bartosz *et al.*, 1981).

A detailed study of peroxidized rat liver microsomes was reported by Eichenberger *et al.* (1982). A close correlation between the production of malondialdehyde, a breakdown product of peroxidized lipids, and the increase of steady-state fluorescence polarization r_s of DPH was observed. An unequivocal proof for an increase in the orientational order of the acyl chains was obtained with time-resolved fluorescence polarization. The time-dependent anisotropy r_t of DPH in both control and peroxidized microsomes decayed with a time constant of a few nanoseconds to a time-dependent residual anisotropy r_∞ within the time range of several tens of nanoseconds. A significant increase in r_∞ from 0.09 (control) to 0.13 (60 nmol malondialdehyde/ mg protein) was seen. These values correspond to a conventional order parameter $S = 0.48$ and $S = 0.59$, respectively. Malondialdehyde did not change r_s when added to membranes, indicating that it does not cause cross-links in the lipid domain. The quantitative determination of the amount of malondialdehyde produced and the amount of acyl chains detected after transesterification and gas chromatography suggested that during peroxidation neighboring acyl side chain radicals form covalent bonds between each other and thus decrease the motional freedom of DPH. (An account of the relation between lipid peroxidation and the rotational mobility of cytochrome P-450 in rat liver microsomes will be given in Section 2.3.)

A single oral dose of polybrominated biphenyls (FireMaster) given to rats led to a persistent decrease of the phospholipid/cholesterol ratio, increased susceptibility to lipid peroxidation, and decreased fluorescence anisotropy in microsomes and liposomes prepared from microsomal lipids (Bernert and Groce, 1984).

An increased fluorescence polarization of DPH was observed in lecithin liposomes upon ascorbate/Fe^{2+}-induced lipid peroxidation (Fukuzawa *et al.*, 1981), both parameters being sensitive to α-tocopherol. In dispersions of polar lipid extracts of *Vicia faba* leaves oxidized by potassium peroxychromate, DPH fluorescence polarization indicated a considerable restriction of probe motion (Galanopoulou *et al.*, 1982). The restriction was apparently not due to the removal of unsaturated residues by peroxidation. Rather, oxidation seemed to destroy the bilayer arrangement as indicated by x-ray diffraction studies. In another model for oxidant stress, alterations in the structure and integrity of the membrane after exposure of erythrocytes and "ghosts" to t-butyl hydroperoxide were investigated (Rice-Evans *et al.*, 1985). DPH fluorescence polarization showed that the hydrophobic region of the membrane became less "fluid" upon peroxidation although the overall cholesterol/phospholipid ratio was unaffected. The surface probe ANS showed a pronounced decrease in fluorescence with no shift in the emission maximum in peroxidized membranes, indicative of an increased negative charge density on the membrane surface. Time-resolved and steady-state measurements of the rotational mobility of pyrene and n-(9-anthroyloxy) stearic acid ($n = 2$ or 12) and the accessibility to fluorescence quenchers have recently been reported for porcine intestinal brush-border membranes (Ohyashiki *et al.*, 1986a,b). Again, lipid peroxidation resulted in a reduction of the bilayer "fluidity." The different sensitivities of the two anthroyloxy fatty acids to quenchers indicated localized changes in the lipid order of membranes upon peroxidation. The antioxidant α-tocopherol itself induced perturbations of the lipid organization in these membranes (Ohyashiki *et al.*, 1986c). In another study, the environmental oxidant pollutant NO_2 increased the rotational correlation time of DPH concomitant with lipid peroxidation in both pulmonary artery and aortic endothelial cells (Patel and Block, 1986).

Tumor cell membranes are characterized by changes in lipid constituents and membrane "fluidity" (Shinitzky, 1984). Microsomal and plasma membranes isolated from hepatomas with different degrees of differentiation have an increased order in the lipid domain related to their phospholipid content (Galeotti *et al.*, 1984a,b). The more undifferentiated and rapidly growing tumors had a lower membrane lipid content, degree of acyl unsaturation, lipid/cholesterol and lipid/protein ratios. Time-resolved DPH fluorescence polarization studies showed a more ordered lipid phase in hepatoma microsomes which was not affected by superoxide radical-dependent peroxidation (Masotti *et al.*, 1986).

2.2. Mobility of Membrane-Bound Cytochrome P-450

2.2.1. Rotational Mobility Studied by Delayed Fluorescence Polarization

The rotational mobility of P-450 was first measured in the microsomal membrane by the flash photolysis technique of the heme–CO complex (Richter *et al.*, 1979). E (eosin)-type delayed fluorescence depolarization was used to study the rotational relaxation of P-450 reconstituted into a membrane vesicle system (Greinert *et al.*, 1979). The fluorescence decay of the membrane preparation extended to several hundred microseconds and allowed measurements of molecular motion in the millisecond range. Cytochrome P-450 LM-2 purified from phenobarbital-induced rabbits had a rotational correlation time of 111 µsec in proteoliposomes with a lipid/protein ratio of 5 (w/w) and a phospholipid composition of phosphatidylcholine (PC)/phosphatidylethanolamine (PE)/dipalmitoylphosphatidic acid (PA) of 2 : 1 : 0.06 (w/w). On the basis of these measurements the formation of cytochrome clusters comprising 6-60 monomers was suggested (Greinert and Stier, 1980).

While the flash photolysis technique of the heme–CO complex is limited to the investigation of the reduced form of the cytochrome, the E-type delayed fluorescence can be performed with both the oxidized and reduced cytochrome (Greinert *et al.*, 1982a,b). It was found that the substrate benzphetamine retards rotation of the cytochrome whereas reduction of the cytochrome in the presence of the substrate accelerates rotation. The same studies also indicated that the labeled cytochrome exhibits strict uniaxial rotation about the membrane normal. The authors propose a model in which the cytochrome forms disclike hexamers in the membrane. The depth of immersion in the membrane is thought to depend on substrate-induced and redox-state-dependent conformational changes.

A subsequent communication reported interactions of LM-2 and cytochrome P-450 LM-4 with cytochrome P-450 reductase, cytochrome b_5, and epoxide hydrase, as well as the effects of substrates, redox state, and the lipid composition of proteoliposomes (Stier *et al.*, 1985). While LM-2 performs strictly uniaxial rotation, LM-4 undergoes wobbling rotation. When equimolar with reductase in the membrane, the cytochromes form complexes with it which have different mobilities. Reductase–cytochrome complexes dissociate upon reduction. Furthermore, the lipid composition strongly influences the motion of the cytochromes.

These studies confirm and complement the rotational mobility studies of cytochrome P-450 using the flash photolysis technique (Richter *et al.*, 1979; Kawato *et al.*, 1982b; Gut *et al.*, 1982, 1983, 1985) and saturation transfer EPR spectroscopic studies (Schwarz *et al.*, 1982).

2.2.2. Lateral Mobility Studied by Fluorescence Recovery after Photobleaching (FRAP)

Based on the rotational relaxation time measured by the flash photolysis technique, the local (free) lateral diffusion coefficient D of P-450 was estimated to be about 10^{-9} cm^2/sec (Kawato *et al.*, 1982b). Model calculations showed that at least several thousand collisions could occur between cytochromes b_5 and P-450 within the known rate of electron transfer. The lateral diffusion coefficient was subsequently measured with fluorescently labeled

P-450 reconstituted in multi-bilayers formed of dimyristoylphosphatidylcholine (DMPC), egg PC, and their cholesterol mixtures by fluorescence recovery after photobleaching (FRAP) (Wu and Yang, 1984). In the liquid-crystalline phase of egg PC and DMPC, D was about 2×10^{-8} cm^2/sec and dropped to about 5×10^{-10} cm^2/sec in the gel phase of DMPC. Cholesterol reduced the diffusion of P-450 in the liquid-crystalline phase and enhanced it in the gel phase. Also, in this study the estimated rate of collision between the cytochrome and its reductase was two to three orders of magnitude faster than the rate of electron transfer in proteoliposomes (Taniguchi et al., 1980).

2.3. Rotational Mobility of Cytochrome P-450 in Peroxidized Rat Liver Microsomes

Lipid peroxidation alters several biophysical properties of the lipid phase in membranes (see above). However, little is known about its consequences on the mobility of membrane proteins. Rat liver microsomes are the cell organelle of choice to investigate this issue. First, their phospholipids are easily peroxidized. Second, the mobility of P-450, the major microsomal protein, can be measured quantitatively with reasonable accuracy.

When phenobarbital-induced rat liver microsomes were subjected to NADPH- and iron-catalyzed lipid peroxidation (Gut et al., 1985), the formation of about 95 nmol malondialdehyde/mg of protein during 18 min was observed. The order of the lipid phase, measured by steady-state fluorescence of DPH, increased in parallel with the malondialdehyde formation. When peroxidation was stopped after 18 min with EDTA, both the amount of malondialdehyde and the increased r_s value remained constant for 2 hr. Lipid peroxidation led to a time-dependent formation of protein aggregates as revealed by sodium dodecyl sulfate polyacrylamide gel electrophoresis (SDS-PAGE). As compared to the control, significant amounts of proteins in all molecular weight regions had disappeared after 18 min of peroxidation, and a large amount of protein aggregates had formed. Protein aggregation continued during the 2 hr after lipid peroxidation had been stopped.

After 18 min of peroxidation the mobility of P-450 was diminished only moderately. However, the subsequent incubation of the peroxidized microsomal membrane in the presence of EDTA (i.e., when lipid peroxidation had been stopped) resulted in a dramatic increase of the amount of immobilized cytochrome. This drastic immobilization was also seen when peroxidized microsomes were suspended in fresh, nonperoxidizing medium. On the other hand, incubation of nonperoxidized microsomes in the supernatant of peroxidized microsomes did not affect the rotational mobility of P-450. No high molecular weight protein aggregates were observed in SDS-PAGE under these conditions. These results rule out the possibility that the products of lipid peroxidation, once liberated from the membrane, induce cross-links and immobilization of P-450.

It is clear from this study that the rotational mobility of P-450 is largely decreased upon lipid peroxidation. However, the time course of increase in the order of phospholipid acyl residues is not paralleled by the decrease in the mobility of P-450. Thus, its mobility in the membrane is not primarily governed by the order of the lipid phase as measured by DPH. Rather, the decreased cytochrome mobility appears to be due to a slow protein aggregation following lipid peroxidation.

What might be the physiological consequence of the decreased P-450 mobility in peroxidized microsomes? P-450 interacts with its reductase and with cytochrome b_5. Furthermore, there is evidence that it also interacts with the membrane-bound epoxide hydrase (Oesch and Daly, 1972; Guengerich and Davidson, 1982; Stier *et al.*, 1985). A significant rotational and/or lateral mobility of microsomal proteins is probably necessary to overcome enzyme inhibition by protein crowding. A large decrease in protein mobility by lipid peroxidation would clearly decrease the productive interaction of membrane proteins.

2.4. Structure of Free and Membrane-Bound Cytochrome P-450

Fluorescence has been used to investigate the dimension of isolated native and modified P-450 in solution and reconstituted into proteoliposomes.

2.4.1. Tryptophan Fluorescence

Analysis of fluorescence excitation and emission spectra of purified adrenal mitochondrial P-450 (Wang and Kimura, 1976) suggested a larger content of tyrosine relative to tryptophan residues and a nonpolar environment of tryptophan residues in P-450. Tryptophan fluorescence spectra were also recorded from P-450 LM-2, which contains one tryptophan residue, and P-450 LM-4, which contains seven such residues (Chiang and Coon, 1979). When the proteins were excited at 285 nm, tryptophan emission of LM-2 was maximal at 320 nm, whereas that of LM-4 was maximal at 330 nm with similar fluorescence intensity. Under denaturing conditions the maximum for LM-4 was shifted to 350 nm and was greatly enhanced, while a similar change in the peak position but not in the intensity was seen with LM-2. It was concluded that tryptophan fluorescence is largely quenched in the native LM-4 protein presumably due to energy transfer to the heme or other amino acids. After the primary structure of LM-2 had become known, the fluorescence of its single tryptophan residue (position 121) was used to study by energy transfer its distance to the heme (Inouye and Coon, 1985). It was estimated to be less than 40 Å. Since Trp-121 is conserved at or near the same position in mammalian P-450 isozymes, the result may be valid also for these related cytochromes. Tryptophan and tyrosine fluorescence was also used to estimate by energy transfer measurements the distance between these residues and the heme in LM-2 (Schwarze *et al.*, 1983a). It was concluded that there are at least several tyrosine residues that are not localized in the immediate surroundings of the single tryptophan and the heme. The authors speculate that electron transfer between reductase, tyrosines, and tryptophan to the heme could be possible due to overlapping of π-electron clouds.

2.4.2. Covalently Labeled Cytochrome P-450

Fluorescein isothiocyanate (FITC) selectively labels the N-terminal methionine of P-450 LM-2 (Bernhardt *et al.*, 1983; Schwarze *et al.*, 1983a). The overlap of the heme absorption bands between 500 and 600 nm with the emitted fluorescence of fluorescein (peak maximum at 523 nm) makes a determination of the distance between the N terminus and the heme possible. By energy transfer measurements, values of 2.65 and 3.97 nm for the isolated oligomeric and monomeric forms of the cytochrome, respectively, were determined. The estimated distances could in principle be skewed by intermolecular energy transfer in the oligomer. However, measurements in the presence of various detergents proved the intermolecular contribution to be small (Schwarze *et al.*, 1983b).

Fluorescence energy transfer measurements were also used to measure distances between the substrate aflatoxin B_1 and cytochrome P-450 I-d and II-a prepared from hepatic microsomes of polychlorinated-biphenyl-treated rats (Omata and Ueno, 1985). The distances were estimated to be 6.9 and 4.7 nm in I-d and II-a, respectively, which is not compatible with a radius of about 2–2.3 nm of the monomeric P-450. Large distances between benzo[*a*]pyrene bound to cytochrome P-450 I-c and II-d, obtained from identically induced animals, and heme were also reported (Omata *et al.*, 1986). They appeared to be altered by incorporation of the cytochrome into phospholipid or by reduction of the heme.

2.4.3. Distance between Heme and Membrane Bilayer in Proteoliposomes

In proteoliposomes containing cytochrome oxidase and DPH, energy transfer from DPH to heme *a* leads to a decrease of the average lifetime of DPH fluorescence by a factor of 4 (Kinosita *et al.*, 1981). In contrast, energy transfer between cytochrome P-450$_b$ and DPH in proteoliposomes is very small (Kunz *et al.*, 1985). From the available data an average distance of more than 60 Å between the heme of P-450 and DPH was calculated. Since DPH is situated in the lipid bilayer of the membrane, the calculation indicates a large vertical displacement of the heme group from the membrane. This conclusion is corroborated by the finding of Tarr *et al.* (1983) and Heinemann and Ozols (1983) that one of the large hydrophilic segments of P-450 contains the heme group, and by the accessibility of most of the cytochrome's mass to site-directed antibodies in microsomes (Lemos-Chiarandini *et al.*, 1987).

2.5. Structure of NADPH–Cytochrome P-450 Reductase

The cDNA sequences and derived amino acid sequences of NADPH–cytochrome P-450 reductase of rabbit and rat have recently been worked out (Black and Coon, 1982; Porter and Kasper, 1985). The rat enzyme has been titrated with *N*-(1-anilinonaph-4-yl)maleimide (ANM), which becomes fluorescent upon covalent binding to cysteine residues in proteins (Lee and Kaminsky, 1986). Four cysteine residues were accessible and in close proximity to tryptophans. Modified cysteine residues could be assigned to tryptic fragments by peptide analysis by acid mobile phase–reverse phase high-pressure liquid chromatography. They provided a label for function-related, cysteine-containing peptide segments of the reductase.

The fluorogenic monobromobimane was used to alkylate thiols in pig liver NADPH–cytochrome P-450 reductase (Vogel and Lumper, 1983). Loss of activity was caused by alkylation of one single critical cysteine residue, which could be protected by NADP(H). It is localized within a peptide of M_r of about 10 kDa, as shown by cyanogen bromide cleavage and gel permeation chromatography.

2.6. Interaction of Cytochrome P-450 and Its Reductase in Membranes

Rotational motion of P-450 has been studied by delayed fluorescence (see above), by the transient dichroism technique (Richter *et al.*, 1979; McIntosh *et al.*, 1980; Kawato *et al.*, 1982b; Gut *et al.*, 1982, 1983, 1985), and by saturation transfer electron paramagnetic resonance (Schwarz *et al.*, 1982). The formation of specific heterodimeric 1 : 1 complexes between reductase and cytochrome was shown by the combination of spectroscopic and immunochemical techniques (Gut *et al.*, 1983).

The interactions between reductase labeled with one ANM residue and P-450 in proteoliposomes were investigated by the analysis of ANM fluorescence (Nisimoto *et al.*, 1983). A significant decrease in fluorescence intensity attributable to energy transfer and a slight increase of emission anisotropy of ANM were observed and related to significant molecular interactions between the proteins in proteoliposomes. In agreement with this conclusion, the distance between ANM and the heme was calculated to be 35 Å.

2.7. Lipid–Protein Interactions Studied by DPH Fluorescence Anisotropy

2.7.1. Microsomal Membrane of Adrenal Cortex

The influence of proteins on the dynamics of lipids was studied by DPH fluorescence polarization measurements in bovine adrenal cortex microsomal membranes (Gallay *et al.*, 1981, 1982). The r_∞ values of DPH were significantly higher in complete membranes than in liposomes formed from extracted lipids. This indicates that the depolarizing rotation of DPH is more restricted in the presence of proteins. On the other hand, the rotational correlation time values were quite similar in both types of membrane. The results suggest an ordering influence of proteins on the lipid organization in adrenal cortex microsomes, whereas the rotational rate remains unaffected by proteins.

2.7.2. Reconstituted Monoxygenase of Rat Liver

Quite different results were obtained when rat liver microsomal P-450$_b$ and its reductase were reconstituted in unilamellar lipid vesicles prepared by the cholate dialysis technique from DMPC, DPPC, DOPC, and PC/PE/PS (10 : 5 : 1) (Kunz *et al.*, 1985). Incorporation of either protein alone or together increased the steady-state fluorescence anisotropy of DPH in DOPC and PC/PE/PS liposomes. In DMPC and DPPC vesicles, the proteins decreased r_s significantly below the transition temperature T_c of the gel to liquid-crystalline phase transition. Time-resolved fluorescence measurements of DPH performed in reconstituted PC/PE/PS and DMPC proteoliposomes showed that the proteins disordered the bilayer both in the gel and in the liquid-crystalline phase. The effect of P-450 was magnified upon co-reconstitution of reductase. These fluorescence data (and accompanying electron paramagnetic resonance data) show that P-450 and its reductase reconstituted into proteoliposomes decrease the order of the phospholipid acyl chains. The decrease is observed in PC/PE/PS liposomes resembling closely the phospholipid composition of the microsomal membrane and in DMPC vesicles above and below T_c. This sets aside the proteins of the microsomal monoxygenase system from other proteins studied by these techniques, for example, bacteriorhodopsin, cytochrome oxidase, the sarcoplasmic reticulum (Ca^{2+},

Mg^{2+})-ATPase, the M13 virus coat protein, and the glycoprotein of vesicular stomatitis virus. The first four proteins increase the order of their surrounding acyl chains, and they do so only above T_c, while the glycoprotein of vesicular stomatitis virus disorders the bilayer in the gel phase and orders it to a small extent in the liquid-crystalline phase. The disordering of surrounding acyl residues by P-450 and its reductase may reflect an uneven surface of the proteins' membrane domain.

NADPH–cytochrome P-450 reductase increases the disorder introduced by P-450 in PC/PE/PS liposomes (Kunz *et al.*, 1985). P-450 forms self-aggregates in these vesicles when reconstituted in the absence of reductase, whereas P-450 oligomers dissociate in the presence of equimolar amounts of reductase to form heterodimeric complexes of one reductase and one P-450 molecule (Gut *et al.*, 1982, 1983). A larger surface area of P-450 molecules should therefore become available for interaction with phospholipids in this situation. This may account for the increased disorder of the acyl chains in the presence of cytochrome and reductase compared to that found with the cytochrome alone.

ACKNOWLEDGMENTS. The work done in the authors' laboratory was supported by grants from the Schweizerischer Nationalfonds and by the very generous financial help from Solco AG Basel. We thank Prof. Dr. K. H. Winterhalter for his continued interest and support.

3. REFERENCES

Bar-Nun, S., Kreibich, G., Adesnik, M., Alterman, L., Negishi, M., and Sabatini, D. D., 1980, Synthesis and insertion of cytochrome P-450 into endoplasmic reticulum membranes, *Proc. Natl. Acad. Sci. U.S.A.* **77** : 965.

Bernhardt, R., Dao, N. T. N., Stiel, H., Schwarze, W., Friedrich, J., Jänig, G.-R., and Ruckpaul, K., 1983, Modification of cytochrome P-450 with fluorescein isothiocyanate, *Biochim. Biophys. Acta* **745** : 140–148.

Bartosz, G., Szabo, G., Szollosi, J., and Damjanovich, S., 1981, Aging of the erythrocyte. XI. Fluorescence studies on changes in membrane properties, *Mech. Aging Dev.* **16** : 265–274.

Bernert, J. T., Jr., and Groce, D. F., 1984, Acute response of rat liver microsomal lipids, lipid peroxidation, and membrane anisotropy to a single oral dose of polybrominated biphenyls, *J. Toxicol. Environ. Health* **13** : 673–687.

Black, S. D., and Coon, M. J., 1982, Structural features of liver microsomal NADPH–cytochrome P-450 reductase. Hydrophobic domain, hydrophilic domain, and connecting region, *J. Biol. Chem.* **257** : 5929–5938.

Black, S. D., French, J. S., Williams, C. H., Jr., and Coon, M. J., 1979, Role of a hydrophobic polypeptide in the N-terminal region of NADPH–cytochrome P-450 reductase in complex formation with P-450$_{LM}$, *Biochem. Biophys. Res. Commun.* **91** : 1528–1535.

Cherry, R. J., 1978, Measurement of protein rotational diffusion, *Methods Enzymol.* **54** : 447–461.

Chiang, Y.-L., and Coon, M. J., 1979, Comparative study of two highly purified forms of liver microsomal cytochrome P-450: Circular dichroism and other properties, *Arch. Biochem. Biophys.* **195** : 178–187.

Cone, R. A., 1972, Rotational diffusion of rhodopsin in the visual receptor membrane, *Nature New Biol.* **236** : 39–43.

De Pierre, J. W., and Dallner, G., 1975, Structural aspects of the membrane of the endoplasmic reticulum, *Biochim. Biophys. Acta* **415** : 411–472.

De Pierre, J. W., and Ernster, L., 1977, Enzyme topology of intracellular membranes, *Annu. Rev. Biochem.* **46** : 201–262.

Dobretsov, G. E., Borschevskaya, T. A., Petrov, V. A., and Vladimirov, Y. A., 1977, The increase of phospholipid bilayer rigidity after lipid peroxidation, *FEBS Lett.* **84** : 125–128.

Eichenberger, K., Böhni, P. C., Winterhalter, K. H., Kawato, S., and Richter, C., 1982, Microsomal lipid peroxidation causes an increase in the order of the membrane lipid domain, *FEBS Lett.* **142** : 59–62.

Esterbauer, H., Cheeseman, K. H., Dianzani, M. U., Poli, G., and Slater, T. F., 1982, Separation and characterization of the aldehydic products of lipid peroxidation stimulated by ADP-Fe^{2+} in rat liver microsomes, *Biochem. J.* **208** : 129–140.

Ford, R., and Barber, J., 1980, The use of diphenyl hexatriene to monitor the fluidity of the thylakoid membrane, *Photobiochem. Photobiophys.* **1** : 263–270.

Fujii-Kuriyama, Y., Mizukami, Y., Kawajiri, K., Sogawa, K., and Muramatsu, M., 1982, Primary structure of a cytochrome P-450: Coding nucleotide sequence of phenobarbital-inducible cytochrome P-450 cDNA from rat liver, *Proc. Natl. Acad. Sci. U.S.A.* **79** : 2793–2797.

Fukuzawa, K., Chida, H., Tokumura, A., and Tsukatani, H., 1981, Antioxidative effect of alpha-tocopherol incorporation into lecithin liposomes on ascorbic acid–Fe^{2+}-induced lipid peroxidation, *Arch. Biochem. Biophys.* **206** : 173–180.

Galanopoulou, G., Williams, W. P., and Quinn, P. J., 1982, Structural studies of plant membrane lipid dispersions subjected to autoxidation in the presence of decomposing peroxychromate, *Biochim. Biophys. Acta* **713** : 315–322.

Galeotti, T., Borrello, S., Palombini, G., Masotti, L., Ferrari, M. B., Cavatorta, P., Arcioni, A., Stremmenos, C., and Zannoni, C., 1984a, Lipid peroxidation and fluidity of plasma membranes from rat liver and Morris hepatoma 3924A, *FEBS Lett.* **169** : 169–173.

Galeotti, T., Borrello, S., Minotti, G., Palombini, G., Masotti, L., Sartor, G., Cavatorta, P., Arcioni, A., and Zannoni, C., 1984b, Lipid composition, physical state, and lipid peroxidation in tumour membranes, *Toxicol. Pathol.* **12** : 324–330.

Gallay, J., Vincent, M., de Paillerets, C., and Alfsen, A., 1981, Relationship between the activity of the 3β-hydroxysteroid dehydrogenase from bovine adrenal cortex microsomes and membrane structure, *J. Biol. Chem.* **256** : 1235–1241.

Gallay, J., Vincent, M., and Alfsen, A., 1982, Dynamic structure of bovine adrenal cortex microsomal membranes studied by time-resolved fluorescence anisotropy of all-*trans*-1,6-diphenyl-1,3,5-hexatriene, *J. Biol. Chem.* **257** : 4038–4041.

Garfinkel, D., 1958, Studies on pig liver microsomes. I. Enzymic and pigment composition of different microsomal fractions, *Arch. Biochem. Biophys.* **77** : 493–509.

Greinert, R., and Stier, A., 1980, Rotational diffusion of cytochrome P_{450} in a reconstituted system measured by depolarization of delayed fluorescence, in *Biochemistry, Biophysics and Regulation of Cytochrome P-450* (J.-A. Gustafsson *et al.*, eds.), pp. 591–594, Elsevier/North-Holland Biomedical Press, Amsterdam.

Greinert, R., Staerk, H., Stier, A., and Weller, A., 1979, E-type delayed fluorescence depolarization, a technique to probe rotational motion in the microsecond range, *J. Biochem. Biophys. Methods* **1** : 77–83.

Greinert, R., Finch, S. A. E., and Stier, A., 1982a, Conformation and rotational diffusion of cytochrome P-450 changed by substrate binding, *Biosci. Rep.* **2** : 991–994.

Greinert, R., Finch, S. A. E., and Stier, A., 1982b, Cytochrome P-450 rotamers control mixed-function oxygenation in reconstituted membranes. Rotational diffusion studied by delayed fluorescence polarization, *Xenobiotica* **12** : 717–726.

Guengerich, F. P., and Davidson, N. K., 1982, Interaction of epoxide hydrolase with itself and other microsomal proteins, *Arch. Biochem. Biophys.* **215** : 462–477.

Gum, J. R., and Strobel, H. W., 1979, Purified NADPH–cytochrome P-450 reductase. Interaction with hepatic microsomes and phospholipid vesicles, *J. Biol. Chem.* **254** : 4177–4185.

Gut, J., Richter, C., Cherry, R. J., Winterhalter, K. H., and Kawato, S., 1982, Rotation of cytochrome P-450. II. Specific interactions of cytochrome P-450 with NADPH–cytochrome P-450 reductase in phospholipid vesicles, *J. Biol. Chem.* **257** : 7030–7036.

Gut, J., Richter, C., Cherry, R. J., Winterhalter, K. H., and Kawato, S., 1983, Rotation of cytochrome P-450. Complex formation of cytochrome P-450 with NADPH–cytochrome P-450 reductase in liposomes demonstrated by combining protein rotation with antibody-induced crosslinking, *J. Biol. Chem.* **258** : 8588–8594.

Gut, J., Kawato, S., Cherry, R. J., Winterhalter, K. H., and Richter, C., 1985, Lipid peroxidation decreases the rotational mobility of cytochrome P-450 in rat liver microsomes, *Biochim. Biophys. Acta* **817** : 217–228.

Heinemann, F. S., and Ozols, J., 1983, The complete amino acid sequence of rabbit phenobarbital-induced liver microsomal cytochrome P-450, *J. Biol. Chem.* **258** : 4195–4201.

Heyn, M. P., 1979, Determination of lipid order parameters and rotational correlation times from fluorescence depolarization experiments, *FEBS Lett.* **108** : 359–364.

Hogeboom, G. H., 1949, Cytochemical studies of mammalian tissues. The distribution of diphospho-pyridine nucleotide–cytochrome c reductase in rat liver fractions, *J. Biol. Chem.* **177** : 847–858.

Hogeboom, G. H., and Schneider, W. C., 1950, Cytochemical studies of mammalian tissues. Isocitric dehydrogenase and triphosphopyridine nucleotide–cytochrome c reductase of mouse liver, *J. Biol. Chem.* **186** : 417–427.

Horecker, B. L., 1950, Triphosphopyridine nucleotide–cytochrome c reductase in liver, *J. Biol. Chem.* **183** : 593–605.

Inouye, K., and Coon, M. J., 1985, Properties of the tryptophan residue in rabbit liver microsomal cytochrome P-450 isozyme 2 as determined by fluorescence, *Biochem. Biophys. Res. Commun.* **128** : 676–682.

Kapitza, H. G., and Jacobson, K. A., 1986, Lateral motion of membrane proteins, in *Analysis of Membrane Proteins* (C. I. Ragan and R. J. Cherry, eds.), pp. 345–375, Chapman and Hall, London.

Kawato, S., Sigel, E., Carafoli, E., and Cherry, R. J., 1981, Rotation of cytochrome oxidase in phospholipid vesicles. Investigation of interactions between cytochrome oxidase and between cytochrome oxidase and cytochrome bc_1 complex, *J. Biol. Chem.* **256** : 7518–7527.

Kawato, S., Lehner, C., Müller, M., and Cherry, R. J., 1982a, Protein–protein interactions of cytochrome oxidase in inner mitochondrial membranes. The effect of liposome fusion on protein rotational mobility, *J. Biol. Chem.* **257** : 6470–6476.

Kawato, S., Gut, J., Cherry, R. J., Winterhalter, K. H., and Richter, C., 1982b, Rotation of cytochrome P-450. I. Investigation of protein–protein interaction of cytochrome P-450 in phospholipid vesicles and liver microsomes, *J. Biol. Chem.* **257** : 7023–7029.

Kinosita, K., Jr., Kawato, S., Ikegami, S., and Orii, Y., 1981, The effect of cytochrome oxidase on lipid dynamics. A nanosecond fluorescence depolarization study, *Biochim. Biophys. Acta* **647** : 7–17.

Klingenberg, M., 1958, Pigments of rat liver microsomes, *Arch. Biochem. Biophys.* **75** : 376–386.

Koppel, D. E., and Sheetz, M. P., 1983, A localized pattern photobleaching method for the concurrent analysis of rapid and slow diffusion processes, *Biophys. J.* **43** : 175–181.

Kunz, B. C., Rehorek, M., Hauser, H., Winterhalter, K. H., and Richter, C., 1985, Decreased lipid order induced by microsomal cytochrome P-450 and NADPH–cytochrome P-450 reductase in model membranes: Fluorescence and electron spin resonance studies, *Biochemistry* **24** : 2889–2895.

Lee, J. J., and Kaminsky, L. S., 1986, Fluorescence probing of the function-specific cysteines of rat microsomal NADPH–cytochrome P-450 reductase, *Biochem. Biophys. Res. Commun.* **134** : 393–399.

Leighton, J. K., De Brunner-Vossbrinck, B. A., and Kemper, B., 1984, Isolation and sequence analysis of three cloned cDNAs from rabbit liver proteins that are related to rabbit cytochrome P-450 (form 2), the major phenobarbital-inducible form, *Biochemistry* **23** : 204–210.

Lemos-Chiarandini, C. D., Frey, A. B., Sabatini, D. D., and Kreibich, G., 1987, Determination of the membrane topology of the phenobarbital-inducible rat liver cytochrome P-450 isoenzyme PB-4 using site-specific antibodies, *J. Cell Biol.* **104** : 209–219.

Lu, A. Y., Junk, K. W., and Coon, M. J., 1969, Resolution of the cytochrome P-450-containing ω-hydroxylation system of liver microsomes into three components, *J. Biol. Chem.* **244** : 3714–3721.

Masotti, L., Cavatorta, P., Ferrari, M. B., Casali, E., Arcioni, A., Zannoni, C., Borrello, S., Minotti, G., and Galeotti, T., 1986, O_2^--dependent lipid peroxidation does not affect the molecular order in hepatoma microsomes, *FEBS Lett.* **198** : 301–306.

Matsuura, S., Masuda, R., Omori, K., Negishi, M., and Tashito, Y., 1981, Distribution and induction of cytochrome P-450 in rat liver nuclear envelope, *J. Cell Biol.* **91** : 212–220.

McIntosh, P. R., Kawato, S., Freedman, R. B., and Cherry, R. J., 1980, Evidence from cross-linking and rotational diffusion studies that cytochrome P-450 can form molecular aggregates in rabbit-liver microsomal membranes, *FEBS Lett.* **122** : 54–58.

Nisimoto, Y., Kinosita, K., Jr., Ikegami, A., Kawai, N., Ichihara, I., and Shibata, Y., 1983, Possible association of NADPH–cytochrome P-450 reductase and cytochrome P-450 in reconstituted phospholipid vesicles, *Biochemistry* **22** : 3586–3594.

Oesch, F., and Daly, J., 1972, Conversion of naphthalene to *trans*-naphthalene dihydrodiol: Evidence for the presence of a coupled aryl monooxygenase–epoxide hydrase system in hepatic microsomes, *Biochem. Biophys. Res. Commun.* **46** : 1713–1720.

Ohyashiki, T., Ohtsuka, T., and Mohri, T., 1986a, A change in the lipid fluidity of the porcine intestinal brush-border membranes by lipid peroxidation. Studies using pyrene and stearic acid derivatives, *Biochim. Biophys. Acta* **861** : 311–318.

Ohyashiki, T., Ohta, A., Ohtsuka, T., and Mohri, T., 1986b, Effects of lipid peroxidation on the membrane-bound ATPases and lipid fluidity of porcine intestinal brush border membranes, *J. Pharmacobiodyn.* **9** : s-124.

Ohyashiki, T., Ushiro, H., and Mohri, T., 1986c, Effect of α-tocopherol on the lipid peroxidation and fluidity of porcine intestinal brush-border membranes, *Biochim. Biophys. Acta* **858** : 294–300.

Omata, Y., and Ueno, Y., 1985, Fluorescence energy transfer measurements of the complexes of aflatoxin B_1 and cytochrome P-450, *Biochem. Biophys. Res. Commun.* **129** : 493–498.

Omata, T., Ueno, Y., and Aibara, K., 1986, Conformational change of cytochrome P-450 indicated by the measurement of fluorescence-energy transfer, *Biochim. Biophys. Acta* **870**: 392–400.

Ortiz de Montellano, P. R., 1986, *Cytochrome P-450 Structure, Mechanism, and Biochemistry*, Plenum Press, New York.

Palade, G. E., and Siekevitz, P., 1956, Liver microsomes. An integrated morphological and biochemical study, *J. Biophys. Biochem. Cytol.* **2** : 171–200.

Patel, J. M., and Block, E. R., 1986, Nitrogen dioxide-induced changes in cell membrane fluidity and function, *Am. Rev. Respir. Dis.* **134** : 1196–1202.

Peters, R., Peters, J., Tews, K. H., and Bähr, W., 1974, A microfluorimetric study of translational diffusion in erythrocyte membranes, *Biochim. Biophys. Acta* **367** : 282–294.

Phillips, A. H., and Langdon, R. G., 1962, Hepatic triphosphopyridine nucleotide–cytochrome c reductase: Isolation, characterization, and kinetic studies, *J. Biol. Chem.* **237** : 2652–2660.

Porter, T. D., and Kasper, C. B., 1985, Coding nucleotide sequence of rat NADPH–cytochrome P-450 oxidoreductase cDNA and identification of flavin-binding domains, *Proc. Natl. Acad. Sci. U.S.A.* **82** : 973–977.

RaziNaqvi, K., Rodriguez, J. G., Cherry, R. J., and Chapman, D., 1973, Spectroscopic technique for studying protein rotation in membranes, *Nature New Biol.* **245** : 249–254.

Rice-Evans, C., and Hochstein, P., 1981, Alterations in erythrocyte membrane fluidity by phenylhydrazine-induced peroxidation of lipids, *Biochem. Biophys. Res. Commun.* **100** : 1537–1542.

Rice-Evans, C., Baysal, E., Pashby, D. P., and Hochstein, P., 1985, t-Butyl hydroperoxide-induced perturbations of human erythrocytes as a model for oxidant stress, *Biochim. Biophys. Acta* **815** : 426–432.

Richter, C., Winterhalter, K. H., and Cherry, R. J., 1979, Rotational diffusion of cytochrome P-450 in rat liver microsomes, *FEBS Lett.* **102** : 151–154.

Saffman, P. C., and Delbrück, M., 1975, Brownian motion in biological membranes, *Proc. Natl. Acad. Sci. U.S.A.* **72** : 3111–3113.

Sato, R., Nishibayashi, H., and Ito, A., 1969, Characterization of two hemoproteins of liver microsomes, in *Microsomes and Drug Oxidations* (J. R. Gilette, A. H. Conney, G. Cosmides, R. W. Estabrook, J. R. Fouts, and G. J. Mannering, eds.), pp. 111–132, Academic Press, New York.

Schwarz, D., Pirrwitz, J., and Ruckpaul, K., 1982, Rotational diffusion of cytochrome P-450 in the microsomal membrane—Evidence for a clusterlike organization from saturation transfer electron paramagnetic resonance spectroscopy, *Arch. Biochem. Biophys.* **216** : 322–328.

Schwarze, W., Jänig, G. R., Berhardt, R., and Ruckpaul, K., 1983a, Topological studies on cytochrome P-450 with fluorescence methods, *Studia Biophys.* **93** : 233–234.

Schwarze, W., Bernhardt, R., Jänig, G. R., and Ruckpaul, K., 1983b, Fluorescent energy transfer measurements on fluorescein isothiocyanate modified cytochrome P-450 LM2, *Biochem. Biophys. Res. Commun.* **113** : 353–360.

Shinitzky, M., 1984, Membrane fluidity in malignancy, adversative and recuperative, *Biochim. Biophys. Acta* **738** : 251–261.

Shinitzky, M., and Bahrenholz, Y., 1978, Fluidity parameters of lipid regions determined by fluorescence polarization, *Biochim. Biophys. Acta* **515** : 367–394.

Singer, S., and Nicholson, G., 1972, The fluid mosaic model of the structure of cell membranes, *Science* **175** : 720–731.

Stier, A., Finch, S. A. E., Greinert, R., and Taniguchi, H., 1985, Membrane protein interactions, in *Cytochrome P-450, Biochemistry, Biophysics, and Induction* (L. Vereczky and K. Magyar, eds.), pp. 139–146, Akademiai Kiado, Budapest.

Strittmatter, C. F., and Ball, E. G., 1952, A hemochromogen component of liver microsomes, *Proc. Natl. Acad. Sci. U.S.A.* **38** : 19–25.

Strittmatter, P., and Velick, S. F., 1956a, The isolation and properties of microsomal cytochrome, *J. Biol. Chem.* **221** : 253–264.

Strittmatter, P., and Velick, S. F., 1956b, A microsomal cytochrome reductase specific for diphosphopyridine nucleotide, *J. Biol. Chem.* **221** : 277–286.

Taniguchi, H., Imai, Y., and Sato, R., 1980, Protein–protein and lipid–protein interactions in a reconstituted liver microsomal monooxygenase system, in *Microsomes, Drug Oxidations, and Chemical Carcinogenesis* (M. J. Coon, A. H. Conney, R. W. Eastbrook, H. V. Gelboin, J. R. Gilette, and P. J. O'Brien, eds.), pp. 537–540, Academic Press, New York.

Tarr, G. E., Black, S. D., Fujita, V. S., and Coon, M. J., 1983, Complete amino acid sequence and predicted membrane topology of phenobarbital-induced cytochrome P-450 (isozyme 2) from rabbit liver microsomes, *Proc. Natl. Acad. Sci. U.S.A.* **80** : 6552–6556.

Thomas, D. D., 1978, Large-scale rotational motions of proteins detected by electron paramagnetic resonance and fluorescence, *Biophys. J.* **24** : 439–462.

Thomas, D. D., 1986, Rotational diffusion of membrane proteins, in *Techniques for the Analysis of Membrane Proteins* (C. I. Ragan and R. J. Cherry, eds.), pp. 377–431, Chapman and Hall, London.

Thomas, P. E., Lu, A. Y. H., West, S. B., Ryan, D., Miwa, G. T., and Levin, W., 1977, Accessibility of cytochrome P450 in microsomal membranes: Inhibition of metabolism by antibodies to cytochrome P450, *Mol. Pharmacol.* **13** : 819–831.

Vlasuk, G. P., Ghrayeb, J., Ryan, D., Reik, L., Thomas, P. E., Levin, W., and Waltz, F. G., Jr., 1982, Multiplicity, strain differences, and topology of phenobarbital-induced cytochrome P-450 in rat liver microsomes, *Biochemistry* **21** : 789–798.

Vogel, F., and Lumper, L., 1983, Fluorescence labelling of NADPH–cytochrome P-450 reductase with the monobromomethyl derivative of *syn*-9,10-dioxabimane, *Biochem. J.* **215** : 159–166.

Wang, H.-P., and Kimura, T., 1976, Purification and characterization of adrenal cortex mitochondrial cytochrome P-450 specific for cholesterol side chain cleavage activity, *J. Biol. Chem.* **251** : 6068–6074.

Williams, C. H., and Kamin, H., 1962, Microsomal triphosphopyridine nucleotide–cytochrome c reductase of liver, *J. Biol. Chem.* **237** : 587–595.

Wu, E.-S., and Yang, C. S., 1984, Lateral diffusion of cytochrome P-450 in phospholipid bilayers, *Biochemistry* **23** : 28–33.

Chapter 9

Fluorescence Studies on Prokaryotic Membranes

P. Proulx

1. INTRODUCTION

Some of the more common techniques used to study the physical properties of biological membranes include electron microscopy, x-ray and neutron diffraction, Raman spectroscopy, electron spin resonance (ESR), nuclear magnetic resonance (NMR), infrared spectroscopy, fluorescence spectroscopy, and differential scanning calorimetry (DSC). The uses and limitations of these various analytical tools were examined in a number of recent reviews (Axelrod *et al.*, 1976; Andersen, 1978; Seelig and Seelig, 1980; Jacobs and Oldsfield, 1981; Yguerabide and Foster, 1981; Amey and Chapman, 1983; Bach, 1983; Davis, 1983; Devaux, 1983, 1985; Hoffmann and Restall, 1983; Verma and Wallach, 1983; Chapman and Benga, 1984; Makowski and Li, 1984; Bergelson *et al.*, 1985; Blaurock, 1985; Bloom and Smith, 1985; Mühlethaler and Jay, 1985; McElhaney, 1986; Restall and Chapman, 1986).

Abbreviations used in this chapter: ANS, anilino-napthalene-1-sulfonate; *n*-AS, *n*-(9-anthroyloxy) stearic acid; CCCP, carboxyl cyanide *m*-chlorophenylhydrazone; DPH, 1,6-diphenyl-1 *trans*, 3 *trans*, 5 *trans*-hexatriene; EDTA, ethylenediaminetetraacetate; NBD-PE, *N*-(7-nitrobenz-2-oxa-1,3-diazol-4-yl)phosphatidylethanolamine; NPN, *N*-phenyl-1-napthylamine; N-Rh-PE, *N*-(lissamine rhodamine B sulfonyl)phosphatidylethanolamine; *cis*- or *trans*-PA, *cis*- or *trans*-parinaric acid.

P. Proulx   Department of Biochemistry, School of Medicine, Faculty of Health Sciences, University of Ottawa, Ottawa, Ontario K1H 8M5, Canada.

Fluorescence spectroscopy has been very useful as an analytical tool because of its sensitivity, and its different facets involving steady-state, time-resolved, or photobleaching recovery measurements have permitted the study of molecular order as well as molecular motion on a relatively broad time scale. Steady-state fluorescence has been especially popular because it depends on relatively simple instrumentation and makes use of either natural fluorescent probes or easily available exogenous probes. In establishing the structural and dynamic properties of prokaryotic membranes, it has been very valuable as an approach complementing other techniques and in many investigations has served to study the thermotropic behavior and the molecular characteristics of the constituents structuring the membranes of microbes.

Fluorescence spectroscopy in general has been extensively used for detecting changes in membrane structure, permeability, and electrical potential as well as fluctuations in cellular ion concentrations, which result from interactions of the microbial cell surface with various agents including antibiotics, toxins, phages, alcohols, chelating agents, polycations, and other compounds. It has been a technique of choice for studying a variety of problems including those concerned with the mechanism of interaction between liposomes and bacterial membranes and the effects of such interactions on structure–function relations in prokaryotic membranes. The present chapter is concerned with some of the numerous applications of fluorescence spectroscopy to the study of prokaryotic membranes.

2. FLUORESCENT PROBES

The fluorescence studies considered here have all involved the use of extrinsic probes, although in some cases the probe was biosynthetically incorporated into the lipids of the membranes. The validity of the extrinsic probe approach depends on several conditions: (1) a knowledge of its location in the membrane, (2) a measurable response to the membrane properties under study in terms of its fluorescent parameters, and (3) a minimal perturbation caused by its incorporation into the membrane.

The fluorescent probes that have been most common for examining prokaryotic membrane structure are 8-anilino-naphthylene-1-sulfonate (ANS), *N*-phenyl-1-napthylamine (NPN), *cis*- or *trans*-parinaric acid (*cis*- or *trans*-PA), 1,6 diphenyl-1 *trans*, 3 *trans*, 5 *trans*-hexatriene (DPH), and the *n*-(9-anthroyloxy) stearic acids (*n*-AS). The structures of these compounds are given in Figure 1.

The probes preferred in earlier studies to determine phase transitions and assess fluidity were ANS and NPN (Traüble, 1971; Sackmann and Traüble, 1972; Overath and Traüble, 1973; Traüble and Overath, 1973; Thilo and

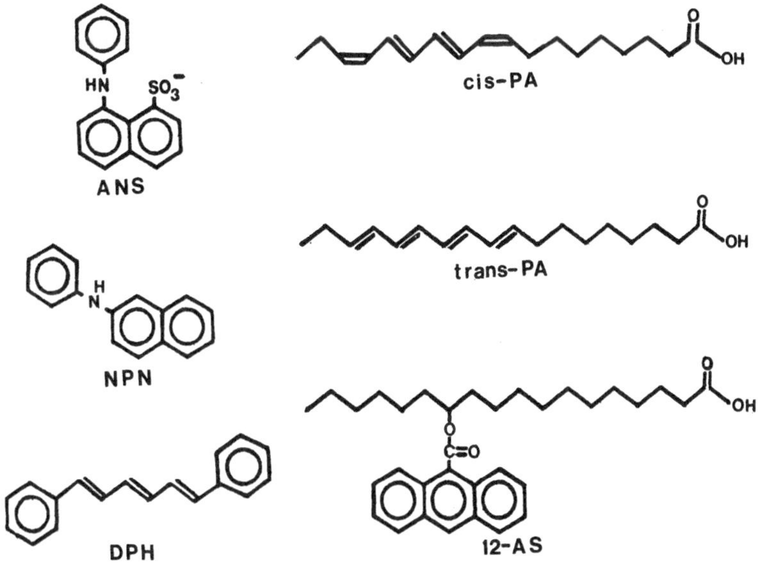

FIGURE 1. Structure of fluorescent membrane probes: ANS, 1-anilino-8-naphthalene sulfonate; NPN, *N*-phenylnapthylamine; DPH, 1,6-diphenyl-1,3,5-hexatriene; *cis*-PA, *cis*-parinaric acid; *trans*-PA, *trans*-parinaric acid; 12-AS, 12-(9-anthroyloxy) stearic acid.

Overath, 1976). ANS is an amphipathic substance and interacts with the lipid bilayer at the polar surface, such that the sulfonate group faces the aqueous phase and the aromatic ring system intercalates a short distance within the hydrocarbon chains. NPN, on the other hand, is only sparingly soluble in water and partitions into the hydrophobic region of the bilayer. These conclusions are based partly on evidence obtained from x-ray diffraction and NMR studies (Lesslauer *et al.*, 1971; Colley and Metcalfe, 1972) and are compatible with the fact that the quantum yields of ANS and NPN in aqueous dispersions of dipalmitoylphosphatidylcholine (DPPC) correspond to solvent dielectric constants of approximately 35 and 10 or less, respectively (Traüble and Overath, 1973). Also in support of these conclusions is the observation that ANS once bound to phospholipid is accessible to water as can be surmised from the fluorescence intensity enhancement caused by D_2O. Such enhancement is not seen with lipid-bound NPN (Radda, 1975). Equally indicative of the hydrophobic location of NPN in membranes is the effects of this dye on lipid phase transitions seen at higher concentrations (Overath and Traüble, 1973).

Both ANS and NPN are polarity-dependent probes and changes in either their emission maximum or fluorescence intensity have been used to report on variations in their environment (Azzi, 1975; Radda, 1975; Waggoner,

1976); whereas the other probes have been used mainly to examine fluorescence polarization changes in membranes. Dyes such as ANS bind to lipids, proteins, and other membrane constituents, which makes interpretations concerning the nature and polarity of their surroundings difficult (Radda, 1975; Waggoner, 1976; Slavik, 1982). However, both ANS and NPN possess fluorescent properties appropriate for the study of certain aspects of membrane structure and accordingly changes in their fluorescence intensity, as a function of temperature, have been frequently measured to determine lipid phase transitions. Good agreement was obtained between the results of such studies and those of others, making or avoiding the use of exogenous probe molecules (Overath *et al.*, 1980; Linden *et al.*, 1973a, b; Overath and Traüble, 1973; Traüble and Overath, 1973; Haest *et al.*, 1974; Overath *et al.*, 1975; Sackmann *et al.*, 1973; Sackmann and Traüble, 1982).

Of the other probes used to study prokaryotic membranes, DPH has been the most popular because of very favorable fluorescent properties; it has a high quantum yield (0.8 hexane at 25°C) and an extinction coefficient of approximately 80,000 $M^{-1}\cdot cm^{-1}$ (Shinitzky and Barenholz, 1978). Being a polyene hydrocarbon, it partitions into the hydrophobic core of the membrane and thereby increases its fluorescence emission 1000-fold as compared to that in aqueous medium. It is equally soluble in the fluid and gel regions of the bilayer (Lentz *et al.*, 1976) and consequently polarization values obtained with this probe represent weight averages for all domains of the membrane.

The location of DPH in the bilayer is relatively well known, being positioned so that its emission dipole is preferentially oriented perpendicular to the surface of the bilayer along the long axis of the molecule (Andrich and Vanderkooi, 1976). Depending on the type of lipids constituting a membrane, the probe may be in close vicinity of the bilayer center or may be broadly spread about this center (Davenport *et al.*, 1985).

Although DPH fluorescence polarization studies have been very valuable for the empirical assessment of lipid fluidity of membranes, the use of this probe has some shortcomings. It has often been observed that polarization values found for membranes are different from those obtained for their lipid extracts. This may be indicative of interactions between lipids and non-lipid elements in the membrane but direct interactions of DPH with proteins (Mely-Goubert and Freedman, 1980; Chapman *et al.*, 1983) could be implicated instead.

Since their introduction by Gunstone and Subbarao (1967) as natural fluorescent analogues of fatty acids, the parinaric acid probes and their phospholipid derivatives have been widely used for polarization studies in model systems (Sklar *et al.*, 1975, 1976, 1977a,b, 1979a,b,c; Gallay and Vincent, 1986) and in membranes (Schroeder, 1985), including bacterial membranes (Tecoma *et al.*, 1977; Fraley *et al.*, 1978; Souzu, 1982, 1986a,b). A number of spectral properties of the PA molecules make them useful probes. These compounds have extinction coefficients greater than $65000 \ M^{-1} \cdot cm^{-1}$ in several organic solvents tried at emission peaks varying between 304 and 310 nm for *cis*-PA and 299 and 305 nm for *trans*-PA as well as high quantum yields in lipid environments (Sklar *et al.*, 1977b). Partitioning experiments involving mixed populations of phospholipid vesicles indicated that *trans*-PA associated preferentially with solid phase lipids, while *cis*-PA showed a more equal distribution between solid and fluid lipids (Sklar, 1977b). This last property has been exploited to demonstrate the coexistence of both fluid and solid phases and to quantify the extent of these domains in membranes (Sklar *et al.*, 1979a,b,c; Schroeder, 1983; Schroeder and Soler-Argilaga, 1983). However, the interpretation of the features of the fluorescence emission kinetics of PA molecules in terms of differential partitioning inside the lipid bilayer has been questioned recently (Parassassi *et al.*, 1984). Because of their favorable fluorescence properties, probes such as DPH and PA can be used in relatively small concentrations compared to the lipid of the membrane. Lipid/probe ratios are usually greater than 100, such that bilayer perturbations due to the presence of probe molecules are expected to be minimal, as has indeed been shown (Sklar *et al.*, 1975, 1977b). However, certain studies have succeeded in incorporating the fluorescent fatty acid derivatives into the membrane lipids of cells to be examined via a biosynthetic route or by introducing the PA-labeled lipid, exploiting exchange reactions (Bergelson *et al.*, 1985). These techniques would be expected to produce even less disturbance in the lipid bilayer. The use of PA probes for examining thermotropic behavior in prokaryotic membranes (Tecoma *et al.*, 1977) is validated at least to the extent that results obtained with these fluorescent fatty acids are similar to those obtained with a variety of other physical methods (Ashe and Steim, 1971; Linden *et al.*, 1973b; Shechter *et al.*, 1974; Morrisett *et al.*, 1975; Overath *et al.*, 1975; Thilo and Overath, 1976).

In the *n*-AS series of probes, the 9-anthroyloxy reporter group is attached at different positions along the acyl chain. The fluorescent properties of the reporter group are similar at the different positions when examined in organic solvents but these change when the probes are intercalated within bilayers. Accordingly, fluorescent lifetimes and quantum yields increase as the probe positions progress from C2 to C9. The fluorescent compounds appear to partition into model membranes with the probe fatty acid chain aligned parallel to the phospholipid acyl chains and the anthroate group positioned at a defined depth in the bilayer (Podo and Blasie, 1977; Thulborn and Sawyer, 1978; Thulborn *et al.*, 1979). These properties have permitted their wide use in the determination of fluidity and polarity gradients of phospholipid bilayers. At high concentrations they may cause perturbation of the membrane, but usually not at concentrations corresponding to less than 1 mol % (Thulborn, 1981). Matayoshi and Kleinfeld (1981) have pointed out complications with the use of these probes but it has nevertheless been shown that n-AS derivatives monitor the same phase transition temperatures as nonperturbing techniques (Thulborn, 1981).

3. STRUCTURAL ASPECTS OF BACTERIAL MEMBRANES

3.1. Outer Membrane of Gram-Negative Bacteria

The surface architecture of gram-negative bacteria differs from that of gram-positive organisms inasmuch as it accommodates, in addition to an inner cytoplasmic membrane and a peptidoglycan layer, an outer membrane — the structure and function of which has been extensively reviewed in the last decade (Inouye 1979; Nikaido and Nakae, 1979; Osborn and Wu, 1980; Lugtenberg and Van Alphen, 1983; Nikaido and Vaara, 1985; Nakae, 1986).

Composed essentially of phospholipids, proteins, lipoprotein, and lipopolysaccharide, the outer membrane possesses the characteristic bilayer arrangement as indicated from the results of electron microscopy, x-ray diffraction, and NMR (Overath *et al.*, 1975; Ueki *et al.*, 1970; Verkleij and Ververgaert, 1978; Smit *et al.*, 1975; Van Gool and Nanninga, 1971; Fiil and Branton, 1968; Van Alphen *et al.*, 1978; Schweizer and Henning, 1977; Nurminen *et al.*, 1976; Burnell *et al.*, 1980; Van Alphen *et al.*, 1980; Gally *et al.*, 1980; Nichol *et al.*, 1980).

The outer membrane appears to be very asymmetric in its architecture, the outer leaflet being composed mainly of lipopolysaccharide and protein, and the inner leaflet containing mainly protein, lipoprotein, and phospholipid (Mühlradt and Golecki, 1975; Kamio and Nikaido, 1976, 1977; Funahara and Nikaido, 1980; Gmeiner and Schlecht, 1980; Lugtenberg and Van Alphen, 1983; Munford and Osborn, 1983; Nikaido and Vaara, 1985). About one-third of the inner leaflet lipoprotein is covalently linked to the peptidoglycan layer (Inouye *et al.*, 1972; Braun, 1975).

3.2. Molecular Interactions

How constituent molecules interact in the outer membrane has been a recurring point of interest in a number of studies as well as the object of some controversy (Lugtenberg and Val Alphen, 1983; Nikaido and Vaara, 1985). That lipids associate with nonlipid constituents was suggested partly on the basis that membranes and their lipid extracts often display differences in thermotropic behavior (Janoff *et al.*, 1979, 1980; Herring *et al.*, 1985; McGibbon *et al.*, 1985; Proulx and Szabo, 1985) and partly because, in several gram-negative bacteria examined, the outer membrane, which is richer in protein and contains all the lipopolysaccharide, displays more rigidity than the inner membrane (Rottem *et al.*, 1975; Rottem and Leive, 1977; Davis *et al.*, 1979; Janoff *et al.*, 1979; Gally *et al.*, 1980; Nichol *et al.*, 1980). Both types of supportive evidence have been obtained from fluorescence studies as well (Cheng *et al.*, 1974; Souzu, 1986a).

^{2}H-NMR revealed in particular that the liquid-crystalline–gel transition region in the outer membrane, when compared to cytoplasmic membrane, was shifted to a higher temperature and that the orientational order was greater in the outer membrane (Davis *et al.*, 1979; Nichol *et al.*, 1980). In a fluorescence study, Cheng *et al.* (1974) examined the dynamics of the lipids of inner and outer membranes of *Escherichia coli* labeled with pyrene. The decay time of the excited state was measured after a 10-msec pulse of ultraviolet light and these measurements as well as steady-state polarization data indicated that the outer membrane was more rigid than the inner membrane. A similar conclusion could be drawn from the results of Souzu (1986a), who labeled the inner and outer membranes of *Escherichia coli* with parinaric acid.

On the other hand, Overath *et al.* (1975), using fluorescence spectroscopy and x-ray diffraction as analytical methods, reported that the two membranes of *Escherichia coli* displayed phase transitions over the same temperature range. Significant differences in the rigidity of these membranes were nevertheless concluded to exist, since in the outer membrane the x-ray data indicated that only 25–40% of the phospholipid could be seen to participate in the phase transition as compared to 60–80% for the cytoplasmic membrane. The remaining phospholipid was suggested to interact with either protein or lipopolysaccharide. Davis *et al.* (1979), however, concluded from their ^{2}H-NMR studies that most of the phospholipid molecules do participate in the phase transition and reconciled the results of Overath *et al.* (1975) on the basis that, at lower temperatures, a large portion of the oriented phospholipid molecules form small domains of a few hundred molecules. Such domains would produce Bragg peaks that are very wide and indistinguishable from incoherently scattered x-rays.

3.2.1. Phospholipid–Lipopolysaccharide Interactions in the Outer Membrane

To explain the increased rigidity of the outer membrane compared to the inner membrane, Rottem and Leive (1977) proposed that lipopolysaccharide might interact with phospholipid to immobilize the fatty acyl chains. This suggestion, however, is not supported by the results of Nikaido *et al.* (1977), who found the ESR spectra of inner and outer membranes of *Salmonella typhimurium* to be remarkably similar when these membranes had been labeled with nitroxide stearate. Further results by Takeuchi and Nikaido (1981) indicated that when suspended together under various conditions, lipopolysaccharide and spin-labeled phospholipids formed bilayers composed of stable segregated domains in which little mixing occurred. On the other hand, Mackay *et al.* (1984) concluded, on the basis of proton and deuterium magnetic resonance studies, that lipopolysaccharide could affect the orientational order of the hydrocarbon chains of DPPC when these two substances were dispersed together. Order was decreased in the gel phase and increased in the liquid phase. The authors argued in favor of lipopolysaccharide–phospholipid interactions as the basis for the greater rigidity of the outer membrane. However, Nikaido and Vaara (1985) dismissed the difference in rigidity between the two membranes as an artifact of the membrane isolation procedure (Osborn *et al.*, 1972) that most investigators used. This procedure, which eliminates peptidoglycan from the membranes, causes mixing of lipopolysaccharide and phospholipid entities normally separated by the asymmetric arrangement of the outer bilayer. Their conclusion, although possibly correct, would have to be reconciled with the earlier report by Takeuchi and Nikaido (1981) indicating the inability of phospholipids and lipopolysaccharide to mix.

To explain the higher degree of rigidity in the outer membrane seen in some studies, one need not evoke lipid–lipolysaccharide interactions, since outer membranes have been shown to be enriched with saturated fatty acids (Lugtenberg and Peters, 1976) and contain less phosphatidylglycerol (Osborn *et al.*, 1972; Lugtenberg and Peters, 1976; Jones and Osborn, 1977a). In fact, fluorescence polarization studies by Souzu (1986a) have indicated quite clearly that lipids extracted from outer membranes are more rigid than those extracted from the inner membrane.

3.2.2. Phospholipid–Protein Interactions

The precise nature and extent of phospholipid–protein interaction in microbial membranes is still a matter of some debate. The conclusions often vary with the type of physical technique used to probe this question. Much of the evidence indicating lipid–protein interactions comes from electron spin resonance spectroscopy and fluorescence spectroscopy. Nikaido *et al.* (1977), working with nitroxide stearic acid-labeled membrane of *Salmonella typhimurium* in a temperature range of 30–45°C, found that in the inner membrane the probe was clearly in a heterogeneous environment, reflecting the presence of free phospholipids as well as protein-immobilized phospholipids. The spectra of the outer membrane in the same temperature range were more homogeneous presumably because of a greater relative contribution of the immobilized signal, reflecting the higher protein/lipid ratio of this membrane. Another study was made by Takeuchi *et al.* (1978) of the cytoplasmic membrane-enriched fractions of *Escherichia coli* B, labeled *in situ* by the incorporation of 12-nitroxide stearic acid chains into phosphatidylglycerol. The electron spin spectrum of the labeled acyl chains consisted again of at least two components: one due to labels undergoing rapid anisotropic motions and the other due to strongly immobilized chains. The spectrum of the membrane was markedly influenced by divalent cations but not that of the extracted lipids, which suggested that divalent cations mediated an interaction between phosphatidylglycerol and protein. A further study by these authors (Takeuchi *et al.*, 1981), using membrane spin-labeled *in situ* in both phosphatidylglycerol and phosphatidylethanolamine, confirmed the previous conclusion that phosphatidylglycerol tended to interact with membrane proteins through the mediation of divalent cations, whereas phosphatidylethanolamine did not display this tendency to the same extent.

Outer membranes from *Escherichia coli* grown at 37°C, labeled with 5-doxyl stearate, and examined by electron spin resonance spectroscopy undergo a transition beginning at 9°C and ending at 42°C, whereas the phospholipid obtained from those same membranes undergoes a transition between 4 and 24°C (Janoff *et al.*, 1980). Since 5-doxyl stearate probes phospholipid domains mainly (Nikaido *et al.*, 1977) and these reside in a distinct monolayer of the outer membrane from which lipopolysaccharide is excluded, it would appear that the association of lipids with nonlipid elements of the outer membrane involved protein rather than lipopolysaccharide. In another study, Coughlin *et al.* (1983) used a cationic spin label to monitor the lipopolysaccharide and a 5-doxyl stearate probe to monitor the phospholipids of the outer membrane. With both probes, a structural transition was seen at 9°C. A similar transition was obtained with the cationic probe when a protein–lipopolysaccharide–peptidoglycan complex was examined. Purified lipopolysaccharide, however, displayed structural transition at 20 and 40°C. The results indicated to the authors that a structural rearrangement of the intact outer membrane occurs at approximately 9°C, involves both the lipopolysaccharide and phospholipids, and is characterized by a phase transition that is highly dependent on lipid–protein interactions.

The involvement of nonlipid (protein)–lipid interactions in microbial membranes is also supported by steady-state fluorescence or combined steady-state fluorescence–electron spin resonance studies (Janoff *et al.*, 1979, 1980; Proulx and Szabo, 1985; Souzu, 1986a). Accordingly, examination of membranes or whole cells of *Escherichia coli* compared to their lipid extracts reveals important differences in fluorescence polarization. In the case of whole cell or outer membranes, these differences might well be attributable to the presence of proteins and/or lipopolysaccharides. However, when DPH is used–a probe that dissolves in all regions of the membrane—differences in fluorescence polarization may simply reflect the influence of lipopolysaccharide acyl chains on the average fluidity of the membrane and not necessarily an interaction of lipid with nonlipid elements.

DPH fluorescence polarization measurements of membranes and extracted lipid of *Micrococcus cryophilus*, a psychrophilic bacterium, were found to be quite different and indicated that protein decreased lipid fluidity in the membrane of this organism (McGibbon *et al.*, 1985). It is interesting to note for this bacterium that, despite its extent of unsaturation as well as its low phase transition temperature revealed by differential scanning calorimetry, the polarization values of the membranes and lipids were elevated, indicating a relatively high degree of order.

Experiments by Janoff *et al.* (1979, 1980) using paramagnetic and fluorescent probes indicated that, with changes in growth temperature, corresponding changes in the phase transition temperature occurred principally in the outer membrane. Little or no change in the phase transition temperature of the cytoplasmic membrane or in the lipids extracted from either membrane occurred as a function of growth temperature. The changes in thermotropic behavior found for the outer membrane were attributed to interactions with nonlipid components. These results are somewhat surprising, however, in view of the fact that saturated hydrocarbon chain content has been reported to increase with growth temperature in both cytoplasmic and outer membranes (Lugtenberg and Peters, 1976; Ishinaga *et al.*, 1979). Also, in another study by Nakayama *et al.* (1980), examination of *Escherichia coli* membranes by x-ray diffraction revealed, as expected, changes in transition temperatures of the membranes and extracted lipids which reflected the alterations in fatty acid composition brought about by growth temperature changes. The reason for the disagreement in results obtained by both groups is not clear.

In a recent study by Herring *et al.* (1985), the fluidity of the lipids in membrane preparations from a mutant strain of *Escherichia coli*, resistant to the uncoupler carbonyl cyanide *m*-chlorophenylhydrazone (CCCP), was examined by electron spin resonance using the spin probe 5-doxyl stearic acid. The fluidity of the inner membrane, in particular, decreased with cells grown in the presence of uncoupler, whereas an opposite effect of the uncoupler was seen with the lipids extracted from such membranes. The effect noted with the extracted lipids was in accord with the increased proportion of unsaturated fatty acids resulting from the CCCP treatment. The anomalous thermotropic behavior of the membranes was explained by an increased protein/phospholipid ratio.

Results obtained with other physical techniques often failed to indicate extensive effects of proteins on lipid fluidity of the membrane. A recent examination by differential scanning calorimetry, coupled with electron spin resonance spectroscopy, of the lipid thermotropic behavior of *Butyrivibrio*

S2 membranes and their extracted lipids (Hauser *et al.*, 1985) indicated no difference in the lipid transition temperature. Also, the correlation time for the motion of 16-doxyl stearic acid in cell membranes and liposomes of the extracted lipids was similar, which indicated a lack of effect of proteins on the fast motion of the label located in the membranes core.

In an investigation by Schindler *et al.* (1980), involving the fluorescence recovery after photobleaching technique, it was shown that the lateral diffusion of *N*-4-nitrobenz-2-oxa-1,3-diazole phosphatidylethanolamine (NBD-PE), in membranes reconstituted from coliform outer membrane constituents, was not influenced significantly by the presence of outer membrane protein or lipopolysaccharide. On the other hand, the diffusion of rhodaminated lipopolysaccharide in this reconstituted system was decreased 10-fold by the presence of 60% protein. A recent study by Ranck *et al.* (1984) involved x-ray analysis of the kinetics of *Escherichia coli* lipid and membrane structural transitions. The results indicated that the relaxation times of the disorder to order transition observed with membranes and their extracts were strongly correlated. The authors concluded that although proteins may possibly affect the amount of lipids taking part in the disorder to order transition, they do not significantly change the kinetics of the conformational transition of the lipids and consequently do not interact with the lipids involved with this transition.

Casal *et al.* (1980) used Fourier transition infrared spectroscopy to study lipid organization and motion in *Acholeplasma laidlawii* membranes and, on the basis of a comparison of the isolated membranes and membrane lipid dispersions, concluded that membrane proteins had only minor effects on the phase transition. They caused a decrease in the rate of acyl chain motion while increasing the population of gauche conformers of the fatty acyl chains in the liquid-crystalline phase of the membranes. These results were in agreement with a previous study by Stockton *et al.* (1977) using *Acholeplasma laidlawii* grown on perdeuterated acid, which revealed a striking agreement between the order profile of the cells and those of the pure phospholipid membranes.

A study by Gally *et al.* (1979) failed to indicate any effect of protein on the order of the acyl chains of the membrane. Using a fatty acid auxotroph of *Escherichia coli*, grown in medium supplemented with deuterated palmitic and oleic acids, they were able to show that the order parameter at different segmental positions was the same whether the deuterated acyl chains were incorporated in the lipids of whole cell samples or aligned as in dioleoylphosphatidylcholine liposomes. In a later study, involving the separated inner and outer membranes from *Escherichia coli* cells supplemented with deuterated fatty acids, Gally *et al.* (1980) were able to show that the membranes were slightly less ordered than artificial membranes prepared from their lipids. This study showed that while ^{2}H-NMR spectra of phosphatidylglycerol containing liposomes was affected by the presence of Mg^{2+}, the spectrum of the membranes was not. The results indicated that either the phosphatidylglycerol was bound to protein in the membrane and inaccesible to Mg^{2+} or that Mg^{2+} binds preferentially to protein sites on the membrane of much larger affinity.

Kang et al. (1979, 1981), using Escherichia coli, observed the ^2H-NMR spectrum of biosynthetically incorporated palmitic acid, specifically deuterated in the terminal methyl groups or in other positions of the chain and found that protein interacted with the lipids of the membrane to cause some disordering. On the other hand, similar experiments with Acholeplasma laidlawii showed no marked effect of protein on ordering of lipids in the membranes of this organism. The authors concluded that protein–lipid interactions in the membrane were of a nature that produced either little disordering or no effect on ordering of the membrane lipid.

The effect of protein on the polar head group region of the lipid molecules was examined by Gally et al. (1981). These authors showed that Escherichia coli, grown on deuterated glycerol, displayed quadruple splittings of its membrane fractions almost identical to those of the corresponding pure lipid vesicles. They concluded that the glycerol backbone conformation is not altered to any appreciable extent by the presence of large amounts of protein. However, examination of their membrane spectra reveals subcomponent resonances of significantly different width, which suggested effects of proteins on the motion of glycerol within its ordered environment (Bloom and Smith, 1985)—conclusions that were confirmed in a later study by Borle and Seelig (1983).

The results obtained from NMR studies have been reviewed and evaluated by a number of authors in recent years (Seelig and Seelig, 1980; Jacobs and Oldfield, 1981; Davis, 1983; Devaux, 1983; Bloom and Smith, 1985) and the general conclusion brought forth has been that protein effects on lipid organization are minor. The question as to why the conclusions from NMR studies do not apparently agree completely with those obtained from other techniques such as ESR and fluorescence spectroscopy has been addressed recently by Jähnig (1979), Smith and Oldfield (1984), and Restall and Chapman (1986). It seems from the present knowledge at hand that the issue of lipid–protein interactions in membranes of prokaryotes and other organisms is not yet entirely resolved.

3.3. Phase Transitions and Homeoviscous Adaptation

Fluorescence techniques, with parameters such as intensity, emission maxima, and polarizations of different types of probes, have been used to follow lipid thermotropic transitions in membranes of prokaryotes (Melchior, 1982), Escherichia coli in particular (Overath and Traüble,1973; Sackmann et al.,1973; Traüble and Overath, 1973; Cheng et al., 1974; Overath et al., 1975; Thilo and Overath 1976; Tecoma et al., 1977; Janoff et al., 1980; Proulx and Szabo, 1985; Souzu, 1986a). The acyl chain gel to liquid-crystalline transitions have also been examined by ESR (Baldassare et al., 1973; Linden et al., 1973a, b; Overath and Traüble 1973; Kleeman et al., 1974; Overath et al., 1975; Rottem and Leive, 1977), by x-ray diffraction (Dupont *et al.*, 1972; Shechter *et al.*, 1972; Sackmann *et al.*, 1973; Overath and Traüble, 1973; Traüble and Overath, 1973; Shechter *et al.*, 1974; Overath *et al.*, 1975; Letellier *et al.*, 1977; Nakayama *et al.*, 1980), by NMR (Davis *et al.*, 1979; Gally *et al.*, 1979; Kang *et al.*, 1979, 1981; Nichol *et al.*, 1980), and by scanning calorimetry (Steim 1970; Haest *et al.*, 1974; Baldassare *et al.*, 1976; Melchior and Steim, 1976; Jackson and Sturtevant, 1977; Jackson and Cronan, 1978). Good agreement between fluorometric methods and other physical techniques was reported (Overath and Thilo, 1978), although a great deal of variation in results from independent laboratories exists because of the differences in strain, growth conditions, and methods of membrane isolation used. Nevertheless, the reports generally indicate for wild-type strains that the membranes undergo a rather broad gel to liquid-crystalline phase transition. This phase transition responds to acyl chain composition, which may be altered by temperature of growth or, particularly in auxotrophs, by fatty acid supplementation (Melchior, 1982; McElhaney, 1984).

There is a tendency for prokaryotes to regulate their acyl chain composition in response to external conditions such that the fluidity of their membranes remains approximately constant. This regulatory phenomenon, termed homeoviscous adaptation, is operative in microbes as different as *Escherichia coli* (Sinensky, 1974) and *Butyrivibrio* S_2 (Hauser *et al.*, 1979, 1985), although its effectiveness may vary considerably in different poikilothermic organisms (Ray *et al.*, 1971; Finne and Matches, 1976; Cossins, 1977; Wakayama and Oshima, 1978; Yang and Haug, 1979; McElhaney, 1984; Russell, 1984). In a study illustrating this point, Reizer *et al.* (1985) examined the effect of increasing the growth temperature from 45 to 65°C on the thermophile *Bacillus stearothermophilus* var. *nondiastaticus*. There resulted from this treatment a substantial decrease in the proportion of branched-chain fatty acids and a significant increase in the amount of saturated acyl chains. The gel to liquid-crystalline phase transition of the membrane lipids was monitored by differential scanning calorimetry and by fluorescence anisotropy of DPH. The apparent microviscosity of the membranes and their extracted lipids decreased with elevation of growth temperatures and was kept within a certain range rather than at a fixed value. It was apparent from the results that this thermophile lacked rigorous homeoviscous control. More detailed information on the thermotropic behavior of prokaryotic organisms and their adaptive responses to growth temperature and other culture conditions is given in recent reviews by Melchior (1982) and Russell (1984).

3.4. Effects of Alcohols

The effects of alcohols on the growth, morphology, and biochemistry of microorganisms were reviewed recently by Ingram and Buttke (1984). The effects of these agents have been studied on bacterial membranes in particular: they are numerous and depend on chain length. The shorter-chain alcohols such as ethanol are hydrophilic and dissolve poorly in the hydrophobic environment of the membrane. Consequently, to cause direct effects of appreciable extent, that is, effects on isolated cells or membranes not mediated by adaptive lipid compositional changes, relatively high concentrations of ethanol are required. The changes that occur may be the result of a chaotropic, hydrophobic bond-weakening effect rather than a deep intercalation of ethanol into the bilayer somewhat akin to the effect of thiocyanate (Ingram, 1982). Examined by using a variety of fluorescent probes, namely, DPH, perylene, and a set of n-(9-anthroyloxy) fatty acids, the decreasing effects of ethanol on fluorescence polarization were small and located near the membrane surface (Dombek and Ingram, 1984). Longer chain analogues such as hexanol are more hydrophobic and partition effectively into membranes (Jain and Wu, 1977). In comparison to ethanol, hexanol decreased the polarization of probes located more deeply in the membrane (Dombek and Ingram, 1984).

Alcohols have profound effects on the lipid metabolism of bacterial cells. Ethanol when added during growth of *Escherichia coli* cells causes preferential inhibition of saturated fatty acid synthesis and an increase in the proportion of *cis*-vaccenic acid (Ingram, 1976; Buttke and Ingram, 1978; Buttke and Ingram, 1980; Ingram, 1982). The synthesis of phosphatidylethanolamine decreases and the proportion of the acidic phospholipids, phosphatidylglycerol, and cardiolipin increases together with the overall decrease in content of phospholipid (Ingram, 1977). Similar changes were reported to occur in other microorganisms (Kates *et al.*, 1962; Taneja and Khuller, 1980).

The ethanol-induced increase in vaccenic acid in the membranes of *Escherichia coli* is paradoxical inasmuch as ethanol is known to cause increased fluidity (Vanderkooi, 1979) and increases in unsaturation would tend to accentuate this effect and destabilize the membrane. However, when plasma membranes of cells grown in the presence of ethanol were examined with the use of various fluorescent probes, these were found to be slightly more rigid than those from control cells (Dombek and Ingram, 1984). In contrast, liposomes prepared from lipid extracts of ethanol-grown cells were more fluid than those from control cells. The decrease in fluidity was explained on the basis of increased protein/lipid ratios, protein exerting a rigidifying effect on lipid fatty acyl chains.

On the other hand, *Escherichia coli* cells grown in the presence of longer chain alcohols, such as hexanol, increase the proportion of palmitic acid and decrease that of vaccenic acid in their lipids (Ingram, 1976; Sullivan *et al.*, 1979). Despite these changes and a diminished phospholipid/protein ratio, membranes from hexanol-grown cells displayed increased fluidity at higher concentrations of alcohol as could be detected by measuring changes in fluorescence polarization following labeling with DPH (Ingram and Vreeland, 1980).

3.5. Permeability of the Outer Membrane to Hydrophobic Substances

The question of permeability of the outer membrane of Gram-negative bacteria to various permeants was addressed in several recent reviews (Lugtenberg and Van Alphen, 1983; Hancock, 1984; Nikaido and Vaara, 1985; Nakae, 1986). It is well established now that the outer membranes of wild-type Gram-negative bacteria are poorly permeable to hydrophobic substances. Accordingly, Nikaido (1976) found that the efficacy of inhibitors against wild-type as compared to deep-rough mutants of *Salmonella typhimurium* was low and decreased with increase in partition coefficient measured in an octanol–phosphate buffer system. Leive *et al.* (1984) determined that resistance of wild-type *Escherichia coli* to a series of tetracyclines decreased as their hydrophobicity increased. They also exploited the fluorescent properties of 13-phenylmercapto-α-6-deoxytetracycline to study penetration into wild-type and mutant strains of *Escherichia coli*. However, there are a number of pitfalls in assuming that fluorescence intensity is an appropriate means of measuring permeability. For instance, it was found that whereas tetracycline incorporation into *Escherichia coli* K12 was nonsaturable when radiolabeled permeant entry was measured, fluorescence increase due to uptake became maximal at an external antibiotic concentration of about 200 µM. The two uptake assays also gave different results for pH optimum (Smith and Chopra, 1983). It can be concluded from these results and those of Samra *et al.* (1979) that fluorescence intensity of incorporated tetracyclines may depend on the hydrophobicity and pH of the milieu as well as on the presence of ions and chelating substances and cannot be related quantitatively to the drug transport in a simple way. Despite these shortcomings, Leive *et al.* (1984) were able to show that penetration of the hydrophobic fluorescent tetracycline derivative occurred most readily in deep-rough, heptoseless strains of *Escherichia coli*. The basis for the barrier against hydrophobic substances, displayed by wild-type strains, is not completely known but appears to involve the highly asymmetric nature of the outer membrane, with all the lipopolysaccharide in the outer leaflet and most if not all of the phospholipid in the inner leaflet (Lugtenberg and Van Alphen, 1983; Hancock, 1984a; Nikaido and Vaara, 1985). The lack of exposed lipid may be one reason why hydrophobic molecules do not readily penetrate wild strains of Gram-negative bacteria in contrast to Gram-positive bacteria, which may be penetrated (Franklin and Snow, 1981). It has also been proposed that bridging of negatively charged lipopolysaccharide molecules afforded by divalent cations (Nikaido and Vaara, 1985) would result in a tight barrier to hydrophobic molecules.

The increased permeability to hydrophobic molecules displayed by the outer membrane of deep-rough mutants may be related to the pleiotropic changes that occur in such strains. Besides possessing lipopolysaccharides of greatly abridged polysaccharide chains, the mutants are often characterized by a reduction of several other membrane proteins and an increased phospholipid content (Ames et al., 1974; Koplow and Goldfine, 1974; Nikaido and Vaara, 1985). It has been proposed that the phospholipid replaces areas of the outer leaflet normally filled by protein (Lugtenberg and Van Alphen, 1983; Nikaido and Vaara, 1985); however, there is some evidence indicating that the phospholipid/lipopolysaccharide ratio may not be greatly increased in some of these mutants (Gmeiner and Schlecht, 1979; Nikaido and Vaara, 1985). Since exposure of such strains to Mg^{2+} decreases their hydrophobic permeability, it was suggested that, in the absence of sufficient Mg^{2+} due to increases in lipopolysaccharide content, much weaker interactions between the lipopolysaccharide would be expected and accessibility to hydrophobic sites would increase (Hancock, 1984; Nikaido and Vaara, 1985). At any rate, it seems that a marked decrease in protein content does not necessarily accompany the usual characteristics of hypersensitivity to hydrophobic agents noted for deep-rough mutants (Nakamura, 1968; Parton, 1975; Henson and Walker, 1982).

Treatment of Gram-negative bacteria with EDTA under appropriate conditions results in the release of large amounts of lipopolysaccharide from the envelope (Gray and Wilkinson, 1965; Leive, 1965). Loss of lipopolysaccharide is accompanied by a marked increase in sensitivity to hydrophobic substances (Leive, 1974). The action of EDTA is dependent on its cation chelating properties. Lipopolysaccharide, stripped of its neutralizing cations, becomes destabilized by electrostatic repulsion between like negative charges of adjacent molecules. This in itself could allow hydrophobic molecules to reach the apolar phase of the membrane even without shedding lipopolysaccharide. In addition, the known loss of lipopolysaccharide, followed by a redistribution of phospholipid to fill the spaces created in the outer leaflet, would certainly favor the entry of exogenous hydrophobic substances (Nikaido and Vaara, 1985).

Tris is usually required to assist the action of EDTA but by itself is also able to cause lipopolysaccharide release at higher concentrations and to permeabilize the outer membrane (Hancock et al., 1981; Irvin et al., 1981; Nikaido and Vaara, 1985). Resembling polycations in its mode of action, Tris probably acts by interacting with lipopolysaccharide (Peterson et al., 1985) and displacing bound cations required to stabilize its aggregate structure (Nikaido and Vaara, 1985). The overall effects of these agents, which can be reversed or blocked by Mg^{2+}, includes loss of lipopolysaccharide, release of periplasmic enzymes, formation of outer membrane blebs, and increased permeability to antibiotics and hydrophobic compounds (Hancock, 1984).

A detailed study of Pseudomonas aeruginosa cells by Hancock and Wong (1984) indicated a parallel increase in lysozyme susceptibility to lysis and permeability to the fluorescent probe NPN, resulting from treatment with EDTA, polycations, and other agents. Gentamicin, and other polycations tested, appeared to involve binding to the outer membrane at sites normally accommodating divalent cations. The kinetics of the permeabilization as studied by measuring fluorescence increase due to NPN uptake was sigmoidal. This prompted Loh *et al.* (1984) to surmise that aminoglycosides promote their own uptake.

3.6. Membrane-Potential-Related Permeability Changes

3.6.1. Fluorescent Probes of Membrane Potential

Many cells are too small to allow direct measurement of membrane potential by microelectrodes and in the case of prokaryotic cells, for example, only indirect determinations can be made. The various techniques used to determine membrane potential have been reviewed by Rottenberg (1975, 1979) and those involving optical probes, in particular, were considered by Bashford and Smith (1979) and Waggoner (1979).

The membrane potential can be determined by measuring the equilibrium distribution of labeled permeant ions across the cell membrane. This distribution responds to the membrane potential difference in a manner that can be calculated from the Nernst equation.

Another current and sensitive method for determining membrane potential involves the use of fluorescence dyes, the association of which with the membrane responds to the potential difference. The nature and degree of association in turn alters the fluorescent properties of the dye. Such changes can be measured and calibrated in terms of potential differences as determined by other methods. Cyanine dyes are among the most often used fluorescent compounds and their absorption by cells results in quenching, the degree of which can be measured. In such a study, for example, Ghazi *et al.* (1981) were able to determine the membrane potential in *Escherichia coli* cells by measuring the quenching of 3,3-dimethylindodicarbocyanine iodide. The degree of quenching was calibrated as a function of potential differences, which were in turn measured by determining the equilibrium distribution of radiolabeled Rb^+, triphenylmethylphosphonium ion and tetraphenylphosphonium ion. For example, with such methods changes in potential resulting from coliphage infection could be studied effectively (Oldmixon and Braun, 1978; Labedan and Lettelier, 1981; Letellier and Labedan, 1984).

3.6.2. Effect of Deenergization on Membrane Structure

The uptake of fluorescent probes has been used extensively to study the effect of deenergization on the structure of bacterial membranes (Hancock, 1984). The energy state is assessed by measuring changes in membrane potential with methods involving equilibrium distribution of labeled permeant ions or the uptake of fluorescent dyes. In one such study, Helgerson and Cramer (1977) examined the changes in *Escherichia coli* envelope structure and fluorescent probe binding caused by the uncoupler carbonyl cyanide *p*-trifluoromethoxy-phenylhydrazone (FCCP) and by EDTA. The probes used were NPN and ANS. It was noted that the fluorescence polarization and the rotational relaxation times of these dyes were increased in cells exposed to uncoupler. EDTA treatment of the cells, which removes 30–50% of the lipopolysaccharide (Leive, 1965), increased the binding, the fluorescence intensity, and the rotational relaxation time of the hydrophobic NPN probe as well as the binding of FCCP. The chelating agent, on the other hand, had no marked effect on the binding and the rotational relaxation time of the amphiphilic ANS probe, but the effect of FCCP on these parameters required a 20-fold smaller concentration after treatment with EDTA. It was concluded that inner membrane deenergization causes structural changes in the outer membrane, which removes a permeability barrier to hydrophobic molecules. The structural changes also appeared to result in an increase in the microviscosity of the environment accommodating the probes. It can be mentioned here that the use of steady-state fluorescence measurements for estimating microviscosity has recently been reevaluated (Heyn, 1979; Pottel *et al.*, 1983; Van der Meer *et al.*, 1986). At any rate, in another investigation, Nieva-Gomez and Gennis (1977), using several types of probes including NPN and ANS, concluded that *Escherichia coli* membrane microviscosity was not affected by deenergization, and fluorescence changes resulting from this process were due solely to quantitative changes in the binding of the fluorescent dyes. In a number of other studies it was amply shown that colicins E and K, uncouplers, and processes that in general deenergize the cytoplasmic membrane also cause an increase in the uptake and in the fluorescence of a variety of dyes by susceptible cells (Helgerson *et al.*, 1974; Brewer, 1976; Helgerson and Cramer, 1977; Nieva-Gomez *et al.*, 1976). In these types of uptake studies, the increase in fluorescence intensity resulting from deenergization does not necessarily reflect the amount of dye bound to the membrane, the fluorescence response being very sensitive to environment (Cramer *et al.*, 1976). The effects of deenergization were also examined by Tecoma and Wu (1980), who showed that treatment of *Escherichia coli* membranes with colicin K dramatically increased the fluorescence of exogenously added *cis*-PA or NPN but did not affect that of *cis*-PA biosynthetically esterified to the phospholipids of the envelope. No fluorescence enhancement was seen in cells centrifuged to remove unbound exogenous *cis*-PA before colicin treatment. The results indicated once again that deenergization leads to structural changes in the outer membrane which cause increased partitioning of the fluorescent probes into the hydrophobic phase of the membrane. No evidence for changes in membrane fluidity could be obtained with the endogenously esterified *cis*-PA, yet this probe, like the nonesterified analogue, can be used effectively to report on lipid structural changes (Sklar *et al.*, 1977b; Tecoma *et al.*, 1977).

Further investigations by Wolf and Konisky (1981) revealed that binding of 1-azidopyrene to strains of *Escherichia coli* derived from K12 was enhanced by conditions that deenergized the cells and decreased the potential across the cytoplasmic membrane. Upon photoactivation of the dye, the nitrene radical formed produces covalent adducts with nearest-neighbor molecules. The changes in binding, as assessed by increased fluorescence intensity, could be reversed by reenergization prior to but not after photoactivation. A comparison of the response by deep-rough and wild-type strains of *Salmonella typhimurium* (Wolf and Konisky, 1984) supported the conclusion of other investigators (Helgerson and Cramer, 1977; Nieva-Gomez and Gennis, 1977; Tecoma and Wu, 1980) that changes in the structure of the outer membrane due to deenergization and decreases in membrane potential lead to increased binding of the probes to hydrophobic domains of the envelope. The nature of the alteration in outer membrane structure and the basis for the reversibility of the binding process are unclear.

3.6.3. Studies on Deenergization by Phage Adsorption

Among the agents that cause deenergization of bacterial cytoplasmic membranes are the phages (Oldmixon and Braun, 1978). Irreversible adsorption of phages T4, T5, BF23, and T4 ghosts to *Escherichia coli* membrane receptors results in an immediate and transient depolarization of the inner membrane which can be followed by the cyanine dye distribution method (Waggoner, 1979; Letellier and Shechter, 1979; Labedan and Letellier, 1981).

Following the immediate depolarization effects of phage adsorption, there is a repolarization response of the host cell which leads to a new steady state of reduced membrane potential. The rate and extent of repolarization depend on the amplitude of depolarization, which in turn is proportional to the number of infecting phages attached. The potential changes are independent of the type of phage and appear to be mediated by a signal that is transmitted to the inner membrane (Labedan and Letellier, 1981). In the case of phage T4 but not of phages T5 and BF23, adsorption appears to trigger the liberation of a membrane-bound pool of calcium. This calcium release, together with induced conformational changes, is required for depolarization. The implication of calcium is indicated by the fact that ethylene glycol-bis(β-aminoethyl ether)-N,N,N',N'-tetraacetic acid (EGTA) prevents T4-induced depolarization (Letellier and Labedan, 1984, 1985). The involvement of Ca^{2+} is further supported by recent evidence indicating that adsorption of phage T4 causes release of the divalent cation from the envelope as could be quantified by the Quin 2 fluorescence assay method (Boulanger *et al.*, 1985).

3.7. Factors Increasing Cell Resistance and Membrane Stability

Freezing and thawing of *Escherichia coli* results in structural and functional damage as well as decreased viability due to loss of membrane and cytoplasmic constituents. These deleterious effects are especially noticeable in exponentially grown cells, which are characterized by marked increases in permeability, whereas stationary phase cells do not display such notable changes in permeability and viability and are characterized by a relatively higher content of protein in their inner and outer membranes (Souzu, 1982).

To compare the structural properties of *Escherichia coli* membranes of cells grown to the exponential and stationary phases, Souzu (1986a) labeled the inner and outer membranes, as well as their corresponding lipids, with *cis*-PA and *trans*-PA and examined fluorescence polarization as a function of temperature. The effects of growth phase were complex. Below the phase transition temperature the increased protein/lipid ratio in the stationary phase was associated with decreased polarization ratios in both cytoplasmic and outer membranes. Above the transition temperature, the elevated protein/lipid ratio was associated with slightly increased polarization ratios in the cytoplasmic membrane but was without effect in the outer membrane. Again, above the transition temperature, the lipid extracts in general displayed lower polarization ratios than did the membranes from which they were prepared. Below the transition temperature, however, the lipid extracts displayed higher polarization ratios compared to the membranes. It can be surmised from these results that proteins and perhaps other constituents associated with the lipids of the membrane affected their thermotropic behavior. On this basis, Souzu (1986a) advanced that stationary phase membranes were in a more highly ordered state because of the increased protein content. Protein–protein and protein–lipid interactions would become reinforced, lipid motion reduced, and stabilization of the membrane increased as a consequence.

The changes in polarization ratios seen below the transition temperature, attributed to the presence of protein, could possibly be explained on the basis that the probes partitioned predominantly into more fluid domains, the appearance of which may have resulted from segregation of protein-rich and lipid-rich phases at the lower temperatures (Kleeman and McConnell, 1974; Letellier *et al.*, 1977; Legendre *et al.*, 1980). When DPH was used as a probe, lipid extracts were generally found to display lower polarization ratios than their corresponding membranes (Janoff *et al.*, 1980; McGibbon *et al.*, 1985; Proulx and Szabo, 1985; Merrill *et al.*, 1987), and this may be because of the more uniform distribution of the probe in the different domains of the membrane.

Related to this study was another by Souzu (1986b), which dealt with the effects of polyamines on membrane resistance to freeze-thawing and on membrane structure. Natural polyamines, such as putrescine and spermidine, which are known to stabilize a variety of membranes (Tabor, 1962; Grossowicz and Ariel, 1963), are present in relatively high concentrations in *Escherichia coli* (Tabor and Tabor, 1976). These substances were found to increase cellular resistance of logarithmic phase *Escherichia coli* cells to freeze-thawing, a phenomenon that could be related to changes in fluidity of the coliform membranes (Souzu, 1986b). These changes detected by parinaric acid polarization measurements resembled those brought about by growing cells to the stationary phase. However, the molecular basis for the stabilizing effect of polyamines on logarithmic phase membranes remains unknown.

4. PERIPLASM

Between the two membranes of Gram-negative bacteria lies the periplasm, which contains specific classes of proteins and oligosaccharides (Nossal and Heppel, 1966; Van Golde et al., 1973). Some of these proteins bind permeants specifically and are important elements of transport systems (Boos, 1972). Recent evidence indicates that these binding proteins may function within a gellike organized environment. Hobot et al. (1984), using conventional and scanning transmission electron microscopy, examined the envelope ultrastructure of Escherichia coli by the progressive lowering of temperature embedding technique and freeze-substitution. From their results, they suggested that the periplasmic space is filled with a viscous gel. To investigate this point further, Brass et al. (1986) introduced biologically active rhodamine B isothiocyanate-labeled maltose-binding protein into the periplasmic space of Escherichia coli following permeabilization of the outer membrane with Ca^{2+}. Lateral diffusion of the fluorescent-labeled protein was measured by a photobleaching recovery method and a diffusion coefficient of 0.9×10^{-10} cm^2/sec was found. This value is 1000-fold lower than would be expected for diffusion in aqueous medium and about 100-fold lower than for a protein of equivalent size in the cytoplasm. The data as a whole suggested that the periplasm is a highly viscous continuum, densely populated with proteins and effectively containing little or no free aqueous space. It was suggested that diffusion of small permeant molecules across the periplasm may involve binding to protein molecules, which may still be free to move or rotate within the limited space available and pass substrate from one binding molecule to another and eventually to appropriate transport sites of the cytoplasmic membrane.

5. INCORPORATION OF EXOGENOUS LIPIDS INTO PROKARYOTIC MEMBRANES

5.1. Gram-Negative Bacteria

Unlike the numerous studies involving the incorporation of exogenous lipids into membranes of mammalian origin, those involving incorporation into prokaryotic membranes are relatively few. Early studies by Jones and Osborn

(1977a,b) indicated that a variety of exogenous lipids including some that are not native to the organism, such as phosphatidylcholine and cholesteryloleate, could be incorporated into galactose- and heptose-deficient mutants of *Salmonella typhimurium*. Incorporations into the wild parent strains, synthesizing normal lipopolysaccharide chains, were much less extensive, as could be expected from the known hydrophobic barrier properties of outer membrane and the role that lipopolysaccharides assume in this barrier. The incorporation of lipid, added to the cells as sonicated lipid vesicles, was found to be time, temperature, and concentration dependent, was greatly stimulated by divalent and polyvalent cations such as Ca^{2+} and spermine, and appeared to involve a bulk uptake by a fusion process with the outer membrane. Once taken up, they could rapidly be translocated to the inner membrane (Jones and Osborn, 1977b) as was elegantly shown by incubating G30A mutant cells with phosphatidylserine (PS) and following its decarboxylation to phosphatidylethanolamine by an enzyme known to be located in the inner membrane exclusively. The kinetics of translocation for other lipids could not be studied in such a manner. However, isolation of the inner and outer membranes established that exogenous lipids equilibrated between both membranes.

Further studies (Hellion *et al.*, 1980; Proulx *et al.*, 1982; Proulx, 1985; Proulx and Szabo, 1985) indicated that *Escherichia coli* was also capable of incorporating lipids into its membranes. The uptake of diacyl phosphoglycerides was, as expected, much greater in the deep-rough strain, D21F2, than in the parent wild-type strain (Proulx, 1985). As in the case of *Salmonella typhimurium* (Jones and Osborn, 1977a,b), the process was time, Ca^{2+} concentration, and lipid concentration dependent and, in the presence of divalent cation at least, appeared to involve a fusion process between the added lipid vesicles and the cell membranes (Proulx *et al.*, 1982; Proulx, 1985).

Another report by Michel *et al.* (1984) indicated that an ethanol-tolerant bacterium, with a lipopolysaccharide lacking ketodeoxyoctonic acid (Tornabene *et al.*, 1982), could incorporate the lipids of liposomes prepared from *Escherichia coli* phospholipids. The process, which again appeared to involve a fusion phenomenon, depended on temperature and the presence of divalent cation and was influenced by pH and the liposome/cell ratio.

The uptake of lysophosphoglycerides was also examined in wild and deep-rough mutant strains of *Escherichia coli* (McIntyre and Bell, 1978; Hellion *et al.*, 1980). In the case of 1-oleoyl-lysophosphatidic acid, the uptake proceeded readily in the absence of added Ca^{2+} although the effect of this ion was not specifically tested (McIntyre and Bell, 1978). Incorporations were considerably greater in the deep-rough, heptoseless strain compared to the parent strain, both strains being *pls* B glycerophosphate auxotrophs. These uptakes, which were found to be time, exogenous lipid concentration, and bacterial cell concentration dependent, were followed by a conversion of the lysoderivative to phosphatidylethanolamine and phosphatidylglycerol at a substantial rate, corresponding to 40% of the rate of *de novo* synthesis of phospholipid under the prevailing experimental conditions. Since the conversion steps are catalyzed by enzymes associated with the inner membrane (Bell *et al.*, 1971; White *et al.*, 1971), it follows that the lysophosphatidic acid translocated to these sites and did not remain associated with the outer membrane only. Quite interestingly, the uptake of 1-oleoyl-lysophosphatidate and its conversion to other phospholipid was followed by the appearance of myelinlike figures inside the cell and at its surface. Very likely these represented outgrowths of the inner and outer membranes due to the substantial uptake of lysophosphatidic acid.

The incorporation of lysophosphatidylethanolamine, on the other hand, was found to be significant even in wild-type strains of *Escherichia coli* (Hellion *et al.*, 1980; Proulx, 1985), and the presence of divalent cations such as Ca^{2+} had only a moderate stimulatory effect on the uptake. In the absence of this cation, a very substantial conversion of the labeled lysoderivative to the diacyl analogue occurred, and since most of the lysophosphoglyceride acylating enzyme is located in the inner membrane fraction, and since separation of the inner and outer membranes revealed the presence of both these labeled lipids, it was concluded that the exogenous lipid could translocate from one membrane to the other.

The larger incorporations of diacyl lipids seen in the deep-rough strains can best be explained on the basis of diminished steric hindrance by polysaccharide chains at the cell surface and of increased access to the bilayer because of missing outer membrane proteins (Jones and Osborn, 1977a, 1977b; Lugtenberg and Van Alphen, 1983; Nikaido and Vaara, 1985). Such mutants display greater permeability to hydrophobic substances in general as indicated in Section 3.5. The uptakes of lysophosphatidic acid (McIntyre and Bell, 1978) and lysophosphatidylethanolamine (Hellion *et al.*, 1980) were comparable in two different strains featuring normal lipopolysaccharide chains; however, enhancement of uptake by deep-rough mutation could not be seen in the case of lysophosphatidylethanolamine, possibly because of the more pronounced rate of lysophosphatidylethanolamine breakdown noticed in the D21F2 strain used for this study (Proulx, 1985).

It appears from these various studies that the amount of exogenous lipid incorporated can be relatively large, especially in the case of deep-rough mutant strains. One would expect that a consequence of such uptake would be noticeable changes in the physical properties of the membranes, and in a recent investigation this is indeed what was found for *Escherichia coli* exposed to vesicles made from saturated and unsaturated phosphatidylcholine and phosphatidylethanolamine species (Proulx and Szabo, 1985). In this investigation, normal and lipid-treated cells or their lipid extracts were labeled with DPH and fluorescence polarization ratios were measured as a function of temperature. Incorporations of dipalmitoyl-, dioleoyl-, and didecanoylphosphatides into the deep-rough mutant D21F2, in the presence of Ca^{2+}, caused significant changes in polarization ratios of the cells over a wide temperature range and the appearance of new phase transitions, none of which corresponded to those of the pure lipids. In the presence of Ca^{2+}, uptake of these lipids into wild-type cells, although much less pronounced, caused changes in the polarization ratios qualitatively similar to those noted for the heptoseless strain. The polarization ratios of the extracted lipids were always reduced and their transitions broadened when compared to those of whole cells. Generally, in the lipid extracts there was also an attenuation of any differences in polarization ratios between normal and phospholipid-treated samples. It was apparent from the results that the presence of nonlipid constituents in the membrane significantly influenced the polarization ratios. Whether this influence was due to lipopolysaccharide or protein with which the probe interacted directly or whether lipid–protein or lipid–lipoprotein interactions were involved could not be ascertained.

Another fluorescence study by Emmerling *et al.* (1977) also showed that isolated envelopes of a deep-rough mutant of *Escherichia coli* K12 could incorporate dimyristoylphosphatidylethanolamine into its membranes. This caused an increase of the transition temperature from 30 to 42°C, as could be assessed by changes in fluorescence intensity in envelopes labeled with NPN. When envelopes from wild-type cells were treated in a similar manner, a somewhat altered transition temperature of 33°C occurred as compared to 25–30°C in untreated cells, but there also appeared another transition at 48°C which is that of the acyl chains of the exogenous lipid. It seemed that in both cases some dimyristoylphosphatidylethanolamine had been incorporated into the envelopes; however, in those of wild-type cells, intercalation of the exogenous lipid into the membrane core structures had not occurred extensively, or if it had, it remained as a separate phase. The origin of the phase transitions in the untreated envelopes could not be ascertained since isolated lipid and lipopolysaccharide constituents melted at significantly lower temperatures. Lipid–lipopolysaccharide interactions in the envelope leading to more compact packing of the hydrocarbon chains of the two components were considered as an explanation. However, attempts to reveal such interactions by mixing lipopolysaccharide and phospholipid were unsuccessful.

5.2. Other Bacteria

In studies involving interaction of exogenous lipids with cell membranes it is sometimes difficult to ascertain whether lipids intercalate within the bilayer structure by a fusion or an exchange process or whether they are stably adsorbed at the surface of the cell. This question has been considered in detail in a previous review (Pagano *et al.*, 1981). When intact cells are used, the failure to observe adsorbed lipid vesicles at the surface of the cell following thorough washing is one means of monitoring contamination or stable adsorption and has been applied in the case of the experiments involving Gram-negative bacteria (Jones and Osborn, 1977a; Proulx *et al.*, 1982; Proulx, 1985). Also, when thermotropic transitions are examined, the failure to observe chain melting characteristic of the added lipids and the appearance of new phase transitions are other means of controlling whether or not mixing with the natural lipids has occurred (Emmerling *et al.*, 1977; Proulx and Szabo, 1985).

Fluorescence resonance energy transfer (RET) monitoring and relief of self-quenching provide alternative methods for distinguishing between fusion and adsorption-endocytosis phenomena. In the RET approach, a donor–acceptor energy pair is incorporated into the lipid vesicle bilayer and transfer of the fluorophores to the cell membrane by a fusion or lipid exchange process is accompanied by a change in RET.

In relief of self-quenching, fluorescence increases as the probe becomes diluted from the concentrated state in the lipid vesicles to the dilute state in the cell membrane following fusion. Such fluorescence techniques have recently been exploited by Dreissen *et al.* (1985) to demonstrate a pH-induced fusion of liposomes with membrane vesicles derived from *Bacillus subtilis* and other bacteria. For the RET approach, they used the probes *N*-(7-nitro-2,1,3-benz oxadiazol-4-yl)phosphatidylethanolamine (N-NBD-PE) as the donor and *N*-(lissamine rhodamine B sulfonylphosphatidylethanolamine) (N-Rh-PE) as the acceptor. Fusion was accompanied by an increase in donor fluorescence. For the relief of self-quenching, parinaroylphosphatidylcholine was used. The incorporation mechanism was also elucidated by other methods involving the measurement of rates of transfer of two distinctly radiolabeled lipids from lipid vesicles to cells, or involving the assessment of changes in buoyant density of lipid-treated membranes.

Recent studies with other Gram-positive bacteria have yielded results similar to those obtained with Enterobacteriaceae. Accordingly, Gilliland *et al.* (1985) were able to show that certain strains of *Lactobacillus* when grown in the presence of bile salts under anaerobic conditions could take up cholesterol from the medium. The basis for the differences in uptake among the various strains was not investigated. In another study, *Bacillus* protoplasts were incubated with egg-yolk liposomes loaded with 6-carboxyfluorescein or with phosphotungstate. The fluorescence of the incubation medium decreased with incorporation of lipid and the interior of the protoplasts became stained with electron-dense marker. The results suggested that the lipid uptake had occurred by a fusion process (Urbaneja *et al.*, 1984).

Cells from the family Mycoplasmataceae are interesting in that they lack a cell wall and require exogenous cholesterol for growth. They have the ability to take up large quantities of this sterol and its ester as well as phospholipids. In contrast, cells from the family Acholeplasmataceae, which do not require cholesterol for growth, do not take up the sterol so readily and are unable to incorporate other lipids (Razin *et al.*, 1980). The mechanism of phospholipid incorporation by *Mycoplasma* cells was not studied but the uptake process was shown to be dependent on growth. The massive uptake of phospholipid by these cells had no effect on cell growth and on *de novo* synthesis of phospholipids. However, a marked increase in the diphosphatidylglycerol/phosphatidylglycerol ratio was observed (Gross *et al.*, 1982). The reason for these changes is not known. The larger uptake of cholesterol by *Mycoplasma* was correlated with their ability to take up phospholipids since the increased phospholipid content of these cells offers additional sites for binding of the sterol (Razin *et al.*, 1980). With respect to this finding, it is interesting to note from the studies of Clejan and Bittman (1984) that sphingomyelin assimilation by the membrane causes firmer binding of cholesterol than does phosphatidylcholine assimilation because the presence of this sphingolipid decreases the rate of sterol exchange with unilamellar vesicles much more so than does the presence of the choline phosphatide.

Freeze-fracture microscopy of the *Mycoplasma gallisepticum* membranes, following uptake of lipids from media containing egg-phosphatidylcholine but not from media containing dioleoylphosphatidylcholine, revealed particle-free patches (Rottem and Verkleij, 1982), the appearance of which could be attributed to the presence of disaturated phosphatidylcholine, formed from the exogenously supplied phospholipid. The occurrence of these smooth areas could be correlated with a decrease in osmotic fragility of the cells and an increase in their permeability to K^+. Fluorescence polarization and electron spin resonance measurements revealed similar fluidities in the protein-rich membranes of the organism and aqueous dispersions of their lipids. Furthermore, the ESR spectra of the membranes indicated no immobilized regions. The authors pointed to the possibility that the osmotic resistance of *Mycoplasma gallisepticum* was related to the presence of particle-free patches rather than to protein–lipid interactions.

Other studies relating to uptake of cholesterol by mycoplasma species indicated that incorporation of the sterol was dependent on the fluidity of the membrane bilayer. The uptake is rapid when the membrane is in the liquid crystalline state and very slow when it is in the gel state (Razin, 1978).

Although mycoplasmas require cholesterol in the growth medium, *Mycoplasma gallisepticum* was found to be adaptable to lower concentrations of cholesterol in the presence of various saturated and unsaturated fatty acids. The fluidity of the membranes from cells adapted to these growth conditions was measured by assessing changes in DPH fluorescence polarization as a function of temperature. Adaptation of low cholesterol resulted in the appearance of distinct lipid phase transitions, which varied in accordance with the supplemented fatty acid (Le Grimellec *et al.*, 1981). The results were in accord with the known effects of cholesterol on lipid phase transitions (Ladbrooke and Chapman, 1969).

5.3. Effect of Lipid Uptake on Membrane Function

Techniques for transferring exogenous lipids to cell membranes during cell growth or involving isolated cells or their membranes offer great potential for elucidation of structure–function relation in membranes. However, to date, very few such studies have been made and the effects of lipid insertion into prokaryotic membranes remain unknown except for the fact that growth of the cells is unaffected by this process (Jones and Osborn, 1977a). Recent investigations by Snozzi and Crofts (1984) and by Takamoto *et al.* (1985) were concerned with the fusion of isolated chromatophore membranes of *Rhodopseudomonas capsulata* with liposomes. The purpose of these studies was to test the effect of lipid dilution on the function of the photosynthetic membranes since such dilutions occur during the cell cycle. Indeed, protein/phospholipid ratios of these intracytoplasmic membranes are known to undergo cyclical fluctuations during synchronous growth, a pattern that was shown to result from large changes in the rate of phospholipid synthesis (Fraley *et al.*, 1979b; Knacker *et al.*, 1985). Results from pulse-chase experiments indicated that the oscillation of the chromatophore protein/phospholipid ratio is dependent on the transfer of phospholipids from a discrete site outside the intracytoplasmic membrane (Cain *et al.*, 1981). Since most of the phospholipid synthesis enzymes are located in the cytoplasmic membrane, it would appear that the transfer is from this latter location (Cain *et al.*, 1984) and might well involve a phospholipid transfer protein (Cohen *et al.*, 1979; Tai *et al.*, 1986). Interestingly, the increase in phospholipid/protein ratio occurring after division affects membrane structure significantly and causes an increase in fluidity of the intracytoplasmic membrane, as could be assessed from the decrease in steady-state fluorescence polarization of membranes labeled with α-parinaric acid (Fraley *et al.*, 1979c).

In *Rhodopseudomonas sphaeroides*, the dilution resulting from interactions with liposomes led to a decrease in the rate of cytochrome b-561 reduction and to an increase in lag between flash and onset of this reduction (Snozzi and Crofts, 1984). In *Rhodopseudomonas capsulata*, dilution led to decreased electron transport rates and an uncoupling of energy transfer between B875 light-harvesting and reaction center complexes (Takemoto *et al.*, 1985).

5.4. Interactions with Vehicle Liposomes

Several studies were performed in which the major interest was to use liposomes or lipid vesicles as vectors for the transfer of macromolecules and plasmids. A report by Fraley *et al.* (1979a) indicated that liposomes containing entrapped pBR322 DNA could be used to transfect competent *Escherichia coli* SF8 cells and produce tetracycline-resistant colonies. A similar finding was reported by Nicolau and Rottem (1982) working with liposome-encapsulated pBR322 plasmid of *Escherichia coli* and *Mycoplasma capricolum* cells. The exposed organism expressed β-lactamase activity and acquired tetracycline resistance following transfer of the plasmid via a probable fusion of the liposomes with the cells.

In a study by Lelkes *et al.* (1984), the general usefulness of liposomes for the transfer of entrapped water-soluble as well as liposomal membrane material to isolated flagellated cell envelopes of *Escherichia coli* (strain RP487) was monitored by electron microscopy using ferritin-containing liposomes, and by fluorescence spectroscopy using encapsulated 5,6-dicarboxyfluorescein and the membrane-intercalated fluorescent S9 (*N*,*N'*-distearyldioxocyanine perchlorate)–S11 (*N*,*N'*-distearyldithiocyanine perchlorate) pair. It was estimated that at least 20% of the envelope-associated liposomes had delivered their content into the envelopes by a mechanism probably involving fusion.

6. CONCLUDING REMARKS

This chapter has illustrated the wide use of fluorescence spectroscopy in the study of bacterial membrane structure. Applied in a coordinated manner with other physical techniques, the various facets of this spectroscopic approach can provide useful information on the physical state of the membrane as well as help define the molecular interactions that form the basis for membrane structure. Particular attention was given to experiments dealing with uptake of exogenous lipids since this approach offers great potential for the elucidation of structure–function relations in membranes. Here again, fluorescence techniques have been quite useful in defining the mechanism of exogenous lipid uptake as well as revealing the resulting changes in membrane structure. Future attempts should be made to correlate in a more systematic manner the uptake of exogenous lipids with changes in membrane fluidity and changes in functional aspects of the membrane.

ACKNOWLEDGMENTS. I wish to thank the Medical Research Council of Canada and the National Science and Engineering Research Council of Canada for their financial support.

7. REFERENCES

Ames, G. F., Spudish, E. N., and Nikaido, H., 1974, Protein composition of the outer membrane of *Salmonella typhimurium*: Effect of lipopolysaccharide mutations, *J. Bacteriol.* **117** : 406–416.

Amey, R. L., and Chapman, D., 1983, Infrared spectroscopic studies of model and natural bio-membranes, in *Biomembrane Structure and Function* (D. Chapman, ed.), pp. 199–256, Macmillan, London.

Andersen, H. C., 1978, Probes of membrane structure, *Annu. Rev. Biochem.* **47** : 359–583.

Andrich, M. P., and Vanderkooi, J. M., 1976, Temperature dependence of 1,6-diphenyl-1,3,5-hexatriene fluorescence in phospholipid artificial membranes, *Biochemistry* **15** : 1257–1261.

Ashe, G. B., and Steim, J. M., 1971, Membrane transitions in Gram-positive bacteria, *Biochim. Biophys. Acta* **233** : 810–814.

Axelrod, D., Koppell, D. E., Schlessinger, J., Elson, E., and Webb, W. W., 1976, Mobility measurement by analysis of fluorescence photobleaching recovery kinetics, *Biophys. J.* **16** : 1055–1069.

Azzi, A., 1975, The application of fluorescent probes in membrane studies, *Q. Rev. Biophys.* **8** : 237–316.

Bach, D., 1983, Calorimetric studies of model and natural biomembranes, in *Biomembrane Structure and Function* (D. Chapman, ed.), pp. 1–41, Macmillan, London.

Baldassare, J. J., McAfee, A. G., and Ho, C., 1973, A spin label study of *Escherichia coli* membrane vesicles, *Biochem. Biophys. Res. Commun.* **53** : 617–623.

Baldassare, J. J., Rhinehart, K. B., and Silbert, D. F., 1976, Modification of membrane lipid: Physical properties in relation to fatty acid structure, *Biochemistry* **15** : 2986–2994.

Bashford, C. L., and Smith, J. C., 1979, The use of optical probes to monitor membrane potential, *Methods Enzymol.* **55** : 569–588.

Bell, R. M., Mavis, R. D., Osborn, M. J., and Vagelos, P. R., 1971, Enzymes of phospholipid metabolism: Localization in the cytoplasmic and outer membrane of the cell envelope of *Escherichia coli* and *Salmonella typhimurium, Biochim. Biophys. Acta* **249** : 628–635.

Bergelson, L. D., Molotkovsky, J. G., and Manevich, Y. M., 1985, Lipid-specific fluorescent probes in studies of biological membranes, *Chem. Phys. Lipids* **37** : 165–195.

Blaurock, A. E., 1985, X-ray and neutron diffraction by membranes: How great is the potential for defining the molecular interactions?, in *Progress in Protein–Lipid Interactions* (A. W. Watts and J. J. H. H. M. Depont, eds.), pp. 1–43, Elsevier Science Publishers, Amsterdam.

Bloom, M., and Smith, I. C. P., 1985, Manifestations of lipid–protein interactions in deuterium NMR, in *Progress in Protein–lipid Interactions* (A. W. Watts and J. J. H. H. M. DePont, eds.), pp. 61–88, Elsevier Science Publishers, Amsterdam.

Boos, W., 1972, Structurally defective galactose-binding protein isolated from a mutant negative in the α-methylgalactoside transport system of *Escherichia coli, J. Biol.* **247** : 5415–5424.

Borle, F., and Seelig, J., 1983, Structure of *Escherichia coli* membranes. Deuterium magnetic resource studies of the phosphoglycerol head group in intact cells and model membranes, *Biochemistry* **22** : 5536–5544.

Boulanger, P., Labedan, B., and Letellier, L., 1985, Involvement of calcium in the transient depolarization of *Escherichia coli* cytoplasmic membrane induced by phase adsorption: A study with the fluorescent calcium indicator Quin 2, *Biochem. Biophys. Res. Commun.* **131** : 856–862.

Brass, J. M., Higgins, C. F., Foley, M., Rugman, P. A., Birmingham, J., and Garland, P., 1986, Lateral diffusion of proteins in the periplasma of *Escherichia coli, J. Bacteriol.* **165** : 789–794.

Braun, V., 1975, Covalent lipoprotein from the outer membrane of *Escherichia coli, Biochim. Biophys. Acta* **415** : 335–377.

Brewer, G. J., 1976, The state of energization of the membrane of *Escherichia coli* as affected by physiological conditions and colicin K, *Biochemistry* **15** : 1387–1392.

Burnell, E. L., Van Alphen, L., Verkleij, A., De Kruijff, B., and Lugtenberg, B., 1980, ^{31}P-nuclear magnetic resonance and freeze-fracture electron microscopy studies on *Escherichia coli.* 1. Cytoplasmic membrane and total phospholipids, *Biochim. Biophys. Acta* **597** : 492–501.

Buttke, J. M., and Ingram, L. O., 1978, Mechanism of ethanol-induced changes in lipid composition of *Escherichia coli*. Inhibition of fatty acid synthesis *in vivo*, *Biochemistry* **17** : 637–644.

Buttke, T. M., and Ingram, L. O., 1980, Ethanol-induced changes in lipid composition of *Escherichia coli*: Inhibition of saturated fatty acid synthesis in vitro, *Arch. Biochem. Biophys.* **203** : 465–471.

Cain, B. D., Deal, C. D., Fraley, R. T., and Kaplan, S., 1981, *In vivo* intermembrane transfer of phospholipids in the photosynthetic bacterium, *Rhodopseudomonas sphaeroides, J. Bacteriol.* **145** : 1154–1166.

Cain, B. D., Donohue, T. J., Shepherd, W. D., and Kaplan, S., 1984, Localization of phospholipid biosynthetic enzyme activities in cell-free fractions derived from *Rhodopseudomonas sphaeroides, J. Biol. Chem.* **259** : 942–948.

Casal, H. L., Cameron, D. G., Smith, I. C. P., and Mantsch, H. H., 1980, *Acholeplasma laidlawii* membranes: A Fourier transform infrared study of the influence of protein on lipid organization and dynamics, *Biochemistry* **19** : 444–451.

Chapman, D., and Benga, G., 1984, Biomembrane fluidity—Studies of model and natural biomembranes, in *Biological Membranes* (D. Chapman, ed.), pp. 1–56, Academic Press, London.

Chapman, D., Goni, F. M., and Pink, D. A., 1983, On the interactions of probes and integral proteins in lipid bilayers, in *Liposomal Letters* (A. D. Bangham, ed.), pp. 169–179, Academic Press, London.

Cheng, S., Thomas, J. K., and Kulpa, C. F., 1974, Dynamics of pyrene fluorescence in *Escherichia coli* membrane vesicles, *Biochemistry* **13** : 1135–1139.

Clejan, S., and Bittman, R., 1984, Decreases in rates of lipid exchange between *Mycoplasma gallisepticum* cells and unilamellar vesicles by incorporation of sphingomyelin, *J. Biol. Chem.* **259** : 10823–10826.

Cohen, L. K., Leuking, D. R., and Kaplan, S., 1979, Intermembrane phospholipid transfer mediated by cell-free extracts of *Rhodopseudomonas sphaeroides, J. Biol. Chem.* **254** : 721–728.

Colley, C. M., and Metcalfe, J. C., 1972, The localization of small molecules in lipid bilayers, *FEBS Lett.* **24** : 241–246.

Cossins, A. R., 1977, Adaptation of biological membranes to temperature. The effect of temperature acclimation of goldfish upon the viscosity of synaptosomal membranes, *Biochim. Biophys. Acta* **470** : 395–411.

Coughlin, R. T., Haug, A., and McGroarty, E. J., 1983, Electron spin resonance probing of lipopolysaccharide domains in the outer membrane of *Escherichia coli*, *Biochim. Biophys. Acta* **729** : 161–166.

Davenport, L., Dale, R. E., Bisby, R. H., and Cundall, R. B., 1985, Transverse location of the fluorescent probe 1,6-diphenyl-1,3,5-hexatriene in model lipid membrane systems by resonance excitation energy transfer, *Biochemistry* **24** : 4097–4108.

Davis, J. H., 1983, The description of membrane lipid conformation, order and dynamics by ^{2}H-NMR, *Biochim. Biophys. Acta* **737** : 117–171.

Davis, J. H., Nichol, C. P., Weeks, G., and Bloom, M., 1979, Study of the cytoplasmic and outer membrane of *Escherichia coli* by deuterium magnetic resonance, *Biochemistry* **18** : 2103–2112.

Devaux, P. F., 1983, ESR and NMR studies of lipid–protein interactions in membranes, in *Biological Magnetic Resonance* (L. J. Berliner and J. Reuben, eds.), Vol. V, pp. 183–299, Plenum Press, New York.

Devaux, P. F., 1985, Conventional ESR spectroscopy of membrane proteins: Recent applications, in *The Enzyme of Biological Membranes*, 2nd ed. (A. N. Martonosi, ed.), Vol. 1, pp. 259–285, Plenum Press, New York.

Dombek, K. M., and Ingram, L. O., 1984, Effect of ethanol on *Escherichia coli* plasma membranes, *J. Bacteriol.* **157** : 233–239.

Driessen, A. J. M., Hoekstra, D., Scherphof, G., Kalicharan, R. D., and Wilchut, J., 1985, Low pH-induced fusion of liposomes with membrane vesicles derived from *Bacillus subtilis*, *J. Biol. Chem.* **260** : 10880–10887.

Dupont, G., Gabriel, A., Chabre, M., Gulik-Krzywicki, T., and Schechter, E., 1972, Use of a new detector for x-ray diffraction and kinetics of the ordering of the lipids of *Escherichia coli*: Membranes and model systems, *Nature* **238** : 331–333.

Emmerling, G., Henning, U., and Gulik-Kzywicki, T., 1977, Order–disorder conformational transition of hydrocarbon chains in lipopolysaccharide from *Escherichia coli*, *Eur. J. Biochem.* **78** : 503–509.

Fiil, A., and Branton, D., 1968, Changes in the plasma membrane of *Escherichia coli* during magnesium starvation, *J. Bacteriol.* **98** : 1320–1327.

Finne, G., and Matches, J. R., 1976, Spin-labelling studies on the lipids of psychrophilic, psychrotropic and mesophilic *Clostridia*, *J. Bacteriol.* **125** : 211–219.

Fraley, R. T., Jameson, D. M., and Kaplan, S., 1978, The use of the fluorescent probe α-parinaric acid to determine the physical state of the intracytoplasmic membranes of the photosynthetic bacterium *R. sphaeroides*, *Biochim. Biophys. Acta* **511** : 52–69.

Fraley, R. T., Jameson, D. M., and Kaplan, S., 1978, The use of the fluorescent probe α-parinaric acid to determine the physical state of the intracytoplasmic membranes of the photosynthetic bacterium *R. sphaeroides*, *Biochim. Biophys. Acta* **511** : 52–69.

Fraley, R. T., Fornari, C. S., and Kaplan, S., 1979a, Entrapment of a bacterial plasmid in phospholipid vesicles: Potential for gene transfer, *Proc. Natl. Acad. Sci. U.S.A.* **76** : 3348–3352.

Fraley, R. T., Lueking, D. R., and Kaplan, S., 1979b, The relationship of intracytoplasmic membrane assembly to the cell division cycle in *Rhodopseudomonas sphaeroides*, *J. Biol. Chem.* **254** : 1980–1986.

Fraley, R. T., Yen, G. S. L., Lueking, D. R., and Kaplan, S., 1979c, The physical state of the intracytoplasmic membrane of *Rhodopseudomonas sphaeroides* and its relationship to the cell division cycle, *J. Biol. Chem.* **245** : 1987–1991.

Franklin, T. J., and Snow, G. A., 1981, *Biochemistry of Antimicrobial Action*, 3rd ed., Chapman and Hall, London.

Funahara, Y., and Nikaido, H., 1980, Asymmetric localization of lipopolysaccharide on the outer membrane of *Salmonella typhimurium*, *J. Bacteriol.* **141** : 1463–1465.

Gallay, J., and Vincent, M., 1986, Cardiolipin–cholesterol interactions in the liquid-crystalline phase: A steady-state and time-resolved fluorescent anisotropy study with *cis*- and *trans*-parinaric acids as probes, *Biochemistry* **25** : 2650–2656.

Gally, H. U., Pluschke, G., Overath, P., and Seelig, J., 1979, Structure of *Escherichia coli* membranes. Phospholipid composition in model membranes and cells as studied by deuterium magnetic resonance, *Biochemistry* **18** : 5605–5609.

Gally, H. U., Pluschke, G., Overath, P., and Seelig, J., 1980, Structure of *Escherichia coli* membranes. Fatty acid chain order parameters of inner and outer membranes and derived liposomes, *Biochemistry* **19** : 1638–1643.

Gally, H. U., Pluschke, G., Overath, P., and Seelig, P., 1981, Structure of *Escherichia coli* membranes. Glycerol auxotrophs as a tool for the analysis of the phospholipid head-group region by deuterium magnetic resonance, *Biochemistry* **20** : 1826–1831.

Ghazi, A., Shechter, E., Letellier, L., and Labedan, B., 1981, Probes of membrane potential in *Escherichia coli* cells, *FEBS Lett.* **125** : 197–200.

Gilliland, S. E., Nelson, C.R., and Maxwell, C., 1985, Assimilation of cholesterol by *Lactobacillus acidophilus, Appl. Environ. Microbiol.* **49** : 377–381.

Gmeiner, J., and Schlecht, S., 1979, Molecular organization of the outer membrane of *Salmonella typhimurium, Eur. J. Biochem.* **93** : 609–620.

Gmeiner, J.,and Schlecht, S., 1980, Molecular composition of the outer membrane of *Escherichia coli* and the importance of protein–lipopolysaccharide interactions, *Arch. Microbiol.* **127** : 81–86.

Gray, G. W., and Wilkinson, S. G., 1965, The action of ethylenediaminetetraacetic acid on *Pseudomonas aeruginase, J. Appl. Bacteriol.* **28** : 153–164.

Gross, Z., Rottem, S., and Bittman, R., 1982, Phospholipid interconversions in *Mycoplasma capricolum, Eur. J. Biochem.* **122** : 169–174.

Grossowicz, N., and Ariel, M., 1963, Mechanism of protection of cells by spermine against lysozyme-induced lysis, *J. Bacteriol.* **85** : 293–300.

Gunstone, F. D., and Subbarao, R., 1967, New tropical seed oils. Part 1. Conjugated trienoic and tetraenoic acids and their oxo derivatives in the seed oils of *C. iaco* and *P. laurinum, Chem. Phys. Lipids* **1** : 349–359.

Haest, C. W. M., Verkleij, A. J., deGier, J., Scheek, R., Ververgaert, P. H. J., and VanDeenen, L. L. M., 1974, The effect of lipid phase transitions in the architecture of bacterial membranes, *Biochim. Biophys. Acta* **356** : 17–26.

Hancock, R. E. W., 1984, Alterations in outer membrane permeability, *Annu. Rev. Microbiol.* **38** : 237–264.

Hancock, R. E. W., and Wong, P. G. W., 1984, Compounds which increase the permeability of the *Pseudomonas aeruginosa* outer membrane, *Antimicrob. Agents Chemother.* **26** : 48–52.

Hancock, R. E. W., Raffle, V. J., and Nicas, T. I., 1981, Involvement of the outer membrane in gentamicin and streptomycin uptake and killing in *Pseudomonas aeruginosa, Antimicrob. Agents Chemother.* **19** : 777–785.

Hauser, H., Hazlewood, G. P., and Dawson, R. M. C., 1979, Membrane fluidity of a fatty acid auxotroph grown with palmitic acid, *Nature* **279** : 536–538.

Hauser, H. Hazlewood, G. P., and Dawson, R. M. C., 1979, Membrane fluidity of a fatty acid auxotroph grown with palmitic acid, *Nature* **279** : 536–538.

Hauser, H., Hazlewood, G. P., and Dawson, R. M. C., 1985, Characterization of membrane lipids of a general fatty acid auxotrophic bacterium by electron spin resonance spectroscopy and differential scanning calorimetry, *Biochemistry* **24** : 5247–5253.

Helgerson, S. L., and Cramer, W. A., 1977, Changes in *Escherichica coli* cell envelope structure and the sites of fluorescence probe binding caused by carbonyl cyanine *p*-trifluoromethoxyphenylhydrazone, *Biochemistry* **16** : 4109–4117.

Helgerson, S. L., Cramer, W. A., Harris, J. M., and Lytle, F. E., 1974, Evidence for a microviscosity increase in the *Escherichia coli* cell envelope caused by colicin E1, *Biochemistry* **13** : 3057–3061.

Hellion, P., Landry, F., Subbaiah, P. V., and Proulx, P., 1980, The uptake and acylation of exogenous lysophosphatidylethanolamine by *Escherichia coli* cells, *Can. J. Biochem.* **58** : 1381–1386.

Henson, J. M., aand Walker, J. R., 1982, Genetic analysis of *acrA* and *lir* mutations of *Escherichia coli, J. Bacteriol.* **152** : 1301–1302.

Herring, F. G., Krisman, A., Sedgwick, E. G., and Bragg, P. D., 1985, Electron spin resonance studies of lipid fluidity changes in membranes of an uncouple-resistant mutant of *Escherichia coli, Biochim. Biophys. Acta* **819** : 231–240.

Heyn, M. P., 1979, Determination of lipid order parameters and rotational correlation times from fluorescence depolarization experiments, *FEBS Lett.* **108** : 359–364.

Hobot, J. A., Carlemalm, E., Villiger, W., and Killenberger, E., 1984, Periplasmic gel: New concept resulting from the reinvestigation of bacterial cell envelope ultrastructure by new methods, *J. Bacteriol.* **160** : 143–152.

Hoffman, W., and Restall, C. J., 1983, Rotational and lateral diffusion of membrane proteins as determined by laser techniques, in *Biomembrane Structure and Function* (D. Chapman, ed.), pp. 257–318, Macmillan, London.

Ingram, L. O., 1976, Adaptation of membrane lipids to alcohols, *J. Bacteriol.* **125** : 670–678.

Ingram, L. O., 1977, Preferential inhibition of phosphatidylethanolamine synthesis in *Escherichia coli* by alcohols, *Can. J. Microbiol.* **23** : 779–789.

Ingram, L. O., 1982, Regulation of fatty acid composition in *Escherichia coli*: A proposed common mechanism for changes induced by ethanol, chaotropic agents and a reduction of growth temperature, *J. Bacteriol.* **149** : 166–172.

Ingram, L. O., and Buttke, J. M., 1984, Effects of alcohols on microorganisms, *Adv. Microbiol. Physiol.* **25** : 253–300.

Ingram, L. O., and Vreeland, N. S., 1980, Differential effects of ethanol and hexanol on the *Escherichia coli* cell envelope, *J. Bacteriol.* **144** : 481–488.

Inouye, M., 1979, *Bacterial Outer Membranes. Biogenesis and Functions*, Wiley-Interscience, New York.

Inouye, M., Shaw, J., and Shen, C., 1972, The assembly of a structural lipoprotein in the envelope of *Escherichia coli*, *J. Biol. Chem.* **247** : 8154–8159.

Irvin, R. T., MacAlister, T. J., Chan, R., and Costerton, J. W., 1981, Citrate-Tris (Hydroxymethyl) aminomethane-mediated release of outer membrane sections from the cell envelope of a deep-rough (heptose-deficient lipopolysaccharide) strain of *Escherichia coli* 08, *J. Bacteriol.* **145** : 1386–1396.

Ishinaga, M., Kanamoto, R., and Kito, M., 1979, Distribution of phospholipid molecular species in outer and cytoplasmic membranes of *Escherichia coli*, *J. Biochem. (Tokyo)* **86** : 161–165.

Jackson, M. B., and Cronan, J. E., 1978, An estimate of the minimum amount of fluid lipid required for the growth of *Escherichia coli*, *Biochim. Biophys. Acta* **512** : 472–479.

Jackson, M. B., and Sturtevant, J. M., 1977, Studies of the lipid phase transitions of *Escherichia coli* by high sensitivity differential scanning calorimetry, *J. Biol. Chem.* **252** : 4749–4751.

Jacobs, R. E., and Oldfield, E., 1981, NMR of membranes, *Progr. NMR Spectrosc.* **14** : 113–136.

Jähnig, F., 1979, Structural order of lipids and proteins in membranes: Evaluation of fluorescence anisotropy data, *Proc. Natl. Acad. Sci. U.S.A.* **76** : 6361–6365.

Jain, M. K., and Wu, N. M., 1977, Effect of small molecules on the dipalmitoyl lecithin liposomal bilayer: III. Phase transition in lipid bilayer, *J. Membr. Biol.* **34** : 157–201.

Janoff, A. S., Haug, A., and McGroarty, E. J., 1979, Relationship of growth temperature and thermotropic lipid phase changes in cytoplasmic and outer membrane from *Escherichia coli* K12, *Biochim. Biophys. Acta* **555** : 56–66.

Janoff, A. S., Gupte, S., and McGroarty, E. J., 1980, Correlation between temperature range of growth and structural transitions in membranes and lipids of *Escherichia coli* K12, *Biochim. Biophys. Acta* **598** : 641–644.

Jones, N. C., and Osborn, M. J., 1977a, Interaction of *Salmonella typhimurium* with phospholipid vesicles: Incorporation of exogenous lipids into intact cells, *J. Biol. Chem.* **252** : 7398–7404.

Jones, N. C., and Osborn, M. J. 1977b, Translocation of phospholipids between the outer and inner membranes of *Salmonella typhimurium*, *J. Biol. Chem.* **252** : 7405–7412.

Kamio, Y., and Nikaido, H., 1976, Outer membrane of *Salmonella typhimurium*: Accessibility of phospholipid head groups to phospholipase C and cyanogen bromide activated dextran in the external medium, *Biochemistry* **15** : 2561–2570.

Kamio, Y., and Nikaido, H., 1977, Outer membranes of *Salmonella typhimurium*. Identification of proteins exposed on cell surface, *Biochim. Biophys. Acta* **464** : 589–601.

Kang, S. Y., Gutowsky, H. S., and Oldfield, E., 1979, Spectroscopic studies of specifically deuterium labelled membrane systems. Nuclear magnetic resonance investigation of protein–lipid interactions of *Escherichia coli* membranes, *Biochemistry* **18** : 3268–3272.

Kang, S. Y., Kinsey, R. A., Rajan, S., Gutowsky, H. S., Gubridge, M. G., and Oldfield, E., 1981, Protein–lipid interactions in biological and model membrane systems. Deuterium NMR of *Acholeplasma laidlawii B, Escherichia coli* and cytochrome oxidase systems containing specifically deuterated lipids, *J. Biol. Chem.* **256** : 1155–1159.

Kates, M., Kushner, D. J., and James, A. T., 1962, The lipid composition of *Bacillus cereus* as influenced by the presence of alcohols in the culture medium, *Can. J. Biochem. Physiol.* **40** : 83–93.

Kleeman, W., and McConnell, H. M., 1974, Lateral phase separations in *Escherichia coli* membranes, *Biochim. Biophys. Acta* **345** : 220–230.

Kleeman, W., Grant, C. W. M., and McConnell, H. M., 1974, Lipid phase separations and protein distribution in membranes, *J. Supramol. Struct.* **2** : 609–616.

Knacker, T., Harwood, J. L., Hunter, C. N., and Russell, N. J., 1985, Lipid biosynthesis in synchronized culture of the photosynthetic bacterium *Rhodopseudomonas sphaeroides, Biochem. J.* **229** : 701–709.

Koplow, J., and Goldfine, H., 1974, Alterations in the outer membrane of the cell envelope of heptose-deficient mutants of *Escherichia coli, J. Bacteriol.* **117** : 527–543.

Labedan, B., and Letellier, L., 1981, Membrane potential changes during the first steps of coliphage infection, *Proc. Natl. Acad. Sci. U.S.A.* **78** : 215–219.

Ladbrooke, M. D., and Chapman, D., 1969, Thermal analysis of lipids, proteins and biological membranes, *Chem. Phys. Lipids* **3** : 304–367.

Legendre, S., Letellier, L., and Shechter, 1980, Influence of lipids with branched-chain fatty acids on the physical, morphological and functional properties of *Escherichia coli* cytoplasmic membrane, *Biochim. Biophys. Acta* **602** : 491–505.

Le Grimellec, C., Cardinal, J., Giocondi, M.-C., and Carriere, S., 1981, Control of membrane lipids of *Mycoplasma gallisepticum*: Effect on lipid order, *J. Bacteriol.* **146** : 155–162.

Leive, L., 1965, Release of lipopolysaccharide by EDTA treatment of *Escherichia coli, Biochem. Biophys. Res. Commun.* **21** : 290–296.

Leive, L., 1974, The barrier function of the Gram-negative envelope, *Ann. N.Y. Acad. Sci.* **235** : 109–127.

Leive, L., Telesetsky, S., Coleman, W. G., Jr., and Carr, D., 1984, Tetracyclines of various hydrophobicities as a probe for permeability of *Escherichia coli* outer membranes, *Antimicrob. Agents Chemother.* **25** : 539–544.

Lelkes, P. I., Klein, L., Marikovsky, Y., and Eisenbach, M., 1984, Liposome-mediated transfer of macromolecules into flagellated cell envelopes from bacteria, *Biochemistry* **23** : 563–568.

Lentz, R. B., Barenholz, Y., and Thompson, T. E., 1976, Fluorescence depolarization studies of phase transition and fluidity in phospholipid bilayers. 1. Single component phosphatidylcholine liposomes, *Biochemistry* **15** : 4521–4528.

Lesslauer, W., Cain, J., and Blaisie, J. K., 1971, On the location of 1-anilino-8-naphthylene-sulfonate in lipid model systems. An x-ray diffraction study, *Biochim. Biophys. Acta* **24** : 547–566.

Letellier, L., and Labedan, B., 1984, Involvement of envelope-bound calcium on the transient depolarization of the *Escherichia coli* cytoplasmic membrane induced by bacteriophage T_4 and T_5 adsorption, *J. Bacteriol.* **157** : 789–794.

Letellier, L., and Labedan, B., 1985, Release of respiratory control in *Escherichia coli* after bacteriophage adsorption: Process independent of DNA injection, *J. Bacteriol.* **16** : 179–182.

Letellier, L., and Shechter, E., 1979, Cyanine dye as monitor of membrane potentials in *Escherichia coli* cells and membrane vesicles, *Eur. J. Biochem.* **102** : 441–447.

Letellier, L., Moudden, H., and Shechter, E., 1977, Lipid and protein segregation in *Escherichia coli* membrane: Morphological and structural study of different cytoplasmic membrane fractions, *Proc. Natl. Acad. Sci. U.S.A.* **74** : 452–456.

Linden, C. D., Keith, A. D., and Fox, C. F., 1973a, Correlations between fatty acid distribution in phospholipids and the temperature dependence of membrane physical state, *J. Supramol. Struct.* **1** : 523–534.

Linden, C. D., Wright, K. L., McConnell, H. M., and Fox, C. F., 1973b, Lateral phase separations in membrane lipids and the mechanism of sugar transport in *Escherichia coli*, *Proc. Natl. Acad. Sci. U.S.A.* **70** : 2271–2275.

Loh, B., Grant, C., and Hancock, R. E. W., 1984, Use of the fluorescent probe 1-*N*-phenylnaphthylamine to study the interactions of aminoglycoside antibiotics with the outer membrane of *Pseudomonas aeruginosa*, *Antimicrob. Agents Chemother.* **26** : 546–551.

Lugtenberg, E. J. J., and Peters, R., 1976, Distribution of lipids in cytoplasmic and outer membranes of *Escherichia coli* K12, *Biochim. Biophys. Acta* **441** : 38–47.

Lugtenberg, B., and Van Alphen, L., 1983, Molecular architecture and functioning of the outer membranes of *Escherichia coli* and Gram-negative bacteria, *Biochim. Biophys. Acta* **737** : 51–115.

Mackay, A. L., Nichol, C. P., Weeks, G., and Davis, J. H., 1984, A proton and deuterium nuclear magnetic resonance study of orientational order in aqueous dispersions of lipopolysaccharide and lipopolysaccharide/dipalmitoylphosphatidylcholine mixtures, *Biochim. Biophys. Acta* **774** : 181–187.

Makowski, L., and Li, J., 1983, X-ray diffraction and electron microscopy studies of the molecular structure of biological membranes, in *Biomembrane Structure and Function* (D. Chapman, ed.), pp. 43–166, Macmillan, London.

Matayoshi, E. D., and Kleinfeld, A. M., 1981, Emission wavelength-dependent decay of the 9-anthroyloxy-fatty acid membrane probes, *Biophys. J.* **35** : 215–235.

McElhaney, R. N., 1984, The structure and function of the *Acholeplasma laidlawii* plasma membrane, *Biochim. Biophys. Acta* **779** : 1–42.

McElhaney, R. N., 1986, Differential scanning calorimetry studies of lipid–protein interactions in model membrane systems, *Biochim. Biophys. Acta* **864** : 361–421.

McGibbon, L., Cossins, A. R., Quinn, P. J., and Russell, N. J. 1985, A differential scanning calorimetry and fluorescence polarization study of membrane lipid fluidity in a psychrophilic bacterium, *Biochim. Biophys. Acta* **820** : 115–121.

McIntyre, T. M., and Bell, R. M., 1978, *Escherichia coli* mutants defective in membrane phospholipid synthesis: Binding and metabolism of 1-oleoylglycerol 3-phosphate by a *pls B* deep rough mutant, *J. Bacteriol.* **135** : 215–226.

Melchior, D. L., 1982, Lipid phase transitions and regulation of membrane fluidity in prokaryotes, in *Current Topics in Membranes and Transport* (S. Razin and S. Rottem, eds.), Vol. 17, pp. 263–316, Academic Press, New York.

Melchior, D. L., and Steim, J. M., 1976, Thermotropic transitions in biomembranes, *Annu. Rev. Biophys. Bioeng.* **5** : 205–238.

Mely-Goubert, B., and Freedman, M. H., 1980, Lipid fluidity and membrane protein monitoring using 1,6-diphenyl-1,3,5-hexatriene, *Biochim. Biophys. Acta* **601** : 315–327.

Merrill, A. R., Aubry, H., Proulx, P., and Szabo, A. G., 1987, Relation between Ca^{2+} uptake and fluidity of brush-border membranes isolated from rabbit small intestine and incubated with fatty acids and methyl oleate, *Biochim. Biophys. Acta* **896** : 89–95.

Michel, S. P. F., Cisse, M., and Starka, J., 1984, Interactions of liposomes with *Zymomonas mobilis* cells, *FEMS Microbiol. Lett.* **24** : 127–131.

Morrisett, J. D., Pownall, H. J., Plumlee, R. T., Smith, L. C., Lehner, Z. E., Esfahani, M., and Wakil, S. J., 1975, Multiple thermotropic phase transitions on *Escherichia coli* membranes

and membrane lipids: Comparison of results obtained by nitroxyl stearate paramagnetic resonance, pyrene excimer fluorescence and enzyme, *J. Biol. Chem.* **250** : 6969–6982.

Mühlethaler, K., and Jay, J., 1985, Electron microscopy of biological membranes, in *The Enzymes
of Biological Membranes*, 2nd ed. (A. N. Martonosi, ed.), Vol. 1, pp. 1–28, Plenum Press,
New York.

Mühlradt, P. F., and Golecki, J. R., 1975, Asymmetrical distribution and artifactual reorientation
of lipopolysaccharide in the outer membrane bilayer of *Salmonella typhimurium*, *Eur. J. Biochem.*
51 : 343–352.

Munford, C. A., and Osborn, M. J., 1983, An intermediate step in translocation of lipopolysaccharide to the outer membrane of *Salmonella typhimurium*, *Proc. Natl. Acad. Sci. U.S.A.*
80 : 1159–1163.

Nakae, T., 1986, Outer membrane permeability of bacteria, *CRC Crit. Rev. Microbiol.* **13** : 1–62.

Nakamura, H., 1968, Genetic determination of resistance to acriflavine, phenylethyl alcohol, and
sodium dodecyl sulfate in *Escherichia coli*, *J. Bacteriol.* **96** : 987–996.

Nakayama, H., Mitsui, T., Nishihara, M., and Kito, M., 1980, Relation between growth temperature of *Escherichia coli* and phase transition temperatures of its cytoplasmic and outer membranes, *Biochim. Biophys. Acta* **601** : 1–10.

Nichol, C. P., Davis, J. H., Weeks, G., and Bloom, M., 1980, Quantitative study of the fluidity
of *Escherichia coli* membranes using deuterium magnetic resonance, *Biochemistry* **19** : 451–
457.

Nicolau, C., and Rottem, S., 1982, Expression of a β-lactamase activity in *Mycoplasma capricolum*
transfected with the liposome-encapsulated *E. coli* pBR 322 plasmid, *Biochem. Biophys. Res.
Commun.* **108** : 982–984.

Nieva-Gomez, D., Konisky, J., and Gennis, R. B., 1976, Membrane changes in *Escherichia coli*
induced by colicin 1a and agents known to disrupt energy transduction, *Biochemistry* **15** :
2747–2753.

Nieva-Gomez, D., and Gennis, R. B., 1977, Affinity of the intact *Escherichia coli* for hydrophobic
membrane probes is a function of the physiological state of the cells, *Proc. Natl. Acad. Sci.
U.S.A.* **74** : 1811–1815.

Nikaido, H., 1976, Outer membrane of *Salmonella typhimurium* transmembrane diffusion of some
hydrophobic substances, *Biochim. Biophys. Acta* **433** : 118–132.

Nikaido, H., and Nakae, T., 1979, The outer membrane of Gram-negative bacteria, *Adv. Microb.
Physiol.* **20** : 163–250.

Nikaido, H., and Vaara, M., 1985, Molecular basis of bacterial outer membrane permeability,
Microbiol. Rev. **49** : 1–31.

Nikaido, H., Takeuchi, Y., Ohnishi, S.-I., and Nakae, T., 1977, Outer membrane of *Salmonella
typhimurium*. Electron spin resonance studies, *Biochim. Biophys. Acta* **465** : 152–154.

Nossal, N. G., and Heppel, L. A., 1966, The release of enzyme by osmotic shock from *Escherichia
coli* in exponential phase, *J. Biol. Chem.* **241** : 3055–3062.

Nurminen, M., Lounatmaa, K., Sarvas, M., Makela, P. H., and Nakae, T., 1976, Bacteriophage-
resistant mutants of *Salmonella typhimurium* deficient in two major outer membrane proteins,
J. Bacteriol. **127** : 941–955.

Oldmixon, E., and Braun, V., 1978, Changes in fluorescence of 8-anilino-1-naphthalene sulfonate
after bacteriophage T5 infection of *Escherichia coli*. Initial fluorescence rise coincides with
onset of rubidium efflux, *Biochim. Biophys. Acta* **506** : 111–118.

Osborn, M. J., and Wu, H. C. P., 1980, Proteins of the outer membrane of Gram-negative bacteria,
Annu. Rev. Microbiol. **34** : 369–422.

Osborn, M. J., Gander, J.-E., Parisi, E., and Carson, J., 1972, Mechanism and assembly of the
outer membrane of *Salmonella typhimurium*. Isolation and characterization of cytoplasmic and
outer membrane, *J. Biol. Chem.* **247** : 3962–3972.

Overath, P., and Thilo, L., 1978, Structural and functional aspects of biological membranes revealed by lipid phase transitions. *MTP Int. Rev. Sci. Biochem. Ser Two* **19** : 1–44.

Overath, P., and Traüble, H., 1973, Phase transitions in cells, membranes and lipids of *Escherichia coli*. Detection by fluorescent probes, light scattering and dilatometry, *Biochemistry* **12** : 2625–2633.

Overath, P., Brenner, M., Gulik-Krzywicki, T., Shechter, E., and Letellier, L., 1975, Lipid phase transitions in cytoplasmic and outer membranes of *Escherichia coli, Biochim. Biophys. Acta* **389** : 358–369.

Overath, P., Schairer, H. U., and Stoffel, W., 1970, Correlations *in vivo* and *in vitro* phase transitions of membrane lipids in *Escherichia coli, Proc. Natl. Acad. Sci. U.S.A.* **67** : 606–612.

Pagano, R. E., Schroit, A. J., and Struck, D. K., 1981, Interactions of phospholipid vesicles with mammalian cells *in vitro*: Studies of mechanism, in *Liposomes: From Physical Structure to Therapeutic Applications* (C. G. Knight, ed.), pp. 323–347, Elsevier/North-Holland Biomedical Press, Amsterdam.

Parassassi, T., Conti, F., and Gratton, E., 1984, Study of heterogeneous emission of parinaric acid isomers using multifrequency phase fluorometry, *Biochemistry* **23** : 5660–5664.

Parton, R., 1975, Envelope proteins in *Salmonella minnesota* mutants, *J. Gen. Microbiol.* **89** : 113–123.

Peterson, A. A., Hancock, R. E. W., and McGroarty, E. J., 1985, Binding of polycationic antibiotics and polyamines to lipopolysaccharides of *Pseudomonas aeruginosa, J. Bacteriol.* **164** : 1256–1261.

Podo, F., and Blasie, J. K., 1977, Nuclear magnetic resonance studies of lecithin bimolecular leaflets with incorporated fluorescence probes, *Proc. Natl. Acad. Sci. U.S.A.* **74** : 1032–1036.

Pottel, H., Van der Meer, W., and Herreman, W., 1983, Correlation between order parameter and the steady-state fluorescent anisotropy of 1,6-diphenyl-1,3,5-hexatriene and an evaluation of membrane fluidity, *Biochim. Biophys. Acta* **730** : 181–186.

Proulx, P., 1985, Interaction of lipid vesicles with a heptoseless strain of *Escherichia coli, Exp. Biol.* **43** : 191–199.

Proulx, P., and Szabo, A. G., 1985, The effect of exogenous glycerophospholipids on the fluorescence polarization ratios of *Escherichia coli* cells labelled with diphenylhexatriene, *Biochim. Biophys. Acta* **815** : 102–108.

Proulx, P., Hellion, P., and Mackenzie, J., 1982, Studies on the uptake of exogenous phosphoglycerides by *Escherichia coli* B cells, *Can. J. Biochem.* **60** : 980–986.

Radda, G., 1975, Fluorescent probes in membrane studies, in *Methods in Membrane Biology* (E. D. Korn, ed.), pp. 97–188, Plenum, New York.

Ranck, J. L., Letellier, L., Shechter, E., Krop, B., Pernot, P., and Tardieu, A., 1984, X-ray analysis of the kinetics of *Escherichia coli* lipid and membrane structural transitions, *Biochemistry* **23** : 4955–4961.

Ray, P. H., White, D. C., and Brock, T. D., 1971, Effect of temperature on the fatty acid composition of *Thermus aquaticus, J. Bacteriol.* **106** : 25–30.

Razin, S., 1973, Cholesterol uptake is dependent on membrane fluidity in mycoplasmas, *Biochim. Biophys. Acta* **513** : 401–404.

Razin, S., Kutner, S., Efrati, H., and Rottem, S., 1980, Phospholipid and cholesterol uptake by *Mycoplasma* cells and membranes, *Biochim. Biophys Acta* **598** : 628–640.

Reizer, J., Grossowicz, N., and Barenholz, Y., 1985, The effect of growth temperature on the thermotropic behavior of the membranes of a thermophilic *Bacillus*. Composition–structure–function relationships, *Biochim. Biophys. Acta* **815** : 268–280.

Restall, C. J., and Chapman, D., 1986, Spectroscopic and calorimetric studies of lipids and biomembranes, in *Lipids and Membranes, Past, Present and Future* (J. A. F. Op den Kamp, B. Roelofsen, and K. W. A. Wirtz, eds.), pp 61–92, Elsevier Science Publishers, Amsterdam.

Rottem, S., and Leive, L., 1977, Effect of variations of lipopolysaccharide on the fluidity of the outer membrane of *Escherichia coli, J. Biol. Chem.* **252** : 2077–2081.

Rottem, S., and Verkleij, A. J., 1982, Possible association of segregated lipid domains of *Mycoplasma gallisepticum* membranes with cell resistance to osmotic lysis, *J. Bacteriol.* **149** : 338–345.

Rottem, S., Hasin, M., and Razin, S., 1975, The outer membrane of *Proteus microbilis* II. The extractable lipid fraction and electron-paramagnetic resonance analysis of the outer and cytoplasmic membranes, *Biochim. Biophys. Acta* **375** : 395–405.

Rottenberg, H., 1975, The measurement of transmembrane electrochemical protein gradients, *Bioenergetics* **7** : 61–64.

Rottenberg, H., 1979, The measurement of membrane potential and pH in cells, organelles, and vesicles, *Methods Enzymol.* **55** : 547–568.

Russell, N. J., 1984, Mechanisms of thermal adaptation in bacteria: Blueprints for survival, *Trends Biochem. Sci.* **9** : 108–112.

Sackmann, E., and Traüble, H., 1972, Studies of the crystalline–liquid crystalline phase transition of lipid model membranes I. Use of spin labels and optical probes as indicators of the phase transition, *J Am. Chem. Soc.* **94** : 4482–4491.

Sackmann, E., Traüble, H., Galla, H., and Overath, P, 1973, Lateral diffusion, protein mobility and phase transitions in *Escherichia coli* membranes. A spin label study, *Biochemistry* **12** : 5360–5369.

Samra, Z., Krausz-Steinmetz, J., and Sompolinaky, 1979, Transport of tetracyclines through the bacteria cell membranes assayed by fluorescence: A study with susceptible and resistant strains of *Staphylococcus aureus* and *Escherichia coli, Microbios* **21** : 7–21.

Schindler, M., Osborn, M. J., and Koppel, D. E., 1980, Lateral mobility in reconstituted membranes—comparisons with diffusion in polymers, *Nature* **283** : 346–350.

Schroeder, F., 1983, Liquid domains in plasma membranes from rat liver, *Eur. J. Biochem.* **132** : 509–516.

Schroeder, F., 1985, Fluorescence probes unravel asymmetric structure of membranes, in *Subcellular Biochemistry* (D. B. Roodyn, ed.), Vol. 11, pp. 51–101, Plenum Press, New York.

Schroeder, F., and Soler-Argilaga, 1983, Ca^{++} modulates fatty acid dynamics in rat liver plasma membranes, *Eur. J. Biochem.* **132** : 517–524.

Schweizer, M., and Henning, U., 1977, Action of a major outer envelope membrane protein in conjugation of *Escherichia coli* K-12, *J. Bacteriol.* **129** : 1651–1652.

Seelig, J., and Seelig, A., 1980, Lipid conformation in model membranes and biological membranes, *Q. Rev. Biophys.* **13** : 19–61.

Schechter, E., Gulik-Krzywicki, T., and Kaback, H. R., 1972, Correlations between fluorescence, x-ray diffraction, and physiological properties in cytoplasmic membrane vesicles isolated from *Escherichia coli, Biochim. Biophys. Acta* **274** : 466–477.

Shechter, E., Letellier, L., and Gulik-Krzywicki, T., 1974, Relations between structure and function in cytoplasmic membrane vesicles isolated from an *Escherichia coli* fatty acid auxotroph, *Eur. J. Biochem.* **49** : 61–76.

Shinitzky, M., and Barenholz, Y., 1978, Fluidity parameters of lipid regions determined by fluorescence polarization, *Biochim. Biophys. Acta* **515** : 367–394.

Sinensky, M., 1974, Homeoviscous adaptation — a homeostatic process that regulates the viscosity of membrane lipids in *Escherichia coli, Proc. Natl. Acad. Sci. U.S.A.* **71** : 522–525.

Sklar, L. A., Hudson, B. S., and Simoni, R. D., 1975, Conjugated polyene fatty acids as membrane probes: Preliminary characterization, *Proc. Natl. Acad. Sci. U.S.A.* **72** : 1649–1654.

Sklar, L. A., Hudson, B. S., and Simoni, R. D., 1976, Conjugated polyene fatty acids as fluorescent membrane probes: Model system studies, *J. Supramol. Struct.* **4** : 449–465.

Sklar, L. A., Hudson, B. S., and Simoni, R. D., 1977b, Conjugated polyene fatty acids as fluorescent probes: Synethetic phospholipid membrane studies, *Biochemistry* 16 : 819–828.

Sklar, L. A., Miljanich, G. P., and Dratz, E. A., 1979a, Phospholipid lateral phase separation and the partition of *cis*-parinaric acid and *trans*-parinaric acid among aqueous, solid lipid and fluid lipid phases, *Biochemistry* 18 : 1707–1716.

Sklar, L. A., Miljanich, G. P., Bursten, S. L., and Dratz, E. A., 1979b, Thermal lateral phase separations in bovine retinal rat outer segment membranes and phospholipids as evidenced by parinaric acid fluorescence polarization and energy transfer, *J. Biol. Chem.* 254 : 9583–9591.

Sklar, L. A., Miljanich, G. P., and Dratz, E. A., 1979c, A comparison of the effects of Ca^{++} on the structure of bovine retinal rod outer segment membranes, phospholipids, and bovine brain phosphatidylserine, *J. Biol. Chem.* 254 : 9592–9597.

Slavik, J., 1982, Anilinonaphthalene sulfonate as a probe of membrane composition and function, *Biochim. Biophys. Acta* 694 : 1–25.

Smit, J., Kamio, Y., and Nikaido, H., 1975, Outer membrane of *Salmonella typhimurium*: Chemical analyses and freeze-fracture studies with lipopolysaccharide mutants, *J. Bacteriol.* 124 : 942–958.

Smith, M. C. M., and Chopra, I., 1983, Limitations of a fluorescence assay for studies on tetracycline transport into *Escherichia coli, Antimicrob. Agents Chemother.* 23 : 175–178.

Smith, R. L., and Oldfield, C., 1984, Dynamic structure of membranes by deuterium NMR, *Science* 225 : 280–288.

Snozzi, M., and Crofts, A. R., 1984, Electron transport in chromatophores from *Rhodopseudomonas sphaeroides* fused with liposomes, *Biochim. Biophys. Acta* 766 : 451–463.

Souzu, H., 1982, *Escherichia coli* B membrane stability related to cell growth phase: Measurement of temperature dependent physical state change of the membrane over a wide range, *Biochim. Biophys. Acta* 691 : 161–170.

Souzu, H., 1986a, Fluorescence polarization studies on *Escherichia coli* membrane stability and its relation to resistance of the cell to freeze-thawing. I. Membrane stability in cells of differing growth phase, *Biochim. Biophys. Acta* 861 : 353–360.

Souzu, H., 1986b, Fluorescence polarization studies in *Escherichia coli* membrane stability and its relation to the resistance of the cell to freeze-thawing. II. Stabilization of the membranes by polyamines, *Biochim. Biophys. Acta* 861 : 361–367.

Steim, J. M., 1970, Thermal phase transitions in biomembranes, *Liq. Cryst. Ordered Fluids* 1 : 1–11.

Stockton, G. W., Johnson, K. G., Buttler, K., Tulloch, A. P., Boulanger, Y., Smith, I. C. P., Davis, J. H., and Bloom, M., 1977, Deuterium NMR study of lipid organization in *Acholeplasma laidlawii* membranes, *Nature* 269 : 268–269.

Sullivan, K. H., Hageman, G. D., and Cordes, E. H., 1979, Alteration of the fatty acid composition of *Escherichia coli* by growth in the presence of normal alcohols, *J. Bacteriol.* 138 : 133–138.

Tabor, C. W., 1962, Stabilization of protoplasts and spheroplasts by spermine and other polyamines, *J. Bacteriol.* 83 : 1101–1111.

Tabor, C. W., and Tabor, H., 1976, 1,4-Diaminobutane (putrescine), spermidine, and spermine, *Annu. Rev. Biochem.* 45 : 285–306.

Tai, S. P., Hoger, J. H., and Kaplan, S., 1986, Phospholipid transfer activity in synchronous populations of *Rhodobacter sphaeroides, Biochim. Biophys. Acta* 859 : 198–208.

Takeuchi, Y., and Nikaido, H., 1981, Persistance of segregated phospholipid domains in phospholipid–lipopolysaccharide mixed bilayers: Studies with spin-labeled phospholipids, *Biochemistry* 20 : 523–529.

Takeuchi, Y., Ohnishi, S.-I., Ishinaga, M., and Kito, M., 1978, Spin-labelling of *Escherichia coli* membrane by enzymatic synthesis of phosphatidylglycerol and divalent cation-induced inter-action of phosphatidylglycerol with membrane proteins, *Biochim. Biophys. Acta* **506** : 54–63.

Takeuchi, Y., Ohnishi, S.-I., Ishivaga, M., and Kito, M., 1981, Dynamic states of phospholipids in *Escherichia coli* B. membranes. Electron spin resonance studies with biosynthetically gen-erated phospholipid spin labels, *Biochim. Biophys. Acta* **646** : 119–125.

Takemoto, J. Y., Schonhardt, T., Golecki, J. R. and Drews, G., 1985, Fusion of liposomes and chromatophores of *Rhodopseudomonas capsulata*: Effect of photosynthetic energy transfer between B875 and reaction center complexes, *J. Bacteriol.* **162** : 1126–1134.

Taneja, R., and Khuller, G. K., 1980. Ethanol-induced alterations in phospholipids and fatty acids of *Mycobacterium smegmatis* ATCC 607, *FEMS Microbiol. Lett.* **8** : 83–85.

Tecoma, E. S., and Wu, D., 1980, Membrane deenergization by colicin K affects fluorescence of exogenously added but not biosynthetically esterified parinaric acid probes in *Escherichia coli*, *J. Bacteriol.* **142** : 931–938.

Tecoma, E. S., Sklar, C. A., Simoni, R. D., and Hudson, B. S., 1977, Conjugated polyene fatty acids as fluorescent probes: Biosynthetic incorporation of parinaric acid by *Escherichia coli* and studies of phase transition, *Biochemistry* **16** : 829–835.

Thilo, L., and Overath, P., 1976, Randomization of membrane lipids in relation to transport system assembly in *Escherichia coli*, *Biochemistry* **15** : 328–334.

Thulborn, K. R., 1981, The use of n-(9-anthroyloxy) fatty acids as fluorescent probes for biom-embranes, in *Fluorescent Probes* (G. S. Beddard and M. A. West, eds.), pp. 113–139, Aca-demic Press, London.

Thulborn, K. R., and Sawyer, W. H., 1978, Properties and the location of a set of fluorescent probes sensitive to the fluidity gradient of the lipid bilayer, *Biochim. Biophys. Acta* **511** : 125–140.

Thulborn, K. R., Tilley, L. M., Sawyer, W. H., and Treloar, F. E., 1979, The use of n-(9-anthroyloxy) fatty acids to determine fluidity and polarity gradients in phospholipid bilayers, *Biochim. Biophys. Acta* **558** : 166–178.

Tornabene, T. G., Holzer, G., Bittnar, A. S., and Grohmann, G., 1982, Characterization of the total extractable lipids of *Zymomonas mobilis* var. *mobilis, Can. J. Microbiol.* **28** : 1107–1118.

Traüble, H., 1971, Phasenumi vandlungen in Lipiden Mogliche Schaltprozesse in biologischen Membranen, *Naturwissenschaften* **58** : 277–284.

Traüble, H., and Overath, P., 1973, The structure of *Escherichia coli* membranes studied by fluorescence measurements of lipid phase transitions, *Biochim. Biophys. Acta* **307** : 491–512.

Ueki, T., Mitsui, T., and Nikaido, H., 1970, X-ray diffraction studies of outer membrane of *Salmonella typhimurium, J. Biochem. (Tokyo)* **85** : 173–182.

Urbaneja, M. A., Villena, A., and Goni, F. M., 1984, The interaction of *Bacillus* properties with sonicated phosphatidylcholine liposomes, *FEBS Lett.* **169** : 40–44.

Van Alphen, L., Verkleij, A., Leunissen-Bijvelt, J., and Lugtenberg, B., 1978, Architecture of the outer membrane of *Escherichia coli*. III. Protein–lipopolysaccharide complexes in intra-membranous particles, *J. Bacteriol.* **134** : 1089–1098.

Van Alphen, L., Verkleij, A., Burnell, F., and Lugtenberg, B., 1980, ^{31}P nuclear magnetic reso-nance and freeze fracture electron microscopy studies on *Escherichia coli*. II. Lipopolysac-charide and lipopolysaccharide complexes, *Biochim. Biophys. Acta* **591** : 502–517.

Vanderkooi, J. M., 1979, Effect of ethanol on membranes: A fluorescent probe study, *Alcohol Clin. Exp. Res.* **3** : 60–63.

Van der Meer, B. W., Van Hoeven, R. P., and Van Blitterswijk, W. J., 1986, Steady-state fluorescence polarization data in membranes. Resolution into physical parameters by an ex-tended Perrin equation for restricted rotation of fluorophores, *Biochim. Biophys. Acta* **854** : 38–44.

Van Golde, L. M. G., Schulmann, H., and Kennedy, E. P., 1973, Metabolism of membrane lipids and its relation to a novel class of oligosaccharides in *Escherichia coli, Proc. Natl. Acad. Sci. U.S.A.* **70** : 1368–1372.

Van Gool, A. P., and Nanninga, N., 1971, Fracture faces in the cell envelope of *Escherichia coli, J. Bacteriol.* **108** : 474–481.

Verkleij, A. J., and Ververgaert, P. H. J. Th., 1978, Freeze-fracture morphology of biological membranes, *Biochim. Biophys. Acta* **515** : 303–327.

Verma, G. P., and Wallach, D. F. H., 1983, Raman spectroscopy of lipids and biomembranes, in *Biomembrane Structure and Function* (D. Chapman, ed.), Macmillan, London.

Waggoner, A., 1976, Fluorescent probes of membranes, in *The Enzymes of Biological Membranes, Vol. 1, Physical and Chemical Techniques* (A. Martonosi, ed.), pp. 119–137, Plenum, New York.

Waggoner, A. S., 1979, The use of cyanine dyes for the determination of membrane potentials in cells, organelles and vesicles, *Methods Enzymol.* **55** : 689–695.

Wakayama, N., and Oshima, T., 1978, Membrane properties of an extreme thermophile I. Detection of phase transition and its dependence on growth temperature, *J. Biochem.* **83** : 1687–1692.

White, D. A., Albright, F. R., Lennarz, W. J., and Schnaitman, C. A., 1971, Distribution of phospholipid-synthesizing enzymes in the wall and membrane subfractions of the envelope of *Escherichia coli, Biochim. Biophys. Acta* **249** : 236–243.

Wolf, M. K., and Konisky, J., 1981, Increased binding of a hydrophobic, photolabile probe to *Escherichia coli* inversely correlates to membrane-potential but not adenosine 5'-triphosphate levels, *J. Bacteriol.* **145** : 341–347.

Wolf, M. K., and Konisky, J., 1984, Membrane potential independent binding of azidopyrene to LPS mutants of *Salmonella typhimurium, FEMS Microbiol. Lett.* **21** : 59–62.

Yang, L. L., and Haug, A., 1979, Structure of membrane lipids and physicobiochemical properties of the plasma membrane from *Thermoplasma acidophilum*, adapted to growth at 37°C, *Biochim. Biophys. Acta* **573** : 308–320.

Yguerabide, J., and Foster, M. C., 1981, Fluorescence spectroscopy of biological membranes, in *Molecular Biology, Biochemistry and Biophysics.* Vol. 31, *Membrane Spectroscopy* (Ernst Grell, ed.), pp. 199–269, Springer-Verlag, Berlin.

Chapter 10

The Study of Cytoskeletal Protein Interactions by Fluorescence Probe Techniques

Edward Blatt and William H. Sawyer

1. INTRODUCTION

The reductionist approach to cell biology aims to explain the functioning of the whole cell in terms of the properties of its individual components. The central place of proteins in all aspects of cell activity justifies the theoretical and experimental attempts that have been made to relate amino acid sequence to molecular structure and function. Since many proteins are present in the cell in small amounts, the biochemist has resorted to microanalytical methods to obtain this information. Recent developments in instrumentation and computer technologies have greatly aided this process.

Fluorescence spectroscopy is an important tool in such endeavors. Improved detection systems now allow determination of concentrations down to the nanomolar level or lower and the availability of image enhancement methods has allowed the use of low illumination intensities in fluorescence microscopy, thus avoiding problems of photobleaching. The fluorescence spectroscopy of proteins began in the 1950s with the pioneering work of Weber (1953). Since that time, the technique has diversified, and there now exist a number of related methods within the technique that can provide information on protein structure and dynamics. It is the intention of this

Edward Blatt   Division of Applied Organic Chemistry, CSIRO, Melbourne, Victoria 3001, Australia.   **William H. Sawyer**   Russell Grimwade School of Biochemistry, University of Melbourne, Parkville, Victoria 3052, Australia.

chapter to review these techniques as related to the study of cytoskeletal protein interactions. In particular, we concentrate on the polymerization properties of actin and its interactions with other cytoskeletal proteins. Section 2 presents a brief description of the dynamic interactions of cytoskeletal components. Section 3 begins by introducing the notions underlying the fluorescence probe technique. This is followed by a detailed examination of specific experimental methods as applied to the study of cytoskeletal interactions. In Section 4, the experimental results of two related luminescence techniques are discussed and shown to complement earlier fluorescence data. Finally, a summary of the chapter is provided, followed by a brief discussion of future directions.

2. THE CYTOSKELETON

2.1. Organization of Cytoskeletal Proteins

There is a great diversity in the number and types of cytoskeletal structures found in cells. It is clear that the major components of the cytoskeleton—namely, actin and tubulin—can be assembled into a large assortment of structural forms involving microfilaments and microtubules, respectively. Intermediate filaments composed of keratinlike proteins add to the diversity of structures. Functionally, these forms control cell shape and motility and may act to position and direct the movements of subcellular organelles. In addition, the cytoskeleton acts as a scaffolding on which is arrayed a number of cytoplasmic enzymes. Thus, components of the cytoplasm, be they small molecules, macromolecules, or subcellular organelles, are not necessarily

Abbreviations used in this chapter: α, fractional amplitude of correlation time; E, efficiency of energy transfer; η, viscosity; I, emission intensity in the presence of quencher or acceptor molecule; I_0, emission intensity in the absence of quencher or acceptor molecule; k_q, bimolecular collisional rate constant; K_{SV}, Stern–Volmer constant; p, steady-state polarization; p_0, limiting polarization at infinite viscosity; r_0, fundamental anisotropy; r_x, residual anisotropy; $r(t)$, decay of anisotropy; R, gas constant; R_0, distance at which energy transfer efficiency is 50%; T, temperature; τ, fluorescence lifetime in the presence of quencher or acceptor molecule; τ_0, fluorescence lifetime in the absence of quencher or acceptor molecule; θ, rotational correlation time; V, effective volume of rotating sphere; ϵ-ATP, 1-N^6-ethenoadenosine 5'-triphosphate; DABMI, 4-dimethylaminophenylazophenyl-4'-maleimide; DDPM, N-(4-dimethylamino-3,5-dinitrophenyl)maleimide; DNS, 5-dimethyl-amino-naphthalene-I-sulfonyl-L-cystine; EIA, 5-iodoacetamide-eosin; EITC, eosin isothiocyanate; FITC, fluorescein isothiocyanate; G3DP, glyceraldehyde phosphate dehydrogenase; IAEDANS, N-(iodoacetyl)-N'-(1-sulfo-5-naphthyl)ethylenediamine; IAF (or 5-AIF), 5-(iodoacetamido)fluorescein; IANDB, 4-(N-(iodoacetoxy)ethyl-N-methyl)amino-7-nitrobenzo-1,2,3-oxadiazole; IATR, (iodoac-etamido)tetramethylrhodamine; NBD-chloride, 7-chloro-4-nitrobenzo-2-oxa-1,3-diazole; NPI, N-(1-pyrenyl)iodoacetamide; NPM, N-(3-pyrenyl)maleimide; TNP, 2'-O-(2,4,6-trinitrophenyl)adenosine 5'-diphosphate; FPR, fluorescence photobleaching recovery; TAA, transient absorption anisotropy.

free to diffuse within the cytoplasmic space. Rather, they may have restricted position and motion by virtue of the high viscosity of the cytoplasmic milieu and the physical restrictions imposed by the cytoskeletal framework. Moreover, it is now recognized that the chemical activity of components in the cytoplasm may be quite high due to the excluded volume effect provided by the macromolecules present (Harris and Winzor, 1985).

In the case of microfilaments, the diversity of structures appears to be determined by the types and amounts of actin-binding proteins present in a cell. Four classes of such proteins have been identified (Weeds, 1982). The first includes those proteins that form complexes with monomeric or G-actin, thereby preventing the polymerization of G-actin into its filamentous form (e.g., profilin, DNase I). The second class of compounds includes those that assist the nucleation of actin complexes, so shortening the lag phase in the association kinetics. Proteins that cap either the barbed or pointed end of the actin filament comprise the third group of proteins (e.g., gelsolin). The fourth group consists of those proteins that can bundle, cross-link, and in some cases sever actin filaments.

Compared to the long-lived intermediate filaments, the microfilaments and microtubules are dynamic structures that can assemble and disassemble according to the needs of the cell. An example is the formation of microtubules to form the mitotic spindle. The change in the arrangement of stress fibers, which is observed when a fibroblast changes from a resting to a mobile cell, provides a further example. The control of these changes is poorly understood at present. Calcium would appear to be one possible modulator controlling the interaction of some proteins with actin.

The structure of the red cell cytoskeleton has been the subject of detailed study and has been reviewed elsewhere (Steck, 1974; Bennett, 1985; Marchesi, 1985). Figure 1 is a schematic representation of some of the major features of the erythrocyte cytoskeleton. Spectrin forms a basic structural element and consists of heterodimers joined head-to-head to form a ribbonlike tetramer. Each end of the tetramer can interact with actin, which in the erythrocyte consists of oligomeric structures of 8–15 monomer units. The multivalency of the spectrin tetramer and the actin oligomer allows the cross-linking of these molecules to form an extensive network that lies on the cytoplasmic face of the membrane. The network is stabilized by protein band 2.1 (ankyrin), which forms a bridge between spectrin and an intrinsic membrane protein (band 3), and by band 4.1, which appears to link spectrin to another intrinsic protein (glycophorin). Actin, spectrin, and band 4.1 form a core structure that is believed to be responsible for maintaining the shape and mechanical properties of the cell by forming the extensive cross-linked network. Evidence now exists that these three proteins form a ternary complex *in vitro* (Cohen and Foley, 1984; Ohanian *et al.*, 1984). Analogues of spectrin

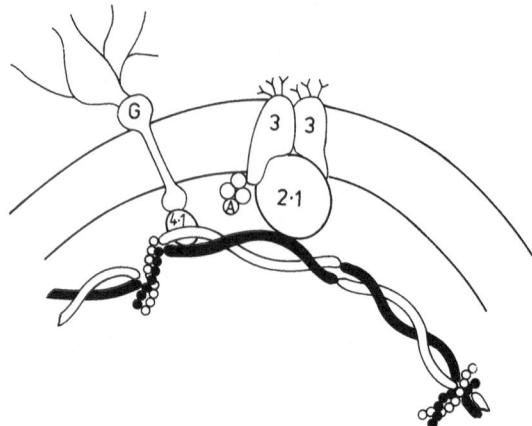

FIGURE 1. Schematic representation of the major proteins associated with the erythrocyte cytoskeleton. The flexible, ribbonlike spectrin dimer forms a head-to-head tetramer that binds actin oligomers at opposite ends of the structure (G = glycophorin; A = aldolase or glyceraldehydephosphate dehydrogenase).

and bands 2.1 and 4.1 have been detected in other nonerythroid cells (Goodman *et al.*, 1981; Cohen *et al.*, 1982; Davis and Bennett, 1983, 1984). Thus, the interactions involved in the erythrocyte cytoskeleton may not be as unique as first thought. Traces of myosin and tropomyosin have also been discovered in the erythrocyte but their role in the organization of the cytoskeleton is not known, although it is possible that they are present in sufficient quantities to form a contractile element capable of controlling cell shape (Fowler and Bennett, 1984; Fowler *et al.*, 1985; Fowler, 1986).

2.2. Assembly of Actin Filaments

Before reviewing applications of fluorescence in the study of actin assembly, it is appropriate to summarize some of the characteristics of this polymerization. G-actin is a globular protein of 375 amino acid residues (MW 41,872). In the presence of Mg^{2+} ions and physiological concentrations of KCl, monomers self-associate in a head-to-tail manner to form double-stranded helical filaments several micrometers in length and with a diameter of about 5 nm. A critical actin concentration exists, below which no association occurs and above which filaments are in equilibrium with the monomeric form. The filament possesses polarity, and "barbed" and "pointed" ends can be distinguished in the electron microscope on the basis of the arrowlike decoration that occurs with heavy meromyosin. The elongation kinetics are complex since monomers can be added at both the barbed and pointed ends, albeit at different rates. One consequence of the difference in assembly rates is that the monomers can cycle through the filaments from the barbed to the pointed end, a phenomenon that has been called treadmilling. ATP hydrolysis is linked to the association of monomer units, although recent experiments have shown that monomer units at the ends of the filament at steady state may bind ATP rather than ADP (Pantaloni *et al.*, 1985).

A number of methods exist to study the mechanism, extent, and rates of polymerization and depolymerization of actin (Cooper and Pollard, 1982). Four reversible steps have been identified: (1) activation, which involves an initial conformational change of the protein; (2) nucleation, involving the association of monomers to form an oligomer; (3) elongation into filaments; and (4) annealing, where filaments break and join to one another end to end. The first two steps are much slower than the elongation process and often cause lag times in experiments that measure rates of polymerization. During elongation, monomers attach to either end of the filament at different rates. At equilibrium, the first-order dissociation rate equals the second-order association rate and the concentration of monomer is at the critical concentration. Several methods are available to characterize individual F-actin molecules and F-actin networks (Pollard and Cooper, 1982).

3. FLUORESCENCE PROBE TECHNIQUES

3.1. Introduction

In its simplest form, the fluorescence probe technique involves measurement of the emission and/or absorption properties of the probe in the host structure of interest. A knowledge of the photophysical properties of the probe may then allow conclusions to be made about the microenvironment of the probe. Fluorescence experiments of this kind in proteins can be performed with three types of probe: extrinsic covalently bonded molecules, extrinsic noncovalently associated molecules (i.e., adsorbed at the surface or bound at specific sites on the protein), or intrinsic fluorophores. The first two approaches have the advantage of enabling the use of fluorescent molecules with particularly suitable spectroscopic properties. The possibility of perturbation or modification of the protein structure and function, however, must be addressed. This is not the case with experiments involving the third approach. Unfortunately, the complex photophysical behavior of intrinsic probes such as tryptophan in solution and within proteins (Cundall and Dale, 1983; Beechem and Brand, 1985) makes interpretation of data difficult.

Apart from the simple observation of the photophysical properties of the probe, a number of complementary techniques are available. These can involve the use of more than one covalently attached extrinsic probe or the addition of nonfluorescent molecules and ions such as quenchers. In this section these techniques are discussed in detail, particularly as they apply to the study of cytoskeletal protein interactions. In each case the type of information obtainable is highlighted by reference to experimental data.

3.2. Energy Transfer

Fluorescence energy transfer is an effective technique for determining proximity relations in biological macromolecules and in molecular structures such as multienzyme complexes and biomembranes. The underlying principle is that electronic excitation energy will be transferred from a donor molecule to an acceptor molecule when the fluorescence emission spectrum of the former significantly overlaps the absorption spectrum of the latter. The efficiency of the energy transfer, E, is related to the distance between the centers of the donor and acceptor chromophores, r, by

$$E = r^{-6}/(r^{-6} + R_0^{-6}) = 1 - I_T/I_0 = 1 - \tau_o/\tau_T \qquad (1)$$

where R_0 is the distance at which the transfer efficiency is 50% and I_T, τ_T and I_0, τ_0 are the fluorescence quantum yields (or steady-state intensities) and the fluorescence lifetimes of the energy donor in the presence and absence of the acceptor, respectively. As a result of the requirement for sufficient spectral overlap, extrinsic covalent chromophores are frequently used in energy transfer experiments, and a number of suitable donor–acceptor pairs are available (Stryer, 1978). An important advantage to the biochemist is that energy transfer experiments can be conducted in reconstituted mixtures of complex composition, where the traditional methods for following solute associations cannot be used (e.g., viscosity measurements, analytical ultra-centrifugation).

Of the many proteins in the cytoskeleton, fluorescence energy transfer experiments with actin have been the most common. The single polypeptide chain of actin contains five sulfhydryl groups per globular monomer at positions 10, 217, 257, 285, and 374 (Elzinga et al., 1973; Vanderkerckhove and Weber, 1979). Of these five, Cys-374 reacts the most readily, and with no apparent biological impairment, with molecules containing the maleimide moiety (Martonosi, 1968). This feature of labeling specificity has been exploited in a number of investigations into the spatial relation between Cys-374 and the nucleotide binding site of actin.

Miki and Mihashi (1978) examined the energy transfer between the fluorescent ATP analogue donor $1-N^6$-ethenoadenosine $5'$-triphosphate (ϵ-ATP), bound to the nucleotide-binding site of G-actin, and the acceptor N-(4-di-methylamino-3,5-dinitrophenyl)maleimide (DDPM), bound to Cys-374. The efficiency of energy transfer from both steady-state intensity and lifetime measurements was approximately 0.57, corresponding to a distance of about 30 Å. In subsequent papers by Miki and Wahl (1984) and Miki and Iio (1984), the energy acceptors 7-chloro-4-nitrobenzo-2-oxa-1,3-diazole (NBD-chloride) and 4-(N-(iodoacetoxy)ethyl-N-methyl)amino-7-nitrobenzo-1,2,3-oxadiazole (IANBD), which also bind to Cys-374, were used. The efficiencies of energy transfer between ϵ-ATP and IANBD or NBD-chloride were greater than that between ϵ-ATP and DDPM because of the greater overlap between the emission and absorption spectra. This resulted in interenergy transfer from adjacent actin monomers upon polymerization of G-actin to F-actin. Subsequent to these investigations, Miki and Wahl (1985) studied the spatial relations between the nucleotide-binding site, the metal-binding site, and the Cys-374 residue of the G-actin molecule. They found that Co^{2+} or Ni^{2+} completely quenched the fluorescence intensity of ϵ-ATP, indicating close proximity of the metal- and nucleotide-binding sites. The transfer efficiency between N-(iodoacetyl)-N'-(1-sulfo-5-naphthyl)ethylenediamine (IAEDANS) bound to Cys-374 and Co^{2+} was approximately 5%, indicating a distance between the metal-binding site and the Cys-374 residue of around 30 Å.

From the above and the more recent energy transfer experiments of Miki and Wahl (1984, 1985), the spatial relations between the nucleotide-binding site and the Cys-374 residue of adjacent actin monomers on an F-actin polymer and the distances between these sites and the metal-binding site of a G-actin monomer have been calculated. Figure 2 shows the results obtained by these authors in diagrammatic form.

Proximity and structural relations have also been sought between F-actin and the biologically active fragments of myosin. Both proteins can be labeled at two sites, a reactive sulfhydryl and a bound nucleotide. Changes in the structural relation between actin and the fast-reacting thiol (SH_1) of myosin subfragment (S-1) are believed to occur during muscle contraction (Geeves et al., 1984). Takashi (1979) investigated the distance between SH_1 and Cys-374 in the rigor complex of acto-S-1 by determining the energy transfer from IAEDANS-labeled actin to 5-(iodoacetamido)fluorescein (IAF)-labeled myosin. A distance of approximately 6.0 nm was obtained. In a later and more detailed study, Dos Remedios and Cooke (1984) determined the distances between different sites on actin and myosin through the use of a number of labels. The results are summarized in Table I. On the basis of the data of Table I and from previous measurements, these authors concluded that the SH_1 and the ATPase sites on myosin are not located adjacent to actin but are probably located in the half of the myosin head that is distal from actin in the acto-myosin complex.

As mentioned earlier, G-actin contains five sulfhydryl groups of which Cys-374 is the most reactive. Recent work by Barden et al. (1986) showed

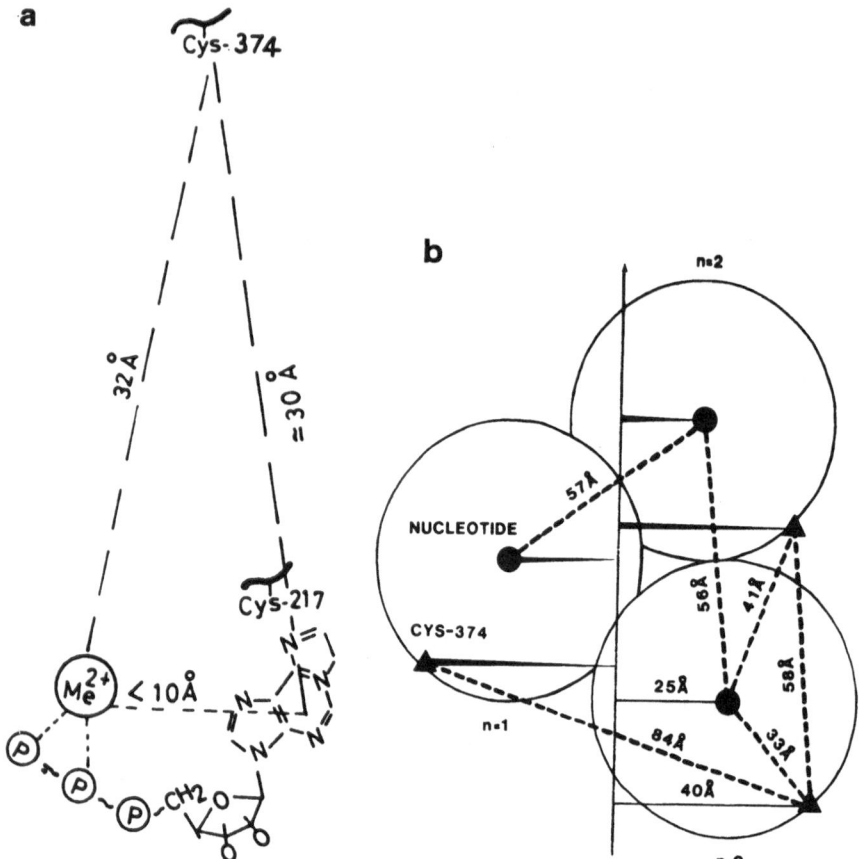

FIGURE 2. Spatial relations among specific sites in G- and F-actin derived from fluorescence energy transfer experiments. (a) Triangulation of the nucleotide and metal ion binding sites with Cys-374 in G-actin. (b) One possible arrangement of Cys-374 and the nucleotide binding sites in F-actin. Reprinted by copyright permission from *Biochim. Biophys. Acta* **828** : 188 (1985) and **871** : 137 (1986).

that upon blocking this reactive site the selective modification of Cys-10 could be achieved without affecting the formation of filaments. The following series of experiments were subsequently performed to elucidate spatial relations between different sites on the actin. In the first, the Cys-10 residues of F-actin were labeled with IAEDANS as the fluorescence energy donor and 4-dimethylaminophenylazophenyl-4′-maleimide (DABMI) as the acceptor (Miki *et al.*, 1986a). The radial coordinate of Cys-10 was calculated as 23 Å, which corresponds to a distance between adjacent sites along the long pitch helix of 56.1 Å and along the genetic helix of 53.3 Å. Secondly, energy transfer experiments performed by Miki *et al.* (1986b) between IAEDANS and Co^{2+}

Table I
Energy Transfer Data between Sites on Actin and Myosin[a]

Experiment[b]	Donor	Acceptor	R_0 (nm)	R (nm)
1	IAEDANS(myosin)	TNP-ADP(F-actin)	4.03	7.5–6.1
2	IAEDANS(myosin)	TNP(F-actin)	3.05	6.6
3	e-ADP(F-actin)	TNP-ADP(myosin)	5.1	10.9–7.7
4	e-ADP(myosin)	TNP-ADP(F-actin)	4.8	6.6–6.2
5	IAEDANS(F-actin)	TNP-ADP(myosin)	4.03	5.8–5.1
6	IAEDANS(F-actin)	TNP-ADP(F-actin)	4.03	3.4

[a]Reprinted with permission from Dos Remedios, C.G., and Cooke, R. *Biochim. Biophys. Acta* **788** : 198 (1984), Table I. Copyright 1984, Elsevier Science.
[b]Experiments refer to: (1) from the SH$_1$ groups of the myosin heads to the nucleotide sites of F-actin; (2) from the SH$_1$ groups of myosin to multiple probes on the surface of the actin filament; (3) from the nucleotide-binding sites of F-actin to the ATPase sites of myosin; (4) from the ATPase sites of myosin to the nucleotide-binding sites of F-actin; (5) from the SH$_1$ sites of myosin to the nucleotide-binding sites of F-actin; (6) from the Cys-374 residues of F-actin to the nucleotide binding sites of F-actin.

bound to the high-affinity metal-binding site indicated a separation distance of approximately 3.0 nm. Finally, Miki *et al.* (1986c) measured the energy transfer between donor ε-ATP bound to the nucleotide site and either DABMI or DDPM conjugated to Cys-10. The donor–acceptor distance was calculated to be approximately 40 Å. Also, intermonomer energy transfer between ε-ADP and DABMI indicated a similar radial coordinate (25 Å) to that obtained earlier by Miki *et al.* (1986b). Corresponding distances separating the donor nucleotide in one monomer from acceptors on Cys-10 in the first and second nearest neighbors in F-actin were calculated as 39–40 Å and 41–43 Å, respectively. A recent review by Dos Remedios *et al.* (1987) discusses in detail the spatial relations obtained by these authors and others between actin monomers and filaments (as well as between myosin and the actomyosin complex).

Apart from proximity relations in actin, fluorescence energy transfer can be used to measure the kinetics of actin polymerization. Figure 3 shows the experimental strategy generally employed to monitor the polymerization of G-actin to F-actin and the exchange of subunits between F-actin filaments. Initially, a predetermined ratio of donor-labeled and acceptor-labeled actin monomers are mixed together and the polymerization initiated by addition of salt. As the monomers coassemble, energy transfer occurs between adjacent donor- and acceptor-labeled monomers. The probability of donor- and acceptor-labeled monomers approaching to within transfer distance depends on the proportion of nonlabeled monomers present. The exchange of monomer units between filaments may then be evaluated by adding unlabeled F-actin.

This procedure was adopted by Taylor *et al.* (1981). Actin was labeled at Cys-374 with the energy donor 5-iodoacetamidofluorescein or the energy acceptors eosin iodoacetamide or tetramethylrhodamine iodoacetamide. In

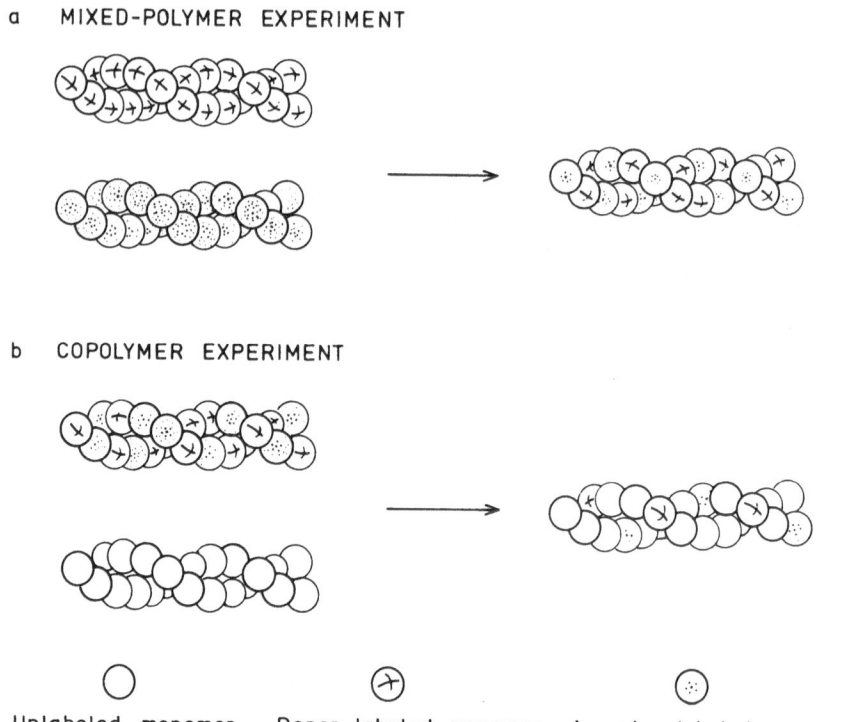

a   MIXED-POLYMER EXPERIMENT

b   COPOLYMER EXPERIMENT

○ Unlabeled monomer   ⊕ Donor–labeled monomer   ⊙ Acceptor–labeled monomer

FIGURE 3. Experimental strategy for measuring the kinetics of polymerization of actin and the exchange of subunits among actin filaments.

Figure 4 the time course of the polymerization shows that in both cases equilibrium was reached after approximately 20 min. Greater energy transfer was observed with eosin because its absorption spectrum overlaps more with the emission spectrum of fluorescein. Values for the spatial relations closely resembled those obtained by Miki and Wahl (1984) discussed above and shown diagrammatically in Figure 2a. Gentle addition of unlabeled filaments to the mixture resulted in only a small decrease in the transfer efficiency; however, upon sonication almost complete loss of energy transfer occurred. These results were interpreted as arising from an increase in the number of ends, and therefore the monomer exchange rate, brought about by mechanical shear, and are consistent with the increase in ATPase activity when filaments are sheared in the same way (Asakura, 1961; Nakaoka and Kasai, 1969).

An important application of the energy transfer method has been the examination of monomer exchange between actin filaments. Two types of experimental design are common: a *mixed polymer* experiment in which donor-labeled filaments are mixed with acceptor-labeled filaments, and a *co-*

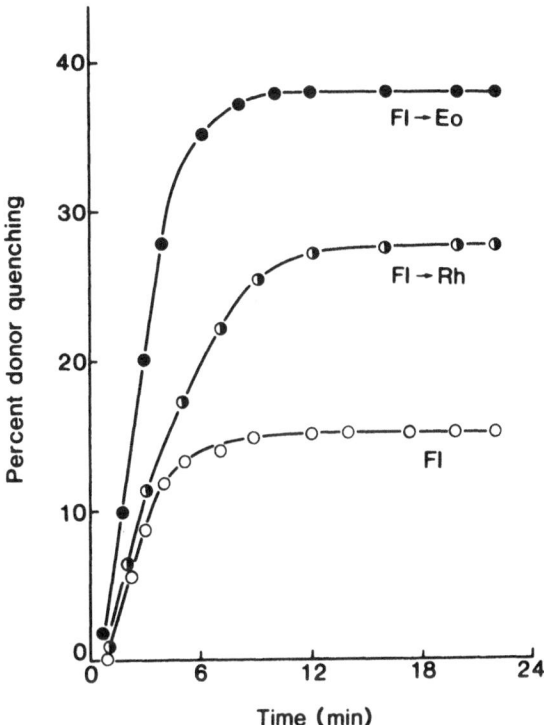

FIGURE 4. The coassembly of fluorescein (Fl)-labeled actin with eosin (Eo)- or rhodamine (Rh)-labeled actin as monitored by fluorescence energy transfer. The efficiencies of energy transfer were calculated taking into account the quenching caused by the formation of F-actin when only fluorescein-labeled actin was used. Reprinted by copyright permission from *J. Cell Biol.* **89** : 362 (1981).

polymer experiment in which donor–acceptor-labeled filaments are mixed with unlabeled filaments. Monomer exchange between filaments results in an increase in energy transfer (i.e., donor quenching) in the mixed polymer experiment, and a decrease in transfer efficiency (i.e., a release of donor quench) in the copolymer experiment. Little monomer exchange is detected if care is taken to mimic physiological conditions as closely as possible (Pardee *et al.*, 1982). The ionic conditions of the buffer can markedly affect the degree of monomer exchange observed (Wang and Taylor, 1981).

The effects of actin-binding molecules on the extent and kinetics of filament assembly and disassembly have also been examined using the copolymer and mixed polymer systems described above (Figure 5). Cytochalasin B decreases the rate of disassembly and prevents monomer exchange under ionic conditions, which favor the exchange (Wang and Taylor, 1981).

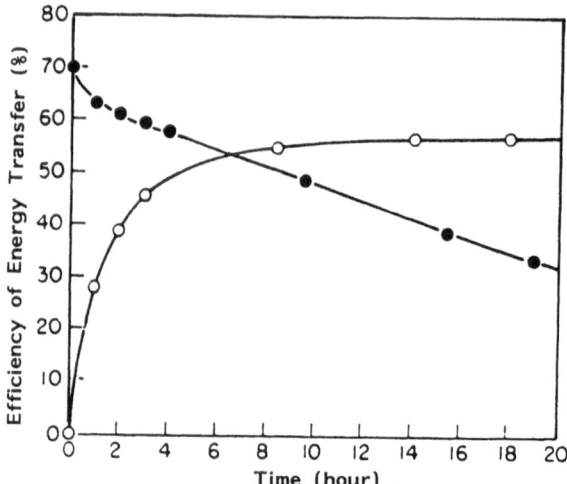

FIGURE 5. Exchange of subunits among actin filaments in the polymerization buffer containing 2.5 mM $MgCl_2$: (O——O) mixed polymer experiment; (●——●) copolymer experiment. The experimental strategies are depicted in Figure 3. Reprinted by copyright permission from *Cell* **27** : 429 (1981).

At the concentration used (1 μM), cytochalasin B occupies a class of high-affinity binding sites. The results were interpreted as indicating that cytochalasin B blocks the fast-growing end of the filament as previously proposed (Brown and Spudich, 1981). The energy transfer method has also been used to examine the effects of an actin-binding protein from *Dictyostelium discoideum* on the fragmentation of actin filaments (Spudich *et al.*, 1982; Yamamoto *et al.*, 1982).

The effect of band 4.1 on the state of cross-linked actin–spectrin complexes has been examined by Husain *et al.* (1983) using IAEDANS-labeled and fluorescein-5-maleimide-labeled actin as donors and acceptors, respectively. A 10% decrease in transfer efficiency was observed irrespective of whether band 4.1 was added to the system before or after polymerization of the actin, indicating that the final state of the cross-linked system is determined by a thermodynamic equilibrium. Further experiments by these authors showed that band 4.1 induces a limited depolymerization of actin filaments and a concomitant rise in the critical actin concentration. A decrease in the filament length could also contribute to these results. Although the mechanism of the depolymerization was not elucidated, the results may explain in part the existence of actin in the erythrocyte cytoskeleton as short oligomers rather than as long filaments.

3.3. Fluorescence Enhancement

Fluorescence molecules whose spectral characteristics are strongly dependent on environmental conditions are commonly used to detect conformational changes of proteins (Edelman and McClure, 1968). Pyrene is one such fluorophore that has been used extensively to investigate the structural and dynamic nature of a diverse range of macromolecular assemblages (Nakajima, 1977; Lee and Meisel, 1985; Hite *et al.*, 1986). The initial discovery that the fluorescence quantum yield and lifetime(s) of pyrene-labeled actin increased sharply upon formation of actin filaments (Kawasaki *et al.*, 1976) led to the notion that the polymerization process could be monitored by the fluorescence enhancement. Experiments using the intrinsic fluorescence of actin showed relatively small changes upon polymerization (Lehrer and Kerwar, 1972).

Early experiments by Kouyama and Mihashi (1980, 1981) involved the labeling of the Cys-374 residue of G-actin with *N*-(1-pyrene)maleimide or *N*-(1-pyrenyl)iodoacetamide (NPI). Significant changes in the fluorescence decay characteristics were observed upon polymerization and the fluorescence intensity increased by a factor of approximately 25 for the latter probe. It was suggested that a conformational change of the actin protomer at the time of association was responsible for the dramatic fluorescence enhancement. Binding of heavy meromyosin after polymerization caused a 25% reduction in intensity; this was interpreted as being due to local structural changes of the F-actin protomer toward a conformation more similar with that of G-actin.

Having established the fluorescence enhancement of the pyrene-labeled G-actin upon filament formation, it was envisaged that the method might allow quantitative determination of the extent and rate constants for the polymerization process in the presence and absence of actin-binding proteins. In a detailed study, Cooper *et al.* (1983) showed that the pyrenyl–actin assay was equivalent to the nonperturbing light-scattering method for measuring the kinetics of actin association. Tellam and Frieden (1982) studied the time course of fluorescence enhancement of NPI-labeled actin in the presence and absence of cytochalasin D under a variety of experimental conditions (e.g., as a function of temperature, concentrations of salt, actin, or cytochalasin D). Figure 6 shows typical data obtained by these authors. In this particular example, the concentration of labeled actin was varied over a range of two orders of magnitude in order to assess the effects of the label. The similarity of the plots indicates that the polymerization characteristics of actin were

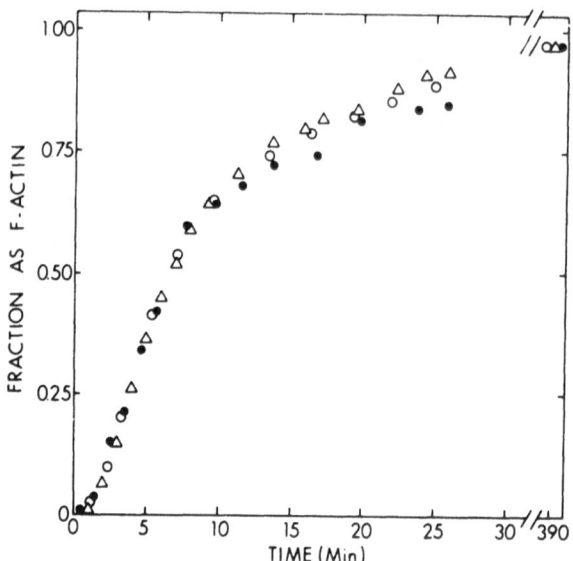

FIGURE 6. Time course of fluorescence enhancement upon polymerization of actin by 2 mM Mg^{2+} at various concentrations of NPI-actin [0.05% (●), 0.50% (○), 5.0% (△)]. Reprinted with permission from *Biochemistry* **21** : 3209 (1982), Figure 2. Copyright 1982, American Chemical Society.

largely unchanged by the label. Under all the experimental conditions these authors found that cytochalasin D accelerated the rate of polymerization as reflected by an enhanced rate of incorporation of monomer into polymer and/ or a decreased lag time. The extent of polymer formation, however, decreased; this was reflected by the increased critical actin concentrations. Since cytochalasin D binds to the fast-growing end of the actin filament, thereby inhibiting elongation, the results suggested that the enhanced rate and decreased lag time occurs at the nucleation step either due to an increased number of nuclei in equilibrium with monomer or to a faster rate of nuclei formation.

Using similar methodology to that presented above, Pinder and Gratzer (1982) investigated the interaction of deoxyribonuclease I (DNase I) with actin. Preincubation of F-actin with DNase at a concentration of 5% or less of that of total subunits caused inhibition of polymerization of additional G-actin onto the filaments. In red cell membranes preincubation of F-actin with DNase resulted in complete inhibition of nucleated polymerization of G-actin. These authors concluded that DNase binds not only to G-actin but also to the preferentially growing ends of F-actin filaments.

The fluorescence enhancement technique has also found application in the study of the interactions between actin and other cytoskeletal components.

Elbaum *et al.* (1984) investigated the effects of the spectrin dimer–band 4.1 complex on the polymerization of actin using the NPI–actin derivative. In the presence of a 1 : 1 molar ratio of spectrin dimer and band 4.1, the rate of actin nucleation was enhanced (i.e., the time of the lag phase decreased) and little change in the rate of elongation in the first 15 min was observed. Consequently, after 75 min, no further changes in the fluorescence occurred and the final relative intensities were identical. The effects of varying the ratio of the complex, spectrin alone, or band 4.1 alone on the length of the lag phase at different actin concentrations were also examined. Of particular interest was the observation that when the stoichiometry of actin, spectrin, and band 4.1 approached the physiological range of 5 : 1 : 1, the spectrin–band 4.1 complex more effectively changed the nucleation process than either spectrin or band 4.1 alone. The spectrin–band 4.1 complex was also shown to stabilize actin filaments but had no effect on the critical actin concentration.

These results are at variance with those of Tilley and Ralston (1987), who observed very little effect of spectrin dimer and tetramer on the kinetics of actin polymerization as measured by the pyrenyl–actin assay. These authors observed a small but consistent decrease in the critical actin concentration caused by spectrin and by a high molecular weight complex of spectrin, actin, and band 4.1. They were able to show that these effects could be explained on thermodynamic grounds and could result from the removal of F-actin into an actin–spectrin complex. Pinder *et al.* (1984) also observed a reduction in the critical actin concentration in the presence of spectrin and band 4.1. However, comparisons of data from different laboratories is frequently difficult because of differences in experimental design and solvent conditions.

The studies so far examined in this section have provided qualitative information on the rate of polymerization of actin and the effects of additives as determined by the pyrenyl–actin assay. Quantitative determination of rate constants by computer simulation and/or extrapolation of specific plots have also found application. In a series of papers by Frieden and co-workers (Frieden, 1983; Frieden and Goddette, 1983; Zimmerle and Frieden, 1986), the temperature-dependent kinetics of Mg^{2+}-induced actin polymerization in the absence and presence of Ca^{2+} was determined by fitting the fluorescence enhancement data of NPI (e.g., see Figure 6) with computer simulations. The mechanism chosen to represent the polymerization process allowed competitive binding between a "tight" metal-binding site on the actin monomer and either Mg^{2+} or Ca^{2+}, and incorporated a second weaker binding site for the Mg^{2+}. The mechanism is represented by the following scheme:

$$1.\ \ A + Ca \underset{K_{Mg}}{\overset{K_{Ca}}{\rightleftharpoons}} ACa$$

2. $A + Mg \underset{k_{-1}}{\overset{k_{+1}}{\rightleftharpoons}} AMg \rightleftharpoons A'Mg$

3. $A'Mg + Mg \overset{K_{Mg}}{\rightleftharpoons} A'(Mg)_2$

4. $A'(Mg)_2 + A'(Mg)_2 \underset{k_{-2}}{\overset{k_{+2}}{\rightleftharpoons}} [A'(Mg)_2]_2$

5. $[A'(Mg)_2]_2 + A'(Mg)_2 \underset{k_{-3}}{\overset{k_{+3}}{\rightleftharpoons}} [A'(Mg)_2]_3$

6. $[A'(Mg)_2]_n + A'(Mg)_2 \underset{k_{-e}}{\overset{k_{+e}}{\rightleftharpoons}} [A'(Mg)_2]_n$

where A is monomeric G-actin, ACa is G-actin containing bound Ca^{2+}, A'Mg is a conformationally altered form of G-actin, and k_e and k_{-e} are the elongation and disassembly rate constants. Table II lists some of the fitted kinetic and thermodynamic constants obtained by these authors. Of particular significance is the result that for polymerization to occur Mg^{2+} must bind to both the strong and weak binding sites of the G-actin. The conformational change induced by the binding of Mg^{2+} is driven by entropy and results in the expulsion of bound water molecules. Also of interest are the results that dimer formation is much less favorable than trimer formation and that elongation originates from the trimer.

An alternate means of measuring the rate constants for actin filament elongation using the fluorescence enhancement of NPI-labeled actin has been pre-

Table II
Kinetic and Thermodynamic Constants for the Polymerization of Actin in the Presence and Absence of Ca^{2+a}

Kinetic constant[b]	ΔH (kcal/mol)	ΔS (cal/mol·K)	ΔG^a (kcal/mol)
$K_{Ca} = 20\ \mu M$	-14.7	-53.7	1.0
$K_{Mg} = 900\ \mu M$	4.3	-0.8	4.5
$k_{+1} = 0.33\ sec^{-1}$	—	—	—
$k_{-1} = 0.01\ sec^{-1}$	—	—	—
k_{+1}/k_{-1}	11.7	45.0	-1.5
$K'_{Mg} = 5\ mM$	-13.0	59.7	-30.5
$k_{+2}/k_{-2} = 1.25\ M^{-1}$	64.3	189.5	8.8
$k_{+3}/k_{-3} = 2 \times 10^5\ M^{-1}$	—	—	—
$k_{+e}/k_{-e} = 6.7$	3.8	16.6	-1.1

[a]Reprinted with permission from Zimmerle, C.T., and Frieden, C. *Biochemistry* **25** : 6435 (1986), Table III. Copyright 1982, American Chemical Society.
[b]Values calculated at 20°C.

sented by Pollard (1983). The method was based on that previously employed by Johnson and Borisy (1977), who examined the self-assembly of microtubules by light scattering techniques. Plots of the initial rate of polymerization against actin concentration allows calculation of the critical concentration, the association rate constant (k_+), and the dissociation rate constant (k_-). Values of the two rate constants varied according to the buffer conditions and ranged from 0.5×10^7 to 2.1×10^7 $M^{-1}sec^{-1}$ for k_+ and from 5 to 13 sec^{-1} for k_-. The method also allows determination of the number of actin monomers per filament: a value of 3100 was obtained, corresponding to a filament length of 8.2 μm. As noted by Pollard (1983), this filament length is much larger than those obtained by other physical techniques. The difference is probably due to the nature of the fluorescence enhancement technique, which minimizes disturbances to the sample under investigation. The drawbacks of other methods such as electron microscopy, flow bifringence, capillary viscometry, and light scattering have been discussed by Cooper *et al.* (1983).

3.4. Anisotropy

The principles underlying the fluorescence anisotropy technique have been reviewed extensively and are dealt with in earlier chapters of this book. The basic approach adopted for the study of actin dynamics has been to label a specific site with a probe with an appropriate lifetime and to measure either the time-resolved anisotropy or the steady-state polarization of the probe under varying experimental conditions. The former measurement gives direct information on the rotational mobility of the probe, which reflects the dynamics of the labeled site and/or the protein as a whole. Unfortunately, the anisotropic environment, physical restrictions, and multiplicity of the fluorescence lifetimes of the probe can lead to complications in the interpretation of experimental data (Heyn, 1979; Jähnig, 1979; Kinosita *et al.*, 1982; Reed *et al.*, 1985; Blatt and Jovin, 1986). Nevertheless, as a qualitative tool the technique offers a powerful means of studying protein dynamics, particularly when comparisons under different conditions are sought. Steady-state polarization measurements provide less information; however, these experiments are simpler and less time-consuming, and the equipment is considerably cheaper.

In general, the time-dependence of the fluorescence anisotropy $r(t)$ may be expressed in the form (Tao, 1969; Blatt and Jovin, 1986)

$$r(t) = (r_0 - r_\infty) \sum_{i=1}^{j} \alpha_i \exp(-t/\theta_i) + r_\infty \qquad (2)$$

where r_0 is the fundamental anisotropy in the absence of rotation (and is dependent only on the angle between the emission and absorption transition moments), r_∞ is the residual anisotropy (Heyn, 1979; Jähnig, 1979), θ_i and α_i are the rotational correlation times and the corresponding fractional amplitudes, respectively, and $j < 6$. For the simplest case of a sphere rotating isotropically, Eq. (2) reduces to (Blatt and Jovin, 1986)

$$r(t) = r_0 \exp(-t/\theta) \tag{3}$$

In Eq. (3), θ is equal to $V\eta/RT$, where V is the volume of the rotating sphere and η is the medium viscosity. The steady-state polarization p of a rigid sphere is given by the Perrin equation:

$$(1/p - 1/3) = (1/p_0 - 1/3)(1 + \tau/\theta) \tag{4}$$

where p_0 is the limiting polarization at infinite viscosity.

Early steady-state polarization data of the intrinsic tryptophans of actin (Weltman et al., 1972) showed that the polarization of F-actin was appreciably greater than that of G-actin, indicating a decrease in tryptophan motion. Cheung et al. (1971) measured the temperature dependence of the polarization of 5-dimethyl-amino-naphthalene-I-sulfonyl-L-cystine (DNS) covalently attached to Cys-374 of G-actin. A rotational relaxation time of $\theta = 44.7$ nsec was obtained, in good agreement to that predicted for a spherical molecule of molecular weight 47,000. These authors also showed that upon polymerization of G-actin or addition of heavy meromyosin, the polarization was temperature independent, indicating a lack of temperature-induced mobilization of the probe. Following these studies, Weltman et al. (1973) synthesized the pyrene derivative N-(3-pyrenyl)maleimide (NPM). Upon conjugation to actin, two fluorescence lifetimes of 24.5 and 100 nsec were obtained, suggesting the use of NPM as a probe of the rotational dynamics of macromolecules.

The first time-resolved anisotropy measurements of actin, reported by Mihashi and Wahl (1975), used ϵ-ATP-labeled G-actin and ϵ-ADP-labeled F-actin. The observed correlation time of around 10 μsec was attributed to an undetermined global rotation of F-actin. In addition to these experiments, Wahl et al. (1975) labeled Cys-374 of actin with DNS and compared the decay of anisotropy in the presence and absence of the regulatory system components tropomyosin, troponin, and Ca^{2+}. The correlation time of $\theta = 411$ nsec obtained in the absence of the additives was 10 times greater than the correlation time expected for the G-actin monomer, but approximately two orders of magnitude lower than that predicted for the global rotation of F-actin (Tawada et al., 1978). The observed correlation time was thus attributed to local motions of the C-terminal end of the actin peptide chain. The binding of heavy meromyosin resulted in greater immobilization of the F-actin conjugate while Ca^{2+} increased the mobility. Kawasaki et al. (1976) measured the time dependence of the anisotropy of NPM conjugated to Cys-374 and obtained a similar correlation time of θ = 560 nsec.

It has been established that at a given pH, salt composition, and temperature the concentration of G-actin in equilibrium with F-actin is determined by a critical concentration that is independent of the total protein concentration. Thus, as the concentration of actin is increased above the critical concentration, the decay of anisotropy contains contributions from both G-actin and F-actin and the relative contribution of the latter should increase. Detailed investigations of this type, performed by Tawada *et al.* (1978) and Ikkai *et al.* (1979) using IAEDANS conjugated to Cys-374 of actin, showed this to be the case. These authors, however, also found that at 3.5°C and with a salt concentration of 0.1 M KCl four exponentials were required to fit the anisotropy decay [i.e., $j = 4$ in Eq. (2) and neglecting any residual anisotropy]. The correlation times obtained were, respectively, θ_j < 6, 45, 100, 900 nsec. The value for θ_1 was attributed to a fast local rotational motion, θ_2 was consistent with that expected for global rotation of G-actin, and θ_3 and θ_4 corresponded to filaments free from intermolecular interaction and those affected by side-by-side filament interaction, respectively. It was noted that θ_3 and θ_4 were very much shorter than a correlation time of 40 μsec expected for the uniaxial rotation of 1-μm F-actin filament as a whole, the authors attributing this to some deformation of the F-actin molecule.

In separate studies using IAEDANS as the conjugate, Miki *et al.* (1982a,b) examined the effects of the divalent cations Ca^{2+} and Mg^{2+} on the interaction between F-actin and myosin heads, and the influence of Ca^{2+} on the flexibility of F-actin. In the latter study, the average fluorescence anisotropy and the longer correlation time of F-actin (in the order of 600–1000 nsec) were found to decrease with increasing concentrations of free Ca^{2+}. These results were interpreted as arising from conformational changes of the F-actin induced by Ca^{2+} binding resulting in increases in the internal Brownian motion of the molecule. Addition of heavy meromyosin or myosin subfragment 1 in the presence of Ca^{2+} caused a decrease in mobility. In the presence of Mg^{2+}, however, the addition of heavy meromyosin or myosin subfragment 1 (to a molar ratio maximum of 1 : 25 and 1 : 50, respectively) caused an increase in mobility. Thus, interaction with myosin heads induces a conformational change of F-actin which is dependent on the divalent cation present.

Fluorescence anisotropy has also been used to probe the interactions between spectrin subunits. Yoshino and Marchesi (1984) used the intrinsic fluorescence of spectrin to follow the reassociation of α and β subunits. The temperature dependence of the anisotropy was strikingly different for each subunit. However, the temperature dependence for the reconstituted dimer was similar to that of the native dimer as was its appearance under the electron microscope. Both the reconstituted and native dimers showed a similar decrease in the anisotropy at temperatures above 48–49°C, indicating an increase in the motional freedom of tryptophan residues as the native structure unfolds.

Spectrin can also be labeled with an extrinsic fluorescence probe. However, our experience has been that it is very prone to aggregation during the labeling process, at least when isothiocyanate or iodoacetamide derivatives of fluorophores are used (W. H. Sawyer and A. G. Woodhouse, unpublished data). The problem can be prevented by labeling the protein *in situ* on ghost membranes before the isolation of the spectrin. Yoshino and Marchesi (1985) labeled the thiol groups of spectrin subunits with *N*-(1-anilinonaphthyl-4)maleimide and were able to follow the kinetics and the stoichiometry of their reassociation by measurements of steady-state anisotropy.

3.5. Fluorescence Photobleaching Recovery

Fluorescence photobleaching recovery (FPR) is a method that has found wide application in determining the lateral diffusion of lipids and proteins in bilayer membranes (Vaz *et al.*, 1982, 1984; Blatt and Vaz, 1986; Peters, 1986). Theoretical aspects of the technique have been treated by Axelrod *et al.* (1976), Koppel (1979), and Vaz *et al.* (1984) and are discussed in earlier chapters of this book. Experimentally, the diffusion coefficient of the labeled lipid or protein is obtained by photobleaching a small fraction of the fluorescent molecules by an intense actinic laser pulse, then monitoring the return of the fluorescence as the unbleached molecules diffuse into the bleached region using an attenuated steady-state laser source. The time-course of the fluorescence recovery is related to the translational diffusion coefficient, while the extent of recovery indicates the fraction of fluorescent molecules that have not diffused over the time scale of the experiment.

Observation of the polymerization of G-actin into filaments by FPR was first reported by Lanni *et al.* (1981) using 5-iodoacetamido fluorescein (5-IAF) or fluorescein isothiocyanate (FITC) conjugated to Cys-374. The rationale underlying the experiments was that as polymerization proceeded a distribution of particle sizes would form, each particle size possessing a unique diffusion coefficient. Sequential measurement of FPR would result in a series of fluorescence recovery curves that changed from an initial expression describing only G-actin monomer diffusion to a weighted integral describing monomers, oligomers, and filaments. These authors found that after addition of 0.1 M KCl a fraction of the recovery did in fact occur with a diffusion coefficient expected for G-actin, while the remainder of the label was immobile and proportionally increased for approximately 30 min after initiation of the polymerization process. An interesting observation was that during the period where the contribution of the immobile fraction increased most rapidly, photobleaching the 5-IAF conjugate resulted in fluorescence recovery followed by a slow increase in the fluorescence emission of the sample well beyond the initial level. Although definitive conclusions on the nature of this photochemically induced polymerization were not possible, these authors hypothesized that when fluorescein is coupled to actin at Cys-374, it appears to be in a position where photooxidation of a residue on the actin monomer or one of its neighbors in a filament can occur with measurable efficiency.

In subsequent studies of actin polymerization using the FPR technique, Tait and Frieden (1982a) labeled the Cys-374 residue of G-actin with (iodoacetamido)tetramethylrhodamine (IATR). The fluorescence characteristics of this probe were previously well characterized (Tait and Frieden, 1982b) and found to be suitable, particularly as the FPR data could be compared with fluorescence enhancement measurements. Figure 7 shows FPR traces obtained by Tait and Frieden (1982a) at specific time intervals after initiation of polymerization. The salient features of Figure 7 are (1) the successive decreases in the diffusion coefficient with time, indicating the formation of filaments; (2) the increasing percent of fluorescence not recovered; and (3) the successive increases in the amount of probe bleached. Figure 8 shows the percent recovery, the amount bleached, and the fluorescence enhancement plotted over the time course of the polymerization. Of particular significance are the observations that (1) the percent recovery drops rapidly to zero before the end of the polymerization reaction,

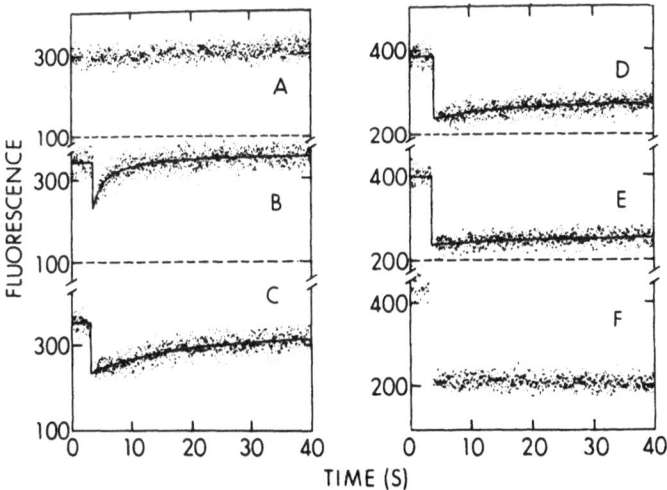

FIGURE 7. Fluorescence photobleaching recovery curves showing the transformation of G-actin into F-actin with IATR as probe. Polymerization was initiated by the addition of 2 mM MgSO$_4$. The profiles were obtained at intervals of (A) 0.7, (B) 2.2, (C) 3.1, (D) 4.1, (E) 5.3, and (F) 30 min after initiation of polymerization. The solid lines are the best fit of the data and allow calculation of diffusion coefficients. Reprinted with permission from *Biochemistry* **21** : 3668 (1982), Figure 2. Copyright 1982, American Chemical Society.

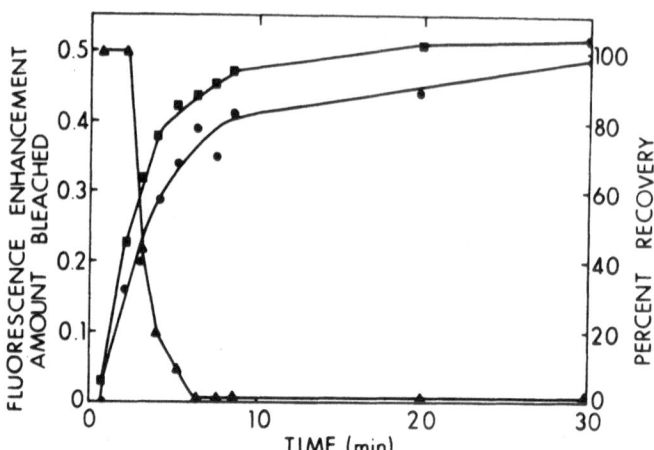

FIGURE 8. Time course of the fluorescence enhancement (●), amount bleached (■), and percent recovery (▲) for polymerization of G-actin using IATR as probe. The data were taken from the results shown in Figure 7. Reprinted with permission from *Biochemistry* **21** : 3669 (1982), Figure 3. Copyright 1982, American Chemical Society.

indicating that early in the reaction polymers are free to diffuse but at a certain point the polymers become immobile; and (2) the calculated diffusion coefficient of polymerized actin 6 min after the initiation of polymerization was much less than 2×10^{-11} cm²/sec, which is at least 100 times less than that predicted for an F-actin filament. These results strongly suggest the formation of cross-linked F-actin networks.

Further experiments by Tait and Frieden (1982a,c) examined the effects of cytochalasin D and chemical modification of actin with *N*-ethylmaleimide, IATR, and 7-chloro-4-nitro-1,2,3-benzoxadiazole, on polymerization. In all cases, polymerization was accelerated compared with unlabeled actin, and lower extents of network formation occurred. Similar results for cytochalasin D and cytochalasin B were obtained by Lanni and Ware (1984). It was hypothesized that these modifications enhance the nucleation of actin. In contrast, addition of very low concentrations of microtubule-associated proteins to rhodamine-labeled actin brought about immobilization of the F-actin (Arakawa and Frieden, 1984). This is most likely the result of the formation of a cross-linked network of F-actin molecules. These results are in conflict with low shear viscosity measurements using the falling ball viscometry method; however, as pointed out by Arakawa and Frieden (1984), the FPR technique is probably a more sensitive measure of network formation and less likely to perturb the system under investigation. An interesting extension of the work was the labeling of the microtubule-associated proteins with iodoacetamidofluorescein. These labeled molecules showed only partial immobilization (compared with complete immobilization for IATR-labeled actin), suggesting the presence of binding and nonbinding forms of the protein.

3.6. Quenching

Since the pioneering work on fluorescence quenching in solution by Stern and Volmer (1919), the technique has found wide application in the study of membrane dynamics (Blatt and Sawyer, 1985) and the accessibility of protein matrices to extrinsically added molecules and ions (Lehrer and Leavis, 1978; Eftink and Ghiron, 1981; Blatt et al., 1986a). In solutions containing proteins, quenching interactions can be measured by monitoring changes in the steady-state fluorescence or lifetime of intrinsic or conjugated probes in the presence and absence of quencher. The quenching efficiencies under different experimental conditions can then indicate the accessibility of the fluorophore to the added quencher (Eftink and Ghiron, 1981).

The quantitative interpretation of fluorescence quenching data in compartmentalized systems or proteins can be complicated by a number of factors, including the type of quenching interaction (static or dynamic), the association of the quencher with the host structure, and the accessibility of multiple emitting chromophores. The effects of these factors have been discussed in earlier chapters of this book and in detail by a number of authors (Lehrer, 1971; Lehrer and Leavis, 1978; Eftink and Ghiron, 1981; Gratton et al., 1984; Somogyi et al., 1985; Blatt et al., 1986a, 1986b; Julien et al., 1986). For the purposes of this section on fluorescence quenching interactions with actin, a brief theoretical background is provided. The quenching of a fluorophore by collisional interactions with a quencher may be described by the Stern–Volmer equation (Stern and Volmer, 1919; Lakowicz, 1983):

$$I_0/I - 1 = \tau_0/\tau - 1 = K_{SV}[Q] = k_q\tau_0[Q] \tag{5}$$

where I_0, I and τ_0, τ are the fluorescence intensities and lifetimes in the presence and absence of quencher Q, respectively, K_{SV} is the Stern–Volmer constant, and k_q is the bimolecular collisional rate constant. Upward curvature in steady-state Stern–Volmer plots can be attributed to a static quenching mechanism either as the result of the formation of a ground-state complex between fluorophore and quencher or according to Perrin's model of quencher–fluorophore interaction (Lakowicz, 1983). A plateau effect at high quencher concentrations can be due to the quenching of more than one population of fluorophores with different quenching constants or accessibilities (Eftink and Ghiron, 1981), although other interpretations may also be valid (Eftink and Ghiron, 1981; Blatt et al., 1986b). Linearity and compatibility between steady-state and lifetime quenching data usually indicates that there is one emitting species and only dynamic interactions are occurring. In cases where more than one species emit, the quenching data can be replotted according to modified forms of Eq. (5) (Lehrer, 1971; Corin *et al.*, 1987) to obtain the fraction of accessible species.

The accessibilities of actin sulfhydryl sites were first investigated by Tao and Cho (1979) using IAEDANS-labeled actin as the fluorophore and acrylamide as the quenching species. Acrylamide is commonly employed in protein quenching studies because it is uncharged, it does not cause significant denaturation up to concentrations of 1 M (Eftink and Ghiron, 1977), and because in solution it quenches indole fluorescence with an efficiency of unity (Eftink and Ghiron, 1976). IAEDANS-labeled G-actin showed two fluorescence decay components with 96% of the total decay having a lifetime of 17.3 nsec and 4% having a lifetime of 33.3 nsec. Lifetime quenching measurements gave linear Stern–Volmer plots corresponding to values of k_q that were approximately 3.5 times greater for the major component. As denatured actin showed only a single decay component, the multiplicity in the active form must be due to heterogeneity of labels rather than because of an intrinsic property of the probe. Thus, from these data (and the result that pretreatment with *N*-ethylmaleimide, which is known to react primarily with Cys-374, blocked conjugation of the IAEDANS derivative to actin), Tao and Cho (1979) concluded that the major component was due to IAEDANS labeled at Cys-374, while the minor component corresponded to a minor labeling site or sites that were less accessible.

Figure 9 shows the steady-state quenching of IAEDANS-G-actin and IAEDANS-F-actin by acrylamide obtained by Tao and Cho (1979). The plateau effects for both quenching plots indicate the presence of multiple emitting species and are consistent with the lifetime data. Polymerization of actin resulted in a substantial decrease in the quenching efficiency and was interpreted as indicating greater shielding of the major component site and deeper burial of

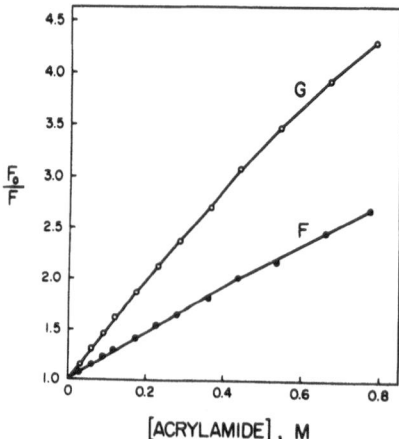

FIGURE 9. Steady-state quenching curves for IAEDANS–G-actin (○) and IAEDANS–F-actin (●). Reprinted with permission from *Biochemistry* **18** : 2762 (1979), Figure 4. Copyright 1979, American Chemical Society.

the minor site(s). Addition of tropomyosin further decreased the accessibility of the major labeling site, indicating that tropomyosin binds to actin at a site near Cys-374.

Subsequent to the work of Tao and Cho (1979), a more detailed investigation of the effects of protein concentration and the cytoskeletal components spectrin and band 4.1 on the fluorescence quenching of IAEDANS-labeled actin by acrylamide was conducted by Husain (1985). Figure 10 shows the effects of protein concentration on steady-state Stern–Volmer plots. For F-actin the quenching efficiencies decreased substantially with increasing protein concentration while a slight decrease was observed for G-actin. Dependencies of these kinds can arise from quencher partitioning or binding (Blatt and Sawyer, 1985; Blatt *et al.*, 1986a); however, analyses of the data using the appropriate partitioning and/or binding models gave unrealistic fitted parameters. Instead, a model was

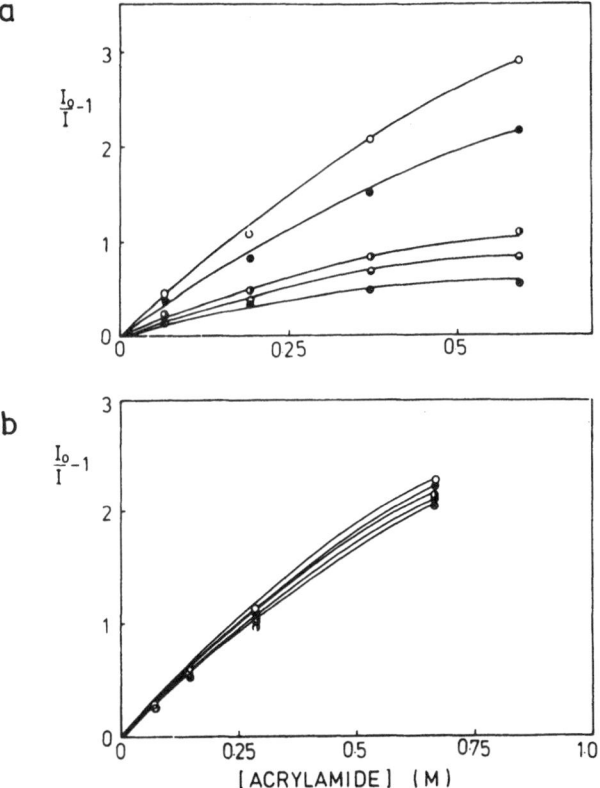

FIGURE 10. Steady-state Stern–Volmer plots for the quenching of (a) F-actin and (b) G-actin by acrylamide as a function of (from top to bottom) increasing actin concentration. From A. Husain, Ph.D. thesis, 1985, University of Melbourne.

chosen for the F-actin data which incorporated contributions from the quenching of both G-actin and F-actin molecules as the total protein concentration increased. Since the initial slopes of the Stern–Volmer plots provided good estimates of K_{SV} for the major component (Tao and Cho, 1979), the slope observed may be treated as reflecting a weighted average of the two actin forms according to

$$S_{obs} = \frac{S_G C_G + S_F C_F}{C_G + C_F} \tag{6}$$

where C_G and C_F are the concentrations of the G and F forms and S_G and S_F are the Stern–Volmer slopes characteristic of pure G and F forms, respectively. The value of S_G was calculated from the data of Figure 11, while S_F was estimated from the ordinate intercept of a plot of slope versus 1/[actin]. The results, shown in Figure 11 as a plot of S_{obs} versus [actin], indicate that at a total protein concentration of 50 μg/ml the quenching efficiency approached that of G-actin. This concentration was in close agreement with the critical actin concentration of 54 μg/ml determined by ultracentrifugation (Husain, 1985).

Husain (1985) also found that addition of spectrin tetramer or dimer decreased the quenching efficiency in a similar, but less dramatic, way to that seen with tropomyosin (Tao and Cho, 1979). The inclusion of band 4.1, however, had no significant effect. The results were interpreted as providing evidence for the binding of spectrin in the vicinity of Cys-374, although the observed changes might also arise from an induced conformational change of the actin filaments.

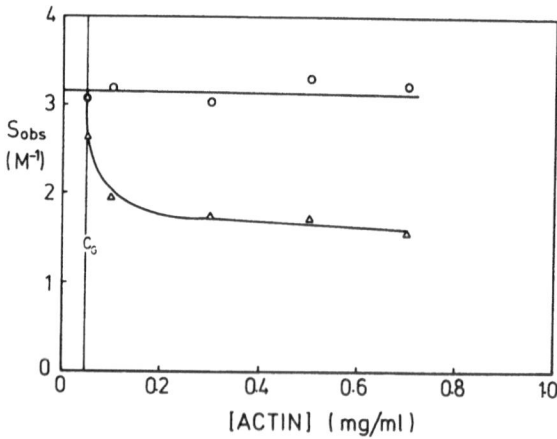

FIGURE 11. Initial slopes of Stern–Volmer plots (S_{obs}) in Figure 10a ($\triangle$) and Figure 10b ($\bigcirc$) plotted as a function of total actin concentration. C_G is the critical actin concentration of 54 μg/ml as determined by ultracentrifugation. From A. Husain, Ph.D. thesis, 1985, University of Melbourne.

3.7. Pressure Relaxation

Pressure relaxation as a technique for studying the interaction of actin with myosin was first reported by Geeves and Gutfreund (1982). The principle underlying the technique is that increased pressure induces a small fraction of the bound S-1 to dissociate from the acto–S-1 complex. Reassociation occurs after rapid release of the pressure and may be monitored by following the changes in light scattering to obtain association constants. The sensitivity of the method, however, limits investigations to association constants in the range 10^4–$10^6 \, M^{-1}$. In order to increase the sensitivity, Coates *et al.* (1985) and Criddle *et al.* (1985) used NPI-labeled actin, which, upon binding to S-1, undergoes a 70% reduction in the fluorescence intensity (Kouyama and Mihashi, 1981).

Pressure-induced changes in the fluorescence of the NPI–acto–S-1 complex resulted in the appearance of two relaxations. Coates *et al.* (1985) proposed the following scheme:

$$A + S \; \underset{}{\overset{K_0}{\rightleftharpoons}} \; AS_0 \; \underset{k_{-a}}{\overset{k_{+a}}{\rightleftharpoons}} \; AS_1 \; \underset{k_{-b}}{\overset{k_{+b}}{\rightleftharpoons}} \; AS_2$$

where A and S refer to actin and myosin subfragment, respectively, and AS_0, AS_1, and AS_2 are isomeric forms of the acto–myosin complex. Table III lists rate and equilibrium constants obtained by Coates *et al.* (1985) under a number of experimental conditions. An interesting aspect of the tabulated data is the dependence of K_0, k_a, and k_b on ionic strength, temperature, and solvent composition, variables that are important to explore in order to relate the data to *in vivo* conditions. The data also showed for the first time the existence of two acto–S1 states which may be implicated in the cross-bridge model of muscle contraction (Geeves *et al.*, 1984).

4. ALTERNATIVE LUMINESCENCE TECHNIQUES

4.1. Introduction

In the preceding sections we discussed investigations of cytoskeletal protein interactions where the fluorescence characteristics of either intrinsic or labeled proteins were employed. An important limitation of fluorescence probe techniques in the study of protein rotational dynamics is that the time window of observation generally does not exceed a few hundred nanoseconds. This time scale is representative of the rotational diffusion of many proteins and macromolecules in solution (Cheung *et al.*, 1971; Munro *et al.*, 1979; Lakowicz *et al.*, 1983) or fluorescence probes in compartmentalized systems (Blatt *et al.*,

Edward Blatt and William H. Sawyer

Table III
Rate and Equilibria Constants for the Interaction of NPI-Actin and S-1[a]

Conditions	Association constant (M^{-1})	$K_0 K_a$ (M^{-1})	K_b	$K_0 k_{+a}$ (M^{-1}·sec^{-1})	Dissociation rate constant (sec^{-1})	k_{-a} (sec^{-1})
Standard[b]	1.7×10^7	5.9×10^4	280	3.63×10^6	0.22	61
2°C	1.5×10^6	7.6×10^3	200	1.48×10^5	0.10	19.5
0.5 M KCl	2.7×10^5	7.4×10^3	37	2.47×10^5	0.9	33
40% (v/v) ethylene glycol	2.5×10^6	1.1×10^5	22	3.7×10^5	0.50	3.4

[a]Reprinted with permission from Coates, J.N., Criddle, A.H., and Greeves', M.A. Biochemical J. 232 : 351–356 (1985). Copyright 1985, The Biochemical Society, London.
[b]0.1 M KCl, 5 mM MgCl$_2$, 20 mM imidazole, pH 7.0, 20°C.

1982, 1983; Vincent *et al.*, 1982a,b). The rotational motions of proteins embedded in cell surfaces and lipid membranes or large proteins in solution, however, occur typically in the microsecond to millisecond time domain (Cherry, 1978; Austin *et al.*, 1979; Coke *et al.*, 1986; Blatt and Jovin, 1986). In order to measure these motions by spectroscopic techniques, it is necessary to use probes with comparable lifetimes. Triplet probes, whose lifetimes are much longer than the fluorescent state due to spin-forbidden singlet–triplet transitions, have thus become useful indicators of these longer rotational motions. In this section we briefly discuss two alternative methods utilizing the triplet state of labels which have found application in the study of the rotational diffusion of erythrocyte proteins.

4.2. Transient Absorption Anisotropy

The principles underlying the method of transient absorption anisotropy (TAA) have been presented by Cherry (1978) and are only briefly described here. In essence, the method utilizes the same notions as those of the fluorescence and phosphorescence emission anisotropy techniques (Ghiggino *et al.*, 1981; Corin *et al.*, 1985; Blatt and Jovin, 1986). Photoselective excitation of the intrinsic or labeled chromophore is achieved with plane-polarized light so that molecules with transition moments for absorption lying parellel or at a small angle to the electric vector of the incident light will be preferentially excited. Following excitation, the initial polarization of the absorption changes as the result of Brownian rotation and other motions. The parallel and perpendicular components of the absorption dichroism are collected and the anisotropy generated as a function of time (Cherry, 1978). The data are usually fitted to functions shown by Eqs. (2) and (3) to extract rotational correlation times.

Band 3, the anion transporter of the human erythrocyte, was the first membrane protein to be investigated by the technique of TAA (Cherry *et al.*, 1976). Eosin isothiocyanate (EITC) was used as the triplet probe due to its preferential labeling of band 3 over other membrane components in intact cells. A rotational correlation time of around 1 msec was obtained, which was essentially independent of the presence of spectrin. These results indicated that if spectrin is linked to band 3 the type of attachment does not hinder the rotation of band 3 about an axis normal to the plane of the membrane. In a further study of band 3 rotation, Mühlebach and Cherry (1982) looked at the effects of cholesterol on band 3 proteins that had been cleaved with trypsin to remove interactions with other cytoskeletal components. It was found that decreasing the mole ratio of cholesterol/phospholipid brought about a more rapid decay of the anisotropy. The authors concluded that addition of cholesterol causes band 3 to aggregate, suggesting a role for cholesterol in regulating protein–protein associations in membranes.

The TAA technique has also found application in the study of the rotational diffusion of F-actin in solution (Mihashi et al., 1983; Yoshimura et al., 1984). In these studies 5-iodoacetamide-eosin (EIA) was used to label the Cys-374 residue of actin. The TAA curves obtained were independent of protein concentration in the range of labeled actin of 7–28 μM. Since the length distribution of F-actin is dependent on protein concentration, it was concluded that the observed decays were not due to global rotation of F-actin but rather due to internal motion. The data were subsequently analyzed according to the theory of anisotropy decay due to torsional motion proposed by Barkley and Zimm (1979); a torsional rigidity of 0.2×10^{-17} dyn·cm^2 and a root mean square fluctuation of the torsional angle between adjacent protomers in the actin helix of 4° at 20°C were obtained. It was concluded that EIA-labeled F-actin behaves as a torsionally flexible thin rod. The addition of tropomyosin, troponin, or phalloidin did not significantly change the TAA curves.

4.3. Phosphorescence

An important limitation in techniques based on absorption is that of sensitivity. In practice, signal-to-noise considerations limit measurements of triplet probes in solution or suspension to concentrations greater than 1 μM (Rigler et al., 1974; Austin et al., 1979). At least two orders of magnitude increase in sensitivity may be obtained by monitoring phosphorescence emission characteristics (Austin et al., 1979). The intrinsic phosphorescence of G-actin has been observed at room temperature (Horie and Vanderkooi, 1982); however, the fundamental emission was composed of more than one band, indicating heterogeneous environments for the various tryptophans. Studies using eosin and erythrosin-labeled actin and band 3 have thus been preferred.

Early work on the phosphorescence anisotropy of EITC-labeled band 3 in human erythrocyte ghosts as a function of temperature were performed by Austin et al. (1979). The rotational correlation times in excess of 1 msec increased as the temperature was lowered. Possible explanations of the data proposed by the authors were that either aggregates of band 3 or complexes with other membrane components were responsible. These studies were extended by Matayoshi et al. (1983), who examined the rotational diffusion of eosin maleimide-labeled band 3 in erythrocyte ghosts and in intact cells as a function of the concentration of aldolase and glyceraldehyde phosphate dehydrogenase (G3PD). These proteins are known to bind to the N-terminal cytoplasmic tail of band 3 near a domain that is involved in the binding of ankyrin, the bridging protein that links spectrin to the membrane proper. The results in Figure 12 show that the binding of G3PD results in the progressive immobilization of band 3. Factors that are known to dissociate the enzyme from the erythrocyte membrane (e.g., high ionic strength, NADH) are able to restore the rotational mobility of band 3. Clearly, the binding

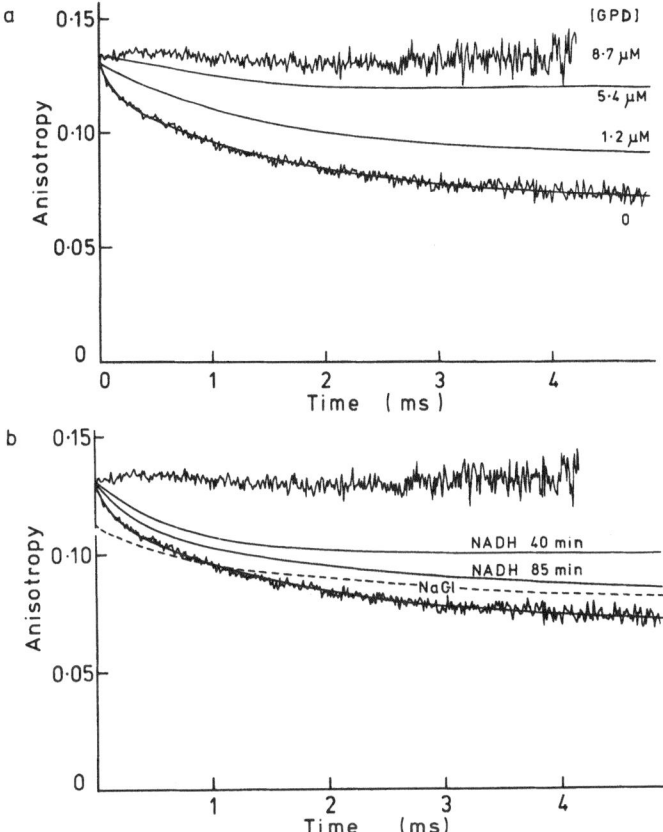

FIGURE 12. (a) Effect of glyceraldehyde phosphate dehydrogenase (GPD) on the anisotropy decay of erythrocyte band 3 labeled with eosin maleimide. The solid lines are computer fits to the experimental data generated by a nonlinear least squares procedure employing Eq. 2. (b) The reversal of the effect of GPD brought about by NADH and NaCl. The experimental data are the same as presented in (a). The figures are adapted in part from *Mobility and Recognition in Cell Biology* (Sund and Veeger, eds.), 1983, pp. 119–134; Figure 4, by copyright permission from Walter de Gruyter Publishers.

of a macromolecule to the cytoplasmic domain of an intrinsic protein can severely limit its rotational diffusion.

5. SUMMARY AND FUTURE PROSPECTS

Fluorescence probe techniques offer a variety of ways of investigating the interaction of cytoskeletal protein interactions. In particular, these techniques have been adapted to study a number of properties of actin. These include the extent and kinetics of polymerization, proximity relation, flexibility, mobility, and accessibility. There is considerable potential for the application of these techniques to other cytoskeletal components and complexes. Table IV summarizes the techniques and the type of information each technique provides and lists important corresponding references and reviews.

It should be emphasized that the fluorescence phenomenon occurs typically in the picosecond hundreds of nanosecond time domain. This is of little concern to experiments that measure steady-state emission characteristics (e.g., energy transfer and fluorescence enhancement). However, with measurements of protein dynamics and diffusion kinetics, the time scale of the emission process becomes more important. Thus, fluorescence anisotropy, photobleaching recovery, and quenching experiments are indicative of localized protein motions over time periods of nanoseconds. Phosphorescence and transient absorption techniques provide information on a time scale more appropriate for global mobilities.

A further technique showing promise as a means of studying protein rotation and mobility in solution is fluorescence recovery spectroscopy (Blatt and Jovin, 1986; Corin et al., 1987). Originally applied in the microscope and termed fluorescence depletion (Johnson and Garland, 1982), this method measures the photoselective depletion of the ground state by monitoring the return of the polarized fluorescence intensity to the steady-state values. That is, the fluorescence is monitored, but the recovery occurs over time scales associated

Table IV
Summary of Fluorescence Probe Techniques

Technique	Information obtained	References
Energy transfer	Kinetics and extent of polymerization, proximity relations	Stryer (1978), Dos Remedios et al. (1987)
Fluorescence enhancement	Kinetics of polymerization, rate constants (quantitative)	Tellam and Frieden (1982), Cooper et al. (1983)
Fluorescence anisotropy	Mobility, flexibility	Rigler and Ehrenberg (1976), Ghiggino et al. (1981), Beecham and Brand (1985)
Fluorescence photobleaching recovery	Mobility, diffusion kinetics, network formation	Tait and Frieden (1982a,b,c), Arakawa and Frieden (1984)
Quenching	Protein dynamics, accessibility, association constants	Lehrer and Leavis (1978), Eftink and Ghiron (1981), Blatt et al. (1986a)
Pressure relaxation	Association constants	Coates et al. (1985), Criddle et al. (1985)

with the triplet state of the probe. This allows the use of a wider variety of probes and extends investigations into the millisecond time domain.

ACKNOWLEDGMENTS. We would like to thank Dr. L. Tilley and Dr. G. Ralston for helpful comments on the manuscript.

6. REFERENCES

Arakawa, T., and Frieden, C., 1984, Interaction of microtubule-associated proteins with actin filaments, *J. Biol Chem.* **259** : 11730–11734.

Asakura, S., 1961, F-actin adenosine triphosphate activated under sonic vibration, *Biochim. Biophys Acta.* **52** : 65–75.

Austin, R. H., Chan, S. S., and Jovin, T. M., 1979, Rotational diffusion of cell surface components by time-resolved phosphorescence anisotropy, *Proc. Natl. Acad. Sci. U.S.A.* **76**: 5650–5654.

Axelrod, D., Koppel, D., Schlessinger, J., Elson, E., and Webb, W. W., 1976, Mobility measurement by analysis of fluorescence photobleaching recovery kinetics, *Biophys. J.* **16** : 1055–1069.

Barden, J. A., Miki, M., and Dos Remedios, C. G., 1986, Selective labelling of Cys-10 on actin, *Biochem. Int.* **12** : 95–101.

Barkley, M. D., and Zimm, B. H., 1979. Theory of twisting and bending of chain macromolecules; analysis of the fluorescence depolarization of DNA, *J. Chem. Phys.* **70** : 2991–3007.

Beechem, J. M., and Brand, L., 1985, Time-resolved fluorescence of proteins, *Annu. Rev. Biochem.* **54** : 43–71.

Bennett, V., 1985, The membrane skeleton of human erythrocytes and its implications for more complex cells, *Annu. Rev. Biochem.* **54** : 273–304.

Blatt, E., and Jovin, T. M., 1986, Rotational dynamics of biological macromolecules, in *Photophysical and Photochemical Tools in Polymer Science* (M. A. Winnik, ed.), pp. 351–370. D. Reidel Publishing, Dordrecht.

Blatt, E., and Sawyer, W. H., 1985, Depth-dependent fluorescence quenching in micelles and membranes, *Biochim. Biophys. Acta* **822** : 43–62.

Blatt, E., and Vaz, W. L. C., 1986, The effects of Ca^{2+} on lipid diffusion, *Chem. Phys. Lipids* **41** : 183–194.

Blatt, E., Ghiggino, K. P., and Sawyer, W. H., 1982, Fluorescence depolarization studies of n-(9-anthroyloxy) fatty acids in cetyltrimethylammonium bromide micelles, *J. Phys. Chem.* **86** : 4461–4464.

Blatt, E., Ghiggino, K. P., and Sawyer, W. H., 1983. The rotational motion of n-(9-anthroyloxy) fatty acids in phospholipid bilayer vesicles, *Aust. J. Chem.* **36** : 1079–1086.

Blatt, E., Husain, A., and Sawyer, W. H., 1986a, The association of acrylamide with proteins. The interpretation of fluorescence quenching experiments, *Biochim. Biophys. Acta.* **871** : 6–13.

Blatt, E., Chatelier, R. C., and Sawyer, W. H., 1986b, Effects of quenching mechanism and type of quencher interaction on Stern–Volmer plots in compartmentalized systems, *Biophys. J.* **50** : 349–356.

Brown, S. S., and Spudich, J. A., 1981, Mechanism of action of cytochalasin B: Evidence that it binds to actin filament ends, *J Cell Biol.* **80** : 487–491.

Cherry, R. J., 1978, Measurement of protein rotational diffusion in membranes by flash photolysis, *Methods Enzymol.* **54** : 47–61.

Cherry, R. J., Bürkli, A., Busslinger, M., Schneider, G., and Parish, G. R., 1976, Rotational diffusion of band 3 proteins in the human erythrocyte membrane, *Nature* **263** : 389–393.

Cheung, H. C., Cooke, R., and Smith, L., 1971, G-actin → F-actin transformation as studied by the fluorescence of bound dansyl cystine, *Arch. Biochem. Biophys.* **142** : 333–339.

Coates, J. H., Criddle, A. H., and Geeves, M. A., 1985, Pressure-relaxation studies of pyrene-labeled actin and myosin subfragment 1 from rabbit skeletal muscle, *Biochem. J.* **232** : 351–356.

Cohen, C. M., and Foley, S. F., 1984, Biochemical characterization of complex formation by human erythrocyte spectrin, protein 4.1, and actin, *Biochemistry* **23** : 6091–6098.

Cohen, C. M., Foley, S. F., and Korsgren, C., 1982, A protein immunologically related to erythrocyte band 4.1 is found on stress fibres of nonerythroid cells, *Nature* **299** : 648–650.

Coke, M., Restall, C. J., Kemp, C. M., and Chapman, D., 1986, Rotational diffusion of rhodopsin in the visual receptor membrane: Effects of temperature and bleaching, *Biochemistry* **25** : 513–518.

Cooper, J. A., and Pollard, T. D., 1982, Methods to measure actin polymerization, *Methods Enzymol.* **85** : 182–210.

Cooper, J. A., Walker, S. B., and Pollard, T. D., 1983, Pyrene actin: Documentation of the validity of a sensitive assay for actin polymerization, *J. Muscle Res. Cell Motil.* **4** : 253–262.

Corin, A. F., Matayoshi, E. D., and Jovin, T. M., 1985, Triplet-state spectroscopy for investigating diffusion and chemical kinetics, in *Spectroscopy and the Dynamics of Biological Systems* (P. M. Bayley and R. E. Dale, eds.), pp. 53–78, Academic Press, London.

Corin, A. F., Blatt, E., and Jovin, T. M., 1987, Triplet-state detection of labeled proteins using fluorescence recovery spectroscopy, *Biochemistry* **26** : 2207–2217.

Criddle, A. H., Geeves, M. A., and Jeffries, T., 1985, The use of actin labeled with N-(1-pyrenyl)iodoacetamide to study the interaction of actin with myosin subfragments and troponin/tropomyosin, *Biochem. J.* **232** : 343–349.

Cundall, R. B., and Dale, R. E., eds. 1983, *Time-Resolved Fluorescence Spectroscopy in Biochemistry and Biology*, Plenum Press, New York.

Davis, J. Q., and Bennett, V., 1983, Brain spectrin. Isolation of subunits and formation of hybrids with ethrythrocyte spectrin subunits, *J. Biol Chem.* **258** : 7757–7766.

Davis, J. Q., and Bennett, V., 1984, Brain ankyrin. A membrane-associated protein with binding sites for spectrin, tubulin and the cytoplasmic domain of the erythrocyte anion channel, *J. Biol. Chem.* **259** : 13550–13559.

Dos Remedios, C. G., and Cooke, R., 1984, Fluorescence energy transfer between probes on actin and probes on myosin, *Biochim. Biophys. Acta.* **788** : 193–205.

Dos Remedios, C. G., Miki, M., and Barden, J. A., 1987, Fluorescence resonance energy transfer measurement of distances in actin and myosin. A critical evaluation, *J. Muscle Res. Cell Motil.* **8** : 97–117.

Edelman, G. M., and McClure, W. O., 1968, Fluorescent probes and the conformation of proteins, *Acct. Chem. Res.* **1** : 65–70.

Eftink, M. R., and Ghiron, C. A., 1976, Fluorescence quenching of indole and model micelle systems, *J. Phys. Chem.* **80** : 486–493.

Eftink, M. R., and Ghiron, C. A., 1977, Exposure of tryptophanyl residues and protein dynamics, *Biochemistry* **16** : 5546–5551.

Eftink, M. R., and Ghiron, C. A., 1981, Fluorescence quenching studies with proteins, *Anal. Biochem.* **114** : 199–227.

Elbaum, D., Mimms, L. T., and Branton, D., 1984, Modulation of actin polymerization by the spectrin–band 4.1 complex, *Biochemistry* **23** : 4813–4816.

Elzinga, M., Collins, J. H., Kuel, W., and Adelstein, R. S., 1973, Complete amino-acid sequence of actin of rabbit skeletal muscle, *Proc. Natl. Acad. Sci. U.S.A.* **70** : 2687–2691.

Fowler, V. M., 1986, An actomyosin contractile mechanism for erythrocyte shape transformations, *J. Cell. Biochem.* **31** : 1–9.

Fowler, V. M., and Bennett, V., 1984, Erythrocyte membrane tropomyosin. Purification and properties, *J. Biol. Chem.* **259** : 5978–5989.

Fowler, V. M., Davis, J. Q., and Bennett, V., 1985, Human erythrocyte myosin: Identification and purification, *J. Cell Biol.* **100** : 47–55.

Frieden, C., 1983, Polymerization of actin: Mechanism of the Mg^{2+}-induced process at pH 8 and 20°C, *Proc. Natl. Acad. Sci. U.S.A.* **80** : 6513–6517.

Frieden, C., and Goddette, D. W., 1983, Polymerization of actin and actin-like systems: Evaluation of the time course of polymerization in relation to the mechanism, *Biochemistry* **22** : 5836–5843.

Geeves, M. A., and Gutfreund, H., 1982, The use of pressure perturbations to investigate the interaction of rabbit muscle myosin subfragment 1 with actin in the presence of MgADP, *FEBS Lett.* **140** : 11–15.

Geeves, M. A., Goody, R. S., and Gutfreund, H., 1984, Kinetics of acto–S1 interaction as a guide to a model for the crossbridge cycle, *J. Muscle Res. Cell Motil.* **5** : 351–361.

Ghiggino, K. P., Roberts, A. J., and Phillips, D., 1981, Time-resolved fluorescence techniques in polymer and biopolymer studies, *Adv. Polym. Sci.* **40** : 69–167.

Goodman, S. R., Zagon, I. S., and Kulikowski, R. R., 1981, Identification of a spectrin-like protein in nonerythroid cells, *Proc. Natl. Acad. Sci. U.S.A.* **78** : 7570–7574.

Gratton, E., Jameson, D. M., Weber, G., and Alpert, B., 1984, A model of dynamic quenching of fluorescence in globular proteins, *Biophys. J.* **45** : 789–794.

Harris, S. J., and Winzor, D. J., 1985, Effect of thermodynamic nonideality on the subcellular distribution of enzymes: Adsorption of aldolase to muscle myofibrils, *Arch. Biochem. Biophys.* **143** : 598–604.

Heyn, M. P., 1979, Determination of lipid order parameters and rotational correlation times from fluorescence depolarization experiments, *FEBS Lett.* **108** : 359–364.

Hite, P., Krasnansky, R., and Thomas, J. K., 1986, Spectroscopic investigations of surfaces using aminopyrene, *J. Phys. Chem.* **90** : 5795–5799.

Horie, T., and Vanderkooi, J. M., 1982, Phosphorescence of tryptophan from parvalbumin and actin in liquid solution, *FEBS Lett.* **147** : 69–73.

Husain, A., 1985, "Protein Interactions in the Erythrocyte Cytoskeleton," PhD thesis, University of Melbourne.

Husain, A., Sawyer, W. H., and Howlett, G. J., 1983, The effect of cross-linking spectrin–actin complexes with band 4.1 on the state of polymerization of the actin, *Biochem. Biophys. Res. Commun.* **111** : 360–365.

Ikkai, T., Wahl, P., and Auchet, J.-C., 1979, Anisotropy decay of labeled actin, *Eur. J. Biochem.* **93** : 397–408.

Jähnig, F., 1979, Structural order of lipids and proteins in membranes: Evaluation of fluorescence anisotropy data, *Proc. Natl. Acad. Sci. U.S.A.* **76** : 6361–6365.

Johnson, K. A., and Borisy, 1977, Kinetic analysis of microtubule self-assembly *in vitro. J. Mol. Biol.* **117** : 1–31.

Johnson, P., and Garland, P. B., 1982, Fluorescent triplet probes for measuring the rotational diffusion of membrane proteins, *Biochem. J.* **203** : 313–321.

Julien, M., Garel, J.-R., Merola, F., and Brochon, J.-C., 1986, Quenching by acrylamide and temperature of a fluorescent probe attached to the active site of ribonuclease, *Eur. Biophys. J.* **13** : 131–137.

Kawasaki, Y., Mihashi, K., Tanaka, H., and Ohnuma, H., 1976, Fluorescence study of *N*-(3-pyrene)maleimide conjugated to F-actin and plasmodium actin polymers, *Biochim. Biophys. Acta* **446** : 166–178.

Kinosita, K., Jr., Ikegami, A., and Kawato, S., 1982, On the wobbling-in-cone analysis of fluorescence anisotropy decay, *Biophys. J.* **37** : 461–464.

Koppel, D., 1979, Fluorescence redistribution after photobleaching, *Biophys. J.* **28** : 281–292.

Kouyama, T., and Mihashi, K., 1980, Pulse-fluorometry study on actin and heavy meromyosin using F-actin labeled with *N*-(1-pyrene)maleimide, *Eur. J. Biochem.* **105** : 279–287.

Kouyama, T., and Mihashi, K., 1981, Fluorimetry study of *N*-(1-pyrenyl)iodacetamide-labeled F-actin, *Eur. J. Biochem.* **114** : 33–38.

Lakowicz, J. R., 1983, *Principles of Fluorescence Spectroscopy*, Plenum Press, New York.

Lakowicz, J. R., Maliwal, B. P., Cherek, H., and Balter, A., 1983, Rotational freedom of tryptophan residues in proteins and peptides, *Biochemistry* **22** : 1741–1752.

Lanni, F., and Ware, B. R., 1984, Detection and characterization of actin monomers, oligomers, and filaments in solution by measurement of fluorescence photobleaching recovery, *Biophys. J.* **46** : 97–110.

Lanni, F., Taylor, D. L., and Ware, B. R., 1981, Fluorescence photobleaching recovery in solutions of labeled actin, *Biophys. J.* **35** : 351–364.

Lee, P. C., and Meisel, D., 1985, Photophysical studies of pyrene incorporated in nafion membranes, *Photochem. Photobiol.* **41** : 21–26.

Lehrer, S. S., 1971, Solute perturbation of protein fluorescence. The quenching of the tryptophyl fluorescence of model compounds and of lysozyme by iodide ion, *Biochemistry* **10** : 3254–3263.

Lehrer, S. S., and Kerwar, G., 1972, Intrinsic fluorescence of actin, *Biochemistry* **11** : 1211–1217.

Lehrer, S. S., and Leavis, P. C., 1978, Solute quenching of protein fluorescence, *Methods Enzymol.* **49** : 222–236.

Marchesi, V. T., 1985, Stabilizing infrastructure of cell membranes, *Annu. Rev. Cell. Biol.* **1** : 531–561.

Martonosi, A., 1968, The sulfhydryl groups of actin, *Arch. Biochem. Biophys.* **123** : 29–40.

Matayoshi, E. D., Corin, A. F., Zidovetzki, R., Sawyer, W. H., and Jovin, T. M., 1983, Rotational dynamics of cell surface proteins by time-resolved phosphorescence anisotropy, in *Mobility and Recognition in Cell Biology* (H. Sund and C. Veeger, eds.), pp. 119–134, Walter de Gruyter, Berlin.

Mihashi, K., and Wahl, P., 1975, Nanosecond pulse fluorometry in polarized light of g-actin–ε-ATP [1,N^6-ethenoadenosine triphosphate] and F-actin–ε-ADP, *FEBS Lett.* **52** : 8–12.

Mihashi, K., Yoshimura, H., Nishio, T., Ikegami, A., and Kinosita, Jr., K., 1983, Internal motion of F-actin in 10^{-6}–10^{-3} s time range studied by transient absorption anisotropy: Detection of torsional motion, *J. Biochem.* **93** : 1705–1707.

Miki, M., and Iio, T., 1984, Fluorescence energy transfer measurements between the nucleotide binding site and Cys-374 in actin and their application to the kinetics of actin polymerization, *Biochim. Biophys. Acta* **790** : 201–207.

Miki, M., and Mihashi, K., 1978, Fluorescence energy transfer between ε-ATP at the nucleotide binding site and *N*-(4-dimethylamino-3,5-dinitrophenyl)-maleimide at Cys-374 of G-actin, *Biochim. Biophys. Acta* **533** : 163–172.

Miki, M., and Wahl, P., 1984, Fluorescence energy transfers in labeled G-actin and F-actin, *Biochim. Biophys. Acta* **786** : 188–196.

Miki, M., and Wahl, P., 1985, Fluorescence energy transfer between points in G-actin: The nucleotide-binding site, the metal-binding site and Cys-374 residue, *Biochim. Biophys. Acta* **828** : 188–195.

Miki, M., Wahl, P., and Auchet, J.-C., 1982a, Fluorescence anisotropy of labelled F-actin: Influence of divalent cations on the interaction between F-actin and myosin heads, *Biochemistry* **21** : 3661–3665.

Miki, M., Wahl, P., and Auchet, J.-C., 1982b, Fluorescence anisotropy of labelled F-actin. Influence of Ca²⁺ on the flexibility of F-actin, *Biophys. Chem.* **16** : 165–172.

Miki, M. Barden. J. A., Hambly, B. D., and Dos Remedios, C. G., 1986a, Fluorescence energy transfer between Cys-10 residues in F-actin filaments, *Biochem. Int.* **12** : 725–731.

Miki, M., Barden, J. A., and Dos Remedios, C. G., 1986b, The distance separating Cys-10 from the high-affinity metal binding site in actin, *Biochem. Int.* **12** : 807–813.

Miki, M., Barden, J. A., and Dos Remedios, C. G., 1986c, Fluorescence resonance energy transfer between the nucleotide binding site and Cys-10 in G-actin and F-actin, *Biochim. Biophys. Acta* **872** : 76–82.

Mühlebach, T., and Cherry, R. J., 1982, Influence of cholesterol on the rotation and self-association of band 3 in the human erythrocyte membrane, *Biochemistry* **21** : 4225–4228.

Munro, I., Pecht, I., and Stryer, L., 1979, Subnanosecond motions of tryptophan residues in proteins, *Proc. Natl. Acad. Sci. U.S.A.* **76** : 56–60.

Nakajima, A., 1977, Fluorescence spectra of condensed aromatic hydrocarbons in water and in aqueous surfactant solution, *Photochem. Photobiol.* **25** : 593–598.

Nakaoka, Y., and Kasai, M., 1969, Behavior of sonicated actin polymers: Adenosine triphosphate splitting polymerization, *J. Mol. Biol.* **44** : 319–332.

Ohanian, V., Wolfe, L. C., John, K. M., Pinder, J. C., Lux, S. E., and Gratzer, W. B., 1984, Analysis of the ternary interaction of the red cell membrane skeletal proteins spectrin, actin, and 4.1, *Biochemistry* **23** : 4416–4420.

Pantaloni, D., Carlier, M.-F., and Korn, E. D., 1985, The interaction between ATP-actin and ADP-actin, *J. Biol. Chem.* **260** : 6572–6578.

Pardee, J. D., Simpson, P. A., Stryer, L., and Spudich, J. A., 1982, Actin filaments undergo limited subunit exchange in physiological salt conditions, *J. Cell Biol.* **94** : 316–324.

Peters, R., 1986, Fluorescence microphotolysis to measure nucleocytoplasmic transport and intracellular mobility, *Biochim. Biophys. Acta* **864** : 305–359.

Pinder, J. C., and Gratzer, W. B., 1982, Investigation of the actin–deoxyribonuclease I interaction using a pyrene-conjugated actin derivative, *Biochemistry* **21** : 4886–4890.

Pinder, J. C., Ohanian, V., and Gratzer, W. B., 1984, Spectrin and band 4.1 as an actin filament-capping complex, *FEBS Lett.* **169** : 161–164.

Pollard, T. D., 1983, Measurement of rate constants for actin filament elongation in solution, *Anal. Biochem.* **134** : 406–412.

Pollard, T. D., and Cooper, J. A., 1982, Methods to characterize actin filament networks, *Methods Enzymol.* **85** : 211–233.

Reed, W., Lasic, D., Hauser, H., and Fendler, J. H., 1985, Effects of photopolymerization on surfactant vesicle surface morphology, *Macromolecules* **18** : 2005–2012.

Rigler, R., and Ehrenberg, M., 19876, Fluorescence relaxation spectroscopy in the analysis of macromolecular structure and motion, *Q. Rev. Biophys.* **9** : 1–19.

Rigler, R., Rabl, C.-R., and Jovin, T.M., 1974, A temperature-jump apparatus for fluorescence measurements, *Rev. Sci. Instrum.* **45** : 580–588.

Somogyi, B., Papp, S., Rosenberg, A., Seres, I., Matko, J., Welch, R., and Nagy, P., 1985, A double-quenching method for studying protein dynamics: Separation of the fluorescence quenching parameters characteristic of solvent-exposed and solvent-masked fluorophors, *Biochemistry* **24** : 6674–6679.

Spudich, J. A., Kuczmarski, E. R., Pardee, J. D., Simpson, P. A., Yamamoto, K., and Stryer, L., 1982, Control of assembly of *Dictyostelium* myosin and actin filaments, *Cold Spring Harbor Symp. Quant. Biol.* **XLVI** : 553–561.

Steck, T. L., 1974, The organization of proteins in the human red blood cell membrane, *J. Cell Biol.* **62** : 1–19.

Stern, V. O., and Volmer, M., 1919, On the quenching-time of fluorescence, *Phys. Z.* **20** : 183–188.

Stryer, L., 1978, Fluorescence energy transfer as a spectroscopic ruler, *Annu. Rev. Biochem.* **47** : 819–846.

Tait, J. F., and Frieden, C., 1982a, Polymerization and gelation of actin studied by fluorescence photobleaching recovery, *Biochemistry* **21** : 3666–3674.

Tait, J. F., and Frieden, C., 1982b, Polymerization-induced changes in the fluorescence of actin labeled with iodoacetamidotetramethylrhodamine, *Arch. Biochem. Biophys.* **216** : 133–141.

Tait, J. F., and Frieden, C., 1982c, Chemical modification of actin. Acceleration of polymerization and reduction of network formation by reaction with *N*-ethylmaleimide, (iodoacetamido)tetramethylrhodamine, or 7-chloro-4-nitro-2,1,3-benzoxadiazole, *Biochemistry* **21** : 6046–6053.

Takashi, R., 1979, Fluorescence energy transfer between subfragment-1 and actin points in the rigor complex of actosubfragment-1, *Biochemistry* **18** : 5164–5169.

Tao, T., 1969, Time-dependent fluorescence depolarization and brownian rotational diffusion coefficients of macromolecules, *Biopolymers* **8** : 609–632.

Tao, T., and Cho, J., 1979, Fluorescence lifetime quenching studies on the accessibilities of actin sulfhydryl sites, *Biochemistry* **18** : 2759–2765.

Tawada, K., Wahl, P., and Auchet, J.-C., 1978, Study of actin and its interactions with heavy meromyosin and the regulatory points by the pulse fluorimetry in polarized light of a fluorescent probe attached to actin cysteine, *Eur. J. Biochem.* **88** : 411–419.

Taylor, D. L., Reidler, J., Spudich, J. A., and Stryler, L., 1981, Detection of actin assembled by fluorescence energy transfer, *J. Cell Biol.* **89** : 362–367.

Tellam, R., and Frieden, C., 1982, Cytochalasin D and platelet gelsolin accelerate actin polymer formation. A model for regulation of the extent of actin formation *in vivo*, *Biochemistry* **21** : 3207–3214.

Tilley, L., and Ralston, G. B., 1987, Effect of erythrocyte spectrin on actin self-association, *Aust. J. Biol. Sci.* **40** : 27–36.

Vanderkerckhove, J., and Weber, K., 1979, The complete amino acid sequence of actins from bovine aorta, bovine heart, bovine fast skeletal muscle and rabbit slow skeletal muscle, *Differentiation* **14** : 123–33.

Vaz, W. L. C., Derzko, Z. I., and Jacobson, K. A., 1982, Photobleaching experiments of the lateral diffusion of lipids and proteins in artificial bilayer membranes, *Cell Surf. Rev.* **8** : 83–135.

Vaz, W. L. C., Goodsaid-Zalduondo, F., and Jacobson, K., 1984, Lateral diffusion of lipids and proteins in bilayer membranes, *FEBS Lett.* **174** : 199–207.

Vincent, M., de Forresta, B., Gallay, J., and Alfsen, A., 1982a, Fluorescence anisotropy decays of *n*-(9-anthroyloxy) fatty acids in dipalmitoyl phosphatidylcholine vesicles. Localization of the effects of cholesterol addition, *Biochim. Biophys. Res. Commun.* **107** : 914–921.

Vincent, M., de Forresta, B., Gallay, J., and Alfsen, A., 1982b, Nanosecond fluorescence anisotropy decays of *n*-(9-anthroyloxy) fatty acids in dipalmitoylphosphatidylcholine vesicles with regard to isotropic solvents, *Biochemistry* **21** : 708–716.

Wahl, P., Mihashi, K., and Auchet, J.-C., 1975, Nanosecond pulse fluorometry in polarized light of dansyl-L-cysteine linked to a unique SH group of F-actin; the influence of regulatory proteins and myosin moiety, *FEBS Lett.* **60** : 164–167.

Wang, Y.-I., and Taylor, D. L., 1981, Probing the dynamic equilibrium of actin polymerization by fluorescence energy transfer, *Cell* **27** : 429–436.

Weber, G., 1953, Rotational Brownian motion and polarization of the fluorescence of solutions, in *Advances in Protein Chemistry* (M. L. Anson, K. Bailey, and J. T. Eskell, eds.), Vol. 8, pp. 415–459, Academic Press, New York.

Weeds, A., 1982, Actin-binding proteins — regulators of cell architecture and motility, *Nature* **296** : 811–816.

Weltman, J. K., Szaro, R. P., Frackelton, A. R., Jr., and Dowben, R. M., 1972, Fluorescence changes associated with G-F transformation of actin, *FEBS Lett.* **22** : 61–63.

Weltman, J. K., Szaro, R. P., Frackelton, A. R., Jr., Dowben, R. M., Bunting, J. R., and Cathou, R.E., 1973, *N*-(3-pyrenyl)maleimide, a long, lifetime fluorescent sulfhydryl reagent, *J. Biol. Chem.* **248** : 3173–3177.

Yamamoto, K., Pardee, J. D., Reidler, J., Stryler, L., and Spudich, J. A., 1982, Mechanism of interaction of *Dictyostelium* severin with actin filaments, *J. Cell Biol.* **95** : 711–719.

Yoshimura, H., Nishio, T., Mihashi, K., Kinosita, K., and Ikegami, A., 1984, Torsional motion of eosin-labeled F-actin as detected in time-resolved anisotropy decay of the probe in the sub-millisecond time range, *J. Mol. Biol.* **179** : 453–467.

Yoshino, H., and Marchesi, V. T., 1984, Isolation of spectrin subunits and reassociation *in vitro*. Analysis by fluorescence polarization, *J. Biol. Chem.* **259** : 4496–4500.

Yoshino, H., and Marchesi, V. T., 1985, Interaction between the subunits of human erythrocyte spectrin using a fluorescence probe, *Biochim. Biophys. Acta* **812** : 786–792.

Zimmerle, C. T., and Frieden, C., 1986, Effect of temperature on the mechanism of actin polymerization, *Biochemistry* **25** : 6432–6438.

Chapter 11

Fluorescent Probes for the Acetylcholine Receptor Surface Environments

Marino Martinez-Carrion, Jeffrey Clarke, Jose-Manuel Gonzalez-Ros, and Jose-Carlos Garcia-Borron

1. INTRODUCTION

The acetylcholine receptor (AchR) was the first neurotransmitter receptor to be identified and purified in an active form. It is a complex transmembrane glycoprotein present in the postsynaptic side of the neuromuscular junctions. When an action potential reaches the motor nerve terminals, acetylcholine is released into the synaptic cleft, where its local concentration can rise transiently to 10^{-4} to 10^{-3} M. Binding of Ach to specific sites located on the extracellular domains of the AchR molecules triggers the opening of short-

Abbreviations used in this chapter: ADS, anthracene-1,5-disulfonic acid; AchR, acetylcholine receptor; ANTS, 8-amino-1,3,6-naphthalene trisulfonate; α-Bgt, α-bungarotoxin; Carb, carbamylcholine; DPH, 1,6-diphenyl-1,3,5-hexatriene; MBTA, 4-(N-maleimide)benzyltrimethylammonium; NEM, N-ethylmaleimide; OG, β-D-octylglucopyranoside; PM, N-(1-pyrene)maleimide; PTSA, 1,3,6,8-pyrene tetrasulfonate; PySA, pyrene-1-sulfonyl azide; PySAH, N-(1-pyrene-sulfonyl)hexadecylamine; TMA-DPH, 1-[4-(trimethylamino)phenyl]-6-phenylhexa-1,3,5-triene.

Marino Martinez-Carrion, Jeffrey Clarke, Jose-Manuel Gonzalez-Ros, and Jose-Carlos Garcia-Borron Division of Molecular Biology and Biochemistry, School of Basic Life Sciences, University of Missouri, Kansas City, Missouri 64110-2499.

lived cation channels, thus increasing the permeability of the postsynaptic membrane and causing the muscle fiber membrane to be depolarized beyond a critical threshold. The final result of this chain of events is muscle contraction. The AchR is present in high amounts in the electric organ of certain fishes. Using *Torpedo* (electric ray) electroplax as the starting material, one can purify milligram quantities of active protein, as well as substantial amounts of its constituent subunits. Moreover, a group of closely related protein toxins (α-neurotoxins) have been isolated from the venom of several Elapid snakes, which bind to the AchR with dissociation constants in the nanomolar to subnanomolar range [for review see Karlsson (1979) and Low (1979)]. The high affinity of α-neurotoxins for AchR, combined with their extreme specificity, has greatly facilitated the purification and characterization of AchR from different sources. Because of these favorable circumstances, the AchR is by far the best characterized neurotransmitter receptor, from both a functional and a structural viewpoint. In some respects, it can be considered an archetype for the study of other less accessible neuroreceptors (Barrantes, 1983) and in general for many transmembrane proteins. Indeed, the favorable circumstances that have made possible our present knowledge of AchR structure and function have also prompted the development of novel approaches that will find a wide application in the study of other neuroreceptor systems. The scope of this chapter is to review several novel applications of fluorescent probes to the study of relevant aspects of AchR structure and function. We first review very briefly some of the salient structural and functional properties of AchR. For a more complete description of AchR structure and function, several recent reviews are available (Conti-Tronconi and Raftery, 1982; Barrantes, 1983; Changeux *et al.*, 1984; Popot and Changeux, 1984). Since our group pioneered the application of a variety of pyrene derivatives to the study of AchR structure and function, most of this chapter is devoted to the description and discussion of the information obtained by such probes and their applicability to monitor the properties of the following environments: the lipid bilayer, the protein–lipid interface, the exocytoplasmic surface, and the endocellular side. However, we also discuss some information obtained by means of other pertinent fluorescent probes.

2. AN OVERVIEW OF AchR PROPERTIES

2.1. Structural Characteristics of *Torpedo californica* AchR

The nicotinic AchR from *Torpedo californica* is a complex transmembrane glycoprotein of molecular weight 270,000 (Martinez-Carrion *et al.*, 1975), comprised of four nonidentical, tightly associated subunits of molecular weights 40,000 (α), 50,000 (β), 60,000 (γ), and 65,000 (δ), in a stoichiometry $\alpha_2\beta\gamma\delta$ (Reynolds and Karlin, 1978). The amino acid sequence of all the subunits has recently been elucidated by sequencing of cDNA clones, and all subunits have been shown to display considerable sequence homology (Noda et al., 1982, 1983a, 1983b; Claudio et al., 1983). This significant homology is generally thought to result in a similar tridimensional arrangement of the four types of polypeptide that are probably disposed pseudosymmetrically around a central pore. This pore, which could correspond to the ionophoretic channel suggested by electrophysiologic data (Huang et al., 1978; Lewis, 1979; Horn and Stevens, 1980), transverses the entire 110 Å of the molecule. Several lines of evidence, including x-ray diffraction studies (Kistler et al., 1982; Klymkowsky and Stroud, 1979), electron imaging techniques (Brisson and Unwin, 1985), and selective proteolysis from the interior and exterior of sealed vesicles (Klymkowsky et al., 1980; Lindstrom et al., 1980; Strader and Raftery, 1980) have clearly established the transmembranous nature of all the AchR subunits. Moreover, it is known that the receptor extends 15 Å on the cytoplasmic side and 55 Å on the synaptic side of the membrane (Kistler et al., 1982). Knowledge of the primary structure of the AchR subunits and of the overall shape of the protein has prompted the proposal of several theoretical models for the secondary structure of the polypeptides (Claudio et al., 1983; Devillers-Thiery et al., 1983; Noda et al., 1983b; Finner-Moore and Stroud, 1984; Guy, 1984; Ratnam et al., 1986). These models differ in the number of transmembrane segments, as well as in the exact location of particular stretches on the individual subunits. The available experimental data support those models with an odd number of transmembrane segments, since the N terminus of the subunits seems to be located on the extracellular side of the membranes (Anderson et al., 1982), while the C terminus appears to be located in the cytoplasmic side (Lindstrom et al., 1984; Ratnam and Lindstrom, 1984; Young et al., 1985); but clearly additional experimental evidence is needed to establish more definite models for AchR secondary and tertiary structures.

2.2. Ligand Binding and Pharmacological Properties

The AchRs from different sources display complex kinetic and pharmacologic properties [reviewed in Conti-Tronconi and Raftery (1982) and Changeux et al. (1984)].

Both purified, solubilized, and membrane-bound AchRs contain a pair of agonist binding sites located on the α subunit. These sites are susceptible to being affinity labeled by 4-(N-maleimide)benzyltrimethylammonium (MBTA) and bromoacetylcholine after pretreatment with reducing agents such as dithiothreitol (Reiter et al., 1972; Damle et al., 1978; Moore and Raftery, 1979). These sites are also responsible for the binding of competitive antagonists such as the α-neurotoxins, and they have recently been mapped to the vicinity of residues Cys-192–Cys-193 in the α subunit, which form the disulfide linkage labeled by MBTA (Kao and Karlin, 1986) and which are comprised in the fragments 173–204 and 185–196, both able to bind α-bungarotoxin (Wilson *et al.*, 1985; Neumann *et al.*, 1986). In addition to these activatory sites, AchR contains specific binding sites for noncompetitive antagonists. These are a heterogeneous class of compounds, including histrionicotoxin and a variety of local anesthetics that may act either allosterically or by direct steric blockade of the ionic channel.

The binding of acetylcholine to the primary activatory sites displays a positive cooperativity. However, the two binding sites must be considerably separated from one another, since one agonist molecule is bound per α subunit. Moreover, kinetic differences between the two sites have been reported (Weber and Changeux, 1974a; Watters and Maelicke, 1983), and monoclonal antibodies can be raised that appear to recognize selectively one of the sites (Mihovilovic and Richman, 1984). Thus, the observed cooperativity must involve molecular rearrangements transmitted over considerable distances. The presence of noncompetitive antagonists induces shifts in the affinity of AchR for acetylcholine and its analogues. On the other hand, prolonged exposure to agonists results in the reversible transition of the receptor to a desensitized state characterized by a higher affinity for agonists (Quast *et al.*, 1978; Weiland and Taylor, 1979) and a low probability for channel opening. This ability to undergo agonist-mediated affinity state transitions seems to be critically dependent on the establishment of certain correct AchR–lipid interactions, since desensitization-related transitions are not observed for the detergent-solubilized receptor (Sugiyama and Changeux, 1975), and the kinetics of affinity changes are modified by a variety of general anesthetics (Young *et al.*, 1978).

The essential need for the establishment of certain critical AchR–lipid interactions for the preservation of AchR functionality has also been established by a variety of experimental approaches, including perturbation of the lipid environment in native membranes with phospholipase (Andreasen and McNamee, 1977; Andreasen *et al.*, 1979) and reconstitution using lipid mixtures of defined composition (Dalziel *et al.*, 1980; Heidmann *et al.*, 1980; Anholt *et al.*, 1981; Criado *et al.*, 1982, 1984; Martinez-Carrion *et al.*, 1982; Ellena *et al.*, 1983; Fong and McNamee, 1986). These studies clearly indicate that the composition and integrity of the lipid medium solvating AchR are critical factors for the manifestation of AchR allosteric properties and gating ability. However, individual experiments should be interpreted with caution since it has been pointed out that not only the type of lipid but also the reconstitution protocol play a role in obtaining the proper AchR configuration (Martinez-Carrion *et al.*, 1982).

Even though considerable progress has recently been made in mapping of the agonist binding site in the α subunit, our understanding of the allosteric behavior of the AchR is still incomplete, mainly owing to the facts that the noncompetitive antagonist sites have not yet been identified and that no spectroscopic probe able to report the conformational changes related to channel opening is available to date.

3. PTSA: A PROBE FOR MEASURING AchR-MEDIATED IONIC FLUXES IN THE PHYSIOLOGICAL TIME SCALE

3.1. Stopped-Flow Assays for AchR-Mediated Ionic Fluxes

In vivo, the series of events leading to receptor activation and nervous signal transmission within the cholinergic synapse takes place in the millisecond time scale. The first methods devised to quantitate the response of AchR to agonists were based on measuring the uptake or the efflux of a radioactive cation tracer, usually $^{22}Na^+$, by a variety of filtration techniques (Kasai and Changeux, 1971). Such methods are still widely used, because of their technical simplicity, speed, and lack of particular requirements in terms of equipment and sample preparation. However, they do not provide an accurate picture of the initial rapid events following receptor activation, since in spite of recent improvements (Paraschos *et al.*, 1983), they can hardly be extended to the subsecond time scale.

To circumvent this problem, a variety of rapid mixing techniques have been proposed. These techniques fall mainly into two categories: quenched-flow techniques (Hess *et al.*, 1979, 1981, 1982; Neubig and Cohen, 1980; Cash and Hess, 1981) and stopped-flow methods (Moore and Raftery, 1980; Karpen *et al.*, 1983; Gonzalez-Ros *et al.*, 1984). The principle of quenched-flow techniques involves the rapid mixing of the reagents (milliseconds), followed by the termination of the reaction by either chemical or physical means. Once the reaction has been stopped, the detection of its extent can be accomplished in a more convenient time scale. On the other hand, stopped-flow techniques provide a more elegant and probably more accurate alternative, since both mixing of reagents and detection of reaction products are achieved in the rapid time scale.

The stopped-flow methods described to date for the determination of AchR-mediated ionic fluxes share a common principle, described in Figure 1. AchR vesicles are loaded with a fluorophore and introduced in one of the stopped-flow driving syringes. A solution of an ion able to quench the fluorescence of the probe is introduced in the other syringe. This solution may contain cholinergic agonists. Upon rapid mixing of both solutions, the influx of the ion can be followed by recording the time-dependent decrease in the

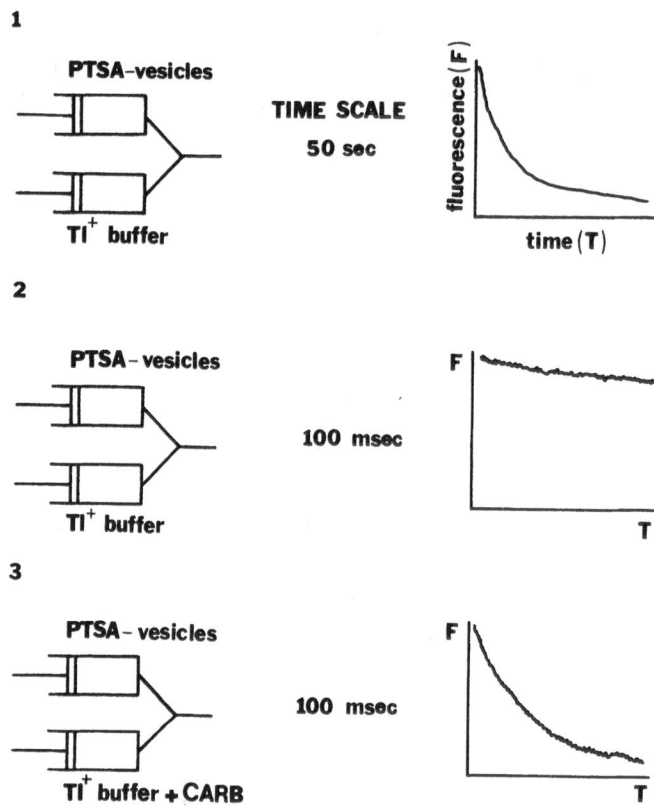

FIGURE 1. Stopped-flow assay for AchR-mediated ionic fluxes. (1) and (2) Fluorophore-loaded AchR membranes and a solution of the quencher are rapidly mixed and the slow decrease in fluorescence intensity due to "passive" quencher in flux is recorded. (3) A rapid flux of the quencher ion is obtained in the presence of cholinergic agonists.

fluorescence of the internally trapped probe. The slow rate of an ion influx in the absence of cholinergic agonists most probably reflects a nonspecific, non-AchR-mediated passive translocation occurring through the membrane by a mechanism not yet well understood. On the other hand, simultaneous mixing of membranes with agonists and the ionic quencher results in the enhancement of the rate of translocation by several orders of magnitude as the AchR channel opens, allowing ions to move according to their concentration gradient.

The available stopped-flow methods differ mainly in the nature of the fluorophore and/or the ion employed as a quencher. The first method reported employed 8-amino-1,3,6-naphthalene trisulfonate (ANTS) and Tl^+, whose ionic radius is similar to that of K^+ (Moore and Raftery, 1980). Subsequently, a similar method was reported employing anthracene-1,5-disulfonic acid (ADS), and Cs^+ (Karpen *et al.*, 1983). We have used 1,3,6,8-pyrene tetrasulfonate (PTSA) as the trapped fluorophore and Tl^+ as the ionic quencher.

3.2. Stopped-Flow Assay with PTSA and Thallous Ion

When compared to ANTS, PTSA presents two definite advantages: a considerably higher quantum yield and a greater stability that eliminates problems associated with photolysis of the probe during the experimental time of exposure to the intense excitation light. Moreover, PTSA is more highly charged than either ADS or ANTS. This is a desirable property since it contributes to minimize leakage of the trapped probe to the exterior of sealed vesicles. In fact, no significant leakage has been detected when sealed vesicles were stored at 4°C for several days. The fluorescent assay is performed in several steps:

1. *Loading of AchR vesicles*: Loading of PTSA into AchR vesicles is efficiently performed by the freeze-thaw technique (Pick, 1981). A 2–3 ml aliquot of AchR membranes containing 3–4 mg/ml protein, 4.3 mM PTSA in Hepes 10 mM, 100 mM $NaNO_3$, pH 7.4 is rapidly frozen in liquid nitrogen and allowed slowly to thaw at 4°C.

2. *Removal of the extravesicular fluorophore*: The freshly thawed vesicles are vigorously homogenized (Polytron, setting 5, 1 min), and the external fluorophore is removed by gel filtration on a Sepharose 6MB column. We have estimated on the basis of α-Bgt binding site recovery that chromatography on Sephadex G-25 or G-50 (coarse grade), or Sepharose 2B or 4B, yields poor recoveries in the 25–30% range. Conversely, a quantitative recovery is approached by means of Sepharose 6MB.

3. *Rapid mixing and recording of the fluorescence decay*: The loaded membranes are placed in one of the stopped-flow syringes. The other is filled with a 30-mM solution of $TlNO_3$ in 70 mM $NaNO_3$, 10 mM Hepes, pH 7.4 with or without varying concentrations of agonist. Upon rapid mixing, the fluorescence decay is recorded in a modified Durrum D-110 stopped-flow spectrophotometer set up in the fluorescence mode. Excitation is performed at 378 nm and emission is monitored using a Corning 3-75 filter. The data are stored in a Physical data model 514 A Transient recorder and visualized in a Tektronix oscilloscope.

4. *Analysis of the data*: Tl^+ quenching occurs through a collisional dynamic mechanism, characterized by a Stern–Volmer constant of 39.4 M^{-1} at room temperature (Gonzalez-Ros *et al.*, 1984). The spectroscopic traces can be computer fitted to the Stern–Volmer equation described by Moore and Raftery (1980) by means of an iterative least-squares algorithm.

Under our experimental conditions, and for a final Tl^+ concentration in the reaction cell of 15 mM, the higher concentrations of cholinergic agonists that can be tested are in the 100-μM range (for carbamylcholine), since the use of higher concentrations results in most of the Tl$^+$ influx occurring within the instrument's "dead time" (~5 msec). The effect of carbamylcholine concentration on the apparent pseudo-first-order rate constant of Tl$^+$ influx is shown in Figure 2.

We have applied the stopped-flow assay described above to a variety of problems including the characterization of AchR inhibition by either polyclonal (Gonzalez-Ros *et al.*, 1984) or monoclonal (Donnelly *et al.*, 1984) antibodies, as well as to the study of the thermal events resulting in AchR denaturation (Soler *et al.*, 1984) and of the effect of alkaline extraction on the properties of AchR and its surrounding lipid bilayer (Martinez-Carrion *et al.*, 1984).

4. PYRENE-1-SULFONYL AZIDE (PySA): A PROBE FOR THE STUDY OF THE AchR–LIPID INTERFACE

Our group started the use of PySA as a selective probe for the transmembrane fragments of AchR in contact with the lipid bilayer. In our first report (Sator *et al.*, 1979a), the probe was synthesized by reaction of sodium azide and pyrene-1-sulfonyl chloride. A radioactive derivative was obtained by tritiation by the Whilzback method. The sulfonyl azide derivative was an extremely lipophilic, hydrophobic probe, very soluble in organic solvents such as chloroform, ether, benzene, or acetone, but sparingly soluble in water (~10^{-5} M, at room temperature). These solubility properties allow for an almost complete incorporation of the probe into the lipid phase of membrane suspensions, and therefore for a selective labeling of the transmembrane fragments of integral proteins. Since our initial report, the probe has become commercially available from Molecular Probes (Roseville, MN).

PySA is extremely sensitive to photolysis. Hence, all operations performed during synthesis and handling of PySA should be carried out in the dark, with all glassware protected from light. Care should also be taken to minimize exposure of the azide to room temperature.

4.1. Optical Properties of PySA and Its Photoproducts

Irradiation of PySA with long-wavelength (>300 nm) UV light produces a rapid photolysis that leads to the generation of reactive nitrenes. When the irradiation is performed after incorporation of PySA into AchR membranes, the PySA–membrane suspensions turn brown from the initial yellow color, upon generation of the photoproducts. However, absorbance changes associated with photolysis are difficult to follow quantitatively because of the

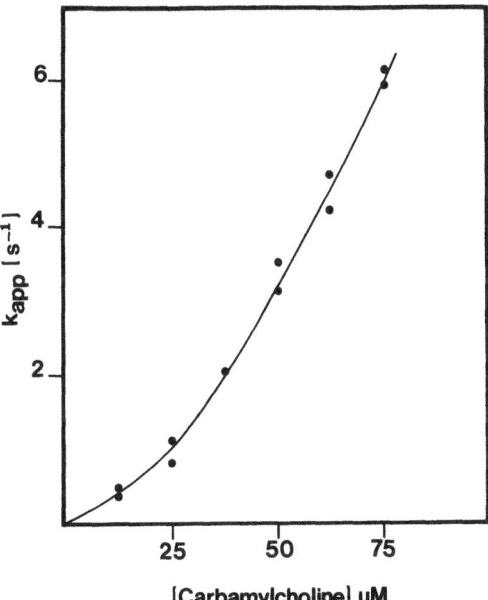

FIGURE 2. Dependence of the apparent Tl$^+$ flux rate on the concentration of carbamylcholine. Apparent rate constants were obtained from computer fitting of the kinetic traces from stopped-flow experiments to the Stern–Volmer equation. Kinetic traces were generated by rapid mixing of PTSA-loaded, AchR-enriched membranes (4 mg/ml protein concentration) with a 30-mM solution of TlNO$_3$ containing varying concentrations of the agonist. From Gonzalez-Ros *et al* (1984).

high light scattering resulting from the presence of large vesicles in suspension. Thus, the appearance of photoproducts and their spectroscopic properties are more conveniently followed using dilute (10 μM) aqueous solutions of PySA. Figure 3 shows the absorption spectra of PySA and its photoproducts. Photolysis proceeds with an increase in the absorbance, which is fourfold increased at 347 nm, after completion of the process.

Upon excitation at 346 nm, the PySA photoproducts, probably a complex mixture of pyrenesulfonylamide derivatives, show three emission maxima, centered around 380, 400, and 415 nm (Figure 4), whose exact positions depend on the polarity of the solvent. The fluorescent properties of PySA photoproducts have been studied employing a pyrenesulfonylamide derivative (PySAH) as a model compound (Gonzalez-Ros *et al.*, 1979a). The fluorescence lifetime values for PySAH are highly dependent on the solvent composition, as illustrated in Table I. In chloroform–methanol mixtures, the lifetime diminishes as the chloroform content is increased, while the emission intensity decreases at higher chloroform/methanol ratios (Figure 5).

4.2. Labeling of the AchR

4.2.1. Labeling Conditions

Labeling of native AchR is performed in two steps. First, PySA is incorporated into AchR-enriched membranes. For this purpose, an aliquot of a

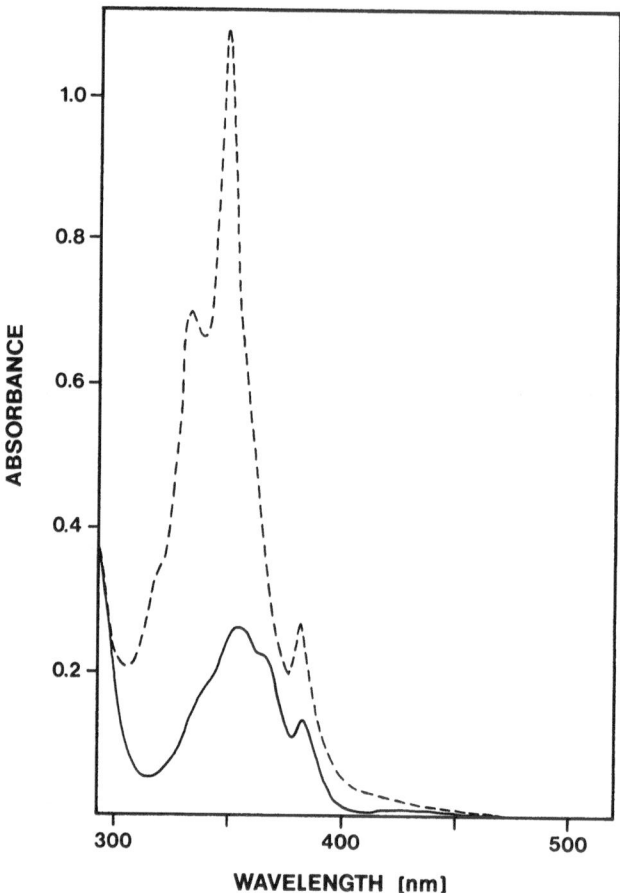

FIGURE 3. Absorption spectra of PySA before (solid line) and after (broken line) irradiation. A 30-μM PySA solution in 10 mM sodium phosphate buffer, pH 7.4, containing 0.03% Triton X-100, 0.02% NaN_3, and 3 mg/ml borine serum albumin, was irradiated with long-wavelength ultraviolet light (>300 nm) through a 1-cm pathlength quartz cuvette for 15 min at room temperature. From Sator *et al.* (1979a).

stock $CHCl_3$ solution is dried onto glass beads (vitris #16-220) under a stream of argon. AchR-enriched membranes, suspended in 10 mM Hepes, 100 mM $NaNO_3$, 1 mM EDTA, and 10 mM NaN_3, pH 7.5 to achieve a final protein concentration of 2 mg/ml, are added in an appropriate amount to achieve a final concentration of 1 mM PySA. The mixture is stirred under argon, for 12 – 17 hr at 4°C. Photoactivation is performed with 2–3 ml of the loaded membranes, placed in a 5-ml quartz cuvette that has been flushed with argon prior to irradiation. Irradiation is carried out for 12 min at room

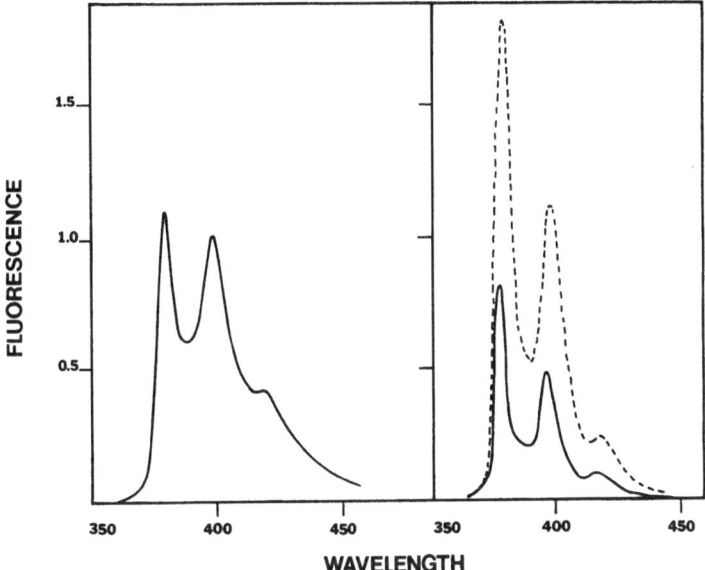

FIGURE 4. (A) Fluorescence emission spectrum of PySA incorporated into AchR membranes, after irradiation with long-wavelength ultraviolet light, for 13 min, in 5 mM Tris buffer, pH 7.4, containing 0.5 M sucrose and 0.02% NaN$_3$. PySA photoproducts concentration was approximately 7 μM and total protein concentration was 0.13 mg/ml. Excitation at 346 nm. (B) Fluorescence emission spectra of a 2 μM solution of PySAH in spectra grade hexane (solid line) and methanol (broken line), upon excitation at 346 nm. From Gonzalez-Ros *et al.* (1979a).

temperature with gentle stirring, by means of a UVL-56 lamp (Ultraviolet Products, Inc.), with a 0-54 Corning filter (50% transmission at 320 nm), and with a distance of 2 cm between the UV source and the cuvette.

Table I
Fluorescence Lifetimes Values of PySAH (2×10^{-6} M) in Different Solvents[a]

Solvent	Fluorescence lifetime (nsec)[b]
Methanol	14.54 ± 0.03[c]
Chloroform/methanol (1 : 9, v/v)	14.15 ± 0.05
Chloroform/methanol (1 : 2, v/v)	13.21 ± 0.09
Chloroform/methanol (1 : 1, v/v)	12.15 ± 0.04
Chloroform/methanol (2 : 1, v/v)	10.91 ± 0.07
Chloroform	9.28 ± 0.06
Hexane	12.36 ± 0.09
Cyclohexane	18.84 ± 0.15

[a] From Gonzalez-Ros *et al.* (1979a).
[b] Emission filters: Corning 4-96 and 3-144.
[c] Standard deviations from four different determinations.

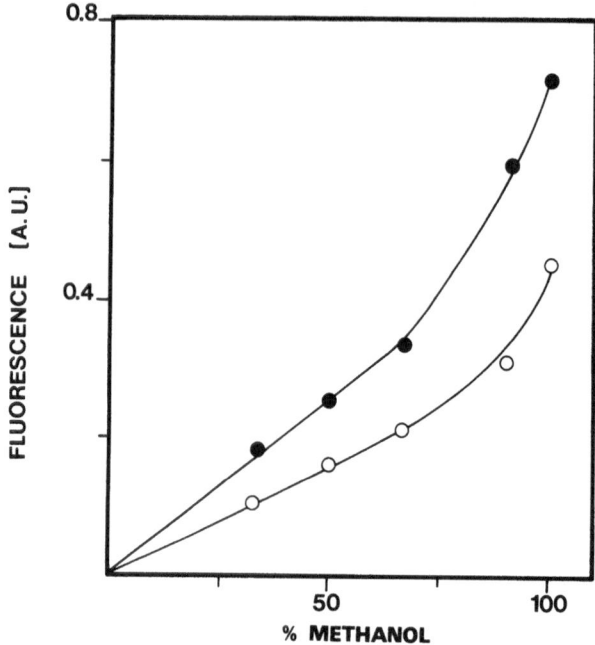

FIGURE 5. Effect of solvent composition on the fluorescence intensity of PySAH at 379 (closed circles) and 396 (open circles) nm. The spectra were recorded for 2 μM PySAH solutions in chloroform–methanol mixtures at different proportions, upon excitation at 346 nm. From Gonzalez-Ros *et al.* (1979a).

4.2.2. Removal of Unincorporated and Noncovalently Associated Photoproducts

The PySA photoproducts are slightly more water soluble than the parent compound. AchR membranes can be separated from the unincorporated photoproducts dissolved in the aqueous medium by several cycles of centrifugation and resuspension in fresh buffer (Sator *et al.*, 1979a), or, preferably, by layering the membrane suspension onto a 45% sucrose solution in Hepes buffer and centrifuging for 1 hr at 29,000 rpm in a Beckman type 65 rotor. The membrane fragments are collected at the sucrose interface, but most of the unincorporated PySA and its photoproducts form large aggregates in aqueous solution and are forced onto the walls of the centrifuge tubes. After this centrifugation step, most of the remaining photoproducts are noncovalently embedded in the lipid bilayer. When intact membranes are required for further work, the majority of the noncovalently bound photoproducts may be removed by several exchanges with lipid vesicles prepared by sonicating *Tor-*

pedo lipid extracts (Gonzalez-Ros *et al.*, 1979a). The effectiveness of such a treatment is shown in Figure 6. It can be seen that as much as 70% of the PySA photoproducts associated with AchR-enriched membranes are removed by four exchange cycles. Solubilized receptor can be freed from the majority of noncovalently bound PySA photoproducts by gel filtration chromatography on Sephacryl S-200 (Sator *et al.*, 1979a) in the presence of 0.03% Triton X-100. However, in this case, some PySA photoproducts may remain associated with the AchR–detergent micelles. A more effective procedure takes advantage of the use of an easily dialyzable detergent, β-D-octylglucopyranoside (OG). Membranes solubilized in the presence of 1% OG, typically overnight at 4°C, are centrifuged and the solubilized supernatant is submitted to several

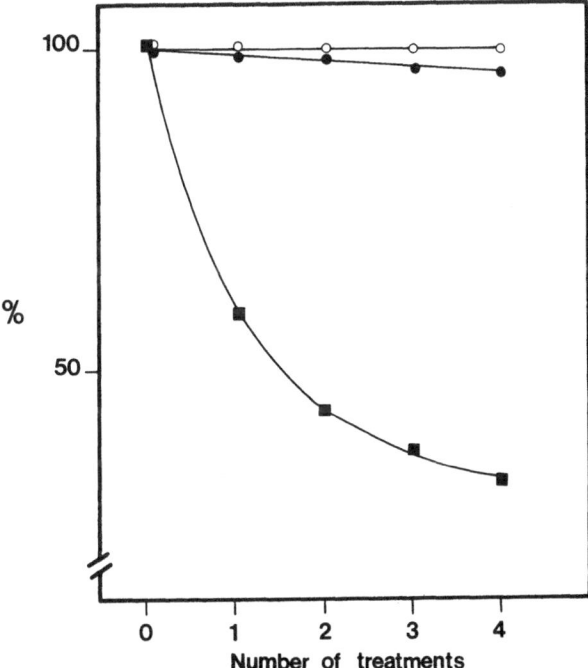

FIGURE 6. Removal of unbound PySA photoproducts from labeled AchR-enriched membranes by treatment with lipid vesicles. PySA-labeled membranes were incubated with lipid vesicles prepared by sonication of total lipid extracts from *Torpedo californica*. The ratio vesicle lipids/AchR membrane lipids was roughly 2 : 1. After 2 hr of incubation at room temperature, plain lipid vesicles and AchR membranes were separated by centrifugation for 1 hr at 35,000 *g* over a 0.5-M sucrose cushion in 5 mM Tris-HCl, pH 7.4. The resulting pellet was resuspended and assayed for total protein (●), α-Bgt binding sites concentration (○), and radioactivity associated with PySA photoproducts (■). From Gonzalez-Ros *et al.* (1979a).

cycles of concentration and redilution with buffer containing 0.1% OG, in an Amicon concentrator equipped with a YM-100 or XMA-100 membrane (Clarke *et al.*, 1987). The latter procedure seems to be the more effective and reliable for stoichiometry determinations.

4.2.3. Labeling Stoichiometry

Both the labeling stoichiometry and the specificity of PySA for the different AchR subunits can be controlled by careful manipulation of the experimental conditions during photolabeling. A 13-min irradiation of a concentrated AchR membrane suspension (8 mg/ml protein concentration) in the presence of a 1-mM final concentration of PySA, at a distance of 3.5 cm from the light source results in the preferential labeling of the β and γ subunits, with a residual tagging of the α and δ subunits (Sator *et al.*, 1979a). This apparent specificity can be abolished by diminishing the protein concentration of the sample and reducing the distance between the sample and the light source, so as to decrease light scattering and increase the amount of light impinging on the sample. Thus, at a protein concentration of 2 mg/ml and a distance from the light source of 2.1 cm, a 12-min irradiation of a sample containing 1 mM PySA results in the labeling of all four types of subunit. Under these conditions, one probe molecule is incorporated per receptor subunit (Clarke *et al.*, 1987).

4.3. Applications of PySA to the Study of AchR Structure and Function

Three properties of PySA make it a particularly promising probe for the study of AchR structure and function.

1. Due to its insolubility in aqueous media, PySA selectively labels the transmembrane portions of the AchR.
2. PySA seems to be a nonperturbing probe for AchR. As opposed to a variety of treatments causing minor perturbations either on the lipid component of AchR or in the transmembrane fragments of AchR itself, labeling with AchR does not impair AchR functionality, as tested by different criteria. PySA-labeled AchR fully retains its ability to bind α-neurotoxins and to undergo reversible affinity state transitions upon incubation with agonists (Gonzalez-Ros *et al.*, 1979b). Moreover, introduction of one probe per receptor subunit does not impair the ion translocating properties of AchR, since similar dose–response curves for agonists can be obtained for labeled and control membranes (Clarke *et al.*, 1987).

3. PySA photoproducts possess favorable fluorescent properties, including the possibility to act as an acceptor of Trp fluorescence through fluorescence energy transfer. Thus, PySA photoproducts are suitable for the analysis of conformational events affecting the portions of the AchR accessible to membrane lipids.

We have taken advantage of these favorable properties of PySA to probe and compare the transmembrane topology of AchR in native, alkaline extracted and reconstituted membranes, to study ligand-induced, long-range effects at the transmembrane regions of AchR, and to detect conformational changes in the receptor molecule associated with treatments resulting in a perturbation of the lipid environment.

4.3.1. Transmembrane Topology of AchR

Labeling of AchR-enriched membranes under conditions of low total protein concentration and low irradiation distance results in the labeling of all four types of subunits, in a stoichiometry 2 : 1 : 1 : 1 relative to the subunit stoichiometry of $\alpha_2\beta\gamma\delta$. This result confirms the transmembrane nature of all the subunits comprising the AchR, postulated by studies involving the selective proteolitic degradation from the interior and the exterior of sealed AchR vesicles (Klymkowsky *et al.*, 1980; Lindstrom *et al.*, 1980; Strader and Raftery, 1980). Moreover, this model is consistent with the results obtained by others with other photoactivatable, lipid-soluble probes like [³H]adamantanediazirine (Middlemas and Raftery, 1983) and arylazido phospholipids (Giraudat *et al.*, 1985). The preferential labeling of the β and γ subunits under conditions of diminished light intensity could be related *a priori* to a greater degree of exposure of these subunits to the lipid environment. However, an alternative explanation based on the presence of amino acid side chains of higher reactivity in the transmembrane portions of these subunits could more likely account for these results, even in the event of a similar overall degree of exposure of all types of subunit to the lipid components of the membrane. In a study comparing the labeling stoichiometries of the AchR complex and AchR subunits in the native, alkaline extracted and reconstituted states, similar stoichiometries relative to the $\alpha_2\beta\gamma\delta$ subunit composition were found for all three types of preparation (Clarke *et al.*, 1987). This observation suggested a similar transmembrane topology for these preparations and therefore lends further support to the use of alkaline extracted and reconstituted membranes as simplified models for the study of AchR function. It should be emphasized that, in this study, total extracts of *Torpedo* lipids were employed in the reconstitution experiments and that an extrapolation of the results obtained to complex mixtures of synthetic lipids is risky.

4.3.2. Ligand-Induced Effects in the Transmembrane Fragments of the AchR

The selectivity of PySa labeling for the transmembrane regions of AchR, together with the nonperturbing character of the probe on the functions of the receptors, has made it possible to examine ligand-induced effects of possible functional importance which may be affecting the transmembrane segments of the protein, or, more precisely, the protein–lipid interface.

The fluorescence lifetime values of PySa-labeled AchR membranes were found to be relatively insensitive to the preincubation of the AchR samples with cholinergic agonists or antagonists (Gonzalez-Ros *et al.*, 1983). No significant changes in the control value of 13.7 nsec were observed upon preincubation of the samples with micromolar concentrations of agonists, some α-neurotoxins, or local anesthetics. However, a consistent decrease in the lifetime values was observed when the concentration of local anesthetics, but not of agonists, was increased to the millimolar range (Table II). Under these conditions, the lifetime value for the probe in membrane-bound AchR

Table II
Fluorescence Lifetime Values of PySA-Labeled AchR-Rich Membrane Fragments in the Presence of Ligands or Membrane Perturbants[a]

Ligand	Fluorescence lifetime (nsec)
None (control)	13.7
Carbamylcholine (10^{-6} M)	13.6
Carbamylcholine (10^{-3} M)	13.7
α-Bungarotoxin (2.5×10^{-7} M)	13.6
α-Bungarotoxin (2.5×10^{-7} M) and then carbamylcholine (10^{-3} M)	13.7
α-Cobratoxin (2.5×10^{-7} M)	13.4
d-Tubocurarine (2.5×10^{-5} M)	13.4
Tetracaine (10^{-6} M)	13.5
Carbamylcholine (10^{-3} M) and then tetracaine (10^{-6} M)	13.0
α-Cobratoxin (2.5×10^{-7} M) and then tetracaine (10^{-6} M)	13.3
Tetracaine (10^{-3} M)	8.7
Carbamylcholine (10^{-3} M) and then tetracaine (10^{-3} M)	8.4
α-Cobratoxin (2.5×10^{-7} M) and then tetracaine (10^{-3} M)	8.3
Triton X-100 (1%)	8.2
Labeled purified AchR (protein alone in a 0.1% Triton X-100 solution)	8.4

[a] From Gonzalez-Ros *et al.* (1983).

approaches that obtained for the probe when attached to solubilized, purified receptor. This suggests that the severe perturbation in the membrane arising from the presence of high concentrations of tetracaine (a local anesthetic) results in a higher degree of exposure to the aqueous environment of regions of the AchR transmembrane fragments otherwise shielded by the lipid bilayer. In accordance with that view, perturbations of the lipid environment of the PySA attached to the AchR by treatment with 1% Triton X-100 or 10^{-3} M tetracaine produce an increase in the ability of nitromethane to quench PySA fluorescence collisionally, suggesting that both treatments result in an increased accessibility of the aqueous medium to the PySA-labeled regions.

Even though the fluorescence lifetime of AchR-bound PySA is not affected by the presence of carbamylcholine at concentrations above the dissociation constant, ligand-induced conformational events affecting the transmembrane fragments of AchR could not be detected by studying the pattern of quenching of PySA fluorescence by nitromethane. As shown in Table III, preincubation of AchR with 1 mM carbamylcholine reduces the accessibility of PySA to the quencher, as indicated by a small but sizable decrease in the apparent second-order rate constant for the collisional quenching process. On the other hand, preincubation with α-Bgt has an opposite effect, leading to an increase in the apparent rate constant for nitromethane quenching, that is, in the accessibility of PySA to the quencher (Gonzales-Ros *et al.*, 1983). It should be emphasized that these results were obtained with AchR preparations preferentially labeled on the β and γ subunits, while the binding of cholinergic ligands occurs in the extracellular domains of the

Table III
Effect of Cholinergic Ligands on the
Apparent Rate Constants of Nitromethane
Quenching of PySA-Labeled AchR
Membranes from *Torpedo californica*[a]

Ligand	$k_{2app} \times 10^{-8}$ $(M^{-1} \cdot sec^{-1})$[b]
None (control)	29.7
Carbamylcholine (10^{-6} M)	29.5
Carbamylcholine (10^{-3} M)	25.9
α-Bgt[c]	35.8
α-Bgt and then carbamylcholine (10^{-3} M)	35.6

[a] From Gonzalez-Ros *et al.* (1983).
[b] Apparent second-order rate constants for the collisional quenching process were obtained from the slopes of Stern–Volmer plots of the quenching curves.
[c] Samples were preincubated for 30 min with a two-fold excess of α-Bgt over total α-Bgt binding sites.

α subunit. Therefore, the finding of altered quenching patterns in the presence of agonists and antagonists should reflect the occurrence of conformational events that must be transmitted over a considerable distance within the receptor–matrix organization.

Since the excitation spectrum of PySA overlaps the fluorescence emission of Trp, PySA can act as the acceptor of resonance energy transfer of Trp. The apparent efficiency of energy transfer from Trp to PySA has been shown to be sensitive to the conformational state of AchR, and it improves after detergent solubilization (Clarke *et al.*, 1987). Interestingly, no significant differences in the resonance energy-transfer efficiency were detected upon preincubation of PySA-labeled AchR with saturating concentrations of carbamylcholine, decamethonium, *d*-tubocuranine, or α-Bgt. This observation suggests that the relative average distances between PySA and the neighboring Trp residues remain constant, even in the presence of a variety of cholinergic effectors. Thus, it is possible that the conformational rearrangements detected by nitromethane quenching of PySA fluorescence would affect regions of the AchR different from those located in the immediate vicinity of the tryptophan residues.

5. PYRENE MALEIMIDE (PM): THE LABELING OF A FUNCTIONALLY RELEVANT SULFHYDRYL GROUP

Pyrene maleimide (PM) is a hydrophobic alkylating agent that reacts preferentially with protein thiols (Weltman *et al.*, 1973; Wu *et al.*, 1976). PM is soluble in organic solvents such as Me_2SO and $CHCl_3$, but the solubility in ethanol is limited (< 2 mM). In aqueous solvents, PM has a saturation concentration lower than 20 μM at 25°C. When the maximal solubility is exceeded, PM forms large aggregates that precipitate out of solution but do partition into liposomes and biological membranes.

The fluorescence of PM is markedly increased upon reaction with proteins. It has been proposed that the olefinic double bond of the maleimide ring effectively quenches the fluorescence of substituents on the imide nitrogen (Kanaoka *et al.*, 1968; Weltman *et al.*, 1973). This quenching could be primarily effected by radiationless transitions resulting from intersystem crossing to the $^3(\pi \rightarrow \pi^*)$ state through the $^1(n \rightarrow \pi^*)$ state of the intact maleimide. Reaction with thiol groups in proteins could render intersystem crossing unfavorable through an increase in the transition energy of the $^1(n \rightarrow \pi)$ state; this would allow $^1(\pi \rightarrow \pi^*)$ transitions to predominate, resulting in fluorescence emission. The fact that PM does not fluoresce prior to adduct formation allows for a convenient assay of the kinetics of protein labeling, which can be monitored fluorometrically.

5.1. Labeling of Solubilized Receptor

In an early report (Sator *et al.*, 1978), PM was shown to react readily with Triton X-100-solubilized AchR. The labeling could be performed on native preparations, but also on AchR pretreated with *N*-ethylmaleimide (NEM), to block the exposed SH groups. Tagging of the AchR with PM did not affect the binding of either α-bungarotoxin or propidium, a cationic site ligand.

The absorption and emission spectra of PM bound to solubilized AchR are shown in Figure 7. The absorption spectrum displays three maxima at 313, 327, and 343 nm, while the emission spectrum is characterized by three emission bands centered at 374, 385, and 394 nm. The lifetime of the bound probe was estimated to be 106 ± 5 nsec (excitation wavelength was 347.1 nm and emission 400 nm). Some insight as to the microenvironment surrounding the bound PM and its accessibility was obtained from analysis of the quenching patterns of PM fluorescence by thallium nitrate, nitromethane, and potassium iodide. It was found that all three compounds quenched the fluorescence of the probe, free in solution, more effectively than that of the receptor-bound PM. This difference in the effectiveness of the quencher was more pronounced for KI, the less-effective quencher of receptor-bound PM fluorescence, suggesting the presence of negatively charged groups in the vicinity of the sites labeled by PM.

The extent of detergent–AchR interaction in Triton X-100-solubilized preparations was analyzed by taking advantage of the high lifetime value of receptor-bound PM, which allows for the calculation of the rotational relaxation time of a large particle from fluorescence polarization data. A value of 1170 ± 150 nsec in 10 mM Tris-HCl, pH 7.4, 0.03% Triton X-100, was

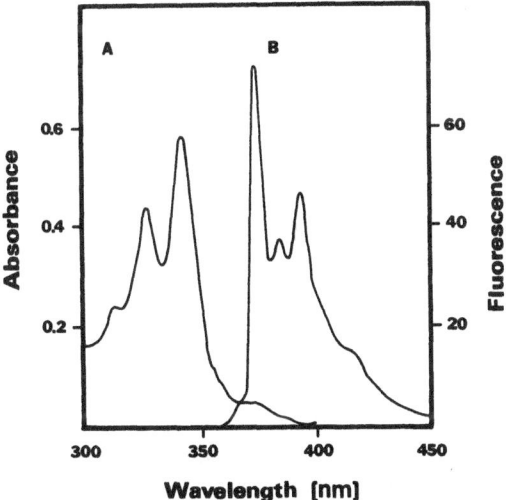

FIGURE 7. Absorption spectrum (A) and fluorescence spectrum (B) of PM covalently bound to solubilized AchR, in 10 mM Tris-HCl, pH 7.4, containing 0.03% Triton X-100 and 0.02% NaN$_3$. The fluorescence spectrum was obtained upon excitation at 343 nm. From Sator *et al.* (1978).

obtained, consistent with a molecular weight of 890,000. Assuming a spherical shape for the receptor–detergent complex, the above value corresponds to a particle with Stokes radius of 72 Å. This value is consistent with the one obtained by chromatography on Sepharose 6B columns (Raftery *et al.*, 1971).

5.2. Labeling of Native Membranes

The reaction of PM with AchR in its native environment has recently been described (Clarke and Martinez-Carrion, 1986). Because of its hydrophobic nature, PM readily partitions into lipid bilayers. The probe can be incorporated into AchR membranes using conditions similar to those described for the incorporation of PySA. The reaction of PM with AchR proceeds rapidly after the incorporation of the probe and can be followed by the increase in the fluorescence of the sample or by means of a radioactive derivative synthesized according to the procedure of Weltman *et al.* (1973).

Alkylation of AchR by PM results in a marked inhibition of the Carb-mediated cation translocation ability of the receptor. The extent of this inhibition is dependent on the concentration of PM and amounts to 75% at the higher PM concentration tested (1 mM). The presence of Carb in the reaction medium does not protect AchR from the PM-mediated inhibition. However, other functional properties of the AchR, namely, the ability to bind bungarotoxin and agonists and the agonist-induced affinity state transitions, seem to be unimpaired by PM labeling. Fluorometric detection of the protein bands after SDS-PAGE showed that each AchR subunit is labeled by PM. Moreover, the relative intensity of each band appeared to be roughly proportional to the intensities after staining with Coomassie blue, and no differences were detected in AchR samples labeled in the presence and absence of agonists. The labeling stoichiometries were determined for each subunit by electroelution from preparative SDS gels of the AchR subunits labeled with a radioactive PM derivative. The amount of PM incorporated was deduced from the radioactivity associated with the band, whereas the protein concentration was determined by amino acid analysis. A labeling stoichiometry of 2 : 1 : 1 : 1, relative to the 2 : 1 : 1 : 1 subunit stoichiometry, was found. Thus, it appears that each polypeptide comprising the AchR complex possesses one PM labeling site.

The ability of a series of substituted maleimide of different hydrophobicity to compete with PM for the alkylation sites was tested. In direct competition experiments with equimolar concentrations of PM, *N*-ethylmaleimide was practically ineffective, whereas phenylmaleimide and naphthylmaleimide decreased both the rate of labeling and the amount of PM incorporated at the completion of the reaction. Since naphthylmaleimide was a more effective inhibitor of PM labeling, the degree of protection against PM afforded by the three alkylating agents showed a good correlation with their hydrophobicity (Figure 8). This observation strongly supports the perception of PM labeling sites being located in a hydrophobic pocket of the AchR displaying limited accessibility to water-soluble compounds.

The site of alkylation of AchR-enriched membranes was shown to exhibit a characteristic pH dependence with an apparent pK_a of 7.5. This pH dependence, together with the observation that reductive methylation of lysine residues does not affect either the extent or the kinetics of PM labeling, supports the view that the labeled residue in each subunit is a cysteine. Additional evidence was obtained by comparison of the amino acid analysis profiles of the labeled subunits with those of appropriate standards.

Even though the positions of the labeled cysteine residues in the primary structure of each subunit have not yet been identified, the following observations clearly suggest that the alkylation sites display a high degree of homology: (1) the pH dependence of the alkylation rate showed only one apparent pK_a; (2) in labeling experiments performed with subsaturating concentrations

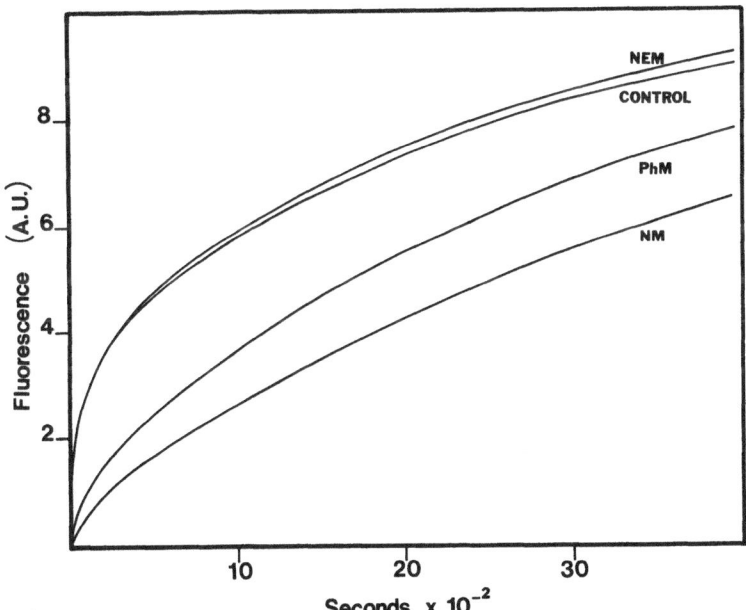

FIGURE 8. Kinetics of PM binding to membrane-bound AchR in the presence of several maleimides. Equimolar concentrations of PM and the competing maleimide derivative (200 μM final concentration) were added simultaneously to AchR-enriched membranes (2.5 mg/ml protein concentration). The reaction of PM with AchR was followed through the increase in PM fluorescence associated with adduct formation. From Clarke and Martinez-Carrion (1986).

of PM, resulting in about 10% of the maximal labeling, each one of the subunits appears to be labeled to a similar extent; (3) the Arrhenius plot for the temperature dependence of PM labeling is linear from 0 to 30°C, with no apparent breaks and suggesting a monophasic behavior; and (4) prelabeling with excess *N*-ethylmaleimide under conditions resulting in a 70% inhibition of PM incorporation affected all four types of subunits equally.

Therefore, PM appears to be probing a homologous class of functionally relevant cysteine residues in the AchR. The role of these residues in the modulation of the ion-gating activity of the AchR will be better understood once the modified cysteine residues are located in the primary structure of the protein.

6. STATE AND ORGANIZATION OF THE LIPID BILAYER IN AchR MEMBRANES

The importance of the nature of the lipids solvating the AchR protein and of the integrity of the lipid bilayer in AchR function has been emphasized by a variety of experimental approaches, including reconstitution of AchR in phospholipid mixtures of defined composition (Criado *et al.*, 1982, 1984; Ellena *et al.*, 1983). We have analyzed several physicochemical features of the lipid bilayer in *Torpedo californica* AchR membranes by means of a series of fluorescent probes, mainly pyrene, DPH, TMA-DPH, and 4-heptadecyl-7-hydrohycoumarine. These probes were employed to characterize the mutual interactions between membrane lipids and the AchR protein, as well as the effects on the state of the lipid bilayer of a variety of cholinergic ligands and other chemicals known to interfere with cholinergic processes.

6.1. Probing of AchR Membranes with Pyrene

Possible changes of fluidity and permeability in the lipid environment of AchR upon binding of cholinergic ligands were investigated in native AchR membranes probed by pyrene (Sator *et al.*, 1979b). The probe can easily be incorporated into the membranes by incubation under mild conditions of a membrane suspension in a test tube whose walls have been coated with pyrene by drying a cyclohexane solution of the fluorophore. Incorporation of pyrene does not impair either the α-Bgt binding ability of the receptor or its ability to undergo affinity state transitions upon incubation with agonists. Moreover, pyrene does not appear to interact directly with AchR in Triton X-100-solubilized preparations.

The emission spectrum of pyrene incorporated into AchR membranes is shown in Figure 9. From the shape of the spectrum and particularly from the ratio of peak heights III (384 nm)/I (373 nm), it was postulated that the

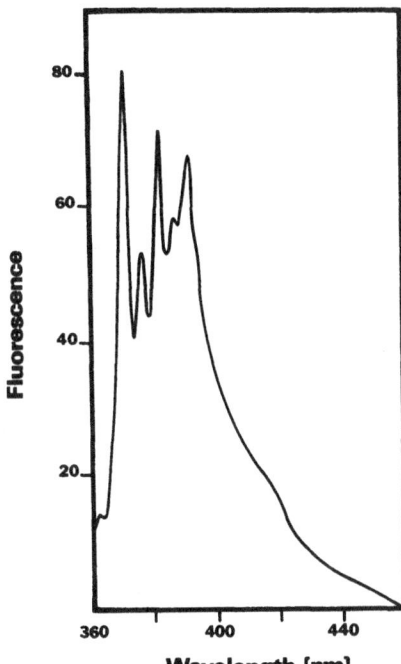

FIGURE 9. Emission spectrum of pyrene (0.1 mM) monomer incorporated into AchR membranes. Excitation wavelength was 335 nm. At concentrations of pyrene higher than 0.5 mM, excimer fluorescence becomes apparent at 475 nm. From Sator *et al.* (1979b).

electroplax membrane lipids should contain a high proportion of unsaturated fatty acids. This was later further substantiated by a detailed analysis of the lipid composition of AchR membranes (Gonzalez-Ros *et al.*, 1982), which showed that AchR-enriched membranes contain high amounts of mono- and polyunsaturated phospholipids.

The lifetime of pyrene singlet excited state was found to be identical in sensitized and desensitized AchR membrane preparations, suggesting that changes in the affinity state of the AchR are not transmitted to the lipid bilayer, or, at least, to the membrane domains probed by pyrene. This lack of effect of cholinergic ligands on the state of the lipid bilayer was further proved by the finding of identical quenching patterns of thallous ion and nitromethane in membranes preincubated with saturating concentrations of either Carb or α-Bgt.

On the other hand, incubation in the presence of the local anesthetics procaine and tetracaine resulted in a marked perturbation of the membrane fluidity as judged by (1) a decrease in pyrene excited-state lifetimes, (2) a change in the absolute quenching constant for thallous ion, and (3) a change in the saturation behavior for the quenching by this ion. Interestingly, tetracaine does not perturb the calorimetric transition assigned to AchR denaturation when present in comparable concentrations (Farach and Martinez-Carrion, 1983).

6.2. Fluidity of the AchR Membranes as Probed by DPH and TMA-DPH

It is generally agreed that the fluorophores DPH and TMA-DPH probe different membrane domains. Above the transition temperature of the lipid bilayer, DPH is assumed to be located within the membrane core, aligned with the phospholipid acyl chains (Shinitzky and Barenholz, 1974; Andrich and Vanderkooi, 1976). On the other hand, the presence of a trimethylammonium head, able to interact with the phospholipid head groups, accounts for a more external location of TMA-DPH in the lipid bilayer, in or near the membrane–water interface (Prendergast *et al.*, 1981).

Both fluorophores have been employed to analyze the fluidity of the lipid environment of AchR (Gonzalez-Ros *et al.*, 1982). The probes were incorporated in plain lipid vesicles, prepared with *Torpedo* electroplax lipids by the detergent dialysis method, and also in native AchR membranes. Comparison of the Arrhenius plots for the temperature dependence of the rotational relaxation times of both probes suggested that the lipid medium surrounding the AchR is a relatively rigid one, owing to the high cholesterol/phospholipid ratio. This rigidity is slightly more pronounced in the surface of the membrane and is increased by the presence of protein, since the rotational relaxation times for both fluorophores were found to be 1.5–2 times higher in the native membranes than in plain lipid vesicles. Thus, the presence of protein in high concentration within the lipid bilayer seems to promote a higher degree of order, no thermal transitions above room temperature were detected for the lipid matrix in any of the samples analyzed, as judged by the linearity of Arrhenius plots for the temperature dependence of the rotational relaxation times.

The studies outlined above provide a general, although still incomplete, picture of the lipid environment surrounding the AchR in native *Torpedo californica* membranes. The high degree of unsaturation of the phospholipid acyl chains provides a fluid medium, with no detectable phase transitions above 20°C. The membrane domain corresponding to the water interface seems to be slightly more rigid than the core of the bilayer. The high cholesterol content limits the fluidity and the permeability of the membrane, and the presence of AchR itself also contributes to mediate a higher degree of order. Thus, AchR seems to influence to some extent the physical state of the surrounding bilayer. This indicates a complementarity between protein and lipid in the membrane since the nature of the lipid is responsible for the functional state of the AchR.

7. SUMMARY

The acetylcholine receptor is a transmembranous protein present at the neuromuscular junction and in the electroplax of electric fishes. As such it contains structural elements in three different environments: exocytoplasmic, endocytoplasmic, and transbilayer (lipid) parts of the membrane. We have used fluorescent probes such as pyrene, pyrene sulfonylazide, pyrene maleimide, and pyrene sulfonic acid to monitor the properties of the following environments: lipid bilayer, protein–lipid interface, exocytoplasmic surface, and endocellular side. Some of these probes bind covalently to the protein, while others sense change in the solvent environment. Through studies of the fluorescence emission spectra, lifetime of fluorescence decay of the singlet excited state, and kinetics of fluorescence quenching, properties of the environment associated with the probes can be detected. Furthermore, the probes are used with receptor in the resting state as well as when neurotransmitter or their pharmacologic analogues are present. The integrated information obtained through the various probes and fluorescence detection techniques provide information regarding environmental changes associated with the state of the lipids in the membrane, as well as those induced by neurotransmitters in native membranes and in those reconstituted with receptor protein and defined phospholipids.

8. REFERENCES

Anderson, D. J., Walter, P., and Blobel, G., 1982, Signal-recognition protein is required for the integration of acetylcholine receptor δ subunit, a transmembrane glycoprotein, into the endoplasmic reticulum membrane, *J. Cell Biol.* **93** : 501–506.

Andreasen, T. J., and McNamee, M. G., 1977, Phospholipase A inhibition of acetylcholine receptor function in *Torpedo californica* membrane vesicles, *Biochem. Biophys. Res. Commun.* **79** : 958–965.

Andreasen, T. J., Doerge, D. R., and McNamee, M. G., 1979, Effects of phospholipase A2 on the binding and ion permeability control properties of the acetylcholine receptor, *Arch. Biochem. Biophys.* **194** : 468–480.

Andrich, M. P., and Vanderkooi, J. M., 1976, Temperature dependence of 1,6-diphenyl-1,3,5-hexatriene fluorescence in phospholipid artifical membranes, *Biochemistry* **15** : 1257–1261.

Anholt, R., Lindstrom, J., and Montal, M., 1981, Stabilization of acetylcholine receptor channels by lipids on cholate solution and during reconstitution in vesicles, *J. Biol. Chem.* **256** : 4377–4387.

Barrantes, F. J., 1983, Recent developments in the structure and function of the acetylcholine receptor, *Int. Rev. Neurobiol.* **24** : 259–341.

Brisson, A., and Unwin, P. M. T., 1985, Quaternary structure of the acetylcholine receptor, *Nature* **315** : 474–477.

Cash, D. J., and Hess, G. P., 1981, Quenched-flow technique with plasma membrane vesicles: Acetylcholine receptor mediated transmembrane ion flux, *Anal. Biochem.* **112** : 39–51.

Changeux, J. P., Devillers-Thiery, A., and Chenouilli, P., 1984, Acetylcholine receptor: An allosteric protein, *Science* **225** : 1335–1345.

Clarke, J. H., and Martinez-Carrion, M., 1986, Labeling of functionally sensitive sulfhydryl containing domains of acetylcholine receptor from *Torpedo californica* membranes, *J. Biol. Chem.* **261** : 10063–10072.

Clarke, J. H., Garcia-Borron, J. C., and Martinez-Carrion, M., 1987, (1-Pyrene)sulfonyl azide is a fluorescent probe for measuring the transmembrane topology of acetylcholine receptor subunits, *Arch. Biochem. Biophys.* **256** : 101–109.

Claudio, T., Ballivet, M., Patrick, J., and Heinemann, S., 1983, Nucleotide and deduced amino acid sequences of *Torpedo californica* acetylcholine receptor γ subunit, *Proc. Natl. Acad. Sci. U.S.A.* **80** : 1111–1115.

Conti-Tronconi, B. M., and Raftery, M. A., 1982, The nicotinic acetylcholine receptor: Correlation of molecular structure with functional properties, *Annu. Rev. Biochem.* **51** : 491–530.

Criado, M., Eibl, H., and Barrantes, F. J., 1982, Effects of lipids on acetylcholine receptor. Essential need of cholesterol for maintenance of agonist induced state transitions in lipid vesicles, *Biochemistry* **21** : 3622–3629.

Criado, M., Eibl, H., and Barrantes, F. J., 1984, Functional properties of the acetylcholine receptor incorporated in model lipid membranes, *J. Biol. Chem.* **259** : 9188–9198.

Dalziel, A. W., Rollins, E. S., and McNamee, M. G., 1980, The effect of cholesterol on agonist induced flux in reconstituted acetylcholine receptor vesicles, *FEBS Lett.* **122** : 193–196.

Damle, V. N., McLaughlin, M., and Karlin, A., 1978, Bromoacetylcholine as an affinity label of the acetylcholine receptor from *Torpedo californica, Biochem. Biophys. Res. Commun.* **84** : 845–851.

Devillers-Thiery, A., Giraudat, J., Bentaboulet, M., and Changeux, J. P., 1983, Complete mRNA sequence of the acetylcholine binding subunit from *Torpedo marmorata* acetylcholine receptor: A model for the transmembrane organization of the polypeptide chain, *Proc. Natl. Acad. Sci. U.S.A.* **80** : 2067–2071.

Donnelly, D., Mihovilovic, M., Gonzalez-Ros, J. M., Ferragut, J. A., Richman, D., and Martinez-Carrion, M., 1984, A noncholinergic site directed monoclonal antibody can impair agonist induced ion flux in *Torpedo californica* acetylcholine receptor, *Proc. Natl. Acad. Sci. U.S.A.* **81** : 7999–8003.

Ellena, J. F., Blazing, M. A., and McNamee, M. G., 1983, Lipid–protein interactions in reconstituted membranes containing acetylcholine receptor, *Biochemistry* **22** : 5523–5535.

Farach, M. C., and Martinez-Carrion, M., 1983, A differential scanning calorimetry study of acetylcholine receptor-rich membranes from *Torpedo californica, J. Biol. Chem.* **258** : 4166–4170.

Finer-Moore, J., and Stroud, R., 1984, Amphipathic analysis and possible formation of the ion channel in an acetylcholine receptor, *Proc. Natl. Acad. Sci. U.S.A.* **81** : 155–159.

Fong, T. M., and McNamee, M. G., 1986, Correlation between acetylcholine receptor function and structural properties of membranes, *Biochemistry* **25** : 830–840.

Giraudat, J., Montecucco, C., Bisson, R., and Changeux, J. P., 1985, Transmembrane topology of acetylcholine receptor subunits probed with photoreactive phospholipids, *Biochemistry* **24** : 3121–3127.

Gonzalez-Ros, J. M., Calvo-Fernandez, P., Sator, V., and Martinez-Carrion, M., 1979a, Pyrenesulfonyl azide as a fluorescent label for the study of protein–lipid boundaries of acetylcholine receptor in membranes, *J. Supramol. Struct.* **11** : 327–338.

Gonzalez-Ros, J. M., Sator, V., Calvo-Fernandez, P., and Martinez-Carrion, M., 1979b, Pyrenesulfonyl azide: A covalent probe permitting *in vitro* desensitization of labeled acetylcholine receptor rich membrane fragments from *Torpedo californica, Biochem. Biophys. Res. Commun.* **87** : 214–220.

Gonzalez-Ros, J. M., Llanillo, M., Paraschos, A., and Martinez-Carrion, M., 1982, Lipid environment of acetylcholine receptor from *Torpedo californica, Biochemistry* **21** : 3467–3474.

Gonzalez-Ros, J. M., Farach, M. C., and Martinez-Carrion, M., 1983, Ligand induced effects at regions of acetylcholine receptor accessible to membrane lipids, *Biochemistry* 22 : 3807–3811.

Gonzalez-Ros, J. M., Ferragut, J. A., and Martinez-Carrion, M., 1984, Binding of anti-acetylcholine receptor antibodies inhibits the acetylcholine receptor mediated cation flux, *Biochem. Biophys. Res. Commun.* 120 : 368–375.

Guy, R., 1984, A structural model of the acetylcholine receptor channel based on partition energy and helix packing calculations, *Biophys. J.* 45 : 249–261.

Heidmann, T., Sobel, A., Popot, J. L., and Changeux, J. P., 1980, Reconstitution of a functional acetylcholine receptor. Conservation of the conformational and allosteric transitions and recovery of the permeability response; role of lipids, *Eur. J. Biochem.* 110 : 35–55.

Hess, G. P., Cash, D. J., and Aoshima, H., 1979, Acetylcholine receptor controlled ion fluxes in membrane vesicles investigated by fast reaction techniques, *Nature* 282 : 329–331.

Hess, G. P., Aoshima, H., Cash, D. J., and Lenchitz, B., 1981, Specific reaction rate of acetylcholine receptor controlled ion translocation: A comparison of measurements with membrane vesicles and with muscle cells, *Proc. Natl. Acad. Sci. U.S.A.* 78 : 1361–1365.

Hess, G. P., Pasquale, F. B., Walker, J. W., and McNamee, M. G., 1982, Comparison of acetylcholine receptor controlled cation flux in membrane vesicles from *Torpedo californica* and *Electrophorus electricus*: Chemical kinetics measurements in the millisecond region, *Proc. Natl. Acad. Sci. U.S.A.* 79 : 963–967.

Horn, R., and Stevens, C. F., 1980, Relation between structure and function of on channels, *Comments Mol. Cell. Biophys.* 1 : 57–68.

Huang, L. Y. M., Catterall, W. A., and Ehrenstein, G., 1978, Selectivity of cations and nonelectrolytes for acetylcholine activated channels in cultured muscle cells, *J. Gen. Physiol.* 71 : 397–410.

Kanaoka, Y., Machida, M., Kokubun, H., and Sekine, T., 1968, Fluorescence and structure of proteins as measured by incorporation of fluorophore. III. Fluorescence characteristics of N-[*p*(2-benzoxazolyl)phenyl] maleimide and the derivatives, *Chem. Pharm. Bull. (Tokyo)* 16 : 1747–1753.

Kao, P. N., and Karlin, A., 1986, Acetylcholine receptor binding site contains a disulfide crosslink between adjacent half-cystinyl residues, *J. Biol. Chem.* 261 : 8085–8088.

Karlsson, E., 1979, Chemistry of protein toxins in snake venoms, in *Handbook of Experimental Pharmacology* (Ch.-Y. Lee, ed.), Vol. 52, pp. 159–212 (Springer-Verlag, Berlin).

Karpen, J. W., Sachs, A. B., Cash, D. J., Pasquale, E. B., and Hess, G. P., 1983, Direct spectrophotometric detection of cation flux in membrane vesicles: Stopped-flow measurements of acetylcholine receptor mediated ion flux, *Anal. Biochem.* 135 : 83–94.

Kasai, M., and Changeux, J. P., 1971, *In vitro* excitation of purified membrane fragments by cholinergic agonists, *J. Membr. Biol.* 6 : 1–23.

Kistler, J., Stroud, R. M., Klymkowsky, M. W., Lalancette, R. A., and Fairclough, R. H., 1982, Structure and function of an acetylcholine receptor, *Biophys. J.* 37 : 371–383.

Klymkowsky, M. W., and Stroud, R. M., 1979, Immunospecific identification and three dimensional structure of a membrane bound acetylcholine receptor from *Torpedo californica*, *J. Mol. Biol.* 128 : 319–334.

Klymkowsky, M. W., Heuser, J. E., and Stroud, R. M., 1980, Protease effects on the structure of acetylcholine receptor membranes from *Torpedo californica*, *J. Cell Biol.* 85 : 823–838.

Lewis, C. A., 1979, Ion concentration dependence of the reversal potential and the single channel conductance of ion channels at the frog neuromuscular junction, *J. Physiol.* 286 : 417–445.

Lindstrom, J., Gullick, W., Conti-Tronconi, B., and Ellisman, M., 1980, Proteolytic nicking of the acetylcholine receptor, *Biochemistry* 19 : 4791–4795.

Lindstrom, J., Criado, M., Hochschwender, S., Fox, J. L., and Sarin, V., 1984, Immunochemical tests of acetylcholine receptor subunit models, *Nature* 311 : 573–575.

Low, B. W., 1979, The three dimensional structure of postsynaptic snake neurotoxins: Consideration of structure and function, in *Handbook of Experimental Pharmacology* (Ch.-Y. Lee, ed.), Vol. 52, pp. 213–257, Springer-Verlag, Berlin.

Martinez-Carrion, M., Sator, V., and Raftery, M. A., 1975, The molecular weight of an acetylcholine receptor isolated from *Torpedo californica, Biochem. Biophys. Res. Commun.* **65** : 129–137.

Martinez-Carrion, M., Gonzalez-Ros, J. M., Llanillo, M., and Paraschos, A., 1982, Acetylcholine receptors from electroplax membranes: *In vitro* and *in situ* properties, in *Advances in Experimental Medicine and Biology* (F. Bossa, E. Chiancone, A. Finnazi-Agro, and R. Strom, eds.), Vol. 148, pp. 209–224, Plenum Press, New York.

Martinez-Carrion, M., Gonzalez-Ros, J. M., Mattingly, J. R., Ferragut, J. A., Farach, M. C., and Donnelly, D., 1984, Fluorescent probes for the study of acetylcholine receptor function, *Biophys. J.* **45** : 141–143.

Middlemas, D. S., and Raftery, M. A., 1983, Exposure of the acetylcholine receptor to the lipid bilayer, *Biochem. Biophys. Res. Commun.* **115** : 1075–1082.

Mihovilovic, M., and Richman, D. P., 1984, Modification of α-bungarotoxin and cholinergic ligand binding properties of *Torpedo* acetylcholine receptor by a monoclonal anti-acetylcholine receptor antibody, *J. Biol. Chem.* **259** : 15051–15059.

Moore, H. P., and Raftery, M. A., 1979, Reversible and irreversible interactions of an alkylating agonist with *Torpedo californica* acetylcholine receptor, *Biochemistry* **10** : 1862–1867.

Moore, H. P., and Raftery, M. A., 1980, Direct spectroscopic studies of cation translocation by *Torpedo* acetylcholine receptor on a time scale of physiological relevance, *Proc. Natl. Acad. Sci. U.S.A.* **77** : 4509–4513.

Neubig, R. R., and Cohen, J. B., 1980, Permeability control by cholinergic receptors in *Torpedo* postsynaptic membranes: Agonist dose response relations measured at second and millisecond times, *Biochemistry* **19** : 2770–2779.

Neumann, D., Barchan, D., Safran, A., Gershoni, J. M., and Fuchs, S., 1986, Mapping of the α-bungarotoxin binding site within the subunit of acetylcholine receptor, *Proc. Natl. Acad. Sci. U.S.A.* **83** : 3008–3011.

Noda, M., Takahashi, H., Tanabe, T., Toyosato, M., Furutani, Y., Hirose, T., Asai, M., Inayama, S., Miyata, T., and Numa, S., 1982, Primary structure of α subunit precursor of *Torpedo californica* acetylcholine receptor deduced from cDNA sequence, *Nature* **299** : 793–797.

Noda, M., Takahashi, H., Tanabe, T., Toyosato, M., Kikyotani, S., Hirose, T., Asai, M., Takashima, H., Inayama, S., Miyata, T., and Numa, S., 1983a, Primary structure of the β and δ subunit precursors of *Torpedo californica* acetylcholine receptor deduced from cDNA sequences, *Nature* **301** : 251–255.

Noda, M., Takahashi, H., Tanabe, T., Toyosato, M., Kikyotani, S., Furutani, Y., Hirose, T., Takashima, H., Inayama, S., Miyata, T., and Numa, S., 1983b, Structural homology of *Torpedo californica* acetylcholine receptor subunits, *Nature* **302** : 528–532.

Paraschos, A., Gonzalez-Ros, J. M., and Martinez-Carrion, M., 1983, Absorption filtration: A tool for the measurement of ion tracer flux in native membranes and reconstituted lipid vesicles, *Biochem. Biophys. Acta* **733** : 223–233.

Pick, U., 1981, Liposomes with a large trapping capacity prepared by freezing and thawing of sonicated phospholipid mixtures, *Arch. Biochem. Biophys.* **212** : 186–194.

Popot, J. L., and Changeux, J. P., 1984, Nicotinic receptor of acetylcholine: Structure of an oligomeric integral membrane protein, *Physiol. Rev.* **64** : 1162–1239.

Prendergast, F. G., Haugland, R. P., Callahan, P. J., and Bodeau, D., 1981, 1-[4-(Trimethylamino)phenyl]-6-phenylhexa-1,3,5-triene: Synthesis fluorescence properties, and use as a fluorescence probe of lipid bilayers, *Biochemistry* **20** : 7333–7338.

Quast, U., Schimerlik, M., Lee, T., Witzemann, V., Blanchard, S., and Raftery, M. A., 1978,

Ligand induced conformation changes in *Torpedo californica* membrane bound acetylcholine receptor, *Biochemistry* **17** : 2405–2414.

Raftery, M. A., Schmidt, J., Clark, D. G., and Wolcott, R. G., 1971, Demonstration of a specific α-bungarotoxin binding component in *Electrophorus electricus* electroplax membranes, *Biochem. Biophys. Res. Commun.* **65** : 1622–1629.

Ratnam, M., and Lindstrom, J., 1984, Structural features of the nicotinic acetylcholine receptor revealed by antibodies to synthetic peptides, *Biochem. Biophys. Res. Commun.* **122** : 1225–1233.

Ratnam, M., Nguyen, D., Rivier, J., Sargent, P., and Lindstrom, J., 1986, Transmembrane topography of nicotinic acetylcholine receptor: Immunochemical tests contradict theoretical predictions based on hydrophobicity profiles, *Biochemistry* **25** : 2633–2643.

Reiter, M. J., Cowburn, D., Prives, J. M., and Karlin, A., 1972, Affinity labeling of the acetylcholine receptor in the electroplax: Electrophoretic separation in sodium dodecylsulfate, *Proc. Natl. Acad. Sci. U.S.A.* **69** : 1168–1172.

Reynolds, J., and Karlin, A., 1978, Molecular weight in detergent solution of acetylcholine receptor from *Torpedo californica*, *Biochemistry* **17** : 2035–2038.

Sator, V., Raftery, M. A., and Martinez-Carrion, M., 1978, *N*-(3-pyrene)maleimide: A fluorescent probe for acetylcholine receptor–Triton X-100 aggregates, *Arch. Biochem. Biophys*. **190** : 57–66.

Sator, V., Gonzalez-Ros, J. M., Calvo-Fernandez, P., and Martinez-Carrion, M., 1979a, Pyrene sulfonyl azide. A marker of acetylcholine receptor subunits in contact with membrane hydrophobic environment, *Biochemistry* **18** : 1200–1206.

Sator, V., Raftery, M. A., Thomas, J. K., and Martinez-Carrion, M., 1979b, Effect of cholinergic ligands and local anesthetics on acetylcholine receptor enriched preparations from *Torpedo californica* electroplax, *Arch. Biochem. Biophys*. **192** : 250–259.

Shinitzky, M., and Barenholz, Y., 1974, Dynamics of the hydrocarbon layer in liposomes of lecithin and sphingomyelin containing deacetylphosphate, *J. Biol. Chem.* **249** : 2652–2657.

Soler, G., Mattingly, J. R., and Martinez-Carrion, M., 1984, Effects of heating on the ion gating function and structural domains of the acetylcholine receptor, *Biochemistry* **23** : 4630–4636.

Strader, C. D., and Raftery, M. A., 1980, Topographical studies of *Torpedo* acetylcholine receptor subunits as a transmembrane complex, *Proc. Natl. Acad. Sci. U.S.A.* **77** : 5807–5811.

Sugiyama, H., and Changeux, J. P., 1975, Interconversion between different states of affinity for acetylcholine of the cholinergic protein from *Torpedo marmorata*, *Eur. J. Biochem.* **55** : 505–515.

Watters, D., and Maelicke, A., 1983, Organization of ligand binding sites at the acetylcholine receptor. A study with monoclonal antibodies, *Biochemistry* **22** : 1811–1819.

Weber, M., and Changeux, J. P., 1974a, Binding of *Naja nigricollis* [³H]α-toxin to membrane fragments from *Electrophorus* and *Torpedo* electric organs. I. Binding of the tritiated α-neurotoxin in the absence of effector, *Mol. Pharmacol.* **10** : 1–14.

Weber, M., and Changeux, J. P., 1974b, Binding of *Naja nigricollis* [³H]α-toxin to membrane fragments from *Electrophorus* and *Torpedo* electric organs. II. Effect of cholinergic agonists and antagonists on the binding of the tritiated α-neurotoxin, *Mol. Pharmacol.* **10** : 15–34.

Weiland, G., and Taylor, P., 1979, Ligand specificity of state transitions in the cholinergic receptor: Behavior of agonists and antagonists, *Mol. Pharmacol.* **15** : 197–212.

Weltman, J. K., Szaro, R. P., Frackelton, A. R., Jr., Dowben, R. M., Bunting, J. R., and Cathou, R. E., 1973, *N*-(3-pyrene) maleimide: A long lifetime fluorescent sulfhydryl reagent, *J. Biol. Chem.* **248** : 3173–3177.

Wilson, P. T., Lentz, T. L., and Hawrot, E., 1985, Determination of the primary amino-acid sequence specifying the α-bungarotoxin binding site on the α subunit of the acetylcholine receptor from *Torpedo californica*, *Proc. Natl. Acad. Sci. U.S.A.* **82** : 8790–8794.

Wu, C. W., Yarbrough, L. Z., and Hsiuch, Y., 1976, *N*-(1-pyrene) maleimide: A fluorescent crosslinking reagent, *Biochemistry* **15** : 2863–2868.

Young, A. P., Brown, F. F., Halsey, M. J., and Sigman, D. S., 1978, Volatile anesthetic facilitation of *in vitro* desensitization of membrane bound acetylcholine receptor from *Torpedo californica*, *Proc. Natl. Acad. Sci. U.S.A.* **75** : 4563–4567.

Young, E. F., Ralston, E., Blake, J., Ramachandran, J., Hall, Z. W., and Stroud, R. M., 1985, Topological mapping of acetylcholine receptor: Evidence for a model with five transmembrane segments and a cytoplasmic COOH-terminal peptide, *Proc. Natl. Acad. Sci. U.S.A.* **82** : 626–630.

Chapter 12

Structural Basis and Physiological Control of Membrane Fluidity in Normal and Tumor Cells

Wim J. van Blitterswijk

1. INTRODUCTION

Mammalian cell membranes basically consist of a bilayer of lipid molecules interacting with each other and with proteins that have either a transmembranous or a superficial position. These physical interactions determine the structure and molecular motions in the membrane.

The fluidity of the membrane is an operationally defined concept that may constitute both static (structural order) and dynamic (acyl chain mobility) properties of the lipid constituents. Fluorescence polarization using apolar

---

Abbreviations used in this chapter: α, cholesterol ordering coefficient; ALL, acute lymphocytic leukemia; AML, acute myelogenous leukemia; CLL, chronic lymphocytic leukemia; CML, chronic myelogenous leukemia; C/PL, cholesterol/phospholipid; DPH, 1,6-diphenyl-1,3,5-hexatriene; HCL, hairy cell leukemia; HDL, high-density lipoproteins; HMG-CoA, 3-hydroxyl-3-methylglutaryl-CoA; LDL, low-density lipoproteins; P_{DPH}, DPH steady-state fluorescence polarization; P_{plat}, plateau value of P_{DPH} reached at high cholesterol content; P_{zero}, polarization of the liposome without cholesterol; S_{DPH}, membrane order parameter; VLDL, very-low-density lipoproteins; WBC, white blood cells.

Wim J. van Blitterswijk   Division of Cellular Biochemistry, The Netherlands Cancer Institute, Antoni van Leeuwenhoek-Huis, 1066 CX Amsterdam, The Netherlands.

probes such as 1,6-diphenyl-1,3,5-hexatriene (DPH) is one of the most powerful techniques to study the physical properties of membrane lipids. Most information can be obtained from time-resolved fluorescence anisotropy decay measurements by differential phase fluorimetry or nanosecond single-photon counting techniques (Yguerabide, 1972; Kinosita et al., 1984; Stubbs and Smith, 1984). These techniques, however, are technically difficult and time consuming. Steady-state fluorescence anisotropy measurements give less information but are much easier to perform and many samples can be measured within a relatively short time. Yet, steady-state data may yield valuable information on the membrane structural order (a static property), provided that one takes advantage of empirical relations between time-resolved and steady-state fluorescence anisotropy data (Van Blitterswijk et al., 1981; Pottel et al., 1983; Van der Meer et al., 1986). The fluorescence polarization data presented in this chapter are derived only from steady-state measurements using the probe DPH. Membrane fluidity may in this context be considered as the reciprocal of the structural order or of the packing of the various apolar lipid chains (Van Blitterswijk et al., 1981).

Structural order parameters in artificial membranes (liposomes) may vary widely, that is, from highly fluid (liquid-crystalline phase) to rigid (gel phase), depending on the phospholipid composition, fatty acid composition, and, most importantly, the cholesterol content, as will be described in Section 2. In biological membranes, however, the overall values of the lipid structural order fall within a rather narrow range, apparently optimal for proper functioning. Nevertheless, within this narrow range of values, alterations in membrane fluidity by treatment of cells with certain complexes of lipids (e.g., certain liposomes or lipoproteins) or culturing cells in certain lipid-defined media have been shown to modify the cell surface expression or function of membrane proteins. Cellular processes that are influenced by the degree of membrane fluidity are, for example, the activity of membrane-bound enzymes and carrier-mediated transport activities [reviewed by Stubbs and Smith (1984), Shinitzky (1984), and Spector and Yorek (1985)]. Also, the binding characteristics of a variety of cell surface membrane antigens and receptors to their respective ligands have been shown to be altered by modifications in the membrane lipid composition and fluidity [reviewed by Spector and Yorek (1985), Gould and Ginsberg (1985), and Muller and Krueger (1986)]. Especially in cell signal transduction, an interesting and rapidly expanding area of research, the lipid composition and fluidity of the plasma membrane may play an important role. Not only the binding of ligands, such as insulin (Farias, 1987; Simon et al., 1987), thrombin (Tandon et al., 1983), thyrotropin, serotonin, transferrin, and adrenalin, to their respective receptors has been shown to depend on the membrane fluidity (Shinitzky, 1984; Gould and Ginsberg, 1985; Muller and Krueger, 1986), but also the functional coupling of the occupied receptor via G-proteins to effector enzymes such as adenylate cyclase has been shown to be affected by alterations in membrane lipid composition and fluidity (Gordon *et al.*, 1980; Neelands and Clandinin, 1983; Stubbs and Smith, 1984; Friedlander *et al.*, 1987; McMurchie *et al.*, 1987; Needham *et al.*, 1987).

The degree of membrane fluidity may also play a role in tumor–host interactions of both humoral and cellular immunological types (Van Blitterswijk, 1985). In several animal and human types of tumor cells, artificial rigidization of the cell membranes with cholesterol or cholesteryl hemisuccinate has resulted in increased immunogenicity of the tumor cells and, by consequence, in partial regression of the tumor and increased survival (Shinitzky *et al.*, 1979; Skornick *et al.*, 1984, 1986). In view of the relevance of the cell membrane fluidity in tumor–host immunological interactions, Section 3 of this chapter contains a review in which for a certain type of tumor (leukemia) the alterations in membrane fluidity and the underlying physiological mechanisms of these alterations are described.

It is well known that the membrane lipid composition of cultured cells can be significantly modified by specific lipid additions to the culture medium (Spector and Yorek, 1985; Spector and Burns, 1987). These modifications are often large enough to affect the physical properties of the membrane, such as those measured by DPH fluorescence polarization. However, *in vivo* membrane lipid alterations can in principle only be achieved by dietary means, via the plasma lipoproteins. The resulting effects are generally much smaller than those obtained *in vitro*, as will be described in Section 4.

2. QUANTITATIVE CONTRIBUTION OF INDIVIDUAL TYPES OF LIPID TO MEMBRANE FLUIDITY

2.1. Cholesterol, Sphingomyelin, and Fatty Acyl (Un)saturation

It has been known for a long time that the content of cholesterol in a membrane is the major determinant of the membrane fluidity. Cholesterol has a small fluidizing effect below the phase transition temperature (gel phase) of phospholipids and may have a large rigidizing (ordering) effect above this temperature (liquid-crystalline phase). In biological membranes at ambient temperatures, cholesterol thus imposes on the heterogeneous population of phospholipids a condition of "intermediate fluidity" (Oldfield and Chapman, 1972) and causes a smoothing or disappearance of phase transitions. In general, the overall effect of cholesterol in these membranes is rigidization, increased order or molecular packing, or increased mutual affinity of the apolar lipid chains. This condensing effect of cholesterol has been shown to depend on the molecular species of the lipids involved (Demel *et al.*, 1977) as well as on their degree of fatty acyl unsaturation (Demel *et al.*, 1972). The studies by Cooper *et al.* (1978) and Hoffman *et al.* (1981) have indicated that such ordering by cholesterol, as measured by DPH steady-state fluorescence polarization (P_{DPH}), levels off at high cholesterol/phospholipid (C/PL) molar ratios toward a plateau.

Van Blitterswijk *et al.* (1987) have recently described that for liposomes of any composition the fluorescence polarization measured at 25°C varies with the C/PL molar ratio according to the empirical exponential equation:

$$P_{DPH} = P_{plat} - (P_{plat} - P_{zero}) \exp(-\alpha C/PL) \tag{1}$$

where P_{plat} is the plateau value reached at high cholesterol content, usually at C/PL > 1; P_{zero} refers to the polarization of the liposome without cholesterol; α is a parameter describing the variation of P_{DPH} by cholesterol and may be called the *cholesterol ordering coefficient*. This coefficient α was found to increase with the fraction of sphingomyelin or dipalmitoylphosphatidylcholine molecules (both being typically "saturated" phospholipids), indicating that the susceptibility of phospholipids to be ordered by cholesterol is increased by these compounds. It also means that in lipid bilayers in the fluid state cholesterol has a relatively high affinity to sphingomyelin and dipalmitoylphosphatidylcholine. This conclusion is in agreement with published results of calorimetric and cholesterol transfer (equilibrium) studies (Demel *et al.*, 1977; Lange *et al.*, 1979; Nakagawa *et al.*, 1979) and a reinterpretation of the theory of Hoffman *et al.* (1981) regarding DPH fluorescence polarization in cholesterol-containing membranes (Van Blitterswijk *et al.*, 1987). For most phospholipids, P_{zero} was shown to increase curvilinearly with the fraction of either of these molecules (Van Blitterswijk *et al.*, 1987). A typical example of such a sigmoidal curve is shown in Figure 1, left-hand panel. This may reflect an isothermal (at 25°C) liquid-crystalline to gel phase transition determined by the chemical composition. The plateau polarization value, P_{plat}, was shown to increase linearly with the fraction of saturated acyl chains for most phospholipids (Figure 1, right-hand panel). Highly unsaturated fatty acyl chains (e.g., 20 : 4 and 22 : 6) were found to depress P_{plat} strongly but not P_{zero}. The results of Van Blitterswijk *et al.* (1987) suggest that such phospholipids are unlikely to associate with cholesterol and may thus create extremely fluid membrane domains. Equations were given according to which the susceptibility of the various individual phospholipids to be ordered by cholesterol can be calculated.

The above-described rules were also shown to be applicable to more complex mixtures of lipids and even to biological membranes (Van Blitterswijk *et al.*, 1987). Figure 1 is a simplified demonstration of how the DPH fluorescence polarization (P_{DPH}) in a membrane with a given C/PL molar

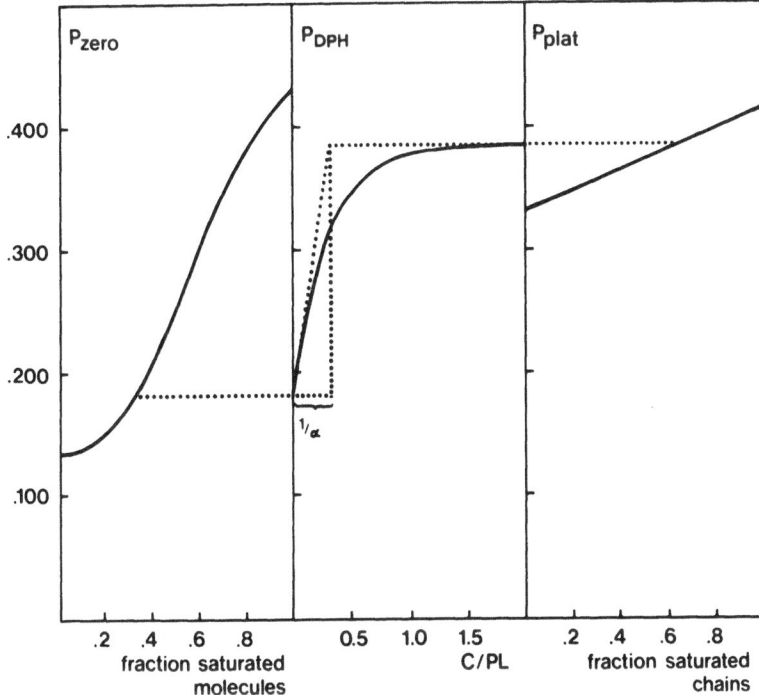

FIGURE 1. Combination of graphs showing how the DPH fluorescence polarization in a membrane with a given cholesterol/phospholipid (C/PL) molar ratio is essentially determined. Left-hand panel: Polarization in the absence of cholesterol (P_{zero}) is determined by the fraction of sphingomyelin and/or dipalmitoylphosphatidylcholine ("saturated" molecules) in the phospholipids. Right-hand panel: The plateau value of fluorescence, P_{plat}, reached at high cholesterol content is essentially determined by the fraction of saturated fatty acyl chains in the phospholipids. Middle panel: Example of P_{DPH} variation with C/PL according to Eq. (1). In addition, α, or rather $1/\alpha$, is visualized here.

ratio can be determined. The P_{DPH} values for liposomes prepared from lipid extracts of natural (purified) membranes of various origins could be more or less predicted (calculated) from the relative contributions of the individual lipid components.

2.2. Cell Biological Implications of Preferential Interactions between Individual Lipids

The systematic quantitative steady-state fluorescence polarization study described above (Van Blitterswijk *et al.*, 1987) on a large variety of liposome compositions has given us much insight in the mutual interactions (affinities) between the individual lipids in biological membranes. The very existence of preferential associations of cholesterol with certain phospholipids while less with others would suggest that, in biomembranes, phase separation into (micro)domains of distinct lipid compositions is likely to occur. Lentz *et al.* (1976) have demonstrated that the DPH probe weighs equally fluid and less fluid regions. Therefore, the steady-state P_{DPH} values generally represent only the overall fluidity and do not resolve possible domains. Nevertheless, the results of Van Blitterswijk *et al.* (1987) have indirectly demonstrated the existence of extremely fluid microdomains devoid of cholesterol in liposomes containing phospholipids with polyunsaturated fatty acyl moieties, such as arachidonic acid (20 : 4) and docosahexaenoic acid (22 : 6) at the *sn-*2 position. In the plasma membrane, such phospholipids, especially of the phosphatidylcholine and phosphatidylethanolamine type, are preferential substrates for the membrane-bound phospholipase A2 *in situ* (Van den Bosch, 1980; Kannagi *et al.*, 1981; Irvine, 1982; Storch and Schachter, 1985; Mahadevappa and Holub, 1987). It may be suggested that this selectivity is (perhaps partly) based on the high lipid fluidity or lipid disarrangement (Kannagi *et al.*, 1981) in the microdomain of the substrate, in which the active site of the enzyme may penetrate more easily to reach the ester bond. This possibility is supported by monolayer studies [reviewed by Verger (1976)] showing that the penetration and activity of phospholipase A2 are inversely related to the tightness of phospholipid packing. Once the enzymatic reaction in the membrane has taken place and the free polyunsaturated fatty acid has subsequently been released, the remaining lyso compound (having much higher P_{zero} and P_{plat} values; Van Blitterswijk *et al.*, 1987) will lead to rigidization of the domain as has indeed been shown experimentally (Storch and Schachter, 1985).

Plasma membranes are generally characterized by a high C/PL molar ratio relative to the cellular endomembranes. Cholesterol is synthesized on intracellular membranes where membrane cholesterol is low and transported to the plasma membrane where cholesterol content is high. This transfer is a unidirectional process (Lange and Matthies, 1984). How and why this occurs have not yet been understood. This problem has been tackled by studies on cholesterol transfer kinetics and equilibrium distribution of cholesterol among the various types of subcellular membranes and artificial membranes of various compositions (Wattenberg and Silbert, 1983; Yeagle and Young, 1986). These studies have suggested that membrane lipid composition may play an important role in the distribution of cholesterol and that the plasma membrane shows the highest "affinity" for cholesterol, attributable at least partly to elevated sphingomyelin levels. The fluorescence polarization study of Van Blitterswijk *et al.* (1987) has confirmed this suggestion and provides a quantitative explanation for the preferential association of cholesterol with the plasma membrane. It is suggested that the disproportionation of cholesterol in the cell is driven and maintained by intracellular segregation (sorting) of sphingomyelin (and probably also glycosphingo-lipids) into vesicles having a distinct destination in the cell. In other words, in the topogenesis of the plasma membrane, cholesterol would be "dragged along" rather than "targeted" specifically.

3. ALTERATIONS IN MEMBRANE FLUIDITY IN LYMPHOID TUMOR CELLS

3.1. Tumor Cell Type and Location

The membrane fluidity of leukemia cells, derived freshly from patients or experimental animals, has been studied extensively by fluorescence polarization and lipid analysis. Figure 2 summarizes the data on purified plasma membranes from various types of human leukemia, in comparison with those of white blood cells (WBC) from normal donors, as compiled in previous reviews (Van Blitterswijk, 1984, 1985). The membrane order parameters (S_{DPH}), derived from the degree of fluorescence polarization using an empirical relation (Van Blitterswijk $et\ al.$, 1981), show a similar trend as the cholesterol/phospholipid molar ratios in the membranes. Acute and chronic lymphocytic leukemias (ALL and CLL) generally show an increased membrane fluidity, whereas hairy cell leukemia (HCL) tends to show increased

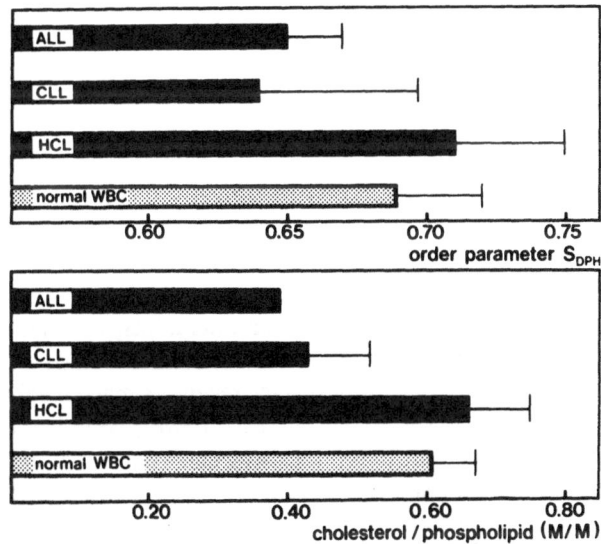

FIGURE 2. Lipid order parameters (S_{DPH}) and C/PL molar ratios in purified plasma membranes from human normal white blood cells (WBC) and leukemic cells, freshly obtained from peripheral blood. ALL and CLL, acute and chronic lymphocytic leukemia; HCL, hairy cell leukemia.

membrane order (rigidity). Similar measurements on whole cells have furthermore indicated an increased fluidity (decreased C/PL ratio) in the myeloid leukemias (Van Blitterswijk, 1984, 1985).

In animal (rodent) systems the tumor cells investigated were generally not taken from the peripheral blood (as in humans), but from the ascites fluid, subcutaneous nodules, or the lymphoid organs (lymph nodes, thymus, or spleen) after transplantation and subsequent outgrowth of the tumor cells. It appears that the type of tumor and the *in situ* location or homing of the tumor cells are important factors determining the membrane lipid fluidity (Van Blitterswijk, 1985). In the murine transplanted GRSL leukemia model (Van Blitterswijk *et al.*, 1984) the fluorescence polarization in isolated plasma membranes of splenic tumor cells (P_{DPH} = 0.316 at 25°C) is slightly higher than those of normal splenic lymphocytes (P_{DPH} = 0.302), while ascites cell membranes are much more fluid (P_{DPH} = 0.261). The chemical parameters responsible for this high fluidity are the extremely low C/PL molar ratio (0.32) and sphingomyelin content (0.9% of the total phospholipids), and the high degree of unsaturation in the phospholipid fatty acyl groups.

A process that substantially contributes to fluidization of the GRSL ascites cell membranes is the shedding of membrane vesicles of high lipid structural order (P_{DPH} = 0.325) (Van Blitterswijk *et al.*, 1982). These extracellular membranes encountered in the ascites fluid in high amounts are highly enriched in cholesterol and sphingomyelin. Extracellular membranes were also found in the thymus and the spleen, but their lipid composition and fluidity differed much less, if at all, from the respective plasma membranes.

3.2. Plasma Lipoproteins and Cholesterol Biosynthesis

The cholesterol content of cells is basically determined by a balance between influx (receptor-mediated uptake of lipoproteins), efflux via high-density lipoproteins (HDL), and/or shed membrane vesicles (see above) and the cell's own biosynthesis, in relation to cell growth. In leukemias these processes are often subjected to significant alterations, as summarized from several studies on human leukemias in Table I: cellular uptake of low-density lipoproteins (LDL) via high-affinity surface receptors may be enhanced, most significantly so in acute myelogenous leukemia (AML) (Heiniger *et al.*, 1976; Vitols *et al.*, 1984). Cellular biosynthesis of cholesterol may be 5–30 times elevated, except in CLL (Heiniger *et al.*, 1976; Ho *et al.*, 1978; Yachnin *et al.*, 1983). Cholesterol levels in blood plasma of leukemic patients are often abnormally low (Heiniger *et al.*, 1976; Ho *et al.*, 1978; Spiegel *et al.*, 1982; W. J. Van Blitterswijk, unpublished data). Spiegel *et al.* (1982) have studied the plasma lipids and lipoproteins in 25 patients with acute leukemia and

Table I

Alterations in Cholesterol Biosynthesis and Receptor-Mediated Uptake
of LDL by Leukemic Cells and in Cholesterol Levels and Fluorescence
Polarization (P_{DPH}) Values of Blood Plasma from Leukemic Patients

Type of human leukemia	Cells[a]		Blood plasma	
	LDL uptake	Cholesterol biosynthesis	Cholesterol level	P_{DPH}
ALL[b]	$0/+$[c]	$++$	$--/+$	$--$
AML	$++$	$++$	$--$	$--$
CLL	$0/+$	0	$-/0$	$-$
CML	$+$		$-$	$--$
HCL		$+$	$-/0$	0

[a]Freshly isolated from peripheral blood.
[b]Abbreviations: see legend to Figure 2. AML and CML, acute and chronic myelogenous leukemia.
[c]$--/-/0/+/++$, range from much decreased to highly increased, as compared to mononuclear cells from healthy volunteers.

non-Hodgkin's lymphoma. Most patients exhibited extremely low levels of HDL cholesterol and elevated levels of very-low-density lipoproteins (VLDL). Patients restudied in remission demonstrated a return to normal values. An increased VLDL (triacylglycerol) level may be a more general trait in leukemias. This lowers the overall structural order of plasma lipids as shown in Table I by the decreased P_{DPH} values, except for hairy cell leukemia (HCL) (Rosenfeld et al., 1979; W. J. Van Blitterswijk, unpublished data).

In mice bearing the GRSL leukemia the cholesterol biosynthesis in the liver and in the splenic and ascitic tumor cells was found to be increased as compared to the liver, spleen cells, and thymocytes of normal control mice, respectively (Van Blitterswijk et al., 1985). This is illustrated in Figure 3 by the activities of 3-hydroxyl-3-methylglutaryl (HMG)-CoA reductase, that is, the rate-determining enzyme of the cellular cholesterol biosynthesis. A similar pattern of cholesterol biosynthetic activities in these organs/cells was found by the incorporation of [^{14}C]acetate into cholesterol. The cholesterol biosynthesis in the GRSL ascites cells was 5–10 times higher than in the splenic tumor cells. Nevertheless, the ascites cells show a strikingly low C/PL ratio of 0.30 in their plasma membrane in comparison to the tumor cells in the spleen (C/PL = 0.55–0.70).

In the transplanted GRSL leukemia, we found significant alterations in the lipoprotein composition and a gradual decrease in cholesterol content of blood and ascites plasma during outgrowth of the tumor (Damen et al., 1984). Studies with density gradient ultracentrifugation generally revealed that HDL$_2$, the major lipoprotein in normal mice, became strongly reduced. At the same time HDL-like lipoproteins accumulated in the plasma low-density fractions. By intravenous injection of [^{14}C]cholesteryl ester-labeled HDL$_2$ into tumor-

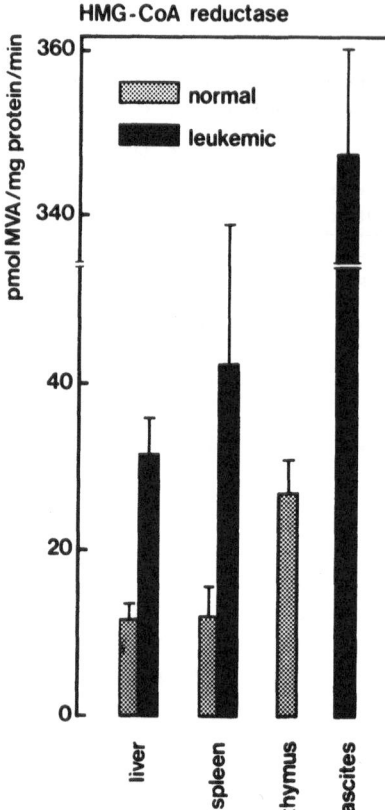

FIGURE 3. Cholesterol biosynthesis in the liver and the various lymphoid cells of leukemic (GRSL-bearing) mice and of normal control mice, estimated by the activity of HMG-CoA reductase, expressed in picomoles of mevalonic acid (MVA) formed per milligram of microsomal protein, per minute.

bearing mice, conversion into lipoproteins of low density was demonstrated directly (Damen *et al.*, 1985). The outset and extent of these alterations in the lipoprotein density profile were found to depend on the malignancy of the particular GRSL line inoculated. For instance, the GRSL 18 line, making the mice moribund within 9 days after ip inoculation, caused more dramatic alterations in the lipoproteins than the GRSL 13 line that grows out more slowly (3 weeks). With the former tumor line we observed 7 days after transplantation extremely low levels of all lipoproteins except VLDL (Damen *et al.*, 1984; Van Blitterswijk *et al.*, 1985). We also found that an occasional transient recovery of the mice during tumor growth was associated with a certain degree of "normalization" of the lipoprotein density profile. Similar observations have been reported in human leukemia (Spiegel *et al.*, 1982). Evidence has been given that the altered lipoprotein density profiles are a direct consequence of decreased lipoprotein lipase and hepatic lipase activities (Damen *et al.*, 1984).

It may be questioned whether the significant alterations in plasma lipoproteins and the plasma cholesterol levels in animals and patients with leukemia determine (perhaps partly) the alterations in cholesterol content and fluidity of cellular membranes described above, and if so, whether experimental *in vivo* manipulation of the lipoprotein composition (e.g., by dietary lipids) may affect membrane lipid composition. Neither in human individuals nor in animal systems could any direct relation be found between membrane cholesterol or fluidity of mononuclear lymphoid cells in blood, spleen, or ascites, and the cholesterol level of their bathing plasma. On the other hand, when the supply of lipoprotein cholesterol to fast growing tumor cells is hampered, this may lower the C/PL ratio in their membranes, in spite of an increased cellular biosynthesis of cholesterol. This seems to be the case in GRSL ascites cells, since the ascites plasma contains a much lower concentration of lipoproteins than the blood plasma, and we have demonstrated a lack of exchange of plasma lipoproteins between the blood and ascites compartments (Van Blitterswijk *et al.*, 1985). Taken together, our results suggest that both the shedding of rigid membrane vesicles (see above) and the limited availability of lipoproteins as a source of cholesterol contribute to the low C/PL ratio in the plasma membrane of GRSL ascites cells.

4. EFFECTS OF DIETARY LIPIDS ON MEMBRANE FLUIDITY

The fatty acid composition of the membrane phospholipids of cultured cells can be modified extensively by changing the type and amount of lipid contained in the growth medium (Spector and Yorek, 1985). This is due to the fact that when fatty acids are available in the extracellular fluid, they are utilized preferentially, and *de novo* fatty acid synthesis is suppressed. Normal and tumor cells *in vivo* also derive most of their fatty acid from the extracellular fluid when an adequate supply of lipid is available in the form of circulating free fatty acids and triacylglycerols contained in the plasma lipoproteins. The fatty acid composition in the plasma can readily be modified by the content and composition of the lipids in the diet.

In experimental animals, the effect of dietary lipids on cell membrane composition is mostly studied by using isocaloric semipurified diets that contain certain plant oils that are enriched in either saturated, monounsaturated, or polyunsaturated (n-6 type) fatty acids or fish oils enriched in n-3 polyunsaturated fatty acids. Using this approach, the fatty acid composition of the phosphoglycerides in a variety of tissues or (intact) cells could be modified to a certain extent [reviewed by Stubbs and Smith (1984)]. Less extensive data are available for isolated plasma membranes, and in only a limited number of studies were these compositional data related to physical parameters, such as membrane fluidity measured by DPH fluorescence polarization.

The effect of a high linoleic acid (18 : 2, *n*-6) content in the diet on the fatty acid composition of rat liver microsomes (Tahin *et al.*, 1981) or plasma membranes (Burns *et al.*, 1983) was clearly detectable by an increased 18 : 2 and/or 20 : 4 (*n*-3) content in these membranes. In the study of Burns *et al.* (1983) the results were compared to those obtained with a coconut-oil-containing diet. The latter diet resulted in a doubling of the 18 : 1 content in the membranes. No membrane fluidity was measured in these studies. Brasitus *et al.* (1985) have determined the lipid composition and fluidity of a number of intestinal membranes after feeding rats nutritionally complete diets enriched in unsaturated (corn oil) or saturated (butter fat) triacylglycerols. The corn oil diet (enriched in 18 : 2) increased the overall unsaturation of the acyl chains and the lipid fluidity (decreased P_{DPH} value) of the membranes. Concomitantly, the cholesterol/phospholipid molar ratio was increased in the microvillus but not in the basolateral membranes. Apparently, rat enterocytes possess regulatory mechanisms that modulate the cholesterol content of the microvillus membranes so as to mitigate changes in lipid fluidity. Burns and Spector (1987) have been able to modify the phospholipid fatty acid composition of L1210 leukemia cells and Ehrlich ascites carcinoma cells by a saturated (coconut oil) or an unsaturated (sunflowerseed oil) fat-containing diet. In the isolated plasma membranes the main effects were found in the 18 : 1 and 18 : 2 fatty acids. There were no changes in the phospholipid or cholesterol content induced by the diets. The differences in the fatty acids were sufficient to yield small differences in the order parameter as measured by electron spin resonance using nitroxystearate spin probes. The growth rate of the tumor cells did not change significantly as a function of the diet.

We ourselves have been investigating the effect of dietary lipids on the composition and fluidity of plasma membranes purified from various lymphoid cells and erythrocytes in normal and leukemic GR/A mice (J. Damen and W. J. Van Blitterswijk, unpublished data). The mice were kept on four different isocaloric semipurified diets, enriched in either corn oil or palm oil or palm oil + 2 wt % cholesterol, or a fat-poor diet containing only a minimal amount of essential fatty acids (18 : 2). The added fat components represented 30 energy % and 7 energy % of the fat-rich and the fat-poor diets, respectively. The major difference between the corn oil and the palm oil diet is the ratio in the fatty acids 16 : 0/18 : 1/18 : 2, being 12.1/24.5/60.5 and 43.0/35.5/16.6, respectively. The results of these studies are as follows.

Table II shows that in spleen lymphocytes of normal mice and in GRSL tumor cells in the enlarged spleens of tumor bearers no alterations in membrane fluidity can be generated by the different diets. Only in the GRSL cells

Table II
DPH Fluorescence Polarization (Lipid Structural Order) and Cholesterol/Phospholipid Molar Ratio in Isolated Plasma Membranes of Lymphoid Cells from Normal and Leukemic Mice ($n = 3$) as a Function of Dietary Lipids

Lipids in diet	Corn oil	Fat-poor	Palm oil	Palm oil + cholesterol
	DPH fluorescence polarization (25°C)			
Normal splenocytes	0.307 ± 0.001	0.306 ± 0.001	0.304 ± 0.001	0.304 ± 0.001
Splenic GRSL cells	0.294 ± 0.002	0.294 ± 0.006	0.292 ± 0.004	0.295 ± 0.001
Ascitic GRSL cells	0.257 ± 0.002	0.254 ± 0.007	0.259 ± 0.007	0.268 ± 0.001
	Cholesterol/phospholipid (M/M)			
Normal splenocytes	0.50 ± 0.01	0.52 ± 0.01	0.52 ± 0.01	0.50 ± 0.01
Splenic GRSL cells	0.51 ± 0.05	0.54 ± 0.07	0.50 ± 0.02	0.51 ± 0.02
Ascitic GRSL cells	0.29 ± 0.02	0.27 ± 0.05	0.31 ± 0.03	0.41 ± 0.02

Wim J. van Blitterswijk

growing in ascites form, does the cholesterol-rich diet induce an increased structural order (decreased fluidity), which can be attributed to an increased cholesterol/phospholipid molar ratio in these cells (Table II). The significant effect of this diet on the ascites cell membranes, rather than on the spleen cell membranes, supports the proposition of Van Blitterswijk *et al.* (1985) that the availability of lipoprotein cholesterol to the GRSL ascites cells is a limiting factor for these cells to obtain a higher cholesterol/phospholipid ratio. The cholesterol-rich diet did not clearly affect the rate of growth of these tumor cells in the peritoneum. This diet also resulted in an increased cholesterol/phospholipid molar ratio (0.75 ± 0.02 compared to 0.67 ± 0.02 for the other three diets) in the blood erythrocyte membranes of the tumor bearers, with concomitant increase in DPH fluorescence polarization (0.313 versus 0.306). The differences in fatty acid composition of the total phospholipids of isolated plasma membranes from the various lymphoid cells and erythrocytes as induced by the different diets were only small and showed a trend as listed for GRSL ascites cells in Table III: the largest differences were seen in the 18 : 1 and 18 : 2 contents, the latter being enhanced by the corn oil diet and the former by the palm-oil-containing diets. Schouten *et al.* (1987) have studied the composition and fluidity of erythrocyte membranes in cholesterol-fed pigs without and with a partial ileal bypass. Dietary cholesterol causes significant increases in the ratios of cholesterol/phospholipids and phosphatidylcholine/sphingomyelin in these membranes. Partial ileal bypass completely nullified the former but not the latter ratio. Neither cholesterol feeding nor the ileal bypass had an effect on the membrane fluidity, as measured by fluorescence polarization.

Currently, there is much interest in the effects of fish oil in the diet on the lipid composition of cellular membranes in relation to biological processes like cardiovascular disease, cancer, and other chronic diseases such as mul-

Table III
Fatty Acid Composition of the Total Phospholipid of Isolated Plasma Membranes of GRSL Ascites Cells as a Function of Dietary Lipids[a]

Fatty acid	Corn oil	Fat-poor	Palm oil	Palm oil + cholesterol
16 : 0	19.8 ± 0.1	17.8 ± 0.6	18.2 ± 1.7	19.4 ± 1.2
18 : 0	21.7 ± 0.7	18.2 ± 2.4	22.5 ± 0.3	18.9 ± 3.0
18 : 1	12.1 ± 0.8	15.2 ± 1.2	23.1 ± 4.3	19.9 ± 0.7
18 : 2	22.1 ± 4.4	23.6 ± 1.0	15.2 ± 1.4	17.5 ± 2.0
20 : 4	15.9 ± 0.9	14.5 ± 2.0	13.9 ± 1.9	13.3 ± 0.2
22 : 6	2.6 ± 0.4	1.3 ± 0.4	1.8 ± 0.5	1.4 ± 0.4

[a]Weight % of total fatty acids $\pm$ SEM, $N = 3$.

tiple sclerosis and arthritis (Leslie *et al.*, 1985; Carroll, 1986). Blood plate-lets play an essential role in arterial thrombosis and atherosclerosis, and dietary fat-induced changes in platelet lipid composition have been demon-strated to affect platelet function (Renaud *et al.*, 1979; Ahmed and Holub, 1984; Nordoy *et al.*, 1985; Rand *et al.*, 1986). Both in rats (Iritani and Narita, 1984; Nordoy *et al.*, 1985) and in humans (Brox *et al.*, 1981; Ahmed and Holub, 1984; Mori *et al.*, 1987) fish oil in the diet generally leads to a substantially increased 20 : 5 (*n*-3) and 22 : 6 (*n*-3) content in the platelet phosphoglycerides, mainly in PC, PE, PS, and PI, at the expense of (mainly) arachidonic acid (20 : 4, *n*-6). Rand *et al.* (1986) have also measured mem-brane fluidity by DPH fluorescence polarization in rat platelets and found a significantly increased fluidity upon feeding a 50 energy % sunflowerseed oil diet ($P_{DPH} = 0.218$) or a 50 energy % sperm-whale oil diet ($P_{DPH} = 0.215$) relative to the control (5 energy % sunflowerseed oil) ($P_{DPH} = 0.223$). The cho-lesterol and phospholipid contents of the platelets showed no difference. Also in humans a small but significant increase in platelet membrane fluidity was found as a result of fish oil in the diet (Hornstra and Rand, 1986).

We ourselves have performed a comparative study on isolated plasma membranes from GR/A mouse spleen lymphocytes (W. J. van Blitterswijk *et al.*, unpublished data), after feeding the mice for 4 weeks with semipurified isocaloric diets, which were provided by Unilever Research Labs (Vlaardin-gen, The Netherlands): (1) a sunflowerseed-oil-containing diet, enriched in linoleic acid (18 : 2, *n*-6; 34%); (2) an olive-oil-containing diet, enriched in oleic acid (18 : 1, 56%); and (3) a fish-oil-containing diet, enriched in the *n*-3 fatty acids (18 : 3, 18 : 4, 20 : 5, 22 : 5, and 22 : 6 (28.5% of the total fatty acids). The content of saturated fatty acids in the three diets (33%) as well as the energy % total lipids (33%) was kept constant. Replacement of 18 : 1 in the diet by 18 : 2 (*n*-6) increased the content of 18 : 2 in phospha-tidylcholine and phosphatidylethanolamine and increased the content of ar-achidonic acid (20 : 4, *n*-6) in phosphatidylinositol at the cost of the contents of 18 : 1 and/or 16 : 0 in these phospholipids (Table IV). The effect of the fish-oil-containing diet was much larger (Table IV): in all phospholipids there was a two- to five-fold reduction in 20 : 4 and a significant increase in the *n*-3 fatty acids 20 : 5, 22 : 5, and, most conspicuously, 22 : 6 (up to 26% in phosphatidylethanolamine and phosphatidylserine). Nevertheless, no sig-nificant differences were found in membrane fluidity as measured by fluo-rescence polarization. Also, the cholesterol/phospholipid ratios in the membranes did not show significant alterations induced by the diets.

A similar study as the one described above was reported by Conroy *et al.* (1986). They analyzed the phosphatidylcholine and phosphatidylethano-lamine fractions of rat spleen (whole) lymphocytes, after feeding the rats for 10 days with an olive oil or a fish oil diet. A similar shift in the fatty acid

Table IV

Fatty Acid Composition of the Phospholipids of Isolated Plasma Membranes of Normal Spleen Lymphocytes as a Function of Diets Enriched in Sunflowerseed Oil (SF), Olive Oil (OV), or Fish Oil (FO)[a]

Fatty acid	Total phospholipids			Phosphatidylcholine			Phosphatidyl-ethanolamine			Phosphatidylserine			Phosphatidylinositol		
	SF	OV	FO	SF	OV	FO	SF	OV	FO	SF	OV	FO	SF	OV	FO
16 : 0	24.3	27.0	28.8	50.2	53.0	60.2	9.9	15.4	11.2	5.7	5.9	6.2	8.0	10.9	10.2
18 : 0	21.2	24.4	20.1	9.9	8.5	6.8	29.2	30.9	24.0	46.8	46.6	40.0	44.7	47.9	41.7
18 : 1	6.0	6.9	4.4	11.6	16.0	9.9	5.4	5.3	4.2	3.4	3.3	3.0	1.8	2.9	1.7
18 : 2	5.9	3.3	3.9	8.5	4.7	4.9	4.3	2.8	2.6	4.5	3.4	3.2	2.0	1.3	1.8
20 : 4	16.5	16.1	4.6	14.5	13.0	3.0	30.7	25.8	6.6	18.6	17.9	4.2	34.2	22.7	15.6
20 : 5	—	—	7.4	—	—	5.8	—	—	9.5	—	—	5.5	—	—	11.2
22 : 5	2.7	2.5	6.3	0.7	0.8	2.1	2.6	2.8	7.0	1.9	2.1	7.5	1.5	1.9	6.2
22 : 6	3.9	3.8	10.4	1.1	1.5	3.5	9.1	9.4	24.5	8.8	10.4	26.2	1.3	2.1	7.2

[a]Expressed in weight % of total fatty acids.

profiles as in Table IV was found. No membrane fluidity measurements were done on these whole lymphocytes, but in rat liver microsomes obtained from these rats no differences in DPH fluorescence polarization were found. However, when lipid motion was examined in vesicles of phosphatidylcholine, isolated from the microsomes from fish-oil-fed animals [21.4% (n-3) fatty acids], the fluorescence polarization was significantly less than the corresponding phosphatidylcholine from olive-oil-fed animals [5.6% (n-3) fatty acids], indicating a more disordered or fluid bilayer in the presence of higher levels of (n-3) fatty acids. Although in native membranes the overall fluidity seems not to be affected, the possibility exists that on the level of membrane (micro)domains enriched in (n-3) polyunsaturated fatty acids fluidity differences are present, in agreement with findings reported by Van Blitterswijk et al. (1987). Apart from fluidity changes in membrane domains, it should be kept in mind that a decrease in the ratio of 20 : 4 (n-6) to (n-3) polyunsaturated fatty acids in the phospholipids induced by the fish oil diet has important consequences for the types of prostaglandins and/or thromboxanes and/or lipoxygenase products produced by the cell and consequently for the cellular behavior (Spector and Yorek,1985; Carroll, 1986).

The effect of dietary (n-3) polyunsaturated fatty acids on human erythrocyte membrane composition and fluidity was studied by Popp-Snijders et al. (1986). An increased unsaturation of phosphatidylcholine and phosphatidylethanolamine was accompanied by a slight decrease in the content of these phospholipids and an increase in sphingomyelin content. The erythrocyte membrane fluidity, measured by electron spin resonance and DPH fluorescence polarization, did not change.

From the above-described studies it may be concluded that dietary lipids may lead to alterations in the fatty acid composition of cellular membranes. The extent of these alterations depends on the type of cell and the type and amount of lipid supplemented to the diet. Fish oils generally cause larger changes than plant oils. In most of the studies these oil supplements to the diet did not alter the phospholipid class composition or the cholesterol content, nor the overall membrane fluidity—intestinal membranes and platelets being exceptions. A cholesterol-enriched diet induced an enhanced cholesterol/phospholipid molar ratio in the plasma membrane of ascites tumor cells and erythrocytes, but not in splenic lymphocyte membranes. Generally, one may conclude that overall membrane fluidity is subject to exquisite homeostatic control. Nevertheless, the relatively small modifications in membrane lipids obtained by dietary means could be important enough to be utilized as a potential adjunct to cancer therapies (Burns and Spector, 1987; Spector and Burns, 1987).

5. REFERENCES

Ahmed, A., and Holub, B. J., 1984, Alteration and recovery of bleeding times, platelet aggregation and fatty acid composition of individual phospholipids in platelets of human subjects receiving a supplement of cod-liver oil, *Lipids* **19** : 617–624.

Brasitus, T. A., Davidson, N. O., and Schachter, D., 1985, Variations in dietary triacylglycerol saturation alter the lipid composition and fluidity of rat intestinal plasma membranes, *Biochim. Biophys. Acta* **812** : 460–472.

Brox, J. H., Killie, J.-E., Gunnes, S., and Nordoy, A., 1981, The effect of cod liver oil and corn oil on platelets and vessel wall in man, *Thromb. Haemost.* **46** : 604–611.

Burns, C. P., and Spector, A. A., 1987, Membrane fatty acid modification in tumor cells: A potential therapeutic adjunct, *Lipids* **22** : 178–184.

Burns, C. P., Rosenberger, J. A., and Luttenegger, D. G., 1983, Selectivity in modification of the fatty acid composition of normal mouse tissues and membranes *in vivo, Ann. Nutr. Metab.* **27** : 268–277.

Carroll, K. K., 1986, Biological effects of fish oils in relation to chronic diseases, *Lipids* **21** : 731–732.

Conroy, D. M., Stubbs, C. D., Belin, J., Pryor, C. L., and Smith, A. D., 1986, The effects of dietary (*n*-3) fatty acid supplementation on lipid dynamics and composition in rat lymphocytes and liver microsomes, *Biochim. Biophys. Acta* **861** : 457–462.

Cooper, R. A., Leslie, M. H., Fischkoff, S., Shinitzky, M., and Shattil, S. J., 1978, Factors influencing the lipid composition and fluidity of red cell membranes *in vitro*: Production of red cells possessing more than two cholesterols per phospholipid, *Biochemistry* **17** : 327–331.

Damen, J., Van Ramshorst, J., Van Hoeven, R. P., and Van Blitterswijk, W. J., 1984, Alterations in plasma lipoproteins and heparin-releasable lipase activities in mice bearing GRSL ascites tumor, *Biochim. Biophys. Acta* **793** : 287–296.

Damen, J., de Widt, J., Hengeveld, T., and Van Blitterswijk, W. J., 1985, Accumulation of HDL-like lipoproteins in the plasma low-density fractions of tumor-bearing mice, *Biochim. Biophys. Acta* **833** : 495–498.

Demel, R. A., Geurts van Kessel, W. S. M., and Van Deenen, L. L. M., 1972, The properties of polyunsaturated lecithins in monolayers and liposomes and the interactions of these lecithins with cholesterol, *Biochim. Biophys. Acta* **266** : 26–40.

Demel, R. A., Jansen, J. W. C. M., Van Dijck, P. W. M., and Van Deenen, L. L. M., 1977, The preferential interactions of cholesterol with different classes of phospholipids, *Biochim. Biophys. Acta* **465** : 1–10.

Farias, R. N., 1987, Insulin-membrane interactions and membrane fluidity changes, *Biochim. Biophys. Acta* **906** : 459–468.

Friedlander, G., Le Grimellec, C., Giocondi, M. C., and Amiel, C., 1987, Benzyl alcohol increases membrane fluidity and modulates cyclic AMP synthesis in intact renal epithelial cells, *Biochim. Biophys. Acta* **903** : 341–348.

Gordon, L. M., Sauerheber, R. D., Esgate, J. A., Dipple, I., Marchmont, R. J., and Houslay, M. D., 1980, The increase in bilayer fluidity of rat liver plasma membranes achieved by the local anesthetic benzyl alcohol affects the activity of intrinsic membrane enzymes, *J. Biol. Chem.* **255** : 4519–4527.

Gould, R. J., and Ginsberg, B. H., 1985, Membrane fluidity and membrane receptor function, in *Membrane Fluidity in Biology* (R. C. Aloia and J. Boggs, eds.), Vol. 3, pp. 257–280, Academic Press, New York.

Heiniger, H. J., Chen, H. W., Applegate, O. L., Jr., Schacter, L. P., Schacter, B. Z., and

Anderson, P. N., 1976, Elevated synthesis of cholesterol in human leukemic cells, *J. Mol. Med.* **1** : 109–116.

Ho, Y. K., Smith, R. G., Brown, M. S., and Goldstein, J. L., 1978, Low-density lipoprotein (LDL) receptor activity in human acute myelogenous leukemia cells, *Blood* **52** : 1099–1114.

Hoffmann, W., Pink, D. A., Restall, C., and Chapman, D., 1981, Intrinsic molecules in fluid phospholipid bilayers: Fluorescence probe studies, *Eur. J. Biochem.* **114** : 585–589.

Hornstra, G., and Rand, M. L., 1986, Effect of dietary *n*-6 and *n*-3 polyunsaturated fatty acids on the fluidity of platelet membranes in rat and man, *Prog. Lipid Res.* **25** : 637–638.

Iritani, N., and Narita, R., 1984, Changes of arachidonic acid and *n*-3 polyunsaturated fatty acids of phospholipid classes of liver, plasma and platelets during dietary fat manipulation, *Biochim. Biophys. Acta* **793** : 441–447.

Irvine, R. F., 1982, How is the level of free arachidonic acid controlled in mammalian cells?, *Biochem. J.* **204** : 3–16.

Kannagi, R., Koizumi, K., and Masuda, T., 1981, Limited hydrolysis of platelet membrane phospholipids; on the proposed phospholipase-susceptible domain in platelet membranes, *J. Biol. Chem.* **256** : 1177–1184.

Kinosita, K., Jr., Kawato, S., and Ikegami, A., 1984, Dynamic structure of biological and model membranes: Analysis by optical anisotropy decay measurement, *Adv. Biophys.* **17** : 147–203.

Lange, Y., and Matthies, H. J. G., 1984, Transfer of cholesterol from its site of synthesis to the plasma membrane, *J. Biol. Chem.* **259** : 14624–14630.

Lange, Y., D'Alessandro, J. S., and Small, D. M., 1979, The affinity of cholesterol for phosphatidylcholine and sphingomyelin, *Biochim. Biophys. Acta* **556** : 388–398.

Lentz, B. R., Barenholz, Y., and Thompson, T. E., 1976, Fluorescence depolarization studies of phase transitions and fluidity in phospholipid bilayers. 2. Two-component phosphatidylcholine liposomes, *Biochemistry* **15** : 4529–4537.

Leslie, C. A., Gonnerman, W. A., Ullman, M. D., Hayes, K. C., Franzblau, C., and Cathcart, E. S., 1985, Dietary fish oil modulates macrophage fatty acids and decreases arthritis susceptibility in mice, *J. Exp. Med.* **162** : 1336–1349.

Mahadevappa, V. G., and Holub, B. J., 1987, Quantitative loss of individual eicosapentaenoyl-relative to arachidonoyl-containing phospholipids in thrombin-stimulated human platelets, *J. Lipid Res.* **28** : 1275–1280.

McMurchie, E. J., Patten, G. S., Charnock, J. S., and McLennan, P. L., 1987, The interaction of dietary fatty acid and cholesterol on catecholamine-stimulated adenylate cyclase activity in the rat heart, *Biochim. Biophys. Acta* **898** : 137–153.

Mori, T. A., Codde, J. P., Van Dongen, R., and Beilin, L. J., 1987, New findings in the fatty acid composition of individual platelet phospholipids in man after dietary fish oil supplementation, *Lipids* **22** : 744–750.

Muller, C. P., and Krueger, G. R. F., 1986, Modulation of membrane proteins by vertical phase separation and membrane lipid fluidity. Basis for a new approach to tumor immunotherapy, *Anticancer Res.* **6** : 1181–1194.

Nakagawa, Y., Inoue, K., and Nojima, S., 1979, Transfer of cholesterol between liposomal membranes, *Biochim. Biophys. Acta* **553** : 307–319.

Needham, L., Dodd, N. J. F., and Houslay, M. D., 1987, Quinidine and melittin both decrease the fluidity of liver plasma membranes and both inhibit hormone-stimulated adenylate cyclase activity, *Biochim. Biophys. Acta* **899** : 44–50.

Neelands, P. J., and Clandinin, M. T., 1983, Diet fat influences liver plasma-membrane lipid composition and glucagon-stimulated adenylate cyclase activity, *Biochem. J.* **212** : 573–583.

Nordoy, A., Davenas, E., Ciavatti, M., and Renaud, S., 1985, Effect of (n-3) fatty acids on platelet function and lipid metabolism in rats, *Biochim. Biophys. Acta* **835** : 491–500.

Oldfield, E., and Chapman, D., 1972, Dynamics of lipids in membranes: Heterogenicity and the role of cholesterol, *FEBS Lett.* **23** : 285–297.

Popp-Snijders, C., Schouten, J. A., Van Blitterswijk, W. J., and van der Veen, E. A., 1986, Changes in membrane lipid composition of human erythrocytes after dietary supplementation of (n-3) polyunsaturated fatty acids. Maintenance of membrane fluidity, *Biochim. Biophys. Acta* **854** : 31–37.

Pottel, H., Van der Meer, B. W., and Herreman, W., 1983, Correlation between the order parameter and the steady-state fluorescence anisotropy of 1,6-diphenyl-1,3,5-hexatriene and an evaluation of membrane fluidity, *Biochim. Biophys. Acta* **730** : 181–186.

Rand, M. L., Hennissen, A. A. H. M., and Hornstra, G., 1986, Effects of dietary sunflowerseed oil and marine oil on platelet membrane fluidity, arterial thrombosis, and platelet responses in rats, *Atherosclerosis* **62** : 267–276.

Renaud, S., Morazain, R., McGregor, L., and Baudier, F., 1979, Dietary fats and platelet function in relation to atherosclerosis and coronary heart disease, *Haemostasis* **8** : 234–251.

Rosenfeld, C., Jasmin, C., Mathe, G., and Inbar, M., 1979, Dynamic and composition of cellular membranes and serum lipids in malignant disorders, *Rec. Results Cancer Res.* **67** : 63–77.

Schouten, J. A., Beynen, A. C., Popp-Snijders, C., Mulder, C., Van Blitterswijk, W. J., and Hoitsma, H. F. W., 1987, The composition of plasma lipoproteins in cholesterol-fed pigs with partial ileal bypass, *Artery* **14** : 165–189.

Shinitzky, M., 1984, Membrane fluidity and cellular functions, in *Physiology of Membrane Fluidity* (M. Shinitzky, ed.), Vol. 1, pp. 1–51, CRC Press, Boca Raton, FL.

Shinitzky, M., Skornick, Y., and Haran-Ghera, N., 1979, Effective tumor immunization induced by cells of elevated membrane lipid microviscosity, *Proc. Natl. Acad. Sci. U.S.A.* **76** : 5313–5316.

Simon, I., Brown, T. J., and Ginsberg, B. H., 1987, Modification of membrane physical properties, biological response and insulin binding in Friend cells by low serum concentration, *Biochim. Biophys. Acta* **896** : 165–172.

Skornick, Y., Kurman, C. C., and Sindelar, W. F., 1984, Active immunization of hamsters against pancreatic carcinoma with lipid-treated cells or their shed antigens, *Cancer Res.* **44** : 946–948.

Skornick, Y. G., Rong, G. H., Sindelar, W. F., Richert, L., Klausner, J. M., Rozin, R. R., and Shinitzky, M., 1986, Active immunotherapy of human solid tumor with autologous cells with cholesteryl hemisuccinate. A phase I study, *Cancer* **58** : 650–654.

Spector, A. A., and Burns, C. P., 1987, Biological and therapeutic potential of membrane lipid modification in tumors, *Cancer Res.* **47** : 4529–4537.

Spector, A. A., and Yorek, M. A., 1985, Membrane lipid composition and cellular function, *J. Lipid Res.* **26** : 1015–1035.

Spiegel, R. J., Schaefer, E. J., Magrath, I. T., and Edwards, B. K., 1982, Plasma lipid alterations in leukemia and lymphoma, *Am. J. Med.* **72** : 775–782.

Storch, J., and Schachter, D., 1985, Calcium alters the acyl chain composition and lipid fluidity of rat hepatocyte plasma membrane *in vitro*, *Biochim. Biophys. Acta* **812** : 473–484.

Stubbs, C. D., and Smith, A. D., 1984, The modification of mammalian membrane polyunsaturated fatty acid composition in relation to membrane fluidity and function, *Biochim. Biophys. Acta* **779** : 89–137.

Tahin, Q. S., Blum, M., and Carafoli, E., 1981, The fatty acid composition of subcellular membranes of rat liver, heart, and brain: Diet-induced modifications, *Eur. J. Biochem.* 121 : 5–13.

Tandon, N., Harmon, J. T., Rodbard, D., and Jamieson, G. A., 1983, Thrombin receptors define responsiveness of cholesterol-modified platelets, *J. Biol. Chem.* 258 : 11840–11845.

Van Blitterswijk, W. J., 1984, Alterations in lipid fluidity in the plasma membrane of tumor cells, in *Physiology of Membrane Fluidity* (M. Shinitzky, ed.), Vol. 2, pp. 53–83, CRC Press, Boca Raton, FL.

Van Blitterswijk, W. J., 1985, Membrane fluidity in normal and malignant lymphoid cells, in *Membrane Fluidity in Biology* (R. C. Aloia and J. Boggs, eds.), Vol. 3, pp. 85–159, Academic Press, New York.

Van Blitterswijk, W. J., Van Hoeven, R. P., and Van der Meer, B. W., 1981, Lipid structural order parameters (reciprocal of fluidity) in biomembranes derived from steady-state fluorescence polarization measurements, *Biochim. Biophys. Acta* 644 : 323–332.

Van Blitterswijk, W. J., DeVeer, G., Krol, H. J., and Emmelot, P., 1982, Comparative lipid analysis of purified plasma membranes and shedded extracellular membrane vesicles from normal murine thymocytes and leukemic GRSL cells, *Biochim. Biophys. Acta* 688 : 495–504.

Van Blitterswijk, W. J., Hilkmann, H., and Hengeveld, T., 1984, Differences in membrane lipid composition and fluidity of transplanted GRSL lymphoma cells, depending on their site of growth in the mouse, *Biochim. Biophys. Acta* 778 : 521–529.

Van Blitterswijk, W. J., Damen, J., Hilkmann, H., and de Widt, J., 1985, Alterations in cholesterol homeostasis and biosynthesis in mice bearing a transplanted lymphoid tumor, *Biochim. Biophys. Acta* 816 : 46–56.

Van Blitterswijk, W. J., Van der Meer, B. W., and Hilkmann, H., 1987, Quantitative contributions of cholesterol and the individual classes of phospholipids and their degree of fatty acyl (un)saturation to membrane fluidity measured by fluorescence polarization, *Biochemistry* 26 : 1746–1756.

Van den Bosch, H., 1980, Intracellular phospholipases A, *Biochim. Biophys. Acta* 604 : 191–246.

Van der Meer, B. W., Van Hoeven, R. P., and Van Blitterswijk, W. J., 1986, Steady-state fluorescence polarization data in membranes. Resolution into physical parameters by an extended Perrin equation for restricted rotation of fluorophores, *Biochim. Biophys. Acta* 854 : 38–44.

Verger, R., 1976, Interfacial enzyme kinetics of lipolysis, *Annu. Rev. Biophys. Bioeng.* 5 : 77–117.

Vitols, S., Gahrton, G., Ost, A., and Peterson, C., 1984, Elevated low density lipoprotein receptor activity in leukemic cells with monocytic differentiation, *Blood* 63 : 1186–1193.

Wattenberg, B. W., and Silbert, D. F., 1983, Sterol partitioning among intracellular membranes; testing a model for cellular sterol distribution, *J. Biol. Chem.* 258 : 2284–2289.

Yachnin, S., Golomb, H. M., West, E. J., and Saffold, C., 1983, Increased cholesterol biosynthesis in leukemic cells from patients with hairy cell leukemia, *Blood* 61 : 50–60.

Yeagle, P. L., and Young, J. E., 1986, Factors contributing to the distribution of cholesterol among phospholipid vesicles, *J. Biol. Chem.* 261 : 8175–8181.

Yguerabide, J., 1972, Nanosecond fluorescence spectroscopy of macromolecules, *Methods Enzymol.* 26 : 498–578.

Chapter 13

Fusion of Enveloped Viruses with Biological Membranes

Fluorescence Dequenching Studies

Nor Chejanovsky, Ofer Nussbaum, Abraham Loyter, and Robert Blumenthal

1. INTRODUCTION

Enveloped virions penetrate eukaryotic cells by two alternative routes (Choppin and Scheid, 1980; White *et al.*, 1983). Envelopes of viruses belonging to the paramyxovirus group fuse with cells' plasma membranes at pH 7.4 and consequently microinject their content, the viral nucleocapsid, directly

---

Abbreviations used in this chapter: $C_{12}E_8$, octaethyleneglycolmono(*n*-dodecyl)ether; CF, carboxyfluorescein; chol, cholesterol; CL, cardiolipin; DPA, dipicolinic acid; DQ, dequenching; DTT, dithiothreitol; EDTA, ethylenediaminetetracetic acid; FRET, fluorescence resonance energy transfer; gang, gangliosides; HA, hemagglutinin; HA_0, precursor of the influenza hemagglutinin glycoprotein; HEG, human erythrocyte ghosts; HSV, herpes simplex virus; HTC, hepatoma tissue cultured; N-NBD, 7-nitro-1,2,3-benzoxydiazole-4-amino; PC, phosphatidylcholine; PE, phosphatidylethanolamine; PMSF, phenylmethylsulfonyl fluoride; PS, phosphatidylserine; R18, octadecylrhodamine B-chloride; RIVE, reconstituted influenza virus envelope; ROVs, right-side-out (erythrocyte membrane) vesicles; RSVE, reconstituted Sendai virus envelope; SFV, Semliki Forest virus; VSV, vesicular stomatitis virus.

Nor Chejanovsky, Ofer Nussbaum, and Abraham Loyter   Institute of Life Sciences, The Hebrew University of Jerusalem, Jerusalem 91904, Israel.   **Robert Blumenthal**   National Cancer Institute, National Institutes of Health, Bethesda, Maryland 20892.

into the cell cytoplasm (Choppin and Scheid, 1980; Loyter and Volsky, 1982; White *et al.*, 1983). A different way of entry has been described for most other enveloped virions such as those belonging to the orthomyxovirus, toga, rhabdo, and herpes groups. Such viruses are taken into cells by endocyticlike processes. Fusion of the viral envelopes with the endosomal or lysosomal membranes is triggered by the intraorganelle low-pH environment and leads to the introduction of the viral content into the intracellular space (Chopin and Scheid, 1980; White *et al.*, 1983). Fusion of the pH-dependent virions with the plasma membrane can be triggered by lowering the pH of the medium containing the virus-associated cells.

A variety of different approaches have been applied to study the complex problem of analyzing intracellular events concomitant with virus penetration. Classically, electron microscopy has been used for such analysis, since individual virions associated with the plasma membrane or localized intracellular can be accounted for with that technique. However, according to Dales (1973), "the methodology of electron microscopy is fraught with artifacts. As a result a number of controversies have arisen regarding the true pathways for penetration." The detailed discussion offered by Dales (1973) of the difficulties and pitfalls in preparation of specimens and evaluation of electron microscopic images to analyze pathways of viral penetration is still very worthwhile reading.

In the ensuing years a number of complementary approaches have been used, and there seems to be a consensus emerging concerning entry mechanisms.

2. RECEPTORS FOR ENVELOPED VIRUSES

The infection of cells by animal viruses involves two main steps: (1) binding of the virus particles to specific cell receptors and (2) introduction of the viral nucleocapsid into the cell cytoplasm (Choppin and Scheid, 1980; Dimmock, 1982). The attachment of the viral particles to the cell surface is mediated by the viral glycoproteins that constitute the spikes in the viral envelope and specifically recognize cellular receptors for the virus on the plasma membrane of the target cells (Rott and Klenk, 1977; White *et al.*, 1983). Thus, the specificity of this recognition process depends on the molecular structure of those viral glycoproteins and their correspondent receptors. The absence of the viral receptors on a specific cell would confer it resistance to viral infection (Rott and Klenk, 1977; Choppin and Scheid, 1980). Some common characteristics can be ascribed to the process of binding of enveloped viruses to the cell surface:

1. The attachment of the viral particle to its cellular receptor is a rela-

tively slow process. For example, binding of the Sendai virus to human erythrocytes is maximal at 15 min (Wolf *et al.*, 1983) and binding of Semliki Forest virus to BHK-21 cells takes about 20–40 min (Fries and Helenius, 1983).

2. The attachment of the viral particles to their cellular receptor is almost irreversible. Elution of viruses bound to the cell surface was achieved by enzymatic digestion of the cellular receptor or, alternatively, by inactivation of the correspondent binding glycoprotein (Helenius *et al.*, 1980; Chejanovsky *et al.*, 1984).

The number of viral particles that are able to penetrate the target cell is mainly controlled by the number of viral receptors available on the cell surface (Helenius *et al.*, 1980; Chejanovsky *et al.*, 1984). The multiple copies of the viral glycoproteins present in the viral envelope contribute to the multivalent binding of the virus to the cell surface (Kielian and Helenius, 1986). The distribution of the bound virus particles on the cell surface is not always even and depends on the cell type and organization of the cell surface. For example, a high concentration of the Semliki Forest virus near the microvilli of BHK-21 cells was reported (Fries and Helenius, 1983).

The next section deals with the different cellular receptors for the different viral groups.

2.1. Myxoviruses

This viral family includes the influenza viruses of humans and lower animals. The viral envelope of viruses belonging to this group comprises three proteins. Two of them are surface glycoproteins, designated HA (hemagglutinin) and NA (neuraminidase), and are responsible for the viral binding and fusion properties and for enzymatic removal of sialic acid residues of the cell surface, respectively (Rott and Klenk, 1977; Choppin and Scheid, 1980; White *et al.*, 1983). Sialic acid residues from glycoproteins and glycolipids were found to serve as the cellular receptor for influenza viruses (Schulze, 1975; Rott and Klenk, 1977). Moreover, different human influenza isolates bearing antigenically different hemagglutinin serotypes were able to differentiate microdomains of the gangliosides such as the sialic acid species (NeuAc, NeuGc) and the sequence of sialic acid linkages (NeuAc alpha 2-3 gal, NeuAc alpha 2-6 gal) preferentially as specific viral receptors (Suzuki *et al.*, 1985). For example, the influenza strain A/PR/8/34, which is serotype H1 for hemagglutinin, recognized IV-3 NeuAc-nLc4Cer containing the NeuAc alpha 2-3 gal sequence preferentially over IV-6 NeuAc-nL c4Cer containing NeuAc alpha 2-6 gal, where the H2 serotype (A/Japan/305/57) recognized the NeuAc alpha 2-6 gal sequence preferentially over NeuAc alpha 2-3 gal (Suzuki *et al.*, 1985).

2.2. Paramyxoviruses

The most representative member of this family of viruses is the Sendai virus. Two glycoproteins are associated with the membrane of the paramyxovirus (Tozawa *et al.*, 1973), namely, the hemagglutinin neuraminidase (HN) and the fusion factor (F) (Homma and Ohuchi, 1973; Scheid and Choppin, 1974). The HN glycoprotein mediates the binding of the viral envelope to its cellular receptor (Homma and Ohuchi, 1973). This glycoprotein specifically recognizes sialic acid containing glycoproteins and glycolipids on the target cell surface (Poste and Pasternak, 1978; Choppin and Scheid, 1980; Dimmock, 1982). Removal of membrane sialic acid residues by treatment with neuraminidase (*Vibrio cholerae* neuraminidase) results in the inhibition of the Sendai virus binding (Rott and Klenk, 1977; Dimmock, 1982). Further insertion of sialoglycolipids into the membranes of neuraminidase-treated cells allowed the subsequent infection of the host cell by Sendai virus particles (Markwell *et al.*, 1981). The neuraminidase activity residing in the HN molecule of Sendai virions seems to play a role in the spreading of the virus in the infected host (Choppin and Scheid, 1980).

2.3. Togaviruses

The knowledge acquired in the penetration of the togaviruses was due to the extensive research performed on the entry of the two alphaviruses Semliki Forest and Sindbis (Helenius *et al.*, 1980; White *et al.*, 1983). The viral nucleocapsid is enclosed in a lipid bilayer membrane in which the viral glycoproteins that constitute the spikes are embedded (Helenius *et al.*, 1980). Three glycoproteins compose the spikes E_1, E_2, and E_3. E_1 and E_2 are transmembrane proteins and E_3 is noncovalently associated with these and is external to the lipid bilayer (Helenius *et al.*, 1980). It has been shown that Semliki Forest viruses were bound preferentially to the major histocompatibility antigens HLA-A and HLA-B present on the surface of JY cells and to the H2-K and H2-L antigens present in peripheral murine lymphocytes (Helenius *et al.*, 1978). However, cells devoid of H2 antigens can be efficiently infected with Semliki Forest virus (Oldstone *et al.*, 1980).

Proteolytic digestion of host cell membranes with the appropriate proteases inhibited Sindbis virus binding to its target cell (Smith and Tignor, 1980). Neuraminidase treatment of the cells marginally affected the binding ability of the Sindbis virus (Smith and Tignor, 1980). Studies performed by chemical cross-linking of the Sindbis virus spike glycoproteins bound to the cell surface of JY and Daudi cells identified an M_r 90,000 protein (Massen and Terhost, 1981). The exact molecular nature of the alphaviruses receptor is still unclear.

2.4. Rhabdoviruses

The prototype of this group is the vesicular stomatitis virus (VSV). The viral envelope contains a single glycoprotein (G) embedded in the viral membrane (Wagner, 1975). The exact nature of the VSV cellular receptor was difficult to establish. The extremely wide host range of the VSV suggested that its cellular receptor is a component present in virtually all animal cells. Competition studies showed saturable binding sites for VSV on Vero cells (Schlegel et al., 1982). Protease treatment of recipient cells did not affect VSV binding but the addition of phosphatidylserine (PS) did. Moreover, PS inhibited VSV plaque formation (Schlegel et al., 1983). Reconstituted VSV envelopes bearing the VSV G-protein were able to bind and fuse at low pH with phosphatidylserine liposomes (Eidelman et al., 1984). However, the specificity of PS as the VSV receptor is still questionable.

Other investigators studied the effect of incubation of VSV particles with gangliosides on the infectivity of the virus to CER cells. In addition, gangliosides from mammalian brains and from CER cells were able to inhibit the VSV attachment to susceptible CER cells (Sinibaldi et al., 1985). Moreover, VSV bound very poorly to CER cells treated with glycosylases. Deglycosylated CER cells reacquired their susceptibility to infection by VSV after coating with gangliosides, suggesting the participation of the gangliosides in the receptorial structure for VSV in the CER cells system (Sinibaldi et al., 1985).

2.5. Retroviruses

The extensive research performed on the biology of the human T-lymphotropic viruses, which are members of the retrovirus family, paved the way to the characterization for the cellular receptor of the human T-lymphotropic virus III (HTLV-III). The HTLV-III virus (AIDS virus) has a selective tropism to a subset of T lymphocytes, which are defined by the expression of the surface glycoprotein T_4 (CD_4 antigen) (Klatzmann et al., 1984). Monoclonal antibodies directed against T_4 molecule were able to block infection of T_4^+ lymphocytes by HTLV-III (Klatzmann et al., 1984). This blocking effect was shown to be specific to these antibodies; monoclonal antibodies against the histocompatibility locus antigens (HLA) class II or anti T-cell natural killer (TNK) target did not prevent HTLV-III infection (Klatzmann et al., 1984). In a parallel study, pseudotypes of vesicular stomatitis virus bearing HTLV-III envelopes were tested for its ability to infect and promote formation of syncytia by mixing virus-producing and cell-bearing receptors for HIV-III (Dalgleish et al., 1984). Receptors were present only on cells. It was found that anti-CD_4 antibodies inhibited syncytia formation and blocked the pseudotypes. When the genes for T_4 were expressed on the surface of T or B lymphocytes or other human cell lines, such cells became susceptible to HTLV-III infection (Maddon *et al.*, 1985; Dewhurst, 1987). The T_4 molecule serves as the cellular receptor to HTLV-III and interacts with the viral envelope glycoprotein gp110 (McDougal *et al.*, 1986).

2.6. Herpesviruses

The envelope of the herpesviruses comprises several glycoproteins embedded in the lipid bilayer (Choppin and Scheid, 1980). The cellular receptor for the Epstein-Barr virus, a member of the herpesvirus family, was recently identified (Fingeroth *et al.*, 1984). The rank order to binding of fluoresceinated EBV to four lymphoblastoid cell lines (SB, JX, Molt-4, and Raji) was identical to the rank order of binding of monoclonal antibodies that recognized the complement receptor type 2 (CR2) as measured by analytical flow cytometry. Transfer of CR2 from SB cells to protein-bearing *Staphylococcus aureus* particles (absorbed with anti-CR2 monoclonal antibody) conferred on them the specific ability to bind ^{125}I-labeled EBV (Fingeroth *et al.*, 1984). Pretreatment of cells with anti-CR2 monoclonal antibody followed by treatment with goat $F(ab)_2$ fragments to mouse IgG blocked the binding of fluoresceinated EBV on SB cells (Fingeroth *et al.*, 1984). Based on these experiments, it was concluded that the complement receptor molecule (C3d receptor CR2) constitutes the cellular receptor for EBV. Further studies had shown that an immunotoxin made from a monoclonal antibody against the C3d receptor and the toxin gelonin was able to inhibit at very low concentrations (10^{-11} M) the ability of EBV to induce polyclonal proliferation of normal B lymphocytes (Tedder *et al.*, 1986).

2.7. Other Enveloped Viruses

The spectrum of cellular receptors for the enveloped virus families of Poxvirus, Iridovirus, Hepadnavirus, Coronavirus, Bunyaveravirus, and Arenavirus is not as yet identified.

3. INTERACTION OF ENVELOPED VIRUSES WITH RECEPTOR-DEPLETED CELLS

A different approach to study the interaction between enveloped animal viruses and target cell membranes was to study the ability of artificial receptors to mediate the viral entry to the host cell. The Sendai virus was mainly used as the model in these studies.

3.1. Use of Antimembrane Antibodies or Polypeptide Hormones to Mediate Virus Attachment

Biologically active Sendai virus envelopes can be obtained by solubilization of the viral membrane in Triton X-100 and subsequent removal of the detergent. Reconstituted Sendai virus envelopes (RSVE) constituted a closed vesicle bearing the two viral glycoproteins, hemagglutinin neuraminidase (HN) and the fusion factor (F), embedded in the lipid bilayer devoid of the nucleocapsid (Volsky and Loyter, 1978). Two alternative methods were used to promote the binding and subsequent fusion of the RSVE with viral-receptor-depleted living cells (viral receptor depletion was performed by neuraminidase digestion of the cellular sialic acid residues):

1. An antibody raised against recipient cell membranes was inserted into the viral envelope. Anti-human erythrocyte IgG was covalently bound to the hydrophobic molecule dodecanethiol and further inserted into the RSVE particles by co-reconstitution (Gitman and Loyter, 1985). The obtained RSVE bearing the anti-human erythrocytes antibody was shown to bind and subsequently fuse with the viral-receptor-depleted erythrocytes. Thus, a cell surface antigen was able to function as an efficient alternative cellular receptor for Sendai virions (Gitman and Loyter, 1984).

2. Anticell specific antibodies or alternatively polypeptide hormones were covalently coupled to the Sendai virus envelope glycoproteins after they had been solubilized with Triton X-100 (Nussbaum et al., 1984). Using this method, RSVE bearing the covalently attached anti-human erythrocyte antibodies or insulin molecules was able to attach and fuse to virus-receptor-depleted human erythrocytes and rat hepatoma cells (Gitman et al., 1985). Again, in this case a cellular antigen or the cell insulin receptor served as functionally active cellular receptors for Sendai virions.

3.2. Implantation of Receptors or Binding Proteins for Enveloped Virions into Recipient Cell Membranes

The ability of different gangliosides to mediate the binding and fusion of the influenza virus with chicken asialoerythrocytes was used as a model system (Suzuki et al., 1985). As mentioned before, insertion of the ganglioside GM3 into the chicken asialoerythrocyte membranes restores their susceptibility to the influenza virus binding and fusion activities (Suzuki et al., 1985). Similarly, insertion of sialoglycolipids into desialyzed cells allowed their infection by Sendai virus particles [see paramyxoviruses in this review and Markwell et al. (1981)].

Receptor negative cells to the Epstein-Barr virus were converted to receptor positive cells by fusion-mediated transfer (Volsky et al., 1980). Membranes rich in receptors for EBV from the human lymphoma cell line Raji were co-reconstituted with RSVE and the formed EBV-receptor-bearing RSVEs were subsequently fused with the EBV receptor negative cells. By this procedure those EBV receptor negative cells became susceptible to EBV infection (Volsky *et al.*, 1980).

The ability of anti-Sendai antibodies to mediate the binding of the virus particles to virus-receptor-depleted cells was also studied (Nussbaum *et al.*, 1984). Anti-Sendai antibodies were chemically coupled to the surface of desialyzed human erythrocytes by the use of the bifunctional cross-linking reagents *N*-succinimidyl-3-2(2-pyridyldithio) propionate or succinimidyl-4-(*p*-maleimidophenyl) butyrate. These antibodies served as attachment sites and were recognized by biologically active Sendai virus particles. Virus erythrocyte fusion was observed following attachment to anti-Sendai antibody-bearing cells (Nussbaum *et al.*, 1984).

A different system was developed by using the binding pair affinity of avidin molecule to biotin (Guyden *et al.*, 1983). Avidin was chemically coupled to surfaces of human red cells and biotin was concomitantly coupled to Sendai virus particles. The couple avidin–biotin served as an alternative recognition system between the treated Sendai virus particles and the target cell. The attachment of the virus particles to the cell surface through the HN binding activity was blocked by preincubation of the particles with an anti-HN monoclonal antibody (Fab preparation) (Guyden *et al.*, 1983).

Targeting of the Sendai virus to desialyzed human erythrocytes was also achieved by utilizing hybrid antibody molecules (Chejanovsky *et al.*, 1985). Anti-Sendai anti-human erythrocyte hybrids [F(ab)$_2$ molecules] were prepared by chemical coupling of anti-Sendai Fab fragments with anti-human erythrocyte Fab fragments. The hybrid molecules obtained were able to mediate the binding of Sendai virus particles to desialyzed human erythrocytes. Incubation of the virus with desialyzed erythrocytes at 37°C resulted in extensive cell fusion (Chejanovsky *et al.*, 1985).

In summary, it has extensively been shown that binding proteins other than membrane receptors can mediate the binding and fusion of virus particles with recipient cell membranes.

4. THEORETICAL ASPECTS OF THE USE OF FLUORESCENCE DEQUENCHING TO MEASURE VIRAL FUSION

In the past, membrane fusion events have been identified and characterized mainly by microscopic techniques (Poste and Pasternak, 1978; Duzgunes, 1985; Blumenthal, 1987). However, since those methods do not readily lend themselves to quantitation of the fusion reaction, new biophysical techniques needed to be developed. In this chapter we focus on techniques using fluorescence.

The first fluorescence dequenching assay was based on self-quenching properties of the water-soluble dye carboxyfluorescein and was originally used to monitor transfer of contents of liposomes to cells (Weinstein *et al.*, 1977). The fluorophore trapped at high concentration inside a liposome emits only a few percent of the fluorescence that it would if released and diluted into the surrounding medium or into the cytoplasmic compartment of a cell. This assay indicated that the amount of liposome–cell fusion was much smaller than originally estimated by other techniques (Blumenthal *et al.*, 1977). On the other hand, the carboxyfluorescein dequenching assay has found a wide range of applications, for example, to monitor release of vesicle contents by a variety of agents such as serum components, enzymes, and cells [for a review see Weinstein *et al.* (1984)]. It has also been used to monitor leakage of vesicle contents during fusion.

Hoekstra *et al.* (1984) was the first to use fluorescence dequenching methods for studying and measuring the kinetics of fusion between enveloped virions and biological membranes. These authors made use of the self-quenching and hydrophobic properties of the fluorescent dye octadecyl rhodamine B-chloride (R18) [originally synthesized by Keller *et al.* (1977) and can be obtained from Molecular Probes, USA]. Because of the hydrophobic properties of the octadecyl chain, this probe can readily be inserted into envelopes of animal virions as well as into biological membranes or liposomes. Hoekstra *et al.* (1984) have shown that incubation of an ethanolic solution of R18 with Sendai virions results in spontaneous incorporation of the fluorescent dye into the viral envelopes. Under the experimental conditions used about 50–70% of the probe became virus-envelope associated, reaching a surface density of 2–3 mol % of the total viral phospholipids and its decrease was shown to be proportional to the fluorescence dequenching (Hoekstra *et al.*, 1984). Experiments by Hoekstra *et al.* (1984) as well as in our laboratories (Citovsky *et al.*, 1985) have shown that the incorporation of R18 into Sendai virus envelopes does not impair the viral biological activities, namely, its ability to induce hemolysis and to promote cell–cell fusion. It has been well established that hemolysis induced by Sendai virus reflects a process of virus–membrane fusion (Maeda *et al.*, 1977).

Quenching that results from collisional encounters between fluorophore and quencher is called collisional or dynamic quenching. The quencher diffuses to the fluorophore during the lifetime of the excited state; upon contact, the fluorophore returns to the ground state without emission of a photon. If the quencher is the same fluorophore it is self-quenched. In case of static quenching a complex is formed between fluorophore and quencher which is nonfluorescent.

An important aspect of collisional quenching is the distance over which fluorophore and quencher travel. The root mean square distance $[(\Delta x^2)^{1/2}]$ over which a quencher can diffuse during the lifetime of the excited state (t) is given by $\Delta x^2 = 2Dt$, where D is the free diffusion coefficient. Consider a typical case of a small, soluble fluorophore with a lifetime of 4 nsec and a diffusion coefficient of 2.5×10^{-5} cm^2/sec (both fluorophore and quencher). The distance over which the fluorophore has to diffuse to produce adequate quenching is 44 Å. To be effective, collisional quenchers must have high efficiency; that is, each collision results in loss of photon emission of the fluorophore. Generally, they are effective in the concentration range of 100 mM.

A simple calculation may determine whether the carboxyfluorescein quenching is consistent with dynamic or static quenching. The dye encapsulated into vesicles at concentrations of 10 mM or greater gives rise to about 50% fluorescence quenching (Weinstein *et al.*, 1984). Since at 10 mM the average distance between fluorophore is about 55 Å, dynamic quenching could possibly be the prevailing mechanism of self-quenching.

The measurement of fluorescent lifetimes is the most definitive method to distinguish static and dynamic quenching. In the case of static quenching no change in fluorescent lifetimes is expected, since the complexed fluorophores are nonfluorescent and the only observed fluorescence is from the uncomplexed fluorophore. Chen and Knutson (1987) found a normal lifetime (> 4 nsec) for over 95% of the fluorescence of carboxyfluorescein encapsulated in liposomes at high concentrations, indicating that the fluorescence decrease may be dominated by a static quenching mechanism. On the other hand, they observe significant dynamic quenching of free CF in the range of 10–50 mM. Depending on liposome size and encapsulation efficiency, both mechanisms may be important.

Absorption spectra of the fluorophore can also be examined to distinguish static and dynamic quenching. Collisional quenching only affects the excited states of fluorophore, and thus no change in the absorption spectra is predicted. On the other hand, ground-state complex formation will frequently result in perturbation of the absorption spectrum of the fluorophore. This is seen in spectra of carboxyfluorescein at high concentrations (Chen and Knutson, 1987).

Other methods to distinguish between static and dynamic quenching are measurements of dependence on temperature and viscosity. Dynamic quenching depends on diffusion and is expected to increase with increasing temperature and decreasing viscosity. In contrast, increased temperature is likely to result in decreased stability of complexes and thus lower the value of the static quenching constants.

In the case of R18 self-quenching, none of the experiments to determine the mechanism of quenching has yet been reported. If we assume a lateral diffusion coefficient of 10^{-7} cm^2/sec (typical of lipid molecules in a bilayer) and a lifetime of 4 nsec, the average distance of lipid probe diffusion to produce collisional quenching is 3 Å. Hoekstra *et al.* (1984) found 50% quenching at a probe surface density of 5.5 mol % R18 in DOPC liposomes. The average distance between probes at this surface density is 33 Å, assuming a lipid head group area of 50 Å^2. Therefore, it is unlikely that R18 quenching is collisional. More likely, the rhodamine molecules will stack up to form nonfluorescent complexes. According to the above calculation the level of R18 quenching at 1% probe surface density indicates that it must be quite self-aggregated. The fatty acyl chain is nonsaturated and in the solid phase below 50°C. Phase separation of solid chains is not uncommon in mixtures of fluid and solid lipids. Labeling of intact Sendai virions with R18 resulted in a 2–3 mol % surface density and 70% quenching of the probe (Hoekstra *et al.*, 1984). Similar results were found with VSV (Blumenthal *et al.*, 1987). This is a higher quenching than in DOPC vesicles. Hoekstra *et al.* (1984) hypothesize that the probe is only on the outer monolayer of the labeled virions, resulting in a surface density of 4–6 mol %, which would explain the higher quenching. Moreover, the probe might be more self-aggregated in the virus membrane or the additional quenching might be due to the interaction of the probe with the more polar environment of the viral spike glycoprotein (see below).

Even if dequenching takes place upon fusion of membranes and diffusion of the probe over a larger surface area, relief of self-quenching might not be 100%. Aggregation of probe molecules even at very low surface densities might lead to some level of quenching. We have found that labeling cells directly with R18 and disrupting with Triton X-100 leads to a fluorescence increase of a factor of about 1.56 (Blumenthal *et al.*, 1987). A practical consequence of this is that measurements of total dequenching should be corrected with a factor of 1.56 (see below).

In static quenching the relation between quenched fluorescence and quencher concentration is given by (Lakowicz, 1983)

$$1 - \frac{F}{F_0} = \frac{K_s Q}{1 + K_s Q} \tag{1}$$

where F is fluorescence, F_0 fluorescence in the absence of quencher, Q concentration (or mole fraction of quencher), and K_s the (self-) association constant for the complex. According to Eq. (1), the quenching range is linear for $K_s Q \ll 1$. Hoekstra *et al.* (1984) found linearity up to 9 mol % R18 in DOPC liposomes, indicating that $K_s < (9 \text{ mol } \%)^{-1}$.

Part of the R18 quenching in the virion might be due to solvent effects;

that is, some of the R18 is bound to the viral spike glycoproteins. The fluorescence emission spectra of many fluorophores are sensitive to the polarity of their surrounding environment. For example, the emission peak wavelength of R18 is shifted to shorter wavelengths (blue shift) as the solvent polarity is decreased. Conversely, increasing solvent polarity generally results in shifts of the emission spectrum to longer wavelength (red shifts). Those red shifts are often accompanied by a decrease in the quantum yield of the fluorophore. The shifts result from both the interaction of the dipole moment of the fluorophore with the reactive fields induced in the surrounding solvent and from specific chemical interactions between the fluorophore with one or more solvent molecules. Localization of membrane-bound fluorophores can be inferred from their emission spectra (Lakowicz, 1983). The emission spectrum of the fluorophore inserted into the membrane is compared with its emission spectra in solvents of varying polarity.

In order to determine the environment of R18 in the virus, excitation and emission spectra of R18 in ethanol, in a lipid bilayer, in a detergent micelle, and R18 bound to protein were compared (Loyter *et al.*, 1988a). The proteins were bovine serum albumin and the extracytoplasmic portion of the VSV G-protein obtained by cathepsin D digestion (Crimmins *et al.*, 1983). R18 bound to proteins was in a much more nonpolar milieu as compared to R18 in the lipid bilayer. Figure 1 shows the emission spectrum of R18 bound to G (cath D) before and after adding detergent. The peak λ_{em} of R18 bound to G (cath D) was 600 nm, whereas its value in the detergent micelle was 586 nm, closer to the value in ethanol of 576 nm. The detergent micelle environment resulted in a fivefold increase in the fluorescence intensity in addition to the blue shift (Figure 1). In the virion the maximal value of λ_{em} was 590 nm, a value between that of the polar environment of the protein and the nonpolar environment of the lipid bilayer. On the other hand, R18-labeled Sendai virus had the same λ_{ex} and λ_{em} peaks as R18 in DOPC vesicles (Hoekstra *et al.*, 1984), indicating that in Sendai the R18 was more lipid bilayer associated.

Methods based on fluorescence resonant energy transfer (FRET) to measure membrane fusion are more sensitive, since they require less probe, which could potentially perturb the fusion system. One such assay was devised for mixing of vesicle contents (Wilschut and Papahadjopoulos, 1979) and involves the fast formation of a chelation complex between Tb^{3+}, encapsulated as a citrate complex in one population of vesicles, and dipicolinic acid in the second population. Fusion and mixing of contents result in the formation of a fluorescent Tb^{3+}–dipicolinic acid complex. Presence of Ca^{2+} and EDTA in the external medium prevents the formation of the fluorescent complex outside the vesicles. Mixing of membrane components can be measured by

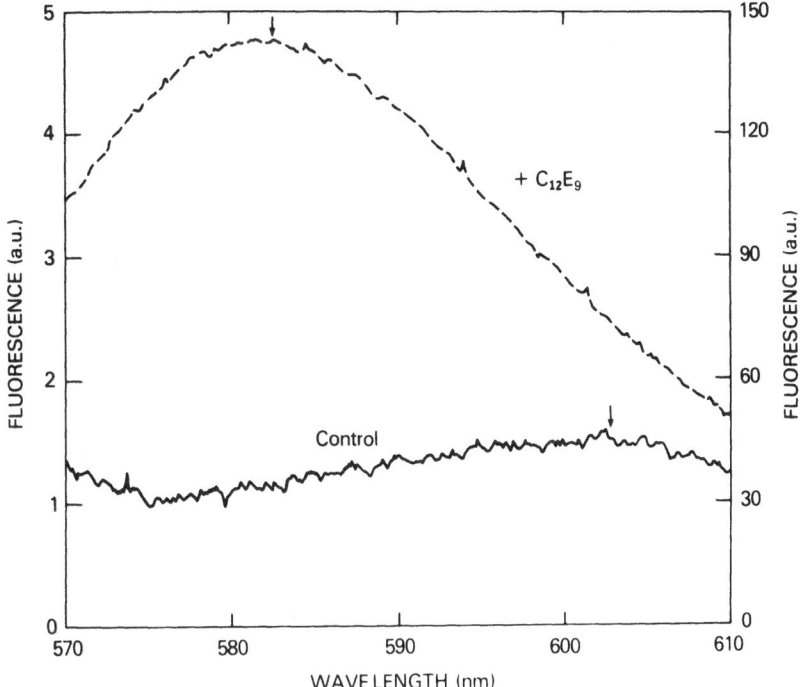

FIGURE 1. Emission spectra of R18 bound to G (cath D) before and after addition of the detergent C12E9. The data were collected using an MPF 44B spectrofluorometer (Perkin–Elmer) at an excitation wavelength of 565 nm. See text for more details.

FRET between a donor and acceptor fluorophore, both of which are attached to a phospholipid (Struck *et al.*, 1981).

Fluorescence energy transfer is the transfer of the excited-state energy from a donor to an acceptor. This transfer occurs without the appearance of a photon and is primarily a result of dipole–dipole interactions between donor and acceptor. The rate of energy transfer depends on the extent of overlap of the emission spectrum of the donor with the absorption spectrum of the acceptor, the relative orientation of the donor and acceptor transition dipoles, and the distance between these molecules. It is this latter dependence on distance that has resulted in the widespread use of energy transfer as a "spectroscopic ruler." Many studies have been performed on fluorescence energy transfer between donor and acceptor pairs, which are lipid fluorophores free to diffuse in the bilayer (Fung and Stryer, 1978). The amount of energy transfer expected for fluorophores randomly distributed on the surface of a membrane has been analyzed extensively. It is a complex problem that requires consideration of the geometric form of the bilayer (planar or spherical) and transfer between donors and acceptors that are on the same side of the bilayer as well as those on opposite sides.

Energy transfer results in quenching of the donor and sensitized emission of the acceptor. A variety of approaches have been used and in some instances solved numerically (Fung and Stryer, 1978; Wolber and Hudson, 1979; Snyder and Freire, 1982). In the case of lipid fluorophores randomly diffusing in the plane of the bilayer, it has been shown that quenching is only dependent on the surface density of acceptors, on the R_0 between donor and acceptor fluorophore, and on the distance of closest approach between donor and acceptor molecules. Only small amounts (<0.1 mol %) of acceptor incorporated into the membrane are needed for quenching. For the energy transfer pair N-NBD-PE and N-Rh-PE, which is widely used in liposome fusion studies, quenching is linear with acceptor concentration up to a surface density of 0.5 mol % (Struck *et al.*, 1981). At this surface density the donor was about 60% quenched. To achieve the same amount of fluorescence quenching, about 10-fold less probe needs to be used in the FRET assay as compared with the R18 dequenching assay. The ability to use less fluorescence probe, which could potentially perturb the membrane, makes the FRET assay more attractive.

It should be noted that membrane mixing assays may generate a multitude of artifactual positive results due to, for example, prefusion interactions, partial fusion, exchange of lipids, and aggregation (see Morris *et al.*, 1988). Therefore, ideally, both core mixing and membrane mixing assays should be done simultaneously or in parallel.

Regardless of what the mechanism of fluorescence quenching is, its relation to the amount of fusion needs to be established. In experiments of fusion between virions and cells, the virus is prebound to the cell surface at 4°C. Subsequently, unbound virus is removed by centrifuging the virus–cell complexes and washing. The virus–cell complexes are then warmed to 37°C and, if necessary, the pH is lowered. For that experimental protocol, the following equation has been derived expressing the relation between measured fluorescence and percentage of fusion (Loyter *et al.*, 1988a):

$$\% \text{ fusion} = \frac{F_x - F_0}{F_t - F_0} \tag{2}$$

where F_x is fluorescence measured at a given time, F_0 is the fluorescence at time 0, and F_t is the maximal fluorescence if all the probe had transferred from virion to cell. The maximum amount of dequenching F_t is determined by disrupting the virion–cell complexes with detergent. As mentioned above, the total fluorescence of the probe in the detergent micelle may differ from that in the cell under conditions of "infinite" dilution. For R18 in Vero cells we have determined a correction factor of 1.56 to account for this difference (Blumenthal *et al.*, 1987). The derivation of Eq. (2) assumes that the fluorophore is diluted from virion into an infinite reservoir of the cell, and that fluorophores that have already entered this reservoir do not affect those coming in.

5. FUSION OF ENVELOPED VIRUSES WITH ANIMAL CELLS AND BIOLOGICAL MEMBRANES: STUDIES WITH INTACT VIRIONS

As mentioned above, the fluorescence probe R18 can readily be inserted into envelopes of intact virus particles (Hoekstra *et al.*, 1984; Citovsky *et al.*, 1985). Incorporation of R18 and at high surface density leads to its self-quenching and its decrease was shown to be proportional to fluorescence dequenching (Hoekstra *et al.*, 1984; Citovsky *et al.*, 1985; Stegmann *et al.*, 1985). Therefore, fusion between enveloped virions and recipient membranes can be studied by determination of R18 fluorescence after the relief from its self-quenching (Hoekstra *et al.*, 1984; Citovsky *et al.*, 1985; Stegmann *et al.*, 1985; Loyter *et al.*, 1988a). The use of fluorescently labeled enveloped virions and the fluorescence dequenching method allowed studies on the various pathways of fusion (i.e., at the plasma membrane or via endocytosis), kinetics of virus–membrane fusion, effect of inhibitors, and fusion with various membrane preparations (Loyter *et al.*, 1988a).

Incubation of fluorescently labeled enveloped virions with living cultured cells resulted in fluorescence dequenching (Table I). This was demonstrated with Sendai (Hoekstra *et al.*, 1984; Citovsky *et al.*, 1985), influenza (Stegmann *et al.*, 1986; Nussbaum and Loyter, 1987), VSV (Blumenthal *et al.*, 1987), Herpes simplex 1 (HSV-1), and Semliki Forest viruses (Nussbaum and Loyter, unpublished data). The view that the fluorescence dequenching observed indeed reflects a process of virus–membrane fusion and results from the dilution of the virus-associated probe in the recipient membrane was supported by studies using inactivated unfusogenic virions (Citovsky *et al.*, 1985; Nussbaum and Loyter, 1987).

Sendai virions can be rendered unfusogenic by treatment with dithiothreitol (DTT) (Ozawa *et al.*, 1979a), trypsin (Ozawa *et al.*, 1979b), phenylmethylsulfonyl fluoride (PMSF) (Israel *et al.*, 1983), or glutaraldehyde (Chejanovsky and Loyter, 1985). No increase in the degree of fluorescence (fluorescence dequenching) was observed following incubation of treated unfusogenic Sendai virions with animal cultured cells (Citovsky *et al.*, 1985; Loyter *et al.*, 1987; Table II). A low level of fluorescence dequenching was

Table I
Interaction of Enveloped Virus with Cultured Cells and HEG:
Fluorescence Dequenching Studies[a]

Virus	Recipient cells	pH of incubation	R18 EQ (%)
Sendai	HeLa	7.4	54
		5.0	38
	HEG	7.4	66
		5.0	43
Influenza	Mouse S_{49} lymphoma	7.4	48
		5.0	51
	HEG	7.4	9
		5.0	51
SFV	Vero cells	7.4	30
		5.0	55
	HEG	7.4	3
		5.0	33
HSV-1	Vero cells	7.4	65
		5.0	63
	HEG	7.4	5
		5.0	25
VSV	Vero cells	7.4	48
		5.9	56
	HEG	n.d.	n.d.

[a]Virus particles were labeled with the fluorescence probe R18 as described for labeling of Sendai or influenza virus and VSV as before (Citovsky *et al.*, 1985; Blumenthal *et al.*, 1987; Nussbaum and Loyter, 1987). Fluorescence-labeled virus particles (2 μg) were incubated with 3×10^6 cultured cells or 100–200 μg protein of erythrocyte membrane either at pH 7.4 or 5.0 (or pH 5.9 for VSV) as described before (Citovsky *et al.*, 1985; Blumenthal *et al.*, 1987; Nussbaum *et al.*, 1987). The degree of fluorescence dequenching (R18 DQ) following incubation with R18-labeled virions was determined as described elsewhere (Citovsky *et al.*, 1985; Nussbaum and Loyter, 1987; Nussbaum *et al.*, 1987).

observed also following incubation of influenza glutaraldehyde-treated virions with living cells. Infectivity as well as the fusogenic ability of influenza virions can be blocked by treatment with hydroxylamine (Schmidt and Lambrecht, 1985) or following incubation at high temperature such as 85°C (Nussbaum and Loyter, 1987) or low pH (Sato *et al.*, 1983). Indeed, no fluorescence dequenching was obtained following incubation of such treated influenza virions and cultured cells (Nussbaum and Loyter, 1987; Table II).

The correlation between the fluorescence dequenching and virus membrane fusion could also be demonstrated by the use of influenza virions bearing an uncleaved HA glycoprotein (HA_0) (Nussbaum and Loyter, 1987). HA_0 influenza virions are neither infective nor fusogenic (Klenk *et al.*, 1975). Indeed, incubation of the fluorescently labeled HA_0 influenza virions with living cells promoted a very low increase in fluorescence dequenching (Nuss-

Table II
Fusion of Enveloped Virions with Animal Cultured Cells: Effect of Inhibitors[a]

System	Virus treated with	Cell treated with	R18 DQ (%)
Sendai + HeLa cells			52
Sendai + HeLa cells	DTT	—	5
Sendai + HeLa cells	Trypsin	—	8
Sendai + HeLa cells	PMSF	—	8
Influenza + mouse S_{49} lymphoma			42
Influenza + mouse S_{49} lymphoma	Hydroxylamine	—	6
Influenza + mouse S_{49} lymphoma	Low pH	—	8
Influenza + mouse S_{49} lymphoma	—	Methylamine	18
Influenza + mouse S_{49} lymphoma	—	Ammonium chloride	21
Influenza + mouse S_{49} lymphoma	—	NaN_3	22
Influenza + mouse S_{49} lymphoma	—	EDTA	16
SFV + Vero cells	—	—	34
SFV + Vero cells	Hydroxylamine	—	6
HSV-1 + Vero cells			58
HSV-1 + Vero cells	Hydroxylamine	—	11

[a]The various virus preparations were labeled with R18 as described for Sendai and influenza before (Citovsky et al., 1985; Nussbaum et al., 1987). Virus particles were incubated with animal cultured cells (3 × 10⁶ cells/system) at pH 7.4 and at the end of the incubation period, the fluorescence dequenching (R18 DQ) was determined as described elsewhere (Citovsky et al., 1985; Nussbaum et al., 1987). Sendai virions were treated with trypsin or PMSF as described by Israel et al. (1983) and with DTT as described by Loyter and Volsky (1982). Influenza, SFV, and HSV-1 were inactivated by treatment with hydroxylamine (Schmidt and Lambrecht, 1985) and pH 5.2 (Sato et al., 1983) as mouse S_{49} lymphoma or Vero cells were treated with methylamine (NH_2CH_3), ammonium chloride (NH_4Cl), sodium azide (NAN_3) (50 mM each), or EDTA (5 mM) as described before (Lapidot et al., 1987; Nussbaum, et al., 1987). All other experimental conditions were as described elsewhere (Citovsky et al., 1987; Lapidot et al., 1987).

baum and Loyter, 1987). It has been well-established that mild trypsinization of the HA_0 virions causes cleavage of the HA_0 viral glycoprotein concomitantly with restoration of the viral fusogenic activity (Klenk et al., 1975). Indeed, a high degree of fluorescence dequenching was observed upon incubation of trypsinized HA_0 influenza virions and living cells (Nussbaum and Loyter, 1987).

Essentially the same results were obtained when human erythrocyte membranes (human erythrocyte ghosts, HEG) were incubated with fluorescently labeled Sendai or influenza virions (Citovsky et al., 1985; Stegmann et al., 1986; Nussbaum and Loyter, 1987; Loyter et al., 1988a; Table I). A high degree of fluorescence dequenching was observed only following incubation with active fusogenic virus particles. Using erythrocyte ghosts as recipient membranes, scientists were able to study aspects of the virus–membrane fusion process. The extent of fluorescence dequenching, which is a direct measure of the percentage of virus particles fused (Chejanovsky and Loyter, 1985; Hoekstra et al., 1985), was highly dependent on the amount of erythrocyte membrane as well as on the medium pH (Citovsky *et al.*, 1985).

Virus–erythrocyte as well as virus–cell fusion appeared to be strongly dependent on the temperature of incubation being stimulated, especially between 23 and 37°C (Hoekstra *et al.*, 1985). Fusion of Sendai virions with erythrocyte ghosts at 37°C was found to be a relatively slow process reaching maximum values within 10–15 min of incubation of 37°C and around pH 7.0–8.0 (Hoekstra *et al.*, 1984). From studies using the fluorescence dequenching method, it was inferred that about 1000 Sendai virus particles can be bound to one erythrocyte ghost as compared to 100–200 particles that actually can fuse with each erythrocyte ghost (Nir *et al.*, 1986). No increase in the degree of fluorescence was observed following incubation of treated unfusogenic Sendai virions and erythrocyte ghosts (Citovsky *et al.*, 1985; Loyter *et al.*, 1988a).

Incubation of influenza virions with human erythrocyte ghosts resulted in fluorescence dequenching only at pH values between 5.0 and 5.5 but not at pH 7.4 (Stegmann *et al.*, 1986; Nussbaum and Loyter, 1987; Table I). This is consistent with the view that the viral fusion protein, namely the HA glycoprotein, is activated only at low pH values (Huang *et al.*, 1981; Skehel *et al.*, 1982; White *et al.*, 1983). Fluorescence dequenching methods were also used to study the fusogenic properties of HSV-1 and SFV. The fusogenic properties of SFV are similar to those of influenza virions. Fusion of SFV with biological membranes was observed only at low pH values (White *et al.*, 1983). This was confirmed by recent experiments showing that incubation of fluorescently labeled SFV with erythrocyte ghosts at pH 7.4 resulted in very little (3%) fluorescence dequenching while a high degree (33%) was observed following incubation at 5.0 (Loyter *et al.*, 1988a; V. Citovsky and A. Loyter, unpublished data; Table II). Recent experiments have shown (Loyter *et al.*, 1988a) that HSV-1 also prompted fluorescence dequenching—following incubation with human erythrocyte ghosts—only at pH 5.0–5.5 and not at pH 7.4 (Loyter *et al.*, 1987; Table II).

It is noteworthy that the extent of fluorescence dequenching observed upon incubation of SFV or HSV-1 with neuraminidase–pronase-treated erythrocyte membranes was higher than that observed with nontreated membranes (Loyter *et al.*, 1988a). This is in contrast to results obtained with Sendai or influenza virions which failed to fuse—and to promote fluorescence dequenching—with such virus-receptor-depleted erythrocytes (Chejanovsky *et al.*, 1986a; Loyter *et al.*, 1988a).

The interaction of fluorescently labeled influenza virions with cultured cells showed a different pattern from that obtained with erythrocyte ghosts (Nussbaum and Loyter, 1987; Loyter *et al.*, 1988a; Table I). A high degree of fluorescence dequenching was promoted—following incubation of influenza virions and culture cells—either at pH 5.0 or pH 7.4 (Nussbaum and Loyter, 1987). Very low or no fluorescence dequenching was observed at both pH values — following incubation with either inactivated unfusogenic virions or with HA_0 influenza virions (Nussbaum and Loyter, 1987; Table II). Evidently, the dequenching observed at either pH 5.0 or pH 7.4 reflects a process of virus–membrane fusion. However, only the fluorescence dequenching observed at pH 7.4, but not at pH 5.0, was inhibited by lysosomotropic reagents such as ammonium chloride and methylamine (Nussbaum and Loyter, 1987; Table II). Also, chelator of bivalent metals such as EDTA or reagents that deplete culture cells from intracellular ATP inhibited the fluorescence dequenching observed at pH 7.4 but not that obtained at pH 5.0 (Nussbaum and Loyter, 1987; Table II). Ammonium chloride and methylamine are known to increase the endosomal or lysosomal pH values (Matlin *et al.*, 1981; Svensson, 1985) while EDTA and NaN_3 inhibit endocytosis itself (Svensson, 1985). Based on these observations, it should be surmised that the fluorescence dequenching observed upon incubation with cultured cells at pH 7.4 results from fusion of influenza virus particles with membranes of intracellular organelles, while that observed at pH 5.0 is due to fusion with cell plasma membranes.

The same results were obtained following incubation of VSV, SFV, or HSV-1 with cultured cells (Blumenthal *et al.*, 1987; Loyter *et al.*, 1988a; Table I). High degrees of fluorescence dequenching were observed upon incubation at pH 5.0–5.1 and 7.4 but only that observed at pH 7.4 was inhibited by lysosomotropic reagents and EDTA.

Experiments with VSV demonstrated that fusion at the plasma membrane began immediately after lowering the pH below 6 and showed an approximately exponential time course, whereas fusion via the endocytic pathway (pH 7.4) became apparent after a time delay of about 2 min (Blumenthal *et al.*, 1987). A 10-fold excess of unlabeled virus arrested R18 VSV entry via the endocytic pathway, whereas R18 dequenching below pH 6 (fusion at the plasma membrane) was not affected by the presence of unlabeled virus. The temperature dependence for fusion at pH 7.4 (in the endosome) was much steeper than that for fusion at pH 5.9 (with the plasma membrane). Fusion via the endocytic pathway was attenuated at hyperosmotic pressures, whereas fusion at the plasma membrane was not affected by this treatment. The pH profile of Vero–VSV fusion at the plasma membrane, as measured by the dequenching method, paralleled that observed for VSV-induced cell–cell fusion. Fusion was blocked by adding neutralizing antibody to the Vero–VSV complexes (Blumenthal *et al.*, 1987). Activation of the fusion process by lowering the pH was reversible, in that the rate of fusion was arrested by raising the pH back to 7.4. The observation that pH-dependent fusion occurred at similar rates with fragments and with intact cells indicates that pH, voltage, or osmotic gradients are not required for VSV fusion with cells (Blumenthal *et al.*, 1987).

6. USE OF FLUORESCENT DEQUENCHING METHODS TO STUDY FUSION OF ENVELOPED VIRUSES WITH BIOLOGICAL MEMBRANES LACKING VIRUS RECEPTORS

The availability of fluorescently labeled enveloped virions and fluorescence dequenching methods allowed one to study the question of whether animal viruses will interact and especially fuse other biological membranes other than cell plasma membranes. Even fusion with membranes of endocytic vesicles initiates and results from binding of virions to specific receptors on cell surfaces (White *et al.*, 1983).

In our laboratory, we studied the interaction of both Sendai and influenza virions with chromaffin granules of bovine medulla (Citovsky *et al.*, 1987a) and with prokaryotic cells, namely mycoplasmas (Citovsky *et al.*, 1987b).

Incubation of fluorescently labeled Sendai or influenza virions at pH 7.4 and pH 5.0, respectively, with chromaffin granule vesicles under isotonic conditions resulted in a very low degree of fluorescence dequenching. It is noteworthy that under physiological conditions, membranes of such organelles never encounter enveloped virions and lack receptors for these virions (Abbs and Phillips, 1980; Ekerdt *et al.*, 1981). However, when Sendai (or influenza) virions were incubated with these vesicles under hypotonic conditions, a relatively high degree (35%) of fluorescence dequenching was observed (Citovsky *et al.*, 1987a). A significantly lower degree (13%) of fluorescence dequenching was observed when treated unfusogenic (DTT, trypsin, PMSF) Sendai virions were incubated with the chromaffin granule vesicles.

Similar results were obtained when fluorescently labeled Sendai or influenza virions were incubated with right-side-out erythrocyte vesicles (ROV) from which virus receptors have been removed by treatment with neuraminidase and pronase (Citovsky and Loyter, 1985; Nussbaum *et al.*, 1987). Only incubation under hypotonic conditions resulted in fusion (fluorescence dequenching) of either Sendai or influenza virions with the virus-receptor-depleted ROVs. Increase in fluorescence dequenching was observed at pH 7.4 following incubation with Sendai virions and at pH 5.0 upon incubation with influenza virions (Citovsky and Loyter, 1985; Nussbaum *et al.*, 1987). It has been suggested that osmotic swelling of human erythrocytes promotes exposure of the masked membrane phospholipids and renders them susceptible to phospholipases or cross-linking reagents (Laster *et al.*, 1972).

The above-described observations may indicate that in order to fuse with recipient membrane the viral envelope fusion proteins should interact directly with the membrane phospholipid bilayers. Under isotonic conditions, the interaction of enveloped virions with their membrane receptors may induce unmasking of the membrane lipid bilayer, thus making it available to the viral fusion glycoproteins (Citovsky and Loyter, 1985). These assumptions and results raise the possibility that enveloped virions will fuse with any biological membranes whose phospholipid bilayer is exposed and available to interaction with the viral glycoproteins even in the absence of appropriate virus receptors.

This was verified by recent experiments showing that Sendai or influenza virions are able to fuse with the membranes of prokaryotic cells (Citovsky *et al.*, 1987b). Incubation of fluorescently labeled Sendai influenza virions with *Mycoplasma gallisepticum* and *Mycoplasma capricolum*, but not with *Acetoplasma laidlawii*, resulted in a high degree of fluorescence dequenching. Fusion of Sendai and influenza virions with mycoplasmas was also confirmed by electron microscopic observations (Citovsky *et al.*, 1987b). The failure of Sendai and influenza virions to fuse with *A. laidlawii* may be due to the low percentage of cholesterol present in membranes of these cells (Razin and Tully, 1970).

The requirement of cholesterol for allowing virus–mycoplasma fusion was demonstrated by showing that a low degree of fusion was obtained with *M. capricolum* whose cholesterol content was decreased by modifying its growth medium (Citovsky *et al.*, 1987b). Fluorescence dequenching was not observed by incubating unfusogenic Sendai or influenza or HA_0 influenza with mycoplasmas (Citovsky *et al.*, 1987b). These results clearly demonstrated that both Sendai and influenza virions are able to fuse with mycoplasmas in spite of the fact that these prokaryotic cells lack virus receptor, namely, sialoglycolipids or sialoglycoproteins.

The above-described systems are an excellent example of the use of the fluorescence dequenching method for elucidating the molecular mechanism of virus–membrane fusion. It was possible to demonstrate using these methods that fusion of enveloped virions with biological membrane requires a high percentage of cholesterol. Such studies are almost impossible by any other methods such as observation by electron microscopy or virus-induced leakage of infected cells (Bashford *et al.*, 1985; Pasternak *et al.*, 1985).

7. ROLE OF VIRAL GLYCOPROTEINS IN THE PROCESS OF VIRUS MEMBRANE FUSION: STUDIES WITH RECONSTITUTED VIRAL ENVELOPES

The relation between the structure of viral glycoproteins and their biological function can be studied by the use of reconstituted viral envelopes.

These are membrane vesicles bearing only the viral envelope glycoproteins and devoid of the viral nucleocapsid (Loyter and Volsky, 1982; Vainstein *et al.*, 1984; Nussbaum *et al.*, 1987; Stegmann *et al.*, 1987). Studies on the biological activity of the isolated viral glycoprotein are of crucial importance for the elucidation of the as yet unknown, initial steps of virus–membrane fusion, virus penetration, and infection.

Most of the methods that have been used to reconstitute viral envelopes are based essentially on three steps: first, solubilization of intact viruses with a detergent; second, sedimentation of the internal proteins and genetic material; and third, removal of the detergent from the supernatant—a step that in most cases results in the formation of empty viral envelopes (Hosaka and Shimizu, 1972; Loyter and Volsky, 1982; Vainstein *et al.*, 1984; Stegmann *et al.*, 1987).

Two kinds of detergent have been employed for solubilization of intact enveloped virions. Detergents with high critical micelle concentration (CMC), such as octylglucoside (Helenius and Simons, 1975), which can be removed effectively by dialysis, have been used for solubilization of VSV, SFV, influenza, and Sendai virions (White *et al.*, 1983; Harmsen *et al.*, 1985; Stegmann *et al.*, 1987). On the other hand, detergents with a low CMC like Nonidet p-40 or Triton X-100 (Helenius and Simons, 1975), which cannot be removed simply by dialysis and require the addition of a hydrophobic resin such as Bio-beads SM-2, have been reported for Sendai (Loyter and Volsky, 1982; Vainstein *et al.*, 1984) or influenza (Nussbaum *et al.*, 1987). Recently, functional reconstituted influenza viral envelopes have been prepared by the use of the nonionic detergent octaethyleneglycol mono(n-dodecyl) ether ($C_{12}E_8$) (Stegmann *et al.*, 1987). Since this detergent possesses the same features as Triton X-100, its removal necessitated also the use of SM-2 Bio-beads (Stegmann *et al.*, 1987).

Removal of Triton X-100 by direct addition of SM-2 Bio-beads to the clear supernatant containing Sendai or influenza viral envelope phospholipids and glycoproteins results in the formation of resealed reconstituted viral envelopes (Loyter and Volsky, 1982; Vainstein *et al.*, 1984; Nussbaum *et al.*, 1987). Reconstituted Sendai virus envelopes contain only the viral hemagglutinin neuraminidase (HN) and fusion (F) glycoproteins, while those of influenza contain the viral HA (hemagglutinin) and NA (neuraminidase) glycoprotein (Vainstein *et al.*, 1984; Nussbaum *et al.*, 1987).

In Sendai virus that belongs to the paramyxovirus group, the hemagglutinin-binding activity and the neuraminidase are located on the same polypeptide, the HN glycoprotein. The virus fusion activity is located on a different polypeptide—the F glycoprotein (White *et al.*, 1983). In the influenzas that belong to the orthomyxovirus groups, the hemagglutinin and the neuraminidase are located on two different polypeptides, the HA and the NA glycoproteins, respectively. The HA glycoprotein mediates binding to cell surface receptors and is required for virus–membrane fusion. The NA glycoprotein possesses only neuroaminidase activity (White *et al.*, 1983).

Energy transfer and fluorescence dequenching methods are used to study— on a quantitative basis—the fusion ability of reconstituted viral envelopes bearing individual viral glycoproteins or their combinations (Citovsky and Loyter, 1985; Citovsky *et al.*, 1985, 1986a; Lapidot *et al.*, 1987; Nussbaum *et al.*, 1987; Stegmann *et al.*, 1987). By using this method it was possible to compare the fusogenic properties of such envelopes to those of intact virions (Nussbaum *et al.*, 1987; Lapidot *et al.*, 1987; Stegmann *et al.*, 1987).

N-4-nitrobenzo-2-oxa-1,3-diazole phosphatidylethanolamine (N-NBD-PE) was incorporated into reconstituted virus envelopes during the reconstitution procedure at self-quenching concentrations (10 mol %) (Citovsky *et al.*, 1985; Citovsky and Loyter, 1985; Loyter *et al.*, 1987; Stegmann *et al.*, 1987). The R18, on the other hand, can be interred into envelopes of virus particles either by its addition during reconstitution or to already reconstituted viral envelopes (Loyter *et al.*, 1987; Nussbaum *et al.*, 1987). As opposed to R18, N-NBD-PE cannot be incorporated into envelopes of intact virions (or to already reconstituted viral envelopes), thus preventing its use for studies of fusion processes of intact virions.

Studies with fluorescently labeled reconstituted Sendai virus envelopes (RSVEs) or reconstituted influenza virus envelopes (RIVEs) have shown that envelopes obtained following solubilization of intact virions with Triton X-100, but not with octylglucoside, are as fusogenic as intact virions (Loyter *et al.*, 1987, 1988a; Nussbaum *et al.*, 1987; Stegmann *et al.*, 1987). These conclusions were based on experiments showing that the degree of fluorescence dequenching obtained following incubation of either RSVEs or RIVEs with HEG or cultured cells was close to that obtained with intact virions (Table III).

RIVEs were able to fuse with erythrocyte membrane only at pH 5.0 while with cultured cells fusion was demonstrated at pH 5.0 as well as at pH 7.4 (Nussbaum *et al.*, 1987; Table III). Only the fusion (fluorescence dequenching) observed upon incubation with cultured cells at pH 7.4 was inhibited by lysosomotropic agents (methylamine and ammonium chloride), as well as by EDTA, indicating that it is due to fusion of RIVEs taken into cells by endocytosis (Nussbaum *et al.*, 1987). The rate and the pH dependence of fusion with RSVEs and RIVEs were found to be essentially the same as those of intact virions. No increase in the fluorescence dequenching was observed when unfusogenic RSVEs or RIVEs were incubated with either HEG or cultured cells (Loyter *et al.*, 1987, 1988a; Nussbaum *et al.*, 1987; Table III). All these experiments clearly demonstrated — as had been inferred from experiments with intact virions — that

Table III
Fusion of Reconstituted Sendai or Influenza Virus Envelopes with HEG[a]

System	pH of incubation	R18 DQ
A. Sendai	7.4	56
RSVE		50
DTT–RSVE		6
HN vesicles		6
F vesicles		8
HN–F vesicles		52
B. Influenza	7.4	5
	5.0	48
RIVE	7.4	8
	5.0	42
Hydroxylamine–RIVE	5.0	8
HA vesicles	5.0	41
Hydroxylamine–HA vesicles	5.0	11

[a]Reconstituted Sendai virus envelopes (RSVEs) or membrane vesicles bearing purified hemagglutinin/neuraminidase (HN vesicles) or fusion (F vesicles) glycoproteins or both (HN–F vesicles) were prepared and fluorescently labeled as described before (Citovsky and Loyter, 1985; Citovsky *et al.*, 1986b). Reconstituted influenza virus envelopes (RIVEs) or membrane vesicles bearing only the hemagglutinin glycoprotein (HA vesicles) were prepared and fluorescently labeled as described by Nussbaum *et al.* (1987) and Lapidot *et al.* (1987).

the fluorescence dequenching observed is due to fusion between the viral envelope and the cell membranes and not to other processes such as lipid–lipid exchange.

Resealed membrane vesicles containing Sendai or influenza individual glycoproteins can be prepared following separation of the viral polypeptides on ion-exchange columns (Fukami *et al.*, 1980; Nussbaum *et al.*, 1984; Lapidot *et al.*, 1987). Membrane vesicles bearing the Sendai HN or F glycoproteins are nonfusogenic as can be inferred from experiments showing that incubation of such fluorescently labeled vesicles with biological membranes results in very little fluorescence dequenching (Citovsky and Loyter, 1985; Loyter *et al.*, 1987; 1988a; Table III). It is noteworthy that membrane vesicles bearing only the Sendai virus binding polypeptide, namely the HN glycoprotein, readily attach to cell plasma membranes and agglutinate red blood cells (Fukami *et al.*, 1980). Membrane vesicles bearing only the Sendai F glycoprotein neither agglutinate cells nor attach to their membranes (Fukami *et al.*, 1980; Nussbaum *et al.*, 1984). Recent experiments showed that even when the binding of the Sendai F vesicles to recipient membranes was mediated by a nonviral binding ligand, no increase in the degree of fluorescence was observed (Loyter *et al.*, 1987). Fluorescence dequenching, namely vesicle–membrane fusion, was observed only with reconstituted envelopes bearing both Sendai HN and F glycoproteins within the same membrane (Citovsky and Loyter, 1985; Citovsky *et al.*, 1986a; Loyter *et al.*, 1987). These experiments raise the possibility that the Sendai virus HN glycoprotein beside being the viral binding protein also actively participates in the virus–membrane fusion step (Gitman *et al.*, 1985).

Fluorescently labeled (R18) membrane vesicles bearing only the influenza viral hemagglutinin (HA) glycoprotein were used to study its function in the fusion process (Wharton *et al.*, 1986; Lapidot *et al.*, 1987). The viral hemagglutinin glycoprotein was separated from the neuraminidase glycoprotein by agarose sulfanilic acid column and was shown to be homogeneous by gel electrophoresis and devoid of any neuraminidase activity (Lapidot *et al.*, 1987). Incubation of fluorescently labeled HA vesicles with HEG or cultured cells gave the same results as incubation of intact influenza virions or RIVEs (Lapidot *et al.*, 1987). Fluorescence dequenching was observed only with fusogenic but not with treated unfusogenic HA vesicles. Similarly, fusion of HA vesicles with living cultured cells — as opposed to fuson with HEG — was observed not only at pH 5.0 but also at pH 7.4 (Lapidot *et al.*, 1987). These results clearly showed that despite the fact that the HA vesicles are devoid of the NA glycoprotein and of neuraminidase activity they are fusogenic and behave in the same manner as intact virions (Lapidot *et al.*, 1987).

Recently, reconstituted envelopes bearing the VSV G-protein were obtained following solubilization of VSV with octylglucoside or by $C_{12}E_8$ (Metsikko *et al.*, 1986). The fusogenic activity of these vesicles was assayed for polykaryon formation (Metsikko *et al.*, 1986). Only envelopes (virosomes) obtained following the use of $C_{12}E_8$ were fusogenic, while those obtained by the use of octylglucoside did not exhibit any fusogenic activity. Membranes bearing the G-protein fuse with BHK-22 cell plasma membranes at pH 5.7–6.0 with an efficiency of fusion comparable to that of the parent virus (Metsikko *et al.*, 1986).

8. FUSION OF ENVELOPED VIRUSES WITH NEGATIVELY CHARGED AND NEUTRAL LIPOSOMES

Various methods have been employed to study and demonstrate fusion between enveloped viruses and phospholipid vesicles (Haywood, 1974; White and Helenius, 1980; Hsu *et al.*, 1983; Kawasaki *et al.*, 1983; Haywood and Boyer, 1984). Haywood (1974) was the first to demonstrate fusion between Sendai virions and lipid vesicles. Using electron microscopy techniques, Haywood showed that fusion of Sendai virions with liposomes composed of neutral lipids such as phosphatidylcholine (PC), sphingomyelin (Sph), phosphatidylethanolamine (PE), and cholesterol (chol) require the presence of virus receptors, namely, sialoglycolipids.

In more recent work fusion between influenza, SFV, or VSV and liposomes of different composition was demonstrated (White *et al.*, 1983). Virus-induced release of liposome content also has been used to follow virus–liposome fusion processes (Oku *et al.*, 1982; Kundrot *et al.*, 1983). It has been well established that fusion processes between hemolytic enveloped virions such as Sendai and influenza and living cells leads to an increase in the cell membranes' permeability to small as well as to large molecules (Bashford *et al.*, 1985; Pasternak *et al.*, 1985). Sendai virions were shown to induce hemolysis at pH values above 6.0–7.0 while the hemolytic activity of influenza virions is manifested at pH 5.2, a pH at which the viral fusion glycoprotein is activated (Huang *et al.*, 1981).

Fluorescence dequenching methods have also been used to follow — on a quantitative basis — fusion between fluorescently labeled enveloped virions and liposomes. Hoekstra *et al.* (1984) have shown that incubation of R18-labeled influenza virions with liposomes composed of negatively charged phospholipids results in fluorescence dequenching. Fluorescence dequenching was observed following incubation at pH 5.0 but not at 7.4. Recently, by the use of fluorescently labeled liposomes, it has been shown that influenza virions fuse readily with liposomes composed of negatively charged phospholipids such as phosphatidylserine (PS) or cardiolipin (CL) and very poorly with liposomes composed of neutral lipids such as sphingomyelin or sphingomyelin and cholesterol (Stegmann *et al.*, 1985). Fusion with negatively charged liposomes was pH dependent, whereas virus-induced release of calcein from these liposomes was pH independent and was observed at pH 5.0 as well as at pH 7.4 (Stegmann *et al.*, 1985; O. Nussbaum, V. Citovsky, and A. Loyter, unpublished data; Table IV).

Neither virus-induced release of liposome content nor virus–liposome fusion was dependent on the presence of virus receptor in the negatively charged liposomes (Hoekstra *et al.*, 1984; Stegmann *et al.*, 1985; Table IV). Surprisingly, a relatively high degree of fluorescence dequenching was observed following incubation of inactivated unfusogenic influenza virions with negatively charged liposomes such as those composed of PS (O. Nussbaum, V. Citovsky, and A. Loyter, unpublished data; Table IV). Furthermore, fluorescence dequenching and release of liposome content [carboxyfluoresceine (CF) dequenching] were also observed following incubation of the unfusogenic HA_0 influenza virions with PS liposomes (O. Nussbaum, V. Citovsky, and A. Loyter, unpublished data). These results raise the possibility that fusion of influenza virions with liposomes composed of negatively charged phospholipids does not reflect the biological activity of the viral glycoproteins needed for infection and penetration (Loyter and Citovsky, 1987).

Support for this view was also obtained from experiments in which fusion of Sendai virions with negatively charged liposomes was studied (Amselem *et*

Table IV

Interaction of Sendai and Influenza Virions with Phospholipid Vesicles: Virus-Induced Lysis of and Fusion with Negatively Charged and Neutral Liposomes[a]

System	pH	PC R18 DQ (%)	PC/chol R18 DQ (%)	PC/chol CF release (% of total)	PC/chol/gang R18 DQ (%)	PC/chol/gang CF release (% of total)	PS R18 DQ (%)	PS CF release (% of total)
Sendai	7.4	2	43	3	48	21	68	74
DTT-treated Sendai	7.4	3	0	2	2	0	63	81
Influenza	5.0	7	31	3	42	54	50	80
	7.4	3	8	0	9	9	17	70
Hydroxylamine-treated influenza	5.0	n.d.	7	n.d.	7	2	30	68
	7.4	n.d.	5	n.d.	8	3	10	55
HA_0 influenza	5.0	n.d.	8	n.d.	9	3	35	63
	7.4	n.d.	5	n.d.	6	3	15	42

[a]Carboxyfluorescein loaded or empty liposomes composed of PC, PC/chol (1 : 0.5 molar ratio), PC/chol/gang (1 : 0.5 : 0.3 molar ratio), or PS were prepared as described by Citovsky et al., (1986b). All other experimental conditions of determination of fluorescence dequenching and carboxyfluorescein release are as described by Citovsky et al. (1986b) and Loyter et al. (1988a). n.d., not done.

al., 1985, 1986; Chejanovsky *et al.*, 1986b). Incubation of fluorescently labeled Sendai virions with PS liposomes resulted in a high degree of fluorescence dequenching (Chejanovsky *et al.*, 1986b). Sendai virions also induced the release of CF from PS-loaded liposomes (Amselem *et al.*, 1985, 1986). No virus receptors, namely sialoglycolipids, were required to allow fusion with or lysis by Sendai virions of such negatively charged liposomes (Amselem *et al.*, 1985; Chejanovsky *et al.*, 1986b). Inactivated unfusogenic virions were able to fuse with liposomes composed of negatively charged phospholipids (Amselem *et al.*, 1985b; Chejanovsky *et al.*, 1986b). A high degree of fluorescence dequenching was observed following incubation of DTT- or PMSF-treated Sendai virions and PS or CL liposomes (Table IV). Fusion of Sendai virions with negatively charged lipsomes is maximal at low pH values, whereas fusion of Sendai virions with biological membranes was maximal between pH 7.0 and 9.0 (Chejanovsky and Loyter, 1985; Chejanovsky *et al.*, 1986b). Similar to intact virions, RSVEs were also able to induce lysis and to fuse with liposomes containing negatively charged phospholipids. The view that such fusion does not reflect the biological activity of the virus was further strengthened by experiments showing that membrane vesicles bearing only the Sendai-virus-binding protein, namely the HN glycoprotein, are also able to induce lysis and to fuse with liposomes composed of PS (Chejanovsky *et al.*, 1986b). Fusion was maximally expressed, at low pH values such as pH 4.0, and was not inhibited by treatment of the HN vesicles with either DTT or PMSF. Neither induction of lysis nor fusion was observed upon incubation of membrane vesicles containing the viral fusion glycoprotein (F vesicles) with the negatively charged liposomes (Chejanovsky *et al.*, 1986b).

It is noteworthy that fusion of Sendai virions or RSVEs with biological membranes is absolutely dependent on the presence of the viral fusion (F) glycoprotein, and HN vesicles neither induce lysis nor fuse with HEG or living cells (Loyter and Volsky, 1982; Citovsky *et al.*, 1986b).

Incubation of fluorescently labeled influenza or Sendai virions with liposomes composed of neutral phospholipids such as PC — as opposed to those composed of PS — resulted in very little fluorescence dequenching (Citovsky *et al.*, 1985; Loyter and Citovsky, 1987). Essentially, the same results were obtained when HA_0 or trypsinized HA_0 influenza virions were incubated with PC liposomes at either pH 5.0 or 7.4 (Loyter and Citovsky, 1987; Table IV). Incorporation of cholesterol into the PC liposomes renders them susceptible to the fusogenic activity of either Sendai or influenza virions. It has been claimed previously that fusion of SFV with phospholipid vesicles is absolutely dependent on the presence of cholesterol (White *et al.*, 1983).

Fusion of Sendai virions with PC/chol liposomes was observed at a wide range of pH values, reaching a maximal degree at pH 7.0–8.0, while that of influenza virions showed maximal values between pH 5.0 and 5.5 (Table IV).

Inactivated unfusogenic Sendai or influenza virions failed to fuse with PC/chol liposomes (Citovsky *et al.*, 1985; Loyter *et al.*, 1988b). HA_0 influenza virions also did not fuse with PC/chol liposomes while trypsinized HA_0 influenza virions readily fuse with these liposomes (Table IV). Fusion of both influenza and Sendai virions with PC/chol liposomes was found to be an unlikely process (Citovsky and Loyter, 1985; Citovsky *et al.*, 1986; Loyter *et al.*, 1988b; Table IV). Incubation of Sendai virions at pH 7.4 or influenza virions at pH 5.0 and 37°C with CF-loaded PC/chol liposomes did not induce any CF release.

Incorporation of virus receptors, namely the sialoglycolipids (gangliosides, gang), into the PC/chol liposomes renders them susceptible to the Sendai or influenza lytic activity. Thus, lysis of liposomes composed of neutral phospholipids by Sendai or influenza virions was absolutely dependent on the presence of cholesterol and virus receptors (Citovsky and Loyter, 1985; Citovsky *et al.*, 1986b; Loyter and Citovsky, 1987). Such virus-induced lysis of loaded liposomes exhibited the same features as shown by virus–membrane fusion and virus-induced lysis of living cells. It was maximally expressed at pH 7.4 when Sendai virions were used and at pH 5.0 with influenza virions (Loyter and Citovsky, 1987). Lysis of liposomes was not observed with DTT- or PMSF-treated Sendai virions or with hydroxylamine-treated influenza virions. These results support the view that in the presence of sialoglycolipids (gang), which serve as receptors for Sendai and influenza virions, lysis of loaded liposomes indeed reflects a process of virus–membrane fusion.

Essentially the same results were observed when RSVEs or RIVEs were incubated with PC/chol or PC/chol/gang liposomes. However, when membrane vesicles containing either the Sendai HN (HN vesicles) or F (F vesicles) glycoproteins were incubated with PC/chol or PC/chol/gang liposomes neither fusion nor release of the liposome content was observed (Citovsky *et al.*, 1986b). Only vesicles bearing the Sendai virus HN and F glycoproteins within the membrane were able to fuse with and to induce lysis of PC/chol/gang liposomes (Citovsky *et al.*, 1986b). From these experiments it should be inferred that even fusion with PC/chol liposomes lacking virus receptor requires the Sendai virus HN glycoprotein in addition to the F polypeptide. An active role of Sendai virus HN glycoprotein in the virus–membrane fusion step itself is thus suggested.

It appears that the following conclusions can be drawn from studies on fusion between enveloped virions and especially Sendai and influenza viruses with phospholipid vesicles: (1) Fusion with negatively charged liposomes does not reflect the biological activity of the viral envelopes. (2) Fusion with liposomes composed of neutral lipids is absolutely dependent on the presence of cholesterol and shows the same features as fusion with biological membranes. (3) Sialoglycolipids (gang), namely virus receptors, must be present in PC/chol liposomes in order to allow expression of the viral lytic activity.

9. ROLE OF CONFORMATIONAL CHANGES AND COOPERATIVITY OF VIRAL PROTEINS IN MEDIATING MEMBRANE FUSION

Significant advances have been made in recent years in elucidating the role of viral spike proteins in inducing membrane fusion. They include (1) the first high-resolution image of a membrane fusion protein, (2) the elucidation of the primary sequence of a large number of viral membrane proteins using DNA sequencing techniques, and (3) the development of genetic and chemical methods for site-specific alterations of viral protein structure.

The viral spike glycoproteins contain in their structures all the information needed for viral entry: recognition, movement to site, apposition, fusion, and dissociation. Conformational changes and subunit interactions have been studied in viral spike glycoproteins. The hemagglutinin (HA) protein of influenza virus is the best-characterized member of the family of viral spike glycoproteins that mediate membrane fusion (White *et al.*, 1983). Its structure contains the recognition site for cell surface sialic acid residues, as well as the "catalytic site" that mediates membrane fusion. The HA consists of two disulfide-linked glycopolypeptide chains, HA1 and HA2, which are derived by proteolytic cleavage from a precursor glycopolypeptide called HA0 (Klenk *et al.*, 1975). This proteolytic cleavage is absolutely required for pH-induced fusion activity of the virus. That HA is necessary and sufficient for fusion activity was shown by expressing HA in eukaryotic cells by transfection of plasmids containing cloned complementary DNAs encoding those viral proteins and monitoring pH-dependent cell–cell fusion (Gething *et al.*, 1986). Mutations in the N-terminal peptide of HA2 significantly altered fusion.

Treatment of intact influenza virus with the enzyme bromelain results in the release of nearly the entire N-terminal ectodomain of HA (95% of its mass) in a water-soluble form. The cleavage of the HA2 chain occurs close to the point where it emerges from the viral membrane (Brand and Skehel, 1972). The resulting fragment, termed BHA, has been crystallized and its three-dimensional structure determined to a resolution of 3 Å (Wilson *et al.*, 1981).

The protein is a trimeric rod-shaped molecule 135 Å in length, consisting of a stem and three globular, highly folded domains (stalks) at the top. The globular domains are composed of the HA1 chain, and they contain the binding sites for sialic acid. The head domain rests on the stem, which is composed of HA2, as well as the C-terminal and N-terminal sequences of HA1. The C terminus of HA1 is located close to the viral membrane, 22 Å from the N terminus to which it was originally linked in the HA0 precursor. The stem domain is a complex of three 76-Å-long α-helixes, which form a triple-stranded coil, stabilized by salt bridges and hydrophobic interactions. The trimer is very stable; it does not dissociate even after treatment with SDS (Doms and Helenius, 1986). The hydrophobic HA2 N-terminal peptide implicated in fusion (fusion peptide) is tucked between the long α-helixes near the base of the molecule. Its position is stabilized by noncovalent interactions with HA1 and with residues on the same and adjoining HA2 subunits. The single HA1–HA2 interchain disulfide bond is also located in the stem region, near the base of the molecule.

Electron microscopy using negative stain reveals the HA spikes in the intact virion at neutral pH as well-ordered rectangular projections about 135 Å in length. Acid treatment results in a disordered appearance. Ruigrok et al. (1986) found that virus spikes become thinner and longer after acid treatment in specimens containing isolated BHA and HA, as well as in HA reconstituted in liposomes. Their interpretation is that the trimeric contacts in the head domain are broken and the HA2 stem is elongated. Doms and Helenius (1986) obtained virus particles with decreasing spike densities by partial digestion of intact virions with bromelain. Acid treatment of those particles resulted in star-shaped aggregates on the surface of the virus.

One of the most powerful methods to examine conformational transitions of HA has been the use of poly- and monoclonal antibodies that recognize known regions of the protein. The neutral form of HA has three major epitopes, termed the loop, hinge, and tip/interface, all located in the globular head domain (Webster et al., 1983; Yewdell et al., 1983). Following acid treatment, antibodies against the tip/interface epitope, especially those that bind close to the trimer interface, no longer interact with HA, whereas binding of antibodies to the loop and hinge epitopes is unaffected. On the other hand, monoclonal antibodies have been raised specific to the acid conformation, which bind sites in HA1 and HA2. The pH-dependent appearance of these epitopes is similar to that of fusion of the intact virion, and BHA and HA show the same changes in antigenic structure.

Other differences between neutral and acid conformations have been examined mainly using BHA. Although incapable of mediating fusion, BHA undergoes many of the conformational changes that are observed in intact HA. For instance, whereas the neutral BHA and HA are resistant to a variety of proteases, the acid forms are susceptible to digestion with trypsin, proteinase K, and other proteases (Skehel et al., 1982; Doms et al., 1985). The pH dependence for conversion to their protease-sensitive forms is similar to that for fusion of the intact virion from which the hemagglutinin was extracted.

pH-induced conformational changes of BHA have been monitored by circular dichroism (Skehel et al., 1982), accessibility of the interchain disulfide bridge to reducing agents (Graves et al., 1983), and increase in hydrophobicity as indicated by aggregation in aqueous solution, interaction with nonionic detergents, and binding to liposomes (Skehel et al., 1982; Doms et al., 1985).

Those irreversible changes are presumably caused by exposure of the hydrophobic fusion peptide. The uncleaved HA0 does not become hydrophobic after low-pH treatment.

In view of the subunit structure and conformational changes discussed above, it is reasonable to view the viral spike glycoproteins as allosteric proteins (Blumenthal, 1988). A model for protein function based on cooperativity through conformational changes was introduced to offer a physical interpretation for heme–heme interactions in hemoglobin (Wyman 1948). A further analysis of the control of activity of a large number of enzymes led to the conclusion that in most of them indirect interactions between distinct binding sites are responsible for the performance of their regulatory function. This led to the model of regulation of protein function by allosteric transitions (Monod *et al.*, 1963, 1965).

The model is described by the following characteristics as formulated by Monod *et al.* (1965): (1) Allosteric proteins are arranged as oligomers whose subunits are associated in such a way that they all occupy equivalent positions. This implies that the molecule possesses at least one axis of symmetry. (2) To each ligand capable of forming a stereospecific complex with the protein there corresponds one site on each protomer. In other words, the symmetry of each set of stereospecific receptors is the same as the symmetry of the molecule. (3) The conformation of each protomer is constrained by its association with the other protomers. (4) Two (at least two) states are accessible to allosteric oligomers. These states differ by the distribution and/or energy of intersubunit bonds, and therefore also by the conformational constraints imposed on the subunits. (5) As a result, the affinity of one (or several) of the stereospecific sites toward the corresponding ligand is altered when a transition occurs from one to the other state. (6) When the protein goes from one state to another state, its molecular symmetry is conserved.

A variety of models have been proposed to describe how viral proteins might interact with the target membrane to induce the lipid rearrangements required for fusion. It is reasonable to assume that the virus binds to the target membrane at many points. The nature of the binding might be electrostatic, specific site–site recognition, or penetration into the acyl chain region of the target membrane. As a result of the multiple binding and presumably following a protein conformational change, the two membranes could be physically deformed and thus brought into close contact. This latter event might precipitate fusion between the two membranes either by overcoming the repulsion forces (steric, electrostatic, hydration) or by inducing an excessive radius of curvature (Blumenthal, 1987).

On the basis of the above discussion, it appears that the HA of influenza virus fits most of those characteristics of an allosteric protein. It therefore seems reasonable to think of the induction of fusion in terms of an allosteric model

(Blumenthal, 1988; Blumenthal *et al.*, 1988). In the proposed model, the viral spike glycoproteins are considered as regulatory proteins involved in catalysis of fusion. Initially, the model does not deal with the specific mechanism of catalysis of fusion, but rather with consequences of ligand–protein interactions on the regulation of fusion. An analogy may be drawn to allosteric enzymes discussed by Monod *et al.* (1965), where not the mechanism of catalysis but the effects of regulatory ligands, conformational changes, and/or cooperativity on the functioning of the protein are considered. The allosteric model is not dependent on mechanism, but analysis of experimental observations according to the model will provide constraints on any proposed mechanism of fusion mediated by regulatory proteins. According to the model the viral spike gly-coproteins are assumed to be arranged as oligomers, consisting of a number (n) of subunits linked through quarternary interactions. In Figure 2 only two or four of the n subunits are drawn as separate entities. This oligomer is capable of undergoing a "concerted" conformational change from what is conventionally known as the *tense* or T state (inactive) to the *relaxed* or R state which is active (Monod *et al.*, 1965). Each subunit in the oligomer contains a regulatory site for a ligand (H^+) with a dissociation constant K_d. Once the protein has undergone a conformational transition from the T state to the R state, the fusion process is catalyzed, leading to the melting of the two linked membranes into the R_f state. Although all the spike proteins of a particular virion can undergo the conformational change, only those involved in the virus–target interaction are relevant. Indeed, according to the model shown in Figure 2, the transition of a single oligomer is considered sufficient to precipitate the fusion event.

In the experiments with the R18-labeled virus a single fusion event brings about the movement of the 1500 R18 molecules from the viral membrane to the target membrane where they diffuse over a larger surface area, resulting in an

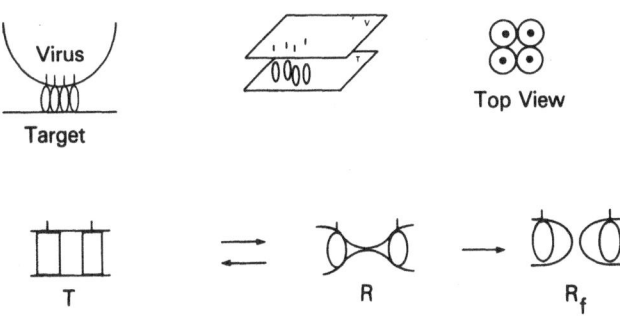

FIGURE 2. Allosteric model for membrane fusion mediated by viral spike glycoproteins. See text for further details. From Blumenthal (1988) with permission.

increased fluorescence signal. A single fusion event is induced by the transition of an oligomer in the virus–target complex from T to R to R_f (Figure 2). The total fluorescence increase is equal to the fluorescence increase per fusion event times the number of virions fused. This situation is similar to ion channel opening, which involves the conformational transition of channel protein oligomers resulting in movement of 1000 ions/msec through the channel, measured as a single channel conductance (Ehrenstein *et al.*, 1974).

By using the allosteric formalism, a simple expression for the relative rate constant for fusion can be derived (Blumenthal, 1988):

$$\text{relative rate constant} = \frac{(1 + \alpha)^n}{L + (1 + \alpha)^n} \tag{3}$$

where L is the equilibrium constant for the conformational transition between T and R states in the absence of ligand, and $\alpha = H^+/K_d$. Equation (3) is very similar to expressions derived by Monod *et al.* (1965) for the activity of allosteric enzymes.

In experiments on the interaction of VSV with Vero cells (a monkey kidney cell line), the rate of fusion was measured as a function of pH using R18 dequenching (Blumenthal *et al.*, 1987). From these data relative rate constants were calculated and plotted as a function of pH. A reasonable fit to Eq. (3) was obtained for pK_d of 6.3, $L = 1000$, and $n = 6$ (Blumenthal, 1988).

Recently, it was found that preincubation of VSV at low pH prior to binding to cells led to significant *enhancement* of fusion (Puri *et al.*, 1988). The data were analyzed in the framework of an allosteric model according to which viral spike glycoproteins undergo a pH-dependent conformational transition to an active (fusion-competent) state. Based on that analysis, it was concluded that the conformational transition to the active state is rate limiting for fusion and that the viral spike glycoproteins are fusion-competent only in their protonated form.

An alternative method to determine the size of the functional unit of the viral spike glycoprotein involved in membrane fusion is radiation inactivation analysis. Radiation inactivation of virus-induced fusion activity was measured by fluorescence dequenching of fluorescent lipid probes incorporated into liposomes (Gibson *et al.*, 1986; Bundo-Morita and Lenard, 1988), as well as by virus-induced hemolysis. With VSV it was found that the calculated functional units for both activities were similar, equivalent to about 15 viral spike glycoproteins (Bundo-Morita and Lenard, 1988). Surprisingly, however, similar studies with influenza and Sendai viruses resulted in a functional unit corresponding in size to a single protein monomer (Gibson *et al.*, 1986).

Examination of the hypotheses presented here regarding the regulatory properties of the viral spike glycoproteins in mediating membrane fusion requires further detailed studies of the kinetics of fusion mediated by viral spike glyco-proteins and analysis based on the allosteric model, in addition to further ex-amination of the physicochemical states of those proteins in the intact virus, as well as in the isolated and reconstituted proteins using a variety of biophysical techniques.

10. CONCLUSIONS

The detailed mechanism of the process of virus–membrane fusion is still obscure. Very little is known about the molecular events that allow viral en-velope glycoproteins of various groups to promote fusion with biological mem-branes and liposomes. Furthermore, the question of whether viral binding proteins and their membrane receptors serve only as a passive bonding pair or alterna-tively play an active role in the process of virus–membrane fusion is as yet unknown. From experiments showing that various binding proteins such as an-tibodies, polypeptide hormones and their receptors (Gitman et al., 1985) or the pair avidin–biotin (Guyden et al., 1983) may mediate virus binding and fusion with receptor-depleted cells should lead to the conclusion that viral binding proteins and their receptors are required only to allow attachment of viruses to recipient membranes. On the other hand, the experiments showing that hemo-lytic viruses such as Sendai or influenza may fuse with virus-receptor-depleted membranes only under hypotonic conditions may indicate an active role or virus receptors under physiological conditions.

This and other questions will have to be answered in future studies and answers obtained may have important clinical implications especially in devel-oping drugs that will inhibit the initial stage of virus infection.

The availability of active fusogenic fluorescently labeled enveloped viruses and the fluorescence dequenching method (Loyter et al., 1988a) made it possible to study by a simple reproducible and quantitative way the fusion of intact virions or the reconstituted envelopes with various membrane preparations.

Studies using the fluorescence dequenching method clearly showed that fusion of enveloped viruses of different groups with PC liposomes requires the presence of cholesterol (Loyter et al., 1988a) and show the same feature as fusion with biological membranes. On the other hand, it appears from these studies that fusion with negatively charged liposomes does not reflect viral biological activity needed for its penetration and fusion of living cells. The interaction between enveloped viruses and liposomes of different lipid compo-sition bearing or lacking virus receptors and the use of fluorescence dequenching methods may serve as an excellent experimental system for elucidating the yet unknown mechanism of virus–membrane fusion and the function of the viral glycoproteins and membrane receptors with this process.

ACKNOWLEDGMENT. The authors wish to express their gratitude to Ms. Mira Laron and Ms. Sherry Kisos for their help in preparing this manuscript.

11. REFERENCES

Abbs, M. T., and Phillips, J. H., 1980, Organization of the proteins of the chromaffin granule membrane, *Biochim. Biophys. Acta* **595** : 200–221.

Amselem, S., Loyter, A., Lichtenberg, D., and Barenholtz, Y., 1985, The interaction of Sendai virus with negative charged liposomes: Virus induced lysis of carboxyfluorescein-loaded small unilamellar vesicles, *Biochim. Biophys. Acta* **820** : 1–10.

Amselem, S., Barenholtz, Y., Loyter, A., and Lichtenberg, D., 1986, Fusion of Sendai virus with negative charged liposomes as studied by pyrene-labeled phospholipid liposomes, *Biochim. Biophys. Acta* **860** : 303–313.

Bashford, C. L., Micklin, K. J., and Pasternak, C. A., 1985, Sequential onset of permeability change in mouse ascite cells induced by Sendai virus, *Biochim. Biophys. Acta* **814** : 242–255.

Blumenthal, R., 1987, Membrane fusion, *Curr. Top. Membr. Transport* **29** : 203–254.

Blumenthal, R., 1988, Cooperativity in viral fusion, *Cell Biophys.* **12** : 1–12.

Blumenthal, R., Weinstein, J. N., Sharrow, S. O., and Henkart, P., 1977, Liposome–lymphocyte interactions: Saturable sites for transfer and intracellular release of liposome contents, *Proc. Natl. Acad. Sci. U.S.A.* **74** : 5603–5607.

Blumenthal, R., Bali-Puri, A., Walter, A., Covell, D., and Eidelman, O., 1987, pH-Dependent fusion of vesicular stomatitis virus with Vero cells: Measurement by dequenching of octadecylrhodamine fluorescence, *J. Biol. Chem.* **262** : 13614–13619.

Blumenthal, R., Puri, A., Walter, A., and Eidelman, O., 1988, pH-Dependent fusion of vesicular stomatitis virus with cells: Studies of mechanism based on an allosteric model, in *Molecular Mechanisms of Membrane Fusion* (S. Ohki, D. Doyle, T. Flanagan, S. W. Hui, and E. Mayhew, eds.), pp. 367–383, Plenum Press, New York.

Brand, C. M., and Skehel, J. J., 1972, Crystalline antigen from the influenza virus envelope, *Nature New Biol.* **238** (83) : 145–147.

Bundo-Morita, K., Gibson, S., and Lenard, J., 1988, Radiation inactivation analysis of fusion and hemolysis by vesicular stomatitis virus, *Virology* **163** : 622–624.

Chejanovsky, N., Beigel, M., and Loyter, A., 1984, Attachment of Sendai virus particles to cell membranes: Dissociation of adsorbed particles by dithiothreitol, *J. Virol* **49** : 1009–1013.

Chejanovsky, N., and Loyter, A., 1985, Fusion between envelopes and biological membranes, *J. Biol. Chem.* **260** : 7911–7918.

Chejanovsky, N., Fridlender, B., and Loyter, A., 1985, Affinity targeting of Sendai virions to desialized human erythrocytes using hybrid antibody molecules, *Biochem. Biophys. Acta* **812** : 353–360.

Chejanovsky, N., Henis, Y. I., and Loyter, A., 1986a, Fusion of fluorescently labeled Sendai virus envelope with living cultured cells as monitored by fluorescence dequenching, *Exp. Cell Res.* **164** : 353–365.

Chejanovsky, N., Amselem, S., Zakai, N., Barenholtz, Y., and Loyter, A., 1986b, Membrane vesicles containing the Sendai virus binding glycoprotein, but not the viral fusion protein, fuse with phosphatidylserine liposomes at low pH, *Biochemistry* **25** : 4810–4817.

Chen, R. F., and Knutson, J. R., 1987, Fluorescence dyes encapsulated in liposomes: Mechanisms of fluorescence changes, *Biophys. J.* **51** : 539a.

Choppin, P. W., and Scheid, A., 1980, The role of viral glycoproteins in adsorption, penetration, and pathogenicity of viruses, *Rev. Infect. Dis.* **2** : 40–58.

Citovsky, V., and Loyter, A., 1985, Fusion of Sendai virions or reconstituted Sendai virus envelopes with liposomes or erythrocyte membranes lacking virus receptors, *J. Biol. Chem.* **260** : 12072–12077.

Citovsky, V., Blumenthal, R., and Loyter, A., 1985, Fusion of Sendai virions with phosphatidylcholine-cholesterol liposomes reflects the viral activity required for fusion with biological membranes, *FEBS Lett.* **193** : 135–140.

Citovsky, V., Yanai, P., and Loyter, A., 1986a, The use of circular dichroism to study conformational changes induced in Sendai virus envelope glycoproteins, *J. Biol. Chem.* **261** : 2235–2239.

Citovsky, V., Zakai, N., and Loyter, A., 1986b, Specific requirement for liposome-associated sialoglycolipids, but not sialoglycoproteins, to allow lysis of phospholipid vesicles by Sendai virions, *Exp. Cell Res.* **166** : 279–294.

Citovsky, V., Schuldiner, S., and Loyter, A., 1987, Osmotic swelling allows fusion of Sendai virions with chromaffin granula and desialized erythrocyte membranes, *Biochemistry* **26** : 3856–3862.

Citovsky, V., Rotem, S., Nussbaum, I., Laster, Y., Rott, R., and Loyter, A., 1988, Animal viruses are able to fuse with prokaryotic cells: Fusion between Sendai influenza virions and mycoplasma, *J. Biol. Chem.* **263** : 461–467.

Crimmins, D. L., Mehard, W. B., and Schlesinger, S., 1983, Physical properties of a soluble form of the glycoprotein of vesicular stomatitis virus at neutral acidic pH, *Biochemistry* **22** : 5790–5796.

Dales, S., 1973, Early events in cell–animal virus interactions, *Bacteriol. Rev.* **37**(2) : 103–135.

Dalgleish, A. G., Beverley, P. C. L., Clapham, P. R., Crawford, D. H., Greaves, M. F., and Weiss, R. A., 1984, The CD4 (T4) antigen is an essential component of the receptor for the AIDS retrovirus, *Nature* **312** : 763–766.

Dewhurst, S., Stevenson, M., and Volsky, D. J., 1987, Expression of the T4 molecule (AIDS virus receptor) by human brain-derived cells, *FEBS Lett.* **213**(1) : 133–137.

Dimmock, N. J., 1982, Initial stages in infection with animal viruses, *J. Gen. Virol* **59** : 1–22.

Doms, R. W., and Helenius, A., 1986, Quarternary structure of influenza virus hemagglutinin after acid treatment, *J. Virol.* **60**(3) : 833–899.

Doms, R. W., Helenius, A., and White, J., 1985, Membrane fusion activity of the influenza virus hemagglutinin. The low pH-induced conformational change, *J. Biol. Chem.* **260**(5) : 2973–2981.

Duzgunes, N., 1985, Membrane fusion, *Subcell. Biochem* **11** : 195–286.

Eidelman, I., Schlegel, R., Tralka, T. S., and Blumenthal, R., 1984, pH-Dependent fusion induced by vesicular stomatitis virus glycoprotein reconstituted into phospholipid vesicles, *J. Biol. Chem.* **259** : 4622–4628.

Ehrenstein, G., Blumenthal, R., Latorre, R., and Lecar, H., 1974, The kinetics of the opening and closing of individual eim channels in a lipid bilayer, *J. Gen. Physiol.* **63** : 707.

Ekerdt, R., Dahl, G., and Gratzl, M., 1981, Membrane fusion secretory vesicles and liposomes for different types of fusion, *Biochem. Biophys. Acta* **646** : 10–22.

Fingeroth, J. D., Weis, J. J., Tedeer, T. F., Straminger, J. L., Biro, P. A., and Fearon, D. T., 1984, Epstein-Barr virus receptor of human B lymphocytes is the C3d receptor CR2, *Proc. Natl. Acad. Sci. U.S.A.* **81** : 4510–4514.

Fries, E., and Helenius, A., 1983, Binding of Semliki Forest Virus and its spike glycoproteins to cells, *Eur. J. Biochem.* **97** : 213–220.

Fukami, Y., Hasaka, Y., and Yamamoto, K., 1980, Separation of Sendai virus glycoproteins by CM-Sepharose column chromatography, *FEBS Lett.* **114** : 342–347.

Fung, B. K., and Stryer, L., 1978, Surface density determination in membranes by fluorescence energy transfer, *Biochemistry* **17** : 5241–5248.

Gething, M. J., Doms, R. W., York, D., and White, J., 1986, Studies on the mechanism of membrane fusion: Site-specific mutagenesis of the hemagglutinin of influenza virus, *J. Cell Biol.* **102**(1) : 11–23.

Gibson, S., Jung, C. Y., Takashi, M., and Lenard, J., 1986, Radiation inactivation analysis of influenza virus reveals different target sizes for fusion, leakage, and neuraminidase activities, *Biochemistry* **25**(20) : 6264–6268.

Gitman, A. G., and Loyter, A., 1984, Construction of fusogenic vesicles bearing specific antibodies, *J. Biol. Chem.* **259**(15) : 9813–9820.

Gitman, A. G., Khanae, I., and Loyter, A., 1985, Use of virus-attached antibodies or insulin molecules to mediate fusion between Sendai virus envelopes and neuraminidase-treated cells, *Biochemistry* **24** : 2762–2768.

Graves, P. N., Schulman, J. L., Young, J. F., and Palese, P., 1983, Preparation of influenza virus subviral particles lacking the HA1 subunit of hemagglutinin: Unmasking of cross-reactive HA2 determinants, *Virology* **126**(1) : 106–116.

Guyden, J., Godfrey, W., Doe, B., Ousley, F., and Wofsky, L., 1983, Immunospecific vesicles targeting facilitates fusion with selected cell populations, in *Cell Fusion*, Vol. 103, pp. 239–253, CIBA Foundation Symposium, Pitman, London.

Harmsen, M. L., Wilsschut, J., Scherphof, G., Hulstuert, C., and Hoekstra, D., 1985, Reconstitution and fusogenic properties of Sendai virus envelopes, *Eur. J. Biochem.* **149** : 591–599.

Haywood, A. M., 1974, Characteristics of Sendai virus receptor in a model membrane, *J. Mol. Biol.* **83** : 427–436.

Haywood, A. M., and Boyer, B. P., 1984, Effect of lipid composition upon fusion of liposomes with Sendai virus membrane, *Biochemistry* **29** : 4161–4166.

Helenius, A., and Simons, K., 1975, Solubilization of membranes by detergents, *Biochim. Biophys. Acta* **415** : 29–79.

Helenius, A., Morrein, B., Fries, E., Simons, K., Robinson, P., Schirrmacher, V., Terhost, C., and Strominger, J. L., 1978, Human (HLA-A and HLA-B) and murine (H-2K and H-2D) histocompatibility antigens are cell surface receptors for Semliki Forest virus, *Proc. Natl. Acad. Sci. U.S.A.* **75** : 3846–3850.

Helenius, A., Kartenbeck, J., Simons, K., and Fries, E., 1980, On entry of Semliki Forest virus into BHK-21 cells, *J. Cell Biol.* **84** : 404–420.

Hoekstra, D., de Boer, T., Klappe, K., and Wilschut, J., 1984, Fluorescence method for measuring the kinetics of fusion between biological membranes, *Biochemistry* **23** : 5675–5681.

Hoekstra, D., Klappe, K., de Boer, T., and Wilschut, J., 1985, Characterization of the fusogenic properties of Sendai virus: Kinetics of fusion with erythrocyte membranes, *Biochemistry* **24** : 4739–4745.

Homma, M., and Ohuchi, M., 1973, Trypsin action on the growth of Sendai virus in tissue culture cells, *J. Virol.* **12** : 1457–1465.

Hosaka, Y., and Schimizu, K., 1972, Artificial assembly of envelope particles of HVI (Sendai virus). I. Assembly of hemolysis and fusion factors from envelope stabilized with Nonidet P-40, *Virology* **49** : 627–639.

Hsu, M., Scheid, A., and Choppin, P. W., 1983, Fusion of Sendai virus with liposomes: Dependence on the viral fusion protein (F) and the lipid composition of liposomes, *Virology* **126** : 361–369.

Huang, R. T. C., Rott, R., and Klenk, H. D., 1981, Influenza viruses cause hemolysis and fusion of cells, *Virology* **110** : 243–247.

Israel, S., Ginsberg, D., Laster, Y., Zakai, N., Milner, Y., and Loyter, A., 1983, Fusion of Sendai virus envelopes with human erythrocytes: A possible involvement of virus-associated protease, *Biochim. Biophys. Acta* **732** : 337–346.

Kawasaki, K., Sato, S. B., and Ohnishi, S. I., 1983, Membrane fusion activity of reconstituted vesicles of influenza virus hemagglutinin glycoproteins, *Biochim. Biophys. Acta* **733** : 286–290.

Keller, P. M., Person, S., and Snipes, W., 1977, A fluorescence enhancement assay of cell fusion, *J. Cell Sci.* **28** : 167–177.

Kielian, M., and Helenius, A., 1986, Alpha viruses, in *The Togaviridae and Flaviridae* (S. Schlesinger and M. J. Schlesinger, eds.), pp. 91–119, Plenum Press, New York.

Klatzmann, D., Champagne, E., Chamaret, S., Gruest, J., Getard, D., Herrend, T., Gluckman, J.-C., and Montagnier, L., 1984, T-lymphocyte T4 molecule behaves as the receptor for human retrovirus LAV, *Nature* **312** : 767–768.

Klenk, H. D., Rott, R., Orlich, M., and Blodorn, J., 1975, Activation of influenza A viruses by trypsin treatment, *Virology* **68**(2) : 426–439.

Kundrot, L. E., Springer, E. A., Kendall, B. A., McDonald, R. L., and McDonald, R. I., 1983, Sendai virus-induced lysis of liposomes requires cholesterol, *Proc. Natl. Acad. Sci. U.S.A.* **80** : 1608–1612.

Lakowicz, J. R., 1983, *Principles of Fluorescence Spectroscopy*, Plenum Press, New York.

Lapidot, M., Nussbaum, O., and Loyter, A., 1987, Fusion of membrane vesicles bearing only the influenza hemagglutinin with erythrocytes, living cultured cells and liposomes, *J. Biol. Chem.* **262** : 13736.

Laster, Y., Sabban, E., and Loyter, A., 1972, Susceptibility of membrane phospholipids in erythrocyte ghosts to phospholipase C and their refractiveness in the intact cell, *FEBS Lett.* **20** : 307–310.

Loyter, A., and Citovsky, V., 1987, The role of Sendai virus envelope glycoproteins in fusion with negatively-charged and neutral liposomes, in *Cellular Membrane Fusion: Fundamental Mechanisms and Application of Membrane Fusion Techniques* (J. Wilschut and D. Hoekstra, eds.), Marcel Dekker, New York.

Loyter, A., and Volsky, D. J., 1982, Reconstituted Sendai virus envelopes as carriers for the introduction of biological materials into animal cells, in *Membrane Reconstitution* (G. Poste and G. L. Nicolson, eds.), pp. 215–266, Elsevier/North-Holland Biomedical Press, Amsterdam.

Loyter, A., Citovsky, V., and Ballas, N., 1987, Sendai virus envelopes as a biological carrier: Reconstitution, targeting and application, in *Cellular Membrane Fusion: Fundamental Mechanisms and Application of Membrane Fusion Techniques* (J. Wilschut and D. Hoekstra, eds.), Marcel Dekker, New York.

Loyter, A., Citovsky, V., and Blumenthal, R., 1988a, The use of fluorescence dequenching methods to follow viral membrane fusion events, in *Methods Biochem. Anal.* **33** : 128–164.

Loyter, A., Nussbaum, O., and Citovsky, V., 1988b, Active function of membrane receptors in fusion of enveloped viruses with cell plasma membranes, in *Molecular Mechanisms of Membrane Fusion* (S. Ohki, D. Doyle, T. Flanagan, S. W. Hui, and E. Mayhew, eds.), pp. 413–426, Plenum Press, New York.

Maddon, P. J., Littman, D. R., Godfrey, M., Maddon, D. E., Chess, L., and Axel, R., 1985, The isolation and nucleotide sequence of a cDNA encoding the T cell surface protein T4: A new member of the immunoglobulin gene family, *Cell* **42**(1) : 93–104.

Maeda, Y., Asano, A., Okada, Y., and Ohnishi, S. I., 1977, Transmembrane phospholipid motions induced by F glycoprotein in hemagglutinating virus of Japan, *J. Virol.* **21** : 232–241.

Markwell, M. A., Svennerholm, L., and Paulson, J. C., 1981, Specific gangliosides function as a host cell receptor for Sendai virus, *Proc. Natl. Acad. Sci. U.S.A.* **78** : 5406–5410.

Massen, J. A., and Terhorst, C., 1981, Identification of a cell-surface protein involved in the binding site of Sindbis virus on human lymphoblastic cell lines using a heterobifunctional cross-linker, *Eur. J. Biochem.* **115** : 153–158.

Matlin, K. S., Reggio, H., Helenius, A., and Simons, K., 1981, Infectious entry pathway of influenza virus in a canine kidney cell line, *J. Cell Biol.* **91** : 601–613.

McDougal, J. S., Kennedy, M. S., Stigh, J. M., Cort, S. P., Mawle, A., and Nicholson, J. K. A., 1986, Binding of HTLV-III/LAV to T4$^+$ T cells by a complex of the 110 K viral protein and the T4 molecule, *Science* **231** : 382–385.

Metsikko, K., van Meer, G., and Simons, K., 1986, Reconstitution of fusogenic activity vesicular stomatitis, *EMBO J.* **5** : 3429–3435.

Monod, J., Changeux, J.-P., and Jacob, F., 1963, Allosteric proteins and cellular control systems, *J. Mol. Biol.* **6** : 306–329.

Monod, J., Wyman, J., and Changuez, J.-P., 1965, On the nature of allosteric transitions: A plausible model, *J. Mol. Biol.* **12** : 88–118.

Morris, S. J., Bradley, D., Gibson, G. C., Smith, P. D., and Blumenthal, R., 1988, Use of membrane-associated fluorescence probes to monitor fusion of vesicles: Rapid kinetics of aggregation and fusion using pyrene excimer/monomer fluorescence, in *Spectroscopic Membrane Probes* (L. Loew, ed.), Vol. I., pp. 161–191, Boca Raton, FL.

Nir, S., Klappe, K., and Hoekstra, D., 1986, Kinetics and extent of fusion between Sendai virus and erythrocyte ghosts: Application of a mass action kinetic model, *Biochemistry* **25** : 2155–2161.

Nussbaum, O., and Loyter, A., 1987, Quantitative determination of virus–membrane fusion events: Fusion of influenza virions with plasma membranes and membranes of endocyte vesicles in living cultured cells, *FEBS Lett.* **221** : 61–67.

Nussbaum, O., Zakai, N., and Loyter, A., 1984, Membrane-bound antiviral antibodies as receptors for Sendai virions in receptor-depleted erythrocytes, *Virology* **138** : 185–197.

Nussbaum, O., Lapidot, M., and Loyter, A., 1987, Reconstitution of functional influenza virus envelopes and fusion with membrane and liposomes lacking virus receptors, *J. Virol.* **61** : 2245–2252.

Oku, M., Nujima, S., and Inoue, K., 1982, Studies on the interactions of HVJ (Sendai virus) with liposomal membranes induced permeability of liposomes containing glycoprotein, *Virology* **116** : 419–427.

Oldstone, M. B. A., Tishon, A., Dukto, F., Kennedy, S. I. T., Holland, J. J., and Lampert, P. W., 1980, Does the major histocompatibility complex serve as a specific receptor for Semliki Forest virus? *J. Virol.* **34** : 256–265.

Ozawa, M., Asano, A., and Okada, Y., 1979a, The presence and cleavage of interpolated disulfide bonds in viral glycoproteins, *J. Biochem. (Tokyo)* **86** : 1361–1364.

Ozawa, M., Asano, A., and Okada, Y., 1979b, Biological activities of glycoproteins of HVJ (Sendai virus) studied by reconstitution of hybrid envelope and by concanavalin A-mediated binding: A new function of HANA protein and structural requirement for F protein hemolysis, *Virology* **99** : 197–202.

Pasternak, C. A., Alder, G. M., Bashford, C. L., Buckley, C. D., Micklem, K. J., and Patel, K., 1985, Cell damage by viruses, toxins and complement: common features of pore-formation and its inhibition by Ca^{2+}, *Biochem. Soc. Symp.* **50** : 247–264.

Poste, G., and Pasternak, C. A., 1978, Membrane fusion, in *Cell Surface Reviews : Membrane Fusion* (G. Poste and G. L. Nicolson, eds.), Vol. 5, pp. 306–321, North-Holland Publishing, Amsterdam.

Puri, A., Winick, J., Lowy, R. J., Covell, D., Eidelman, O., Walterm, A., and Blumenthal, R., 1988, Activation of vesicular stomatitis virus fusion with cells by pretreatment at low pH, *J. Biol. Chem.* **263** : 4749–4763.

Razin, S., and Tully, J., 1970, Cholesterol requirements of mycoplasma, *J. Bacteriol.* **102** : 306–310.

Rott, R., and Klenk, H. D., 1977, Structure and growth of viral envelopes, in *Virus Infection and the Cell Surface* (G. Poste and G. L. Nicolson, eds.), Vol. 2, pp. 47–48, North-Holland Publishing, Amsterdam.

Ruigrok, R. W., Martin, S. R., Wharton, S. A., Skehel, J. J., Bayley, P. M., and Wiley, D. C., 1986, Conformational changes in the hemagglutinin of influenza virus which accompany heat-induced fusion of virus with liposomes, *Virology* **155**(2) : 484–497.

Sato, S. B., Kawasaki, K., and Ohnishi, S., 1983, Hemolytic activity of influenza virus hemagglutinin glycoproteins activated in mildly acidic environments, *Proc. Natl. Acad. Sci. U.S.A.* **80** : 3153–3157.

Scheid, A., and Choppin, P. W., 1974, Identification of biological activities of paramyoxvirus glycoproteins: Activation of cell fusion hemolysis and infectivity by proteolytic change of an active precursor protein of Sendai virus, *Virology* **57** : 475–479.

Schlegel, R., Willingham, M. C., and Pastan, I. H., 1982, Saturable binding sites for vesicular stomatitis virus on the surface of Vero cells, *J. Virol.* **43** : 871–875.

Schlegel, R., Tralka, T. S., Willingham, M. C., and Pastan, I. H., 1983, Inhibition of VSV binding and infectivity by phosphatidylserine: Is phosphatidylserine a VSV-binding site?, *Cell* **32** : 639–646.

Schmidt, M. F. G., and Lambrecht, B., 1985, On the structure of the acyl linkage and the function of fatty acyl chains in the influenza virus hemagglutinin and the glycoproteins of Semliki Forest virus, *J. Gen. Virol.* **66** : 2635–2647.

Schulze, I. T., 1975, The biologically active proteins of influenza virus: The hemagglutinins, in *The Influenza Viruses and Influenza* (E. D. Kilbourne, ed.), pp. 53–82, Academic Press, New York.

Sinibaldi, L., Gordoni, P., Seganti, L., Superti, F., Tsaing, H., and Orsi, N., 1985, Gangliosides in early interactions between vesicular stomatitis virus and CER cells, *Microbiologia* **8**(4) : 355–365.

Skehel, J., Bayley, P., Brown, E., Martin, S., Waterfield, M., White, J., Wilson, I., and Wiley, D., 1982, Changes in conformation of influenza virus hemagglutinin at the pH optima of virus-mediated membrane fusion, *Proc. Natl. Acad. Sci. U.S.A.* **79** : 968–972.

Smith, A. L., and Tignor, G. H., 1980, Host cell receptors for two strains of Sindbis virus, *Arch. Virol.* **66**(1) : 11–26.

Snyder, B., and Freire, E., 1982, Fluorescence energy transfer in two dimensions. A numeric solution for random and nonrandom distributions, *Biophys. J.* **40** : 137–148.

Stegmann, T., Hoekstra, O., Senerphob, G., and Wilschut, J., 1985, Kinetics of pH-dependent fusion between influenza virus and liposomes, *Biochemistry* **24** : 3107–3113.

Stegmann, T., Hoekstra, G., Scherpol, G., and Wilschut, J., 1986, Fusion activity of influenza virus, *J. Biol. Chem.* **261** : 10966–10969.

Stegmann, T., Morselt, W. M., Booy, F. P., van Breemen, J. F. L., Scherpol, G., and Wilschut, J., 1987, Functional reconstitution of influenza virus envelopes, *EMBO J.* **6** : 2651–2659.

Struck, D. K., Hoekstra, D., and Pagano, R. E., 1981, Use of resonance energy transfer to monitor membrane fusion, *Biochemistry* **29** : 4093–4099.

Suzuki, Y., Matsunaga, M., and Matsumoto, M., 1985, N-Acetyl neuraminyl lactosyl ceramide, GM3-NeuAc, a new influenza A virus receptor which mediates the adsorption–fusion process of viral infection. Binding specificity of influenza virus A/Aichi/2/68 (H3N2) to membrane-associated GM3 with different molecular species of sialic acid, *J. Biol. Chem.* **260**(3) : 1362–1365.

Svensson, V., 1985, Role of vesicles bearing adenovirus 2: Internalization into HeLa cells, *J. Virol.* **55** : 442–449.

Tedder, T. F., Goldmacher, V. S., Lambert, J. M., and Schlossman, S. F., 1986, Epstein-Barr virus binding induces internalization of the C3d receptor: A novel immunotoxin delivery system, *J. Immunol.* **137** : 1387–1391.

Tozawa, H., Watanabe, M., and Ishida, N., 1973, Structural components of Sendai virus, *Virology* **55** : 242–253.

Vainstein, A., Hershkovitz, M., Israel, S., Rabin, S., and Loyter, A., 1984, A new method for reconstitution of highly fusogenic Sendai virus envelopes, *Biochim. Biophys. Acta* **773** : 181–188.

Volsky, D. J., and Loyter, A., 1978, An efficient method for reassembly of fusogenic Sendai virus envelopes after solubilization of intact virion with Triton X-100, *FEBS Lett.* **92** : 190–194.

Volsky, D. J., Shapiro, I. M., and Klein, G., 1980, Transfer of Epstein-Barr virus receptors to receptor-negative cells permits virus penetration and antigen expression, *Proc. Natl. Acad. Sci. U.S.A.* **77** : 5453–5455.

Wagner, R. G., 1975, Reproduction of rhabdoviruses, in *Comprehensive Virology* (H. Fraenkel-Conrat and R. R. Wagner, eds.), Vol. 4, pp. 1–93, Plenum Press, New York.

Webster, R. G., Brown, L. E., and Jackson, D. C., 1983, Changes in the antigenicity of the hemagglutinin molecule of H3 influenza virus at acidic pH, *Virology* **126**(2) : 587–599.

Weinstein, J. N., Yoshikami, S., Henkart, P., Blumenthal, R., and Hagins, W. A., 1977, Liposome–cell interaction: Transfer and intracellular release of a trapped fluorescent marker, *Science* **195** : 489–492.

Weinstein, J. N., Ralston, E., Leserman, L. D., Klausner, R. D., Dragsten, P., Henkart, P., and Blumenthal, R., 1984, Self-quenching of carboxyfluorescein fluorescence: Uses in studying liposome stability and liposome–cell interaction, in *Liposome Technology* (G. Gregoriadis, ed.), Vol. 3, pp. 183–204, CRC Press, Boca Raton, FL.

Wharton, S. A., Skehel, J. J., and Wiley, D. C., 1986, Studies of influenza hemagglutinin-mediated membrane fusion, *Virology* **144** : 27–35.

White, J., and Helenius, A., 1980, pH-Dependent fusion between the Semliki Forest virus membrane and liposomes, *Proc. Natl. Acad. Sci. U.S.A.* **77** : 3273–3277.

White, J., Kielian, M., and Helenius, A., 1983, Membrane fusion proteins of enveloped animal viruses, *Q. Rev. Biophys.* **16** : 151–195.

Wilschut, J., and Pahahadjopoulos, D., 1979, Ca^{2+}-induced fusion of phospholipid vesicles monitored by mixing of aqueous contents, *Nature* **281** : 690–692.

Wilson, I., Skehel, J. J., and Wiley, D. C., 1981, Structure of the haemagglutinin membrane glycoprotein of influenza virus at 3 Å resolution, *Nature* **289** : 366–373.

Wolber, P. K., and Hudson, B. S., 1979, An analytic solution of the Förster energy transfer problem in two dimensions, *Biophys. J.* **28** : 197–210.

Wolf, D., Kahana, I., Nir, S., and Loyter, A., 1983, The interaction between Sendai virus and cell membranes, *Exp. Cell Res.* **130** : 361–369.

Wyman, 1948, Heme proteins, *Adv. Protein Chem.* **4** : 407–531.

Yewdell, J. W., Gerhard, W., and Bachi, T., 1983, Monoclonal anti-hemagglutinin antibodies detect irreversible antigenic alterations that coincide with the acid activation of influenza virus A/PR/834-mediated hemolysis, *J. Virol.* **48** (1) : 239–248.

Index

Absorption
anisotropy decay, 58
anisotropy probe, 58
transition moment, 243
Acetoplasma laidlawii, 435
Acetylcholine, 363
receptor, 363–387
Acholeplasmaceae, 306
Acholeplasma laidlaiwii, 291–292
Acousto-optic modulator, 93, 122
Acrylamide fluorescence quencher, 26, 121,
346–347
Actin, 324
F, 323–355
filaments, 80–83, 323–355
G, 325
Action potential, 363
Adamantanediazirine, 377
Adenylate cyclase, 167, 196, 197, 209–231,
394–395
ϵ-ADP (1-N^6-ethenoadenosine 5'-
diphosphate), 329, 340
Adrenal cortex, 273
Adrenalin, 394
β-Adrenergic receptor, 214
ADS, 369
Aflatoxin B_1, 271
AIDS virus (HTLV-III), 419
Alcohol, 294–295
Aldolase, 352
Alkaline phosphatase, 196
Allostery, 446
Aminophospholipids, 9
1-Aminopyrine (1-AP), 178
Ammonium chloride, 433, 438
Androstanol, 145
Anesthetics, 223–231, 380

Angle-resolved fluorescence depolarization,
25
N-(1-Anilinonaphytyl-4)maleimide (ANM),
272, 342, 383
Anisotropy, 11–26, 56–66, 339–342
Ankyrin, 325, 353
ANM [N-(1-anilinonaph-4-tyl)maleimide],
272, 342, 383
Annular lipids, 229
ANS (1-anilinonaphtalene-8-sulfonate), 26,
201, 224, 266, 282, 298
Anthracene, 131
Anthracene-1,5-disulfonic acid, 369
Anthroyloxy fatty acids, 26–28, 131, 134,
138, 201, 217, 224, 267, 282
9-(2-Anthryl)-nonanoic acid, 132
Antibody, 74–75, 421, 445
ANTS (8-amino-1,3,6-naphtalene
trisulfonate), 369
Arachidonic acid, 398, 407
Arc lamps, 122
Argon, 118, 374
Arterial thrombosis, 406
Arthritis, 406
Arylazidophospholipids, 377
Ascites
GRSL, 400, 403
ascitic tumor, 401
Ascorbate, 263
Asymmetry of membranes, 9
Atherosclerosis, 407
ATP, 82, 327
ϵ-ATP (1-N^6-ethenoadenosine 5'-
triphosphate), 329, 340
ATPase
Na^+–K^+ dependent, 167, 183
Mg^{2+}–Ca^{2+} dependent, 273

Autofluorescence, 15
Avidin, 422
Axons, 183
1-Azidopyrene, 299

Bacillus stearothermophilus var.
    nondiastaticus, 293
Bacillus subtilis, 306
Bacteriophage
    λ, 79
    T4, 79, 299–300
Bacteriorhodopsin, 31, 37, 58, 73, 241–256,
    273
Band 4.1., 323–355
Barium, 183
Bending of DNA, 77
Benzo[α]pyrene, 271
Benzphetamine, 268
Benzyl alcohol, 161, 224–227
Bio-beads SM-2, 437
Biomembranes, 129, 259–262, 394
Biotin, 422
Black lipid membranes (BLM), 26
Bleaching, periodic pattern, 35
Blue shift, 426
Bragg peaks, 287
Bromelain, 444
Bromoacetylcholine, 365
Brownian motion, 5, 159
α-Bungarotoxin, 366, 369, 379, 385
Butter fat, 404
Butyrivibrio S_2, 291, 293

Cadmium, 159–186
Calcein, 440
Calcium ion, Ca^{2+}, 10, 11, 81, 82, 159–
    186, 300, 302, 304, 325, 337, 340,
    426
Calorimetry, differential scanning, 137, 218,
    293
Carbamylcholine, 370, 379, 382, 385
Carboxyfluorescein, 423, 442
Carboxyfluorescence dequenching, 442
Cardiolipin, 142, 168, 440
Cardiovascular disease, 406
Carrier protein, cadmium–zinc, 167
Cathepsin D, 193
CCCP (uncoupler), 290
C = C stretching mode, 250
Cell signal transduction, 394

CER cells, 419
Cesium, 367–369
Chloroform, 181
7-Chloro-4-nitro-2,1,3-benzoxadiazole, 344
7-Chloro-4-nitrobenzo-2-oxa-1,3-diazole,
    250–254, 329
Chlorpromazine, 229–231
Cholera toxin, 207, 213, 217, 227
Cholesterol, 9, 69, 71, 127–158, 160, 179,
    269, 307, 336, 386, 395, 396, 398,
    435, 440, 443
    ordering coefficient, 396
Cholesteryl hemisuccinate, 395
Cholinergic agonists, 379, 385
Cholinergic antagonists, 379
Chromaffin granules, 434
Chromatin, 78
Chromophore, 72
Circular dichroism, 446
cis-vaccenic acid, 296
Coated pits, 10
Coat protein, M13 virus, 273
Cobalt ion (Co^{2+}), 329
Co-EDTA (Cobalt-
    ethylenediaminetetraacetate), 246
Colicin, 299
Coliform membranes, 291, 301
Coliphage, 298
Collagen, 160
Complement receptor type 2 (CR2), 420
Con A-Sepharose, 198
Cone model, 23
Corn oil, 404
Correlation time, 122
Cross-correlation detection, 94
9-Cyanoanthracene, 93
Cysteine, 272, 384
Cytochalazin, 333, 335, 344
Cytochrome
    b5, 28, 33
    c, 249
    reductase, 263–265
    oxidase 69, 72, 262, 271, 273
    p-420, 264
    p-450, 264–274
        reductase (NADPH dependent), 272
Cytoskeleton, 5, 10, 73, 160, 325–357

DABMI (4-dimethylaminophenylazophenyl-
    4'-maleimide), 330

Dansyl-PE, 113
DDA (dansyldodecylamine), 33
DDPM [N-(4-dimethylamino-3,5-
      dinitrophenyl)maleimide], 329
Decamethonium, 380
Deenergization, 298–300
Dehydroergosterol, 147
Depolarization, fluorescence, 11–26
Desensitization, 366
Dextran, 175
Diazonium salt of sulfanilic acid, 217
Dictyostelium discoideum, 334
Didecanoyl-phosphatides, 304
Diet, 395, 403–409
Differential phase fluorimetry, 394
Differential scanning calorimetry, 137, 218,
      290
Diffusion, 5, 6, 7, 128
      lateral, 7, 128
      limit model, 7, 40
      rotational, 7, 128, 261
4-Dimethylaminochalcone, 265
Dipicolinate, 34
Dipicolinic acid, 426
Dipole–dipole coupling, 243, 427
Dipole–dipole interactions, 243, 427
Dipole moment, 102
Discocyte, 9
Discontinuous sucrose gradient
      centrifugation, 195–197
Dithiothreitol (DTT), 429, 442, 443
DLPC (dilinoleoylphosphatidylcholine), 142
DMPC (dimyristoylphosphatidylcholine), 23,
      102, 269, 273
DMPE (dimyristoylphosphatidylethanolamine),
      305
DNA, 76–80
      torsional rigidity, 78
DNase I, 325, 336
c-DNA sequencing, 265
DNS (5-dimethyl-amino-naphtalene-I-sulfonyl-
      L-cystine), 340
Docosahexaenoic acid, 398
Dolichol, 218–223
Dolichylphosphate, 218–223
DOPC (dioleylphosphatidylcholine), 104, 110,
      273, 304
Double bond, effects of, 137
5-Doxyl stearic acid, 232, 289, 290
DPH (1,6-diphenyl-1,3,5-hexatriene), 11–26,

66–72, 98–101, 129, 134, 169, 172–
      177, 201–226, 266, 282, 290, 304,
      384–387, 393
DPPC (dipalmitoylphosphatidylcholine), 23,
      100, 104, 109, 135, 221, 273, 304,
      396
      melittin complexes, 98
      dns (dansylated
          dipalmitolylphosphatidylethanolamine),
          178
Duchenne type muscular dystrophy, 183

Echynocyte, 9, 145
EDTA, 296, 298, 426, 433
Ehrlich ascites carcinoma, 404
Electro-optic modulators, 93
Electron diffraction analysis, 241
Electron microscopy, 55, 416
Electron paramagnetic resonance (EPR), 260
      spectroscopy, 269
Electron spin resonance (ESR), 70, 130, 135,
      163, 289, 293
Emission transition moment, 243
Endocytosis, 195, 207, 416, 433
Endomembranes, 26
Endoplasmic reticulum, 222
Endosomes, 433
Energy transfer
      fluorescence, 28–35, 112–117
      in membranes, 112–117, 271, 329–334
Enterobacteriaceae, 306
Enterocytes, from rat, 404
Entropy of mixing, 10
Eosin, 80, 83, 262, 352
      iodoacetamide (EIA), 331, 352
      isothiocyanate (EITC), 351, 352
      phosphatidylethanolamine (EPE), 34
Epithelial cells from kidney, 209
Epitope, 445
Epoxide hydrase, 269, 270
EPR spectroscopy: *see* Electron paramagnetic
      resonance EPR, spectroscopy
Epstein-Barr virus, 420
Erythrocyte(s), 9, 209, 211, 406, 432, 435
      chicken, 78
      ghosts, 26, 352, 432
      plasma membranes of, 170, 174, 179, 406
      rat, 209, 432
Erythrosin, 262, 352
Escherichia coli, 287–294

Ethanol, 183, 294
Ethidium, 76–80
N-Ethylmaleimide, 344, 346, 381, 383
Excimer fluorescence, 38–42, 129
Exocytosis, 195

Falling ball viscometry, 344
F-actin, 81–83, 323–355
FCCP (carbonylcyanide p-
    trifluoromethoxyphenylhydrazone), 298
Ferritin, 249
Fibronectin, 160
Fibroblasts, 9
Fish oil, 407
FITC (fluorescein isothiocyanate), 271, 342
    lectin, 232
Flash photolysis of heme-CO, 262, 267
Flipases, 9
Flip-flop, 9, 160, 260
Fluidity
    gradient, 209
    of membranes, 4, 129, 163–172, 200–209,
        265, 393
Fluid mosaic model, 4
Fluorescein-5-maleimide, 334
Fluorescence, 1–42, 133–134
    anisotropy, 11–26, 129, 133
    depolarization, 11–26, 56–58, 242, 287
        angle-resolved, 25
    dequenching, 422–434
    energy transfer, 28–35, 112–116, 242–244,
        305, 426
    enhancement, 334–339
    lifetime, 3, 72
    microscopy, 11, 89
    polarization, 260, 396
    recovery after photobleaching (FRAP), 35–
        38, 129, 268–269, 291, 342–345
Fluorimetry
    total internal reflection, 11
    phase, 25
    frequency-domain, 90–98, 116–121
    modulation, 25
Fluorophores, 1–3, 11–26, 35, 89, 94, 120
Forskolin, 213, 216
Fourier transform, 117
FRAP: see Fluorescence, recovery after
        photobleaching FRAP
Free radicals, 263

Frequency-domain fluorimetry, 90–98, 116–
    121
Frequency response, 92

Gangliosides, 207, 214–218, 419, 440–444
Gelsolin, 325
Gentamicin, 297
Glucose-6-phosphatase, 198
Glutaraldehyde, 429
Glutathione, 167, 263
Glyceraldehyde phosphate dehydrogenase
    (G_3PD), 352
Glycerin, 174
Glycerol moiety, 132
Glycocalix, 5–6
Glycophorin, 325
Glycosaminoglycans, 160
Gram-negative bacteria
    outer membrane, 286
    incorporation of exogenous lipids, 302–305
GRSL leukemia, 400–402

HA_0-influenza virion, 430
Hemoglobin, 446
Hairy cell leukemia, 399
Halobacterium halobium, 241
Hapten, 74–75
Hemagglutinin, 417, 444
Hemoglobin, 249
Hemolysis, 440
Hepatic lipase, 402
Hepatocyte, 162, 171, 174–180
Hepatotoxin, 164
4-Heptadecyl-7-hydroxycoumarine, 384–387
Herpes simplex virus HSV-1, 432
Herpes virus, 420
Hexagonal II phase, 221
Hexanol, 294, 295
β-Hexosaminidase, 198
High density lipoprotein HDL, 400
Histones, 77–78, 119
Histrionicotoxin, 366
HMG-CoA (hydroxymethylglutaryl-CoA), 401
^2H-NMR, 130, 281–309
Homeoviscous adaptation, 293–294
HSV-1: see Herpes simplex virus HSV-1
HTLV-III virus: See AIDS virus HTLV-III
Hydrodynamic model, 37
Hydrogen belt, 224

Hydroxylamine, 429, 443
Hyperactivation, 226

IAEDANS [N-iodoacetyl-N'-(1-sulfo-5-naphthyl)ethylenediamine], 329, 341, 346
IAF [5-(iodoacetamido)fluorescein], 329, 331, 342, 345
IANBD [4-(N-iodoacetoxy)ethyl-N-methyl)amino-7-nitrobenzo-2,1,3-oxadiazole], 329
IATR [(iodoacetamideo)tetramethylrhodamine], 343, 344
Immunorecognition, 4, 10, 11
Indole: see Tryptophan
Influenza virus, 417–446
Infrared spectroscopy, 291
In-plane mode of rotation, 131, 139
Inside-out vesicles, 217
Insulin, 207, 394
Intensity-modulated light source, 92
Iodine, as fluorescence quencher, 26

Jablonski diagram, 2
JY cells, 418

K$^+$: see Potassium, ion K$^+$
Keratinlike proteins, 324

Lactobacillus, 306
Laser, 92, 116–118, 122, 133
Lateral diffusion coefficient, 5, 128, 209, 260
Legendre polynomials, 7, 17
LDL: see Low-density lipoprotein (LDL)
Leucine aminopeptidase, 200
Leukocyte, chemotaxis, 145
Leukemia, 395, 399, 401
  hairy cell, 399
  myeloid, 400
Light scattering, 339
Linoleic acid, 403
Lipase, hepatic, 402
Lipid
  annular, 227
  brominated, 28
  domains, 10, 25, 167
  hydroperoxides, 262–263
  peroxidation, 265–267
  polymorphism, 5

protein interaction, 69–71
  transfer protein, 210
Lipopolysaccharide, 287–288
Lipoprotein, 395, 401
  lipase, 402
Liposomes, 308, 394, 396, 440–444
Lipoxygenase, 409
Liquid-crystalline phase, 135
Lithium, 183
Liver, 204, 401
Low-density lipoprotein (LDL), 400
LUV: see Unilamellar lipid vesicles, large
T-Lymphotropic viruses, 419
Lysolecithin, 132
Lysophosphatidylethanolamine, 303
Lysophosphoglycerides, 303
Lysosomes, 433
Lysosomotropic reagent, 433, 439

M13 virus, coat protein, 273
Magnesium ion Mg^{2+}, 181, 292, 296, 326, 337
Magnetic resonance spectroscopy, 66, 130
Maleimide, 329
Malondialdehyde, 266, 269
Maximum entropy method, 149
MBTA [4-(N-maleimide)benzyltrimethylammonium], 365
Medula, 434
Mellitin, DPPC complexes, 98
Membrane
  basolateral, 404
  biological, 71
  erythrocyte, 9, 135, 406
  fluidity, 4, 128, 163–172, 403–409
  inner cytoplasmic, 286
  intestinal, 404
  fusion, 207, 415–450
  oriented, 249
  potential, 11, 297–300
  prokaryotic, 302–309
  reconstituted, 182–183, 202–209
  recycling, 207
  structure, 135
  viscosity, 6
Meromyosin, 335, 340
Methionine, 271
3-Methoxybenzanthracene, 265

Methylamine, 433, 439
α-Methylmannoside, 200
Mg^{2+}: *see* Magnesium ion Mg^{2+}
Microchannel plate photomultiplyer, 94
Micrococcus cryophilus, 292
Micrococcus luteus, 135
Microfilament, 324–327
Microsomes, 9, 259–274, 404
Microtubules, 324–327
Microvilli, 209, 404
Microviscosity, 21–24, 129, 260, 298
Milling crowd model, 7, 37, 40
Mitotic spindle, 325
Mobility
  lateral, 4
  rotational, 4
Modulators
  accousto-optic, 93, 122
  electro-optic, 93
Monobromobimane, 272
Monooxygenase, 259–274
Motion
  in-plane, 131
  out-of-plane, 131
  rotational, 65, 260
  thermal, 65
Multilamellar vesicles (MLV), 26
Multinodular goiter, 231–232
Multiple sclerosis, 406
Murine lymphocytes, 418
Muscle, 80
  contraction of, 364
  skeletal, 81
Mutants, 281–309
Mycoplasma, 306, 435
  capricolum, 308, 435
  gallisepticum, 307, 435
Mycoplasmataceae, 306
Myelin, 4
Myeloid leukemia, 400
Myofibril, 84–85
Myosin, 80–82, 326, 329
  actin interaction, 82, 323–355
  filament, 83–85, 323–355
Myxoviruses, 417

N-(1-anilonaph-4-yl) maleimide, 274, 344, 385
NAD-glycohydrolase, 197, 200
NADPH-cytochrome P-450 reductase, 264, 272–273

NaN_3, 433
NBD (7-chloro-4-nitrobenzo-2-oxa-1,3-diazole), 250–254, 291
  chloride, 329
  -PE, 306, 429, 437
Neuraminidase, 417, 436–439
α-Neurotoxins, 364
Nickel (Ni^{2+}), 329
Nitrocellulose transfer, 231
Nitromethane, 378, 381, 385
Nitroxide stearate, 289
NMR: *see* Nuclear magnetic resonance (NMR)
NO_2, 267
Non-Hodgkin's lymphoma, 400
Nonidet P40, 437
Nuclear magnetic resonance (NMR), 70, 130, 135, 218, 260, 287–292
Nuclei, 78
Nucleic acids, 112
Nucleocapsid, 416
Nucleosome, 78
5'-Nucleotidase, 195, 196, 217

Octadecyl rhodamine B-chloride (R18), 415–450
Octathyleneglycol mono(n-dodecyl)ether ($C_{12}E_8$), 437
β-D-Octylglucopyranoside, 202, 252, 376, 437
Oil, 403, 404
1-Oleoyl lysophosphatidic acid, 303
Olive oil, 407
One-swivel model, 84
Order parameter, 7, 134, 394, 399
Orientational order, 7, 19, 128, 135
Orientation factor κ, 243
OSPC (1-oleoyl-2-stearoyl-*sn*-glycero-3-phosphocholine), 135
Out-of-plane mode of rotation, 131, 139
Oxygen, as fluorescence quencher, 26
Oxysterols, 145
Oxytocin, 119

P-450 LM-2, LM-4, 265, 268, 270
Packing, 7–8
Palmitic acid, 292, 295
Paramyxovirus, 418
Parinaric acid (PA), 26, 131, 134, 282
cis-Parinaric acid (*cis*-PnA) (9, 11, 13, 15-*cis*, *trans*, *trans*, *cis*-octadecatetraenoic acid), 31, 131, 138, 282, 299, 300

trans-Parinaric acid (*trans*-PnA) (9, 11, 13, 15-all *trans*-octadecatetraenoic acid), 31, 131, 138, 282, 299, 300
Parinaroylphosphatidylcholine, 138, 306
PATMAN (6-palmitoyl-2[((2-trimethylammonio)ethyl)methylamino]naphtalene chloride), 105, 109
PC-DPH, 130, 132
Pentobarbital, 181
Peptidoglycan layer, 286
Pentylenetetrazole, 181
Periodic pattern bleaching, 35
Periplasm, 301, 302
Peroxidation, of membranes, 262–263
Persistence length, 77
Perylene, 26, 101–102, 131, 265
Phage adsorption, 299
Phalloidin, 352
Phase
    diagram, 10
    separation, 10, 397
    transition, 67, 165, 185, 287, 293
Phenobarbital, 227
Phenylhydrazine, 266
Phenylmaleimide, 383
Phenylmethylsulfonyl fluoride (PMSF), 429, 442, 443
N-Phenyl-1-naphtylamine (NPN), 282, 297, 298, 305
Phosphatidic acid, 164–165, 185, 211
Phosphatidylcholine (PC), 9, 165, 185, 194, 211, 269, 407
Phosphatidylethanolamine (PE), 211, 221, 407
    dansyl-, 113
    TNP-, 113
Phosphatidylglycerol, 289, 292
Phosphatidylinositol (PI), 168, 211–214
Phosphatidylserine (PS), 162–167, 185, 194, 211, 407, 419, 440
Phospholipase, 210, 398
Phospholipase C, 211
Phospholipids, incorporation of, 210–214
Phospholipid transfer protein, 308
Phosphorescence, 58, 352–353
Photobleaching recovery, 302
Pituitary, 218
Plasma membranes, 26, 74, 159, 193–232, 399, 404
Polarization, 11–26, 301
Polarized resonance Raman scattering, 242

Polarizer, 11
Polyamines, 301
Polyethylene glycol, 175
Polylysine, 10, 249
POPOP, 95
Potassium
    chloride, 326, 342
    iodide, 381
    ion K$^+$, 183, 307, 326, 342
Pressure relaxation, 349
Probe
    amphiphilic, 26
    disk-shaped, 131
    intrinsic, 73
    lipophilic, 26
    rodlike, 22, 135
Procaine, 386
Profilin, 325
Prokaryotic membranes, 282, 302–309
Prop-DPH, 130
Propidium, 381
Propylene glycol, 101
Prostaglandins, 409
Protamine, 79–80
Protein
    aggregation, clustering, 6, 269
    –lipid interaction, 69–71
    –protein interaction, 73–74, 112
Pseudomonas aeruginosa, 297
Pulse fluorescence anisotropy, 129
Purple membrane, 244–254
Putrescine, 301
Pyrene, 38–42, 129, 201, 262, 265, 287, 335, 384–387
N-(1-Pyrene)maleimide (PM), 335, 380–384
Pyrenesulfonylazide, 371
Pyrene-1-sulfonyl azide (PySa), 371–380
N-(1-Pyrene-sulfonyl)hexadecylamine (PySaH), 371
1,3,6,8-Pyrene tetrasulfonate (PTSA), 366–370
N-(3-Pyrenyl)maleimide (NPM), 340
N-Rh-PE [N-lissamine rhodamine B sulfonylphosphatidylethanolamine], 306, 429
N-succinimidyl-3-2(2-pyridyldithio)proprionate, 422

Quantum yield, 3, 30, 345–349, 369, 426
Quenching, 26–28, 305, 345–349, 380–384, 422–429

Quenched-flow techniques, 367

Radiation inactivation, 449
Raman
  scatter, 15, 249
  spectroscopy, 130
  tensor, 250
Random walk model, 7, 37, 40
Rb$^+$: *see* Rubidium ion (Rb$^+$)
Receptor, capping, 10, 74
Reconstituted thyroid membrane, 202–208
Reconstituted viral envelopes, 436–439
Red blood cell, 9, 325, 439
Red shift, 426
Regulatory proteins, 447
Relaxation, spectral, 110
Retinal, 32, 60, 72, 244–250
  outer segment (ROS), 31
Retroretinyllysine, 244, 246
Retrovirus, 419–420
Rhabdovirus, 419
Rhodopseudomonas, 308
  capsulata, sphaeroides, 308
Rhodopsin: *see* Bacteriorhodopsin
ROS: *see* Retinal outer segment (ROS)
Rotational constant, 5, 260
Rotational diffusion coefficient, 5, 18–19, 129
Rotational motion, 4, 260
Rubidium ion (Rb$^+$), 298

Saffman–Delbrück equation, 7, 37, 260–261
Salmonella typhimurium, 289–302
Saturation transfer electron paramagnetic
  resonance (ST-EPR), 261, 272
Scanning calorimetry, 137, 218, 293
SDS-gel electrophoresis, 231
Self-quenching, 305, 422–434
Semliki forest virus (SFV), 417–442
Sendai virus, 417–442
Serotonin, 394
Sialoglycolipids: *see* Gangliosides
Signal transduction, 394
Sindbis virus, 418
Single-photon counting, 91, 394
Shape concept, 147
Sheetz and Singer, bilayer coupling
  hypothesis, 10
Sm-2 Bio-beads, 437
SOPC (1-stearyl-2-oleoyl-*sn*-glycero-3-
  phosphocholine), 135

Sodium, 183
  uptake, 181
  tracer, 367
Solenoid, 78
Solvent relaxation, 11, 103
Spectral relaxation, 110
Spectrin, 325–351
Sperm, 79–80
Spermidine, 301
Spermine, 302
Sphingomyelin, 9, 396, 398, 440
Spikes, in the viral envelope, 416, 444
Spleen, 401, 407
  lymphocytes, 404, 407
Spring constant, 77
Staphylococcus aureus, 420
Steady-state fluorescence
  anisotropy, 19, 22, 71–72, 394
  measurement, 72–74, 298, 394
Stern–Volmer equation, 27, 345, 370
Sterols, 127–151
Stopped-flow methods, 367
Stress fibers of fibroblasts, 325
Strontium, 181
Structural order, 394
Succinate dehydrogenase, 198
Succinimidyl-4-(*p*-maleimidophenyl)butyrate,
  422
N-Succinimidyl-3-2(2-pyridyldithio)propionate,
  422
Sucrose gradient centrifugation
  continuous, 195
  discontinuous, 195
Sulfhydryl group in acetylcholine receptor,
  380–384
SUV: *see* Unilamellar lipid vesicles, small,
  26, 211
Synaptic membranes, 231
Synaptosomes, 178–181
Synchroton radiation, 94, 117, 133

Terbium ion (Tb^{3+}), 34, 426
Tetracaine, 378, 386
Tetracycline, 295, 308
Tetrahymena pyriformis, 184
Tetramethylrhodamine iodoacetamide, 331
Tetraphenyl phosphonium ion, 298
Thallous ion, 369, 381, 385
Thiocyanate, 294
Thiolate heme ligand, 265

Thromboxanes, 409
Thymocyte, 204
Thymus, 78
Thyratron gated flash lamp, 133
Thyroglobulin, 193, 194
Thyroid, 193–240
　plasma membranes, 193–232
Thyroid stimulating hormone (TSH), 193, 207, 394
　receptor, 194, 214
Thyrotropin: *see* Thyroid stimulating hormone (TSH)
Thyroxine (T4), 193
Time-resolved
　emission spectra, 101–110
　fluorescence anisotropy, 25, 66, 72, 121, 394
　spatial photometry (TRSP), 261
TMA (tryptamine myristic acid), 114
　-DPH, 130, 178, 201, 214–217, 224, 384–387
TNP-PE (trinitrophenyl-labeled phosphatidylethanolamine), 33, 111
α-Tocopherol, 266
Toga virus, 418–419
Torpedo californica, 364–365
Torsional rigidity of DNA, 78
Transfer protein, 302
　lipid, 210
Transferrin, 394
Trans-gauche conformational change, 260
Transient absorption anisotropy, 351–352
Transient dichroism, 272
Transient optical anisotropy, 261
Transition temperature, 69, 136, 300
Transmembrane location, 246–249
Trans-PNA-lecithin, 139
Transporter protein, 74
Trapping
　diffusion-mediated, 10
Tread milling in actin filaments, 327
Triiodothyronine, 193
Triplet probe, 58, 72, 80
Triphenylmethyl phosphonium ion, 298
Tris buffer, 297
Tris(2,2′-bipyridyl)ruthenium (II), 246
Tropomyosin, 81, 326, 340, 347, 349
Troponin, 81, 340

Trypsin, 429
Tryptophan, 28, 32–35, 89, 119, 270–272, 340, 346, 352, 380
TSH: *see* Thyroid stimulating hormone (TSH)
d-Tubocuranine, 380
Tubulin, 324–327
Tumor cell, 399–403
Tumor–host interactions, 395
Twisting of DNA, 77
Tyrosine, 117–119, 271

Ubiquinone, 28
Unilamellar lipid vesicles
　large (LUV), 26
　small (SUV), 26, 211
　thyroidal, 209
Urate, 263

Vaccenic acid, 294
Cis-Vaccenic acid, 294
Valinomycin, carrier for, 169
Van der Waals interactions, 135
Vehicle liposomes, 308
Veratridine, 179
Vero cells, 419, 449
Very low density lipoprotein (VLDL), 401
Vesicular stomatitis virus (VSV), 273–274, 419, 433, 449
Viral fusion, 415–450
Virus
　infection, 436
　–cell fusion, 433, 436
　–liposome fusion, 440
　–mycoplasma fusion, 435
　penetration, 436
Viscosity, 6, 75, 265
Vitamin E, 263

Wobbling, 5, 7, 18, 22, 66–70
Wobbling-in-cone model, 135

Xenobiotics, 163
X-ray
　diffraction, 55, 242, 266, 287, 293
　scattering, 218, 249

Zinc, 172, 181
Zonal rotor, 195